BLUEPRINT READING
FOR MANUFACTURING

BLUEPRINT READING FOR MANUFACTURING

Edward G. Hoffman
Paul R. Wallach

Delmar Publishers Inc.®

NOTICE TO THE READER

Cover photos and blueprint courtesy of Cincinnati Milacron Inc.

Delmar Staff
Associate Editor: Marjorie A. Bruce
Managing Editor: Barbara A. Christie
Publications Coordinator: Karen Seebald

For information, address Delmar Publishers Inc.
3 Columbia Circle, PO Box 15015
Albany, New York 12212-5015

Printed in the United States of America
Published simultaneously in Canada
by Nelson Canada,
A division of The Thomson Corporation

10 9 8 7 6 5 4

Library of Congress Cataloging in Publication Data

Hoffman, Edward G.
    Blueprint reading for manufacturing / Edward G. Hoffman,
Paul R. Wallach.
        p.   cm.
    Includes index.
    ISBN 0-8273-3325-0 (pbk.)
    1. Blueprints.   2.   Machinery—Drawings.   3.   Freehand
technical sketching.   I. Ross Wallach, Paul.   II. Title.
T379.H5715 1988
621.8'022'1—dc19   88-14987
                                              CIP

CONTENTS

PREFACE

Accurate communication between designers and shop personnel is essential in industry today. Gone forever are the days of the master mechanic, when the designer and builder were the same person. Now we rely on one another to make sure the products we produce are properly designed and made. In industry, the shop print is the most accurate and efficient means of communicating all of the detailed technical information needed to make today's parts.

One of the first skills any journeyman machinist must master is reading and interpreting shop prints. Before anyone can begin to apply the "hands-on" skills needed to become a proficient machinist, he or she must first understand what is to be made and exactly how it should be fashioned. *Blueprint Reading for Manufacturing* is designed to instruct beginning machinists in the fine points of reading and interpreting prints, and to refresh the skills of the more experienced machinists in interpreting today's prints.

The recent changes in drafting standards, made to bring United States drafting practices more in line with international methods have made many former methods obsolete. However, many companies will maintain their older drawings, while their newer prints will begin to comply with the new standards. During this period of transition, both the older and newer procedures must be completely understood.

The overall intent of this text is to thoroughly acquaint beginning and experienced shop personnel with the modern shop print. Every effort has been made to make this text as current as possible. New standards are incorporated where applicable, and both the new and old methods are discussed thoroughly.

Blueprint Reading for Manufacturing leads the reader from the study of what prints are and how they are made to the alphabet of lines and the development of the actual views in a shop print. The text includes all of the elements necessary to read and interpret the views, notes, dimensions, and special details found on modern prints.

A few features of this text which will prepare the reader for the new prints found in industry today are:

- A complete glossary of terms commonly used in the shop and frequently found on prints.

- Coverage of SI metric prints.

- Coverage and explanation of the newest standards used in drafting.

- Discussion of geometric dimensioning and tolerancing.

- Coverage of sheet metal and welding prints.

- Descriptions and discussion of modern dimensioning practices.

- End-of-chapter reviews and Appendix exercises to assist readers in assessing their mastery of the material.

- Prints that can be used for both reading exercises and for shop projects.

ACKNOWLEDGMENTS

No project of this magnitude could ever be an individual effort or accomplishment. The authors acknowledge the contributions of a number of people who worked diligently to produce this text, including Lisa Clyde Cassidy, Charlene Goins and Jackie Riechers. Other individuals who helped make this text possible are technical illustrators Kathy O. Hoffman, Paul Dalla Guardia, Becky Hehr, Rick Jacobs and Susan Shivley.

Technical contributors played a large part in developing the text and verifying the accuracy of the content. Particular appreciation is expressed to William W. Luggen, Manager of Computer Integrated Manufacturing (CIM) Systems Development for General Elecric Aircraft Engine Business Group, for sharing his time and extensive manufacturing know-how.

For their thorough reviews and recommendations for improvements in the text, the authors wish to express their appreciation to

George Sehi, Chairperson, Mechanical Engineering Technology, Sinclair Community College, Dayton, OH 45402

Larry Lamont, Mechanical Design/CADD Instructor, Moraine Park Technical Institute, Fond du Lac, WI 54935-2897

Dietrich R. Kanzler, Rancho Santiago College, Santa Ana, CA 92643

Benny Locken, Mid-Florida Technical Institute, Orlando, FL 32809

The authors are also indebted to the following companies and organizations who provided illustrations and technical assistance:

American National Standards Institute
American Society of Mechanical Engineers
American Welding Society
Boston Gear, Incom International Inc.
DoALL Company
Fred V. Fowler Company, Inc.
Rank Scherr-Tumico, Inc.
The L. S. Starrett Company
Society of Manufacturing Engineers
U.S. Metric Association

NOTE: The prints in this text are not reproduced to an accurate scale. Always follow the stated dimensions. Do not measure the print.

1

WHAT IS A PRINT?

OBJECTIVES

After studying this chapter, you will be able to:

- Define the term *print*.
- Define the term *engineering drawing*.
- List five ways to care for prints.
- Identify the standard drawing sheet sizes.
- Describe the proper procedures for folding prints.

The development of manufacturing has been rapid and dramatic in the past 200 years. Due to new technology, manufacturing has moved from production of one-of-a-kind, custom items to the mass-produced items we use daily. Throughout this period of rapid change one factor has remained constant: the need for accurate communication.

The simplest and most reliable way to convey detailed product information is through the shop print. *Prints* are the universal graphic language of industry. Prints can provide information that is too detailed for words alone to describe. Prints also control and coordinate all areas of production. From design to marketing, prints provide the necessary instructions and related technical information.

Prints are exact copies of original drawings. These original drawings are called *engineering drawings*. Most engineering drawings are drawn by a drafter at a drawing board. However, in some companies, computers are now used to assist designers in developing and constructing engineering drawings. These computer systems are referred to as *CAD systems*. CAD is the abbreviation for *computer-aided-drafting*. CAD systems greatly reduce the time needed to design parts. They also eliminate much of the repetitive work involved in making engineering drawings. Figure 1-1 shows a typical CAD system.

Engineering drawings represent a large investment in time and money. Therefore, they must be protected. To prevent an engineering drawing from becoming torn or soiled, copies, or prints, are made for use in the shop. This not only protects the original, but also permits changes to be made to the drawing as required. In addition, as many additional copies as are needed can be made. Figure 1-2 shows how a print is made from an engineering drawing.

The oldest form of print copy is the *blueprint*. Blueprinting produces a copy that has white lines on a dark blue background. The actual blueprint form of copy is rarely seen in industry today. However, the term *blueprint* has become so widely known that it is often used to identify all forms of drawing reproductions.

1

Figure 1-1. A typical CAD system used to construct engineering drawings (Courtesy of Computervision)

Modern methods of reproduction produce copies that have dark lines on a white background. These copies, called *whiteprints* or *diazo prints*, are the type of print usually seen today. Regardless of the process used to make the copy, the terms *print*, *blueprint*, *whiteprint*, *diazo print*, *industrial print*, and *drawing* are all commonly used to identify the copies used in the shop.

PRINTS ON THE JOB

The print travels along with the job in most shops. When a job is assigned, the supervisor generally gives the machinist a print along with any special instructions. When the task is completed, the machinist returns the print to the supervisor. To ensure that every machinist who

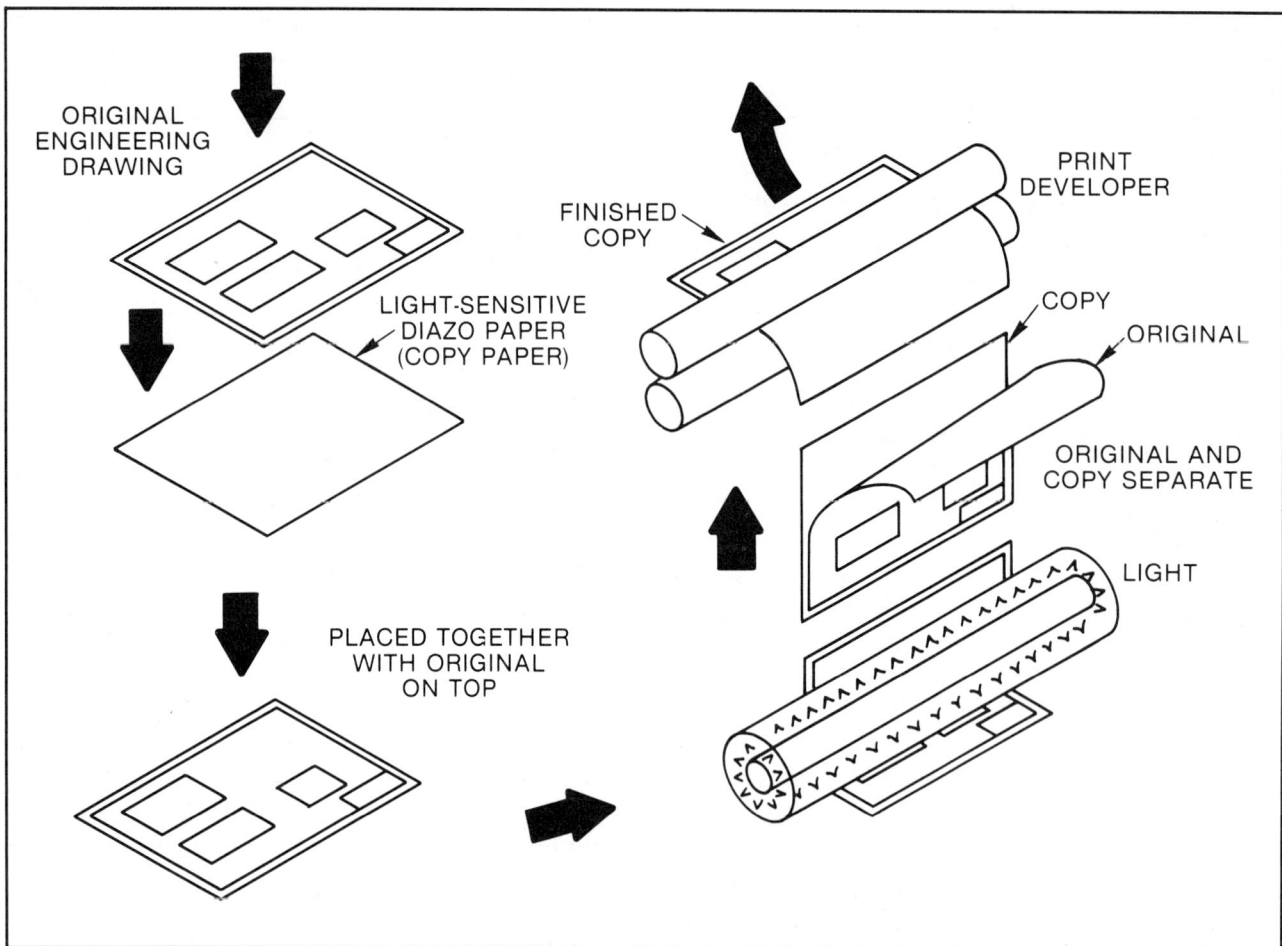

Figure 1-2. Method used to make prints from an engineering drawing

works on each job receives a readable print, the prints must be treated properly.

A print is a valuable document. It must be treated properly. The following simple tips will help to keep prints in good condition while they are handled.

1. Do not treat prints roughly. Fold and unfold prints carefully so they will not tear. Do not stick prints in a toolbox or pocket.

2. Never write on a print. Do not use a print as a pad to calculate sizes or for notes. If there is a mistake on the print or something seems to be missing, show the print to a supervisor.

3. Keep the print clean. Do not allow it to become oily or dirty.

4. Keep tools and chips off the print.

5. Do not carry an unfolded print. Refold the print on the fold lines before taking it anywhere.

FOLDING PRINTS

Prints are made in several standard sizes, as shown in Figure 1-3. Prints are normally folded to make them easier to store and carry. Figure 1-4 shows the methods commonly used to fold the standard print sizes.

When folding prints, the title block should always be on the outside so that each print can be identified easily. If the print has been folded with the title block covered, the flap should be folded back as shown in Figure 1-5. Folding prints in this manner makes every print the same size. This allows prints to be stored in standard letter-size filing cabinets.

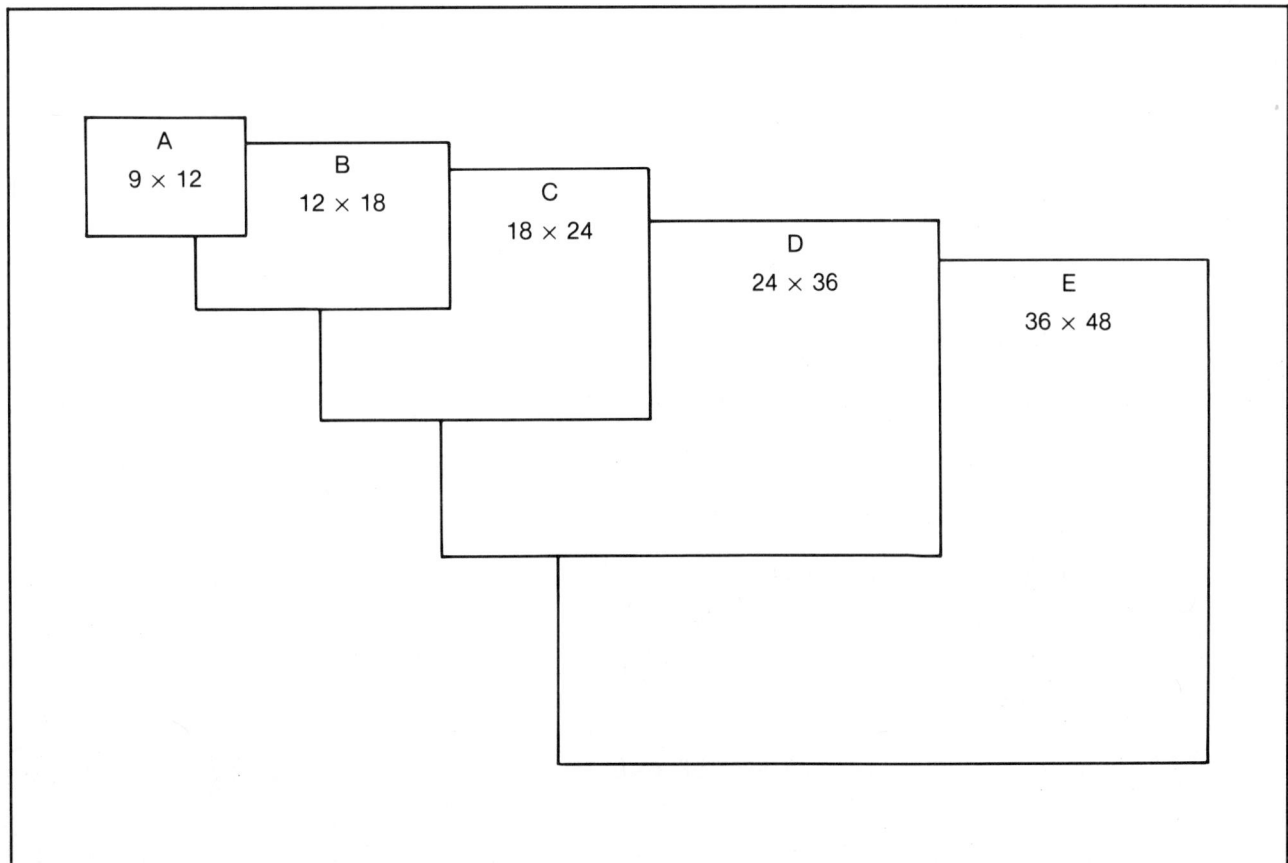

Figure 1-3. Standard drawing sheet sizes, in inches

A — 9 × 12
B — 12 × 18
C — 18 × 24
D — 24 × 36
E — 36 × 48

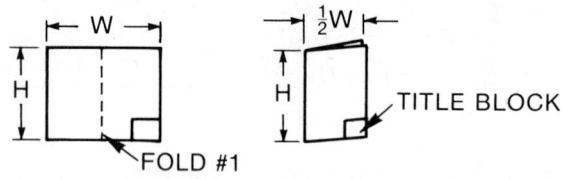

SIZE B SHEET

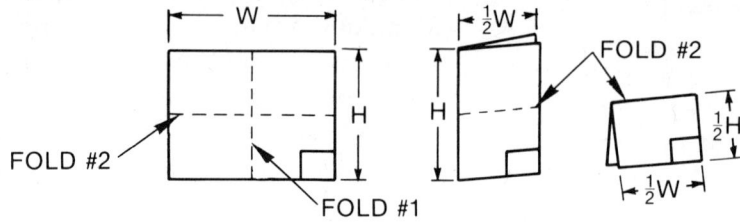

SIZE C SHEET

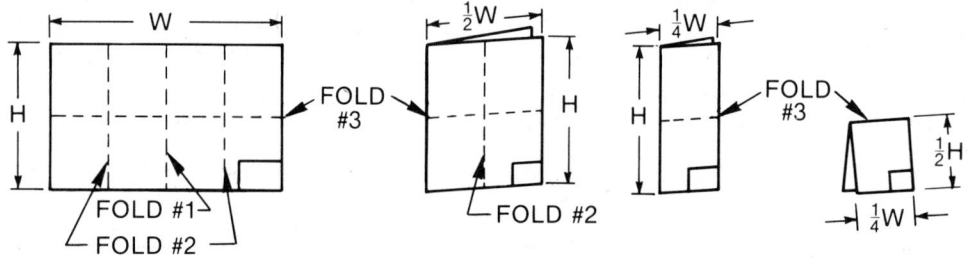

SIZE D SHEET

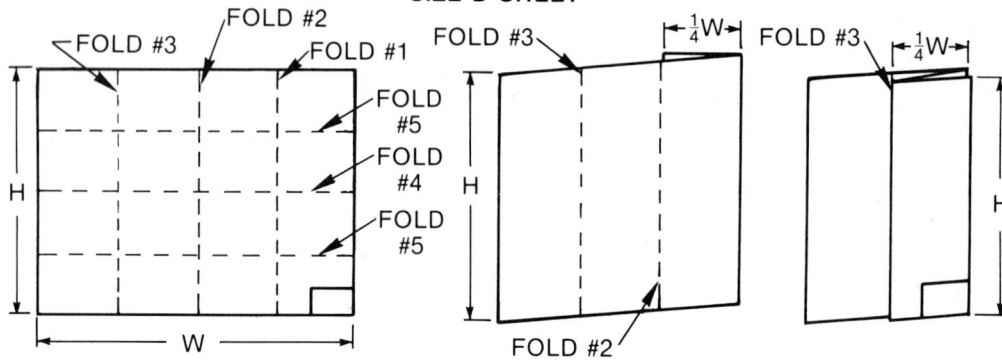

SIZES E AND AO

Figure 1-4. Proper methods of folding prints

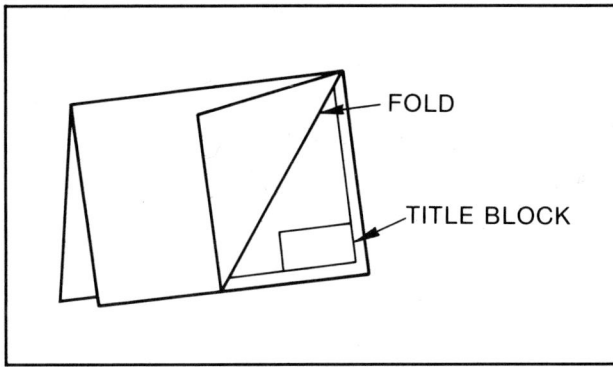

Figure 1-5. The print must be folded so that the title block is visible.

The following key terms were introduced in this chapter. Be sure you know the meaning of each term before proceeding to the review material.

Blueprint
CAD
Diazo Print
Engineering Drawing
Print
Whiteprint

REVIEW

Test your knowledge with this reinforcement study material. Write your answers to the questions in the spaces provided.

1. Explain the difference between a print and an engineering drawing. _____
   _____
   _____
   _____

2. Describe the difference between a whiteprint and a blueprint. _____
   _____
   _____
   _____

3. List five ways to care for prints.

   a. _____
      _____
      _____

   b. _____
      _____
      _____

   c. _____
      _____
      _____

   d. _____
      _____
      _____

   e. _____
      _____
      _____

4. Why are prints folded in a set pattern? _____
   _____
   _____

5. How big is a size B drawing sheet? How big is a size C drawing sheet? _____
   _____
   _____

2

PARTS OF A PRINT

OBJECTIVES

After studying this chapter, you will be able to:

- Identify the parts of a print.
- List the title block entries.
- Identify the elements in the print body.

The term *print reading* refers to the process of studying a shop print to form a mental picture of how a completed object should appear. To correctly read a print, it is necessary to understand the parts of a print. The basic format of a print includes two general parts: the title block and the print body. See Figure 2-1. Each of these parts contains specific information about the object. Both parts should always be used together when reading a print.

TITLE BLOCK

The *title block* is used to organize all general information about the object shown in the print. The size, form, and entries of a title block depend on the requirements of the company that makes the drawing. For these reasons, many types and forms of title blocks are commonly found on prints. The title block is normally located in the lower right corner of the print. Figure 2-2 shows a typical title block with the major sections identified by number. These major sections are described below.

1. NAME AND ADDRESS. This block contains the name and location of the manufacturer or designer that prepared the drawing.

2. TITLE. The title of the drawing is a brief description of the object drawn. Generally, the part name is used as the drawing title.

3. PART NUMBER. A part number is used to identify the specific part.

4. DRAWING NUMBER. The drawing number identifies the specific drawing rather than the part. This number is often used when several parts, with different part numbers, are shown on the same drawing. The part number may be assigned as the drawing number.

5. DRAWN BY/DATE. These blocks identify the drafter who made the drawing and the date it was completed.

6. CHECKED BY/DATE. These blocks identify the checker who inspected the drawing for accuracy and the date it was approved.

7. REPLACES/REPLACED BY. These blocks are used to record the part numbers of the parts that either replace or are replaced by the part in the print. If the object shown is

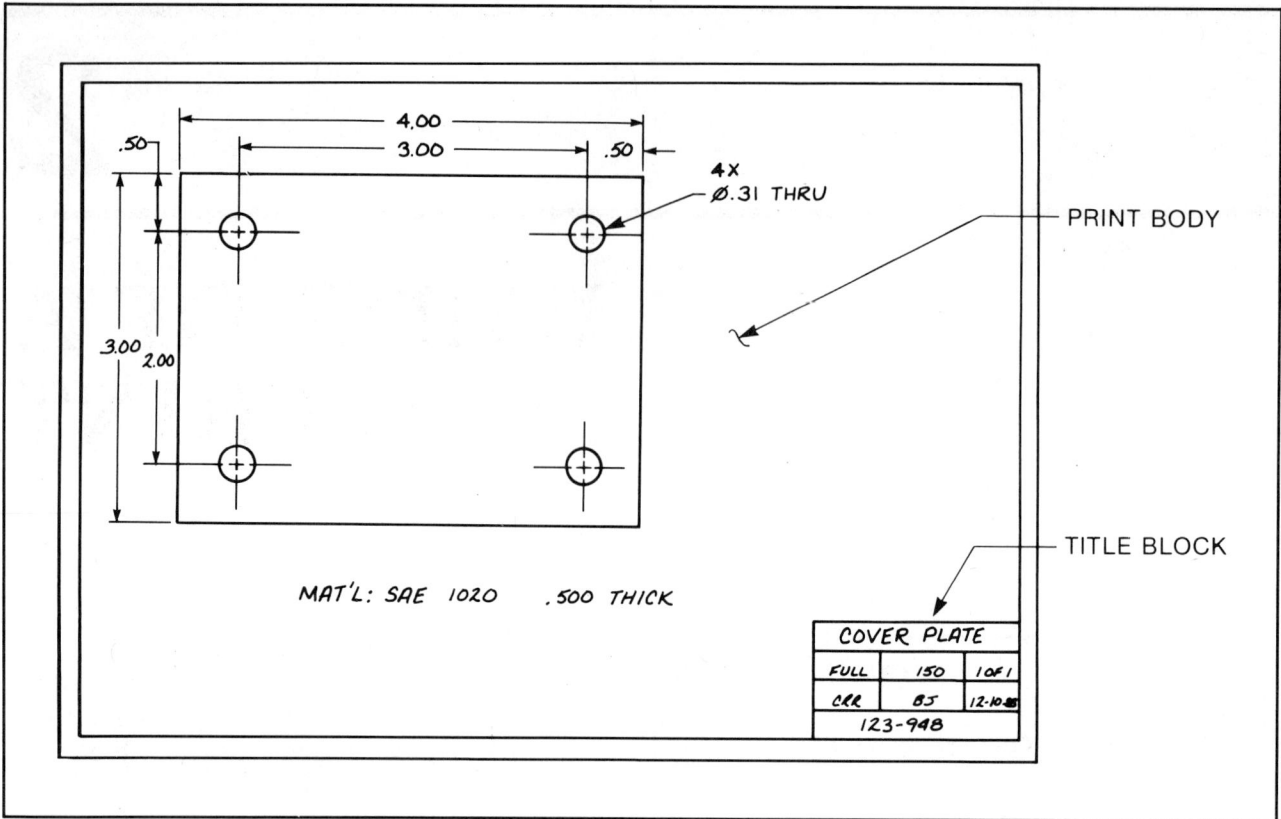

Figure 2-1. The print body shows the views of the object. The title block shows the necessary general information.

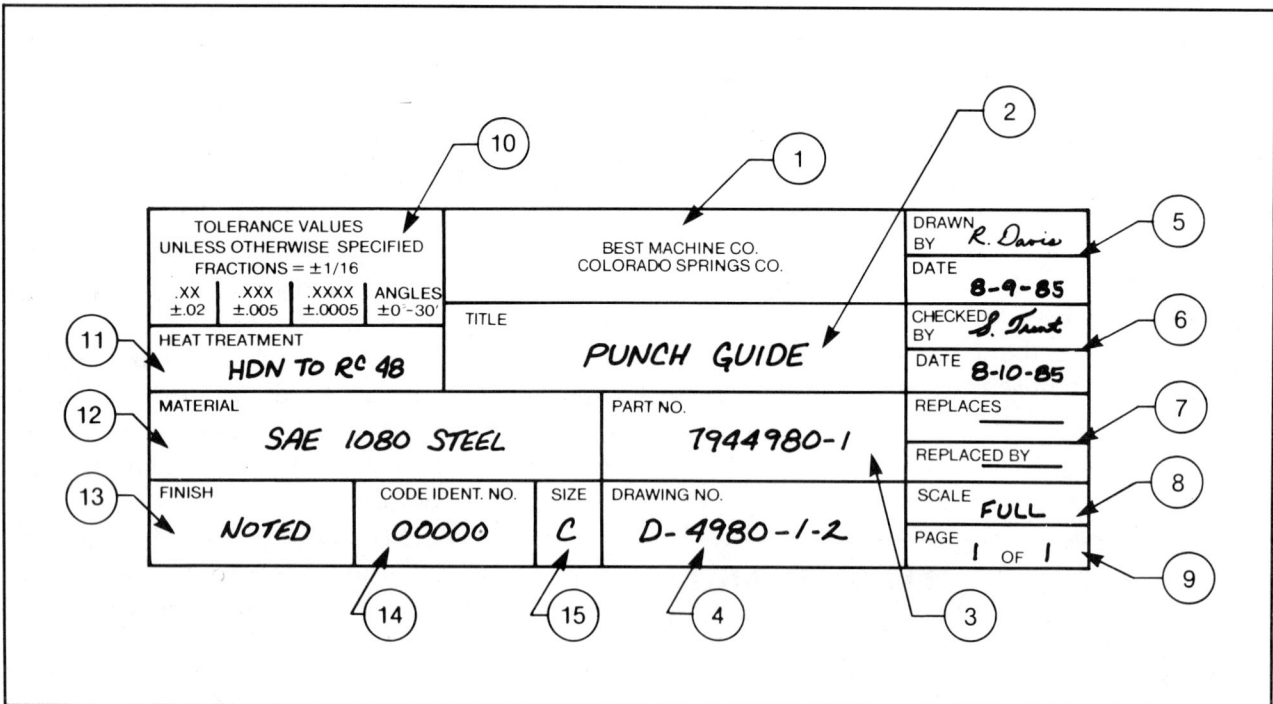

Figure 2-2. Title block

intended to replace another part, the old part number is shown in the REPLACES block. If the object shown is replaced by a newer part, the new part number is shown in the REPLACED BY block. These blocks should always be checked to make sure the print reader is using the most current print available. If the print is not replaced, a line must be drawn through the box.

8. SCALE. The *scale* is the proportional relationship of the part in the print to the actual object. When the word FULL, or 1:1 is used, the part is drawn at the same size as the actual object. A scale of HALF, or $\frac{1}{2}$:1, means that the part is drawn at half the size of the actual object.

9. PAGE. This block is used to record the consecutive order of the drawing sheets and the total number of sheets in the complete set. For example, if a block reads "1 of 4," this means that this page is sheet number 1 of a total set of 4 sheets.

10. TOLERANCES. This block is used to show the general tolerance values for (a) fractional dimensions; (b) two-place, three-place, and four-place decimal dimensions; and (c) angular dimensions. These tolerances are used to control the size limits of dimensions that do not have a tolerance applied directly. If a dimension has a specific tolerance value, then it should always be used in place of the general tolerances given in this block.

11. HEAT TREATMENT. This block is used to record the heat treatment and hardness specifications for the part. If no heat treatment is required, the word NONE should be placed in this block. Tables are available stating different heat treatments for various metals.

12. MATERIAL. The material specification in this block shows the exact material to be used to make the part. Normally, the material will be shown by name and number. If another material must be substituted, the proper authorization must be obtained first.

13. FINISH. The finish specification found in this block refers to the condition of the completed part. General surface roughness designations or finishes such as plating, painting, or chemical treating are indicated here or on the drawing.

14. CODE IDENTIFICATION NUMBER. This is the identification number assigned to a company by the government. This number is used only to identify the specific company or design activity.

15. SIZE. This is the drawing sheet size used for the drawing format.

If the word NOTED is used in any block, it means the information is given near the view, in the print body where it applies. This method is often used when several parts are shown on the same print.

In addition to the entries already discussed, some title blocks also have spaces for the quantity of parts and the part numbers of related parts. The size, form, and content of the title block are determined by the requirements of the company.

PRINT BODY

The *print body*, Figure 2-3, may show the object in a series of one or more drawings, called *views*. An object is shown this way for the purpose of clarity. While a photograph may show the basic shape of the object, for example, it may not be able to show detailed information. Dividing a part into views allows each section of the object to be shown with more detail.

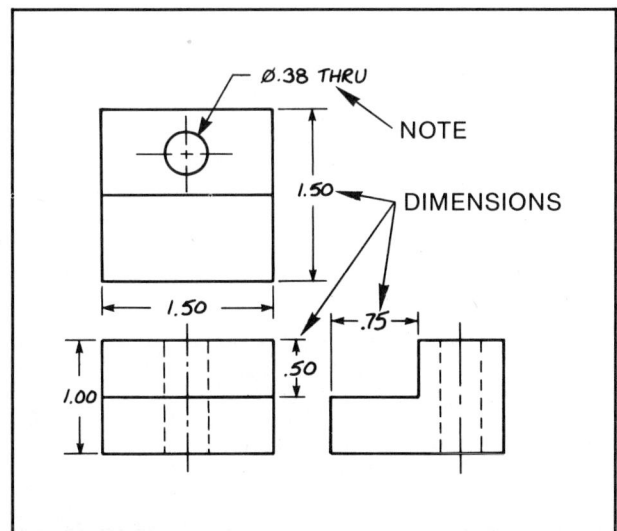

Figure 2-3. Print body with a three-view drawing

In addition to the views, the print body also includes dimensions and notes. *Dimensions* show the size of the object and the location and size of part details such as holes. *Notes* are used to give specific information about making the part.

Each view in a print represents one side of an object. In Figure 2-4 the views show each side in two dimensions. The *top view* shows the width and depth. The *front view* shows the height and width. The *side view* shows the height and depth.

The numbered object and views in Figure 2-5 can help locate and identify the views. By matching the numbers it can be seen how each view is related to the object. A more detailed study of views is covered later in this book. At this point, however, it is only necessary to

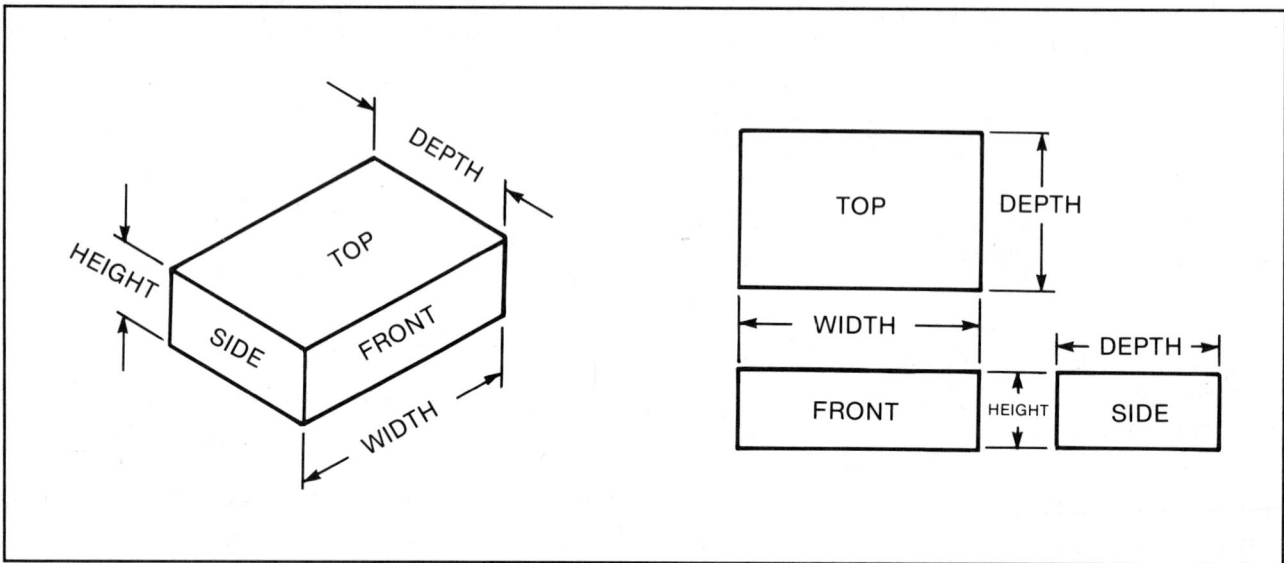

Figure 2-4. Each view shows a side of the object in two dimensions.

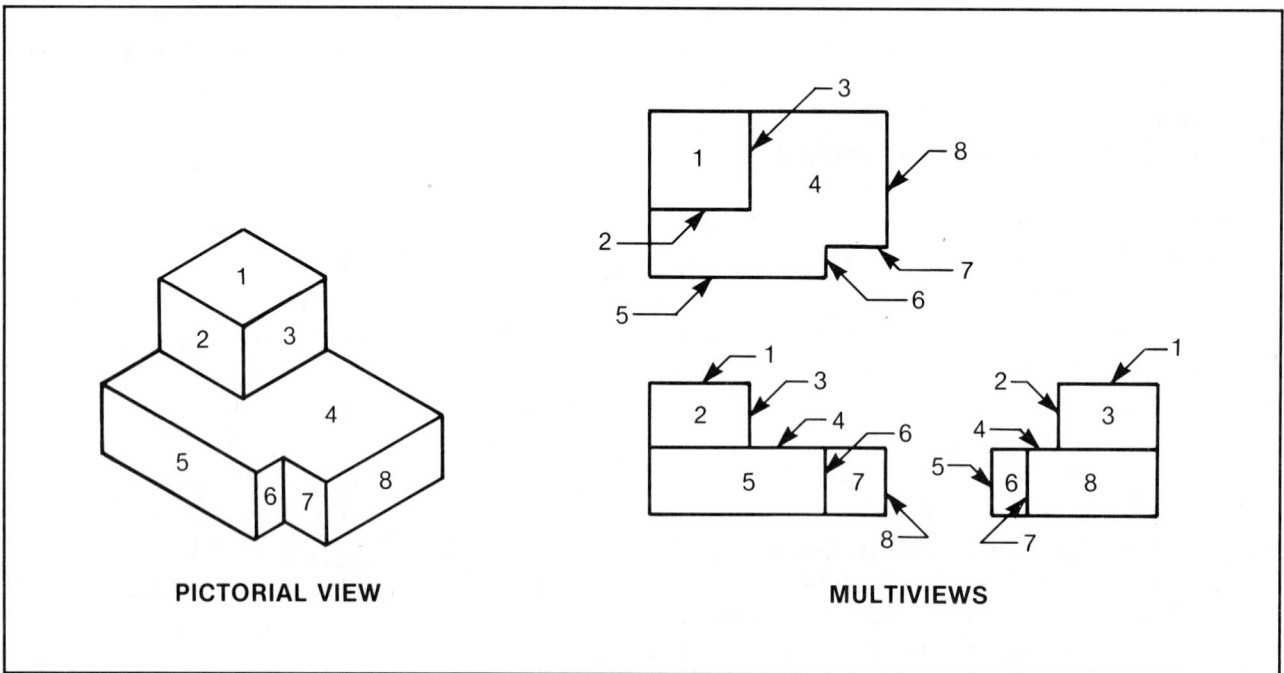

PICTORIAL VIEW   MULTIVIEWS

Figure 2-5. Matching the views with the object they represent

remember the basic relationship between the different views and the object. This is the start of understanding print reading and spatial perceptions.

READING SIMPLE PRINTS

When reading a print, the first step is to study each view separately. Every print uses lines to show edges, corners, and surfaces. Study each line carefully to find the information it shows. After studying all the views, look at how each view relates to the other views. Check the sizes (the dimensions) given to determine the overall size and proportion of the object. Survey all the information in the title block: check the part number, name, drawing scale, dates, and all other relevant information. Make sure the print

is the proper part number and the most current edition available. Following these steps will allow the print reader to form a mental picture of the object.

KEY TERMS

The following key terms were introduced in this chapter. Be sure you know the meaning of each term before proceeding to the review material.

Dimensions
Notes
Print Body
Scale
Title Block
Views

Test your knowledge with this reinforcement study material. Write your answers to the questions in the spaces provided. Refer to Figure R2-1, below, to answer the following questions.

QUANTITY 25	MATERIAL SAE 1020 STEEL	PART NAME BEARING SUPPORT	DRAWN BY E.T. DAVIS	TOLERANCES EXCEPT AS NOTED
REPLACES 2871-49	HEAT TREATMENT NONE		DATE 9-25-85	.XX = ± .02 .XXX = ± .005 .XXXX = ± .0005
REPLACED BY —	FINISH NONE	ABC CO. INC. ANY TOWN ANY STATE	CHECKED BY R. HERZOG	FRACTIONS = ± 1/16 ANGLES = ± 0°30'
CODE IDENT. NO. 00000	SCALE FULL		DATE 9-26-85	ALL DIM. ARE INCHES
PAGE 1 OF 4		SIZE A	PART NO. 39742 -006	

Figure R2-1.

1. What is the name of this part? _____

2. What is the part number? _____

3. What part does this part replace? _____

4. How many parts are required? _____

5. What size drawing sheet was used for this drawing? _____

6. What is the sheet (page) number? _____

7. How many sheets are in the complete set? _____

8. What are the general tolerances?

    a. .XX: _____   **d.** FRACTIONAL: _____

    b. .XXX: _____   **e.** ANGULAR: _____

    c. .XXXX: _____

9. On what date was the drawing completed? _____

10. When was the drawing checked? By whom? _____

11. What material is specified for the part? _____

12. What finish is required? _____

13. What is the scale of the drawing? _____

14. How large is the drawn object in relation to the actual part? _____

15. What heat treatment is specified? _____

16. What is the purpose of the print body? _____

_____

17. List three elements found in the print body.

a. _____

b. _____

c. _____

18. What is the purpose of the title block? _____

_____

EXERCISE 2-1

Refer to Figure E2-1 on page 18 to answer the following questions.

1. What is the part name? _____

2. What is the part number? _____

3. How many parts are required? _____

4. How large is the actual object as compared to the drawn views? _____

5. What material is specified for this part? _____

6. List the overall sizes of the object.
   a. Height = _____
   b. Depth = _____
   c. Width = _____

7. How many views are shown? _____

8. What views are shown? _____

9. What does the note specify? _____

10. Match the surfaces in the top view with those in the front and side views. An example is provided in the chart.

TOP	FRONT	SIDE
A	1	13
B		
C		
D		
E		
F		
G		
H		

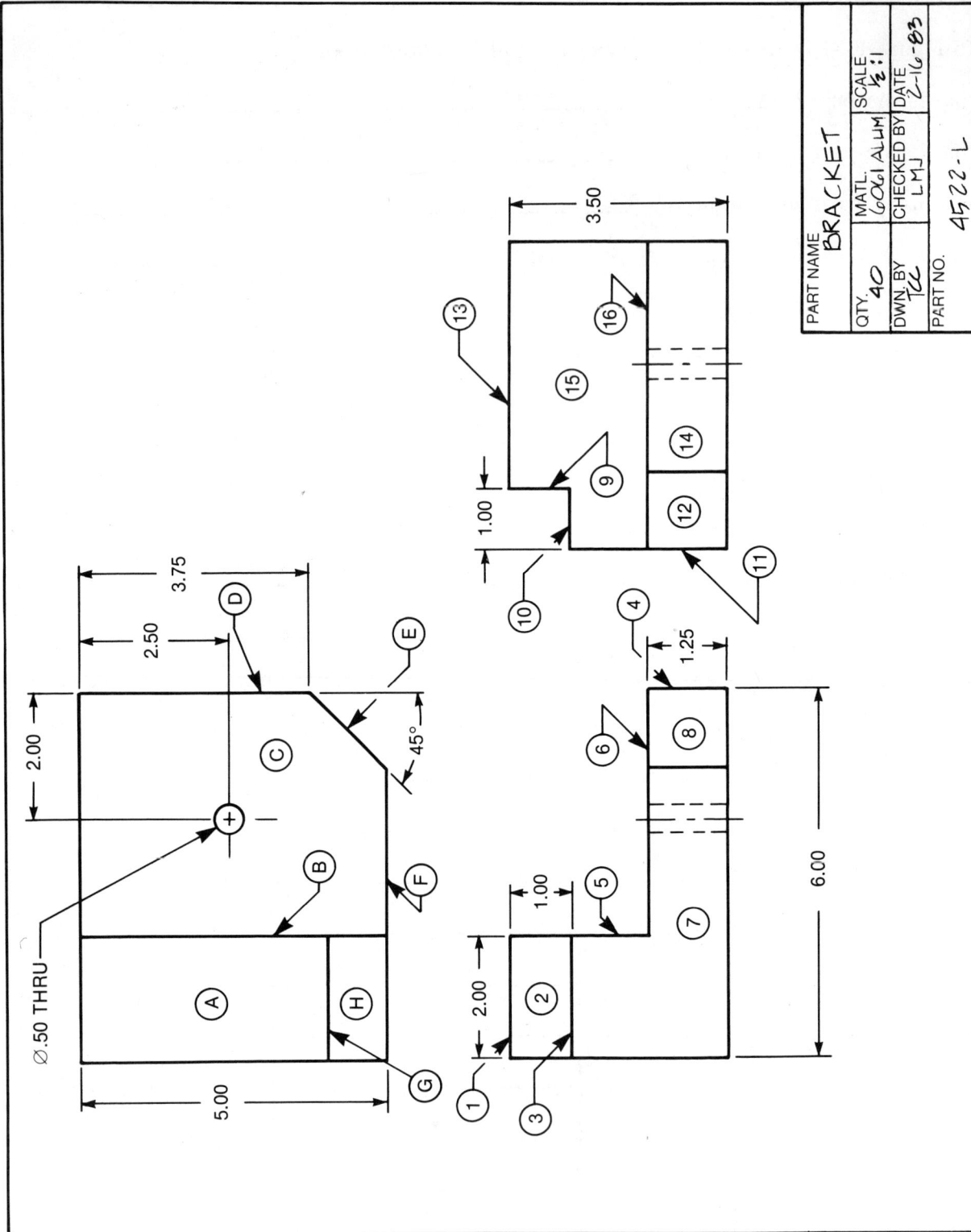

Figure E2-1.

3

LINE
CONVENTIONS

OBJECTIVES

After studying this chapter, you will be able to:

- Define the term *line conventions*.
- Describe the appearance and function of each line variation.
- Identify each line variation on a sample print.
- Determine the meaning of lines.
- Explain and identify the proper precedence of lines.

Every line used on an engineering drawing has a definite meaning and relays specific information about a part. By varying the appearance of these lines, almost any part or complex detail can be described readily.

LINE CONVENTIONS

Specific standards have been developed for the types of lines used on engineering drawings. These standard lines are commonly called *line conventions* or the *alphabet of lines*. They form the common language between the drafter and the machinist. Being able to read the lines and understand the message they convey is essential for proper interpretation of any print. Figure 3-1 shows the standard line conventions.

Visible (Object) Lines

Visible lines (also called *object lines*) are the most common type of lines found on prints. They are thick, solid lines. Visible lines are used to show an object's surfaces, edges, and corners that are visible in the view. See Figure 3-2. Visible lines are purposely drawn thick to stand out on the print. This allows the overall shape and outline of the object to be clearly defined and easily identified.

Hidden Lines

Hidden lines (also called *invisible lines*) are used to show the shape or position of surfaces and details not directly visible in a view. See Figure 3-3. These lines are drawn as evenly spaced dashes and are thinner than visible lines.

Figure 3-4 shows how visible lines and hidden lines are used to describe a part. In the top view, both slots are visible, so visible lines are used for both slots. In the front view, only one slot is shown with visible lines. The other slot is hidden in this view, so hidden lines are used. In the right side view, neither slot is visible, so both slots are indicated with hidden lines.

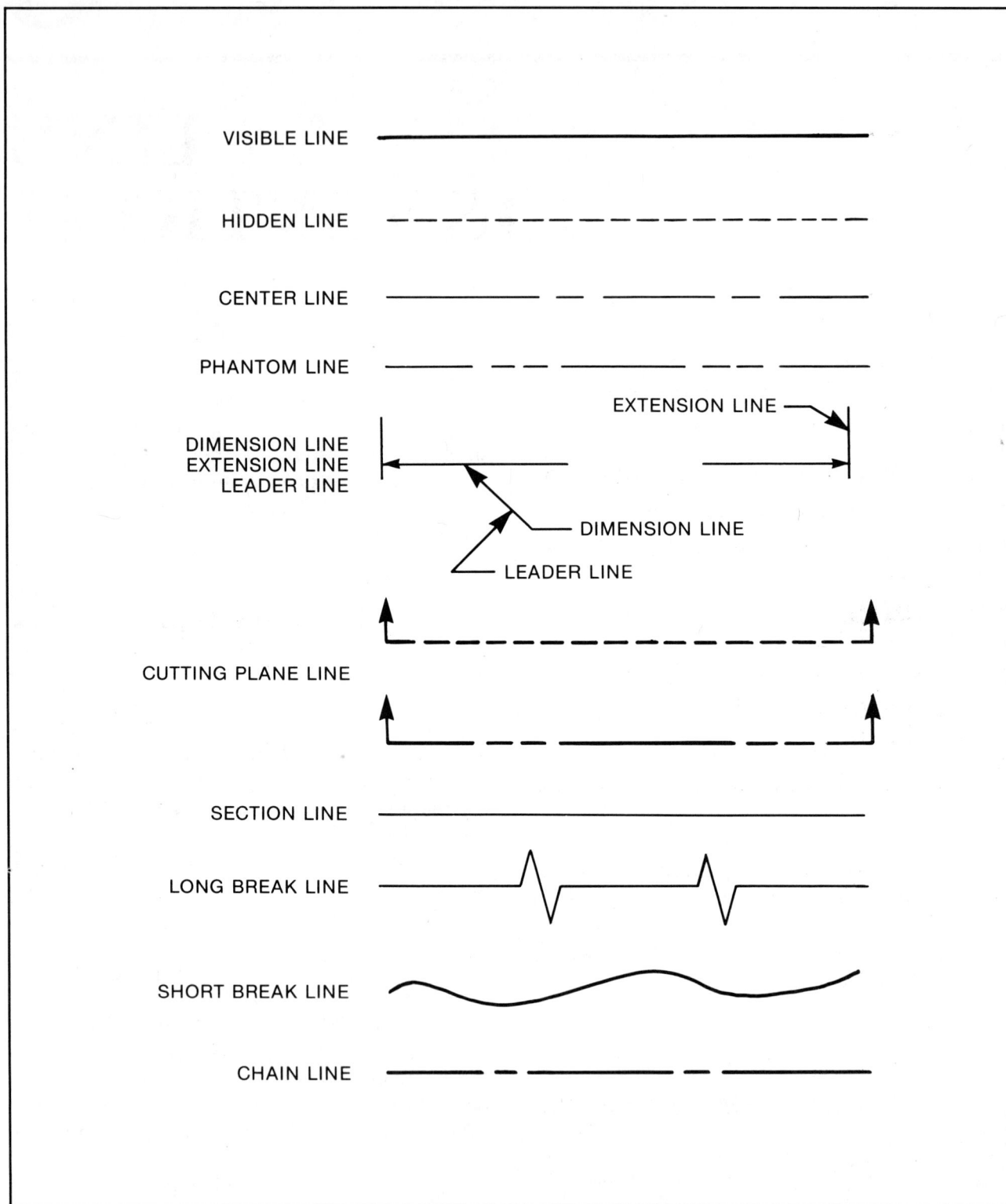

VISIBLE LINE

HIDDEN LINE

CENTER LINE

PHANTOM LINE

DIMENSION LINE
EXTENSION LINE
LEADER LINE

EXTENSION LINE

DIMENSION LINE

LEADER LINE

CUTTING PLANE LINE

SECTION LINE

LONG BREAK LINE

SHORT BREAK LINE

CHAIN LINE

Figure 3-1. Line conventions

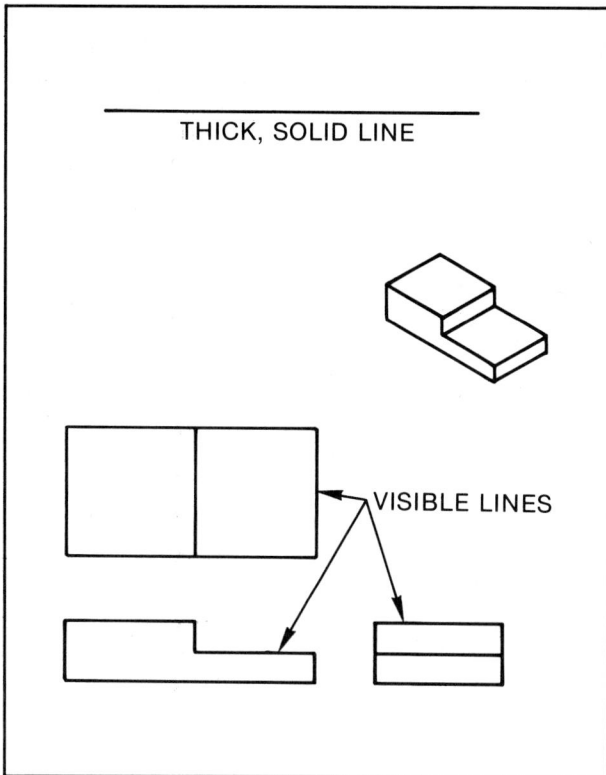

THICK, SOLID LINE

VISIBLE LINES

Figure 3-2. Visible lines are thick, solid lines.

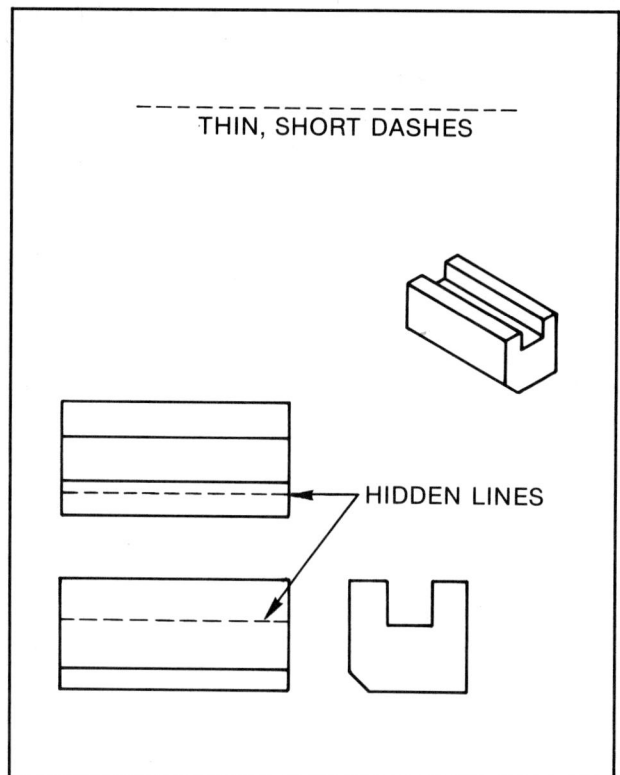

THIN, SHORT DASHES

HIDDEN LINES

Figure 3-3. Hidden lines are thin, dashed lines.

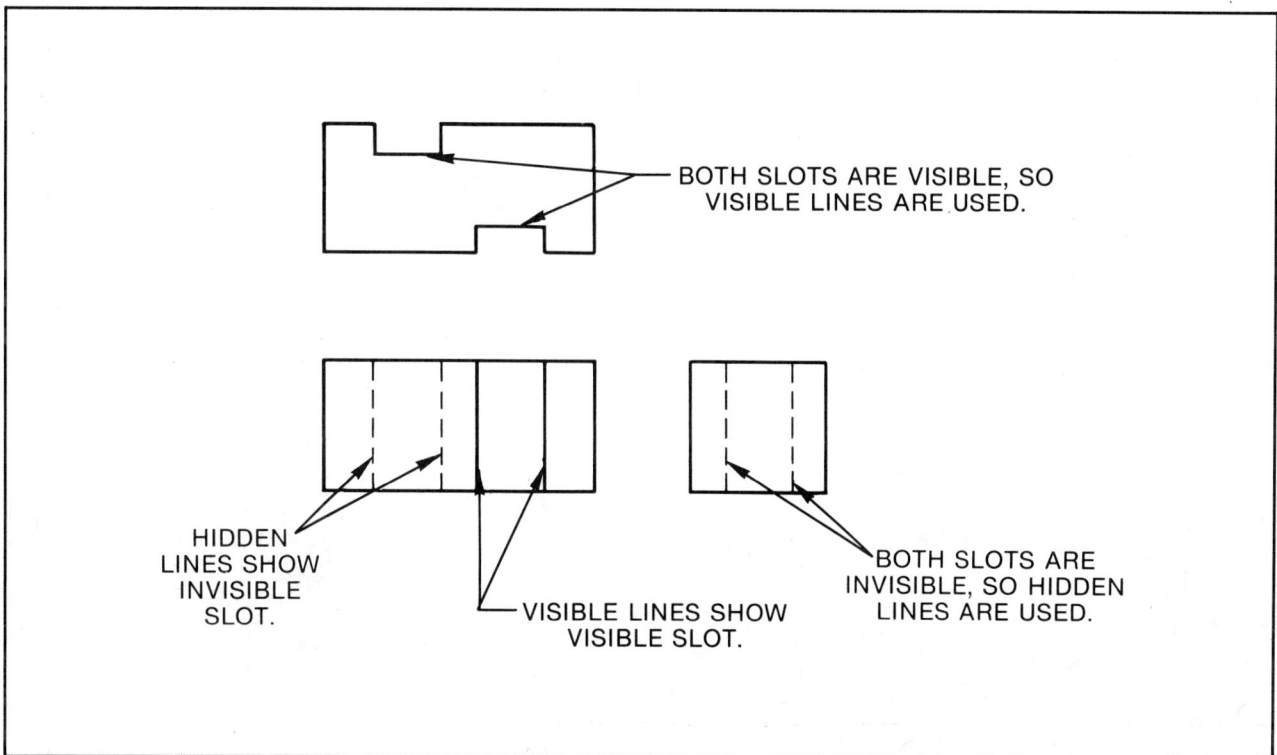

BOTH SLOTS ARE VISIBLE, SO VISIBLE LINES ARE USED.

HIDDEN LINES SHOW INVISIBLE SLOT.

VISIBLE LINES SHOW VISIBLE SLOT.

BOTH SLOTS ARE INVISIBLE, SO HIDDEN LINES ARE USED.

Figure 3-4. Visible and hidden lines used together

Center Lines

Center lines are thin, alternately spaced long and short dashes. Center lines can indicate center positions of holes, rounded edges (called *radii*), or part centers. See Figure 3-5A. In cases where holes are placed in a circular pattern (Figure 3-5B), the location of the bolt circle that locates the hole centers is shown with a center line.

Center lines are also used to show a symmetrical part, or equality on both sides of the center line. When a center line is used for this purpose (Figure 3-5C), only half of the part is normally drawn. The symbol of two short lines on the ends of the center line means that the part is symmetrical about the center line. That is, the part is the same size and shape on both sides of the center line, even though only one side of the part is shown in the drawn view.

Phantom Lines

Phantom lines are thin, alternately spaced long dashes separated by two short dashes. Phantom lines are used: (1) to show alternate positions of moving parts, (2) to show adjacent positions of mating or related parts, or (3) to eliminate unnecessary repeated details. See Figure 3-6. When reading prints, be careful not to confuse phantom lines with center lines.

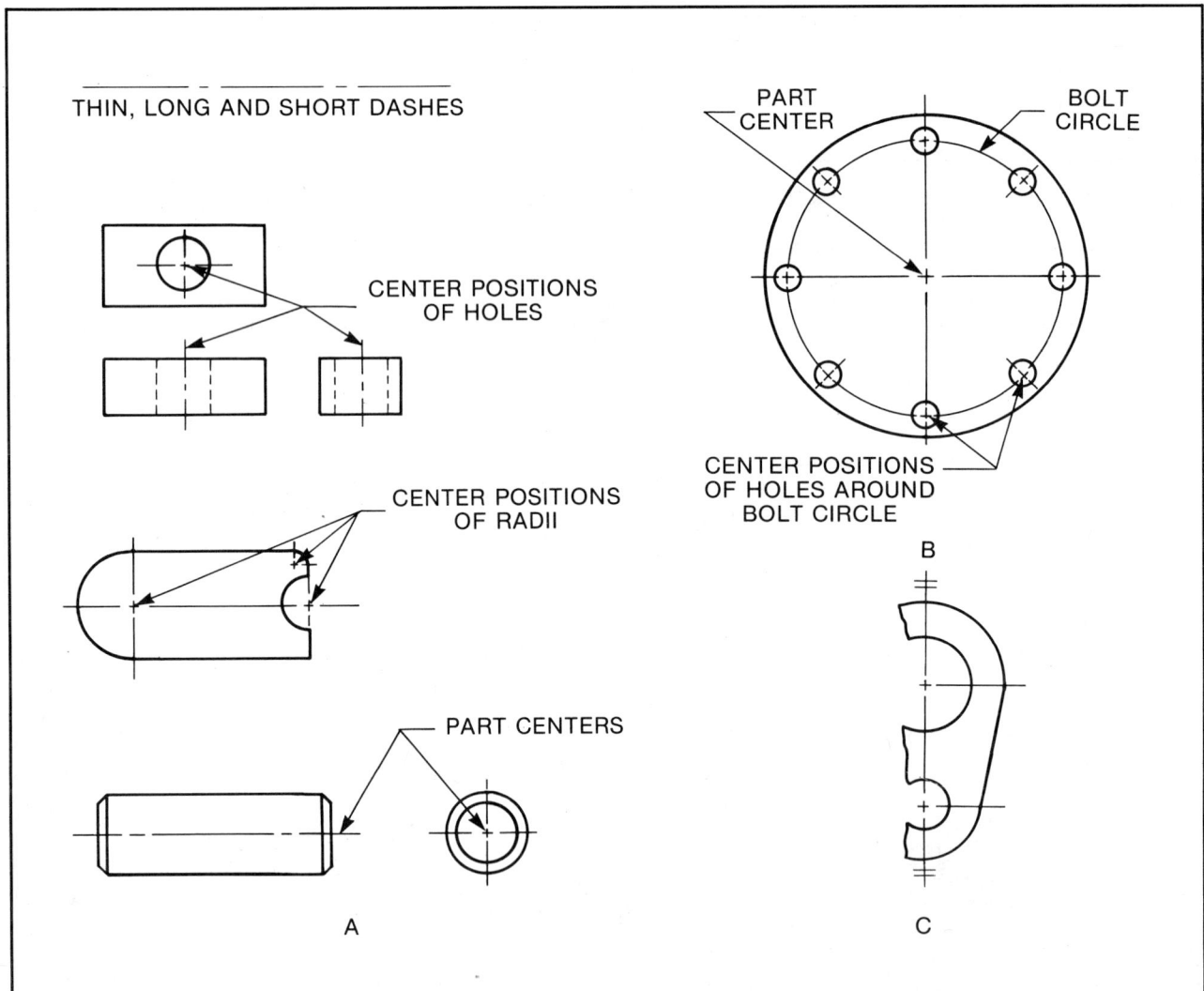

Figure 3-5. Center lines can indicate (A) center positons of holes, radii, or part centers; (B) center positions of holes around a bolt circle; (C) symmetrical parts.

THIN DASHES, ONE LONG, TWO SHORT

PATH OF MOTION SHOWN
WITH A CENTER LINE

PHANTOM
LINES

DOVETAIL
SLIDE

PHANTOM LINES

LEVER IN
OPEN
POSITION

LEVER IN
CLOSED POSITION

POSITION OF
MATING SLIDE

ALTERNATE POSITION

SPUR
GEAR

PHANTOM
LINES

ADJACENT OR RELATED PARTS

CENTER
LINE

CENTER
LINE

PHANTOM
LINES

REPEATED DETAILS

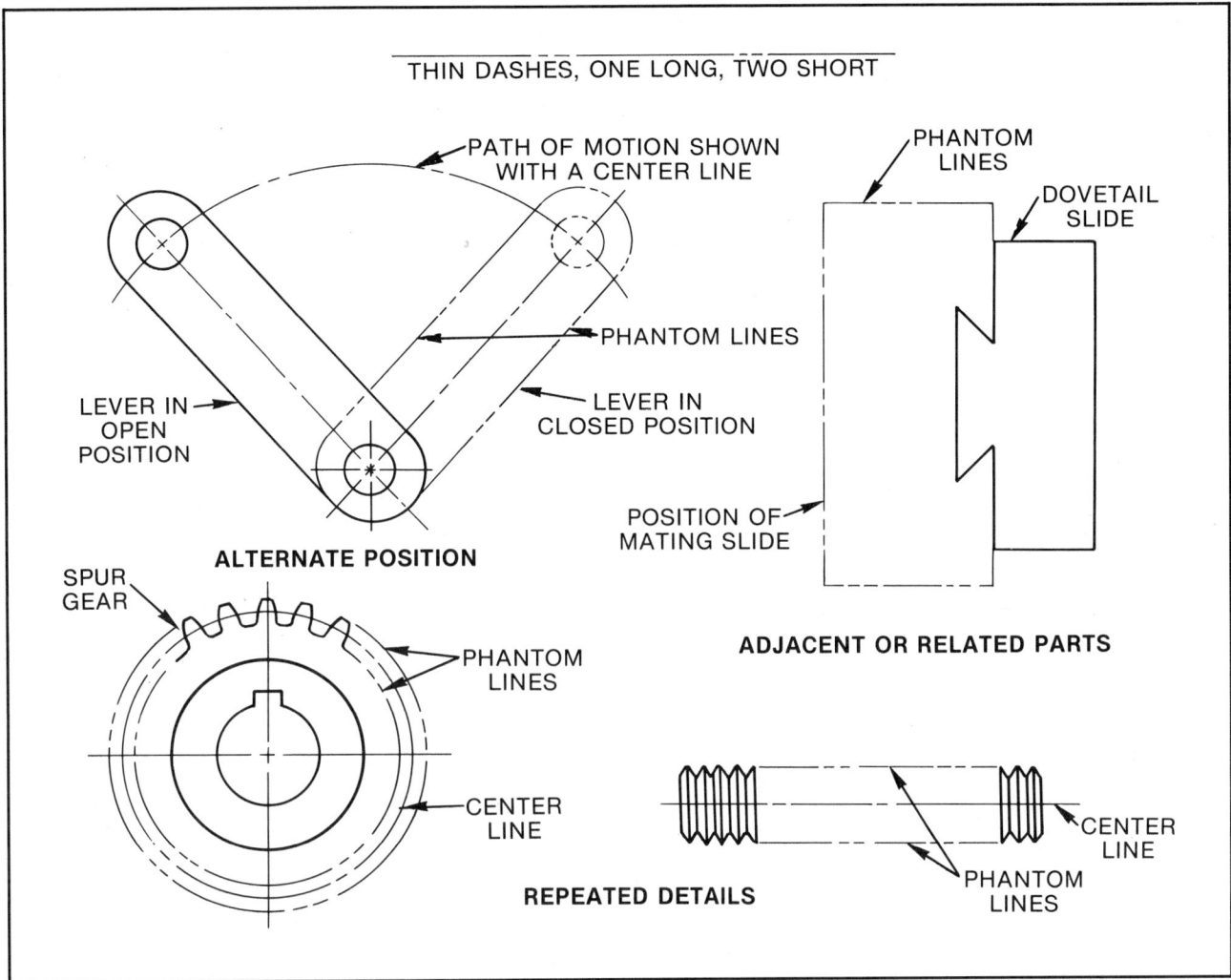

Figure 3-6. Phantom lines can indicate alternate positions, adjacent or related parts, or repeated details.

Dimension Lines

Dimension lines are thin, solid lines that have arrowheads at both ends. Dimension lines are open near the center in order to insert a dimension. See Figure 3-7. The principal use of dimension lines is to show the direction and size of the dimension included within the line.

Extension Lines

Extension lines are used to indicate the limits of a dimension. See Figure 3-8. These thin, solid lines are drawn close to the edges they limit, and extend just past the last dimension line. In cases where a dimension is given from a center line, the center line is used as an extension line.

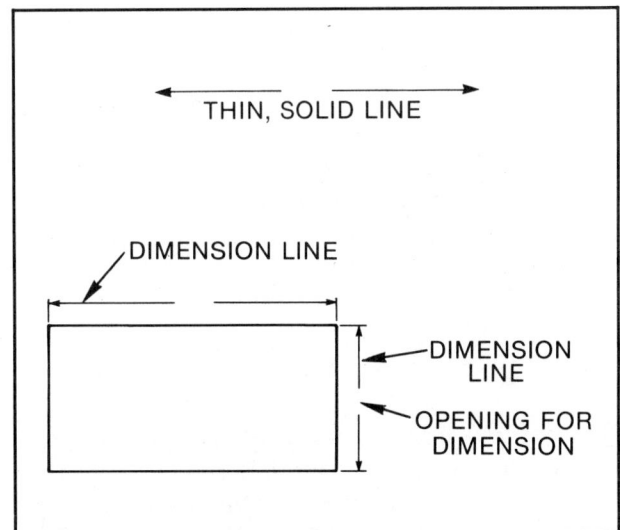

THIN, SOLID LINE

DIMENSION LINE

DIMENSION
LINE

OPENING FOR
DIMENSION

Figure 3-7. Dimension lines

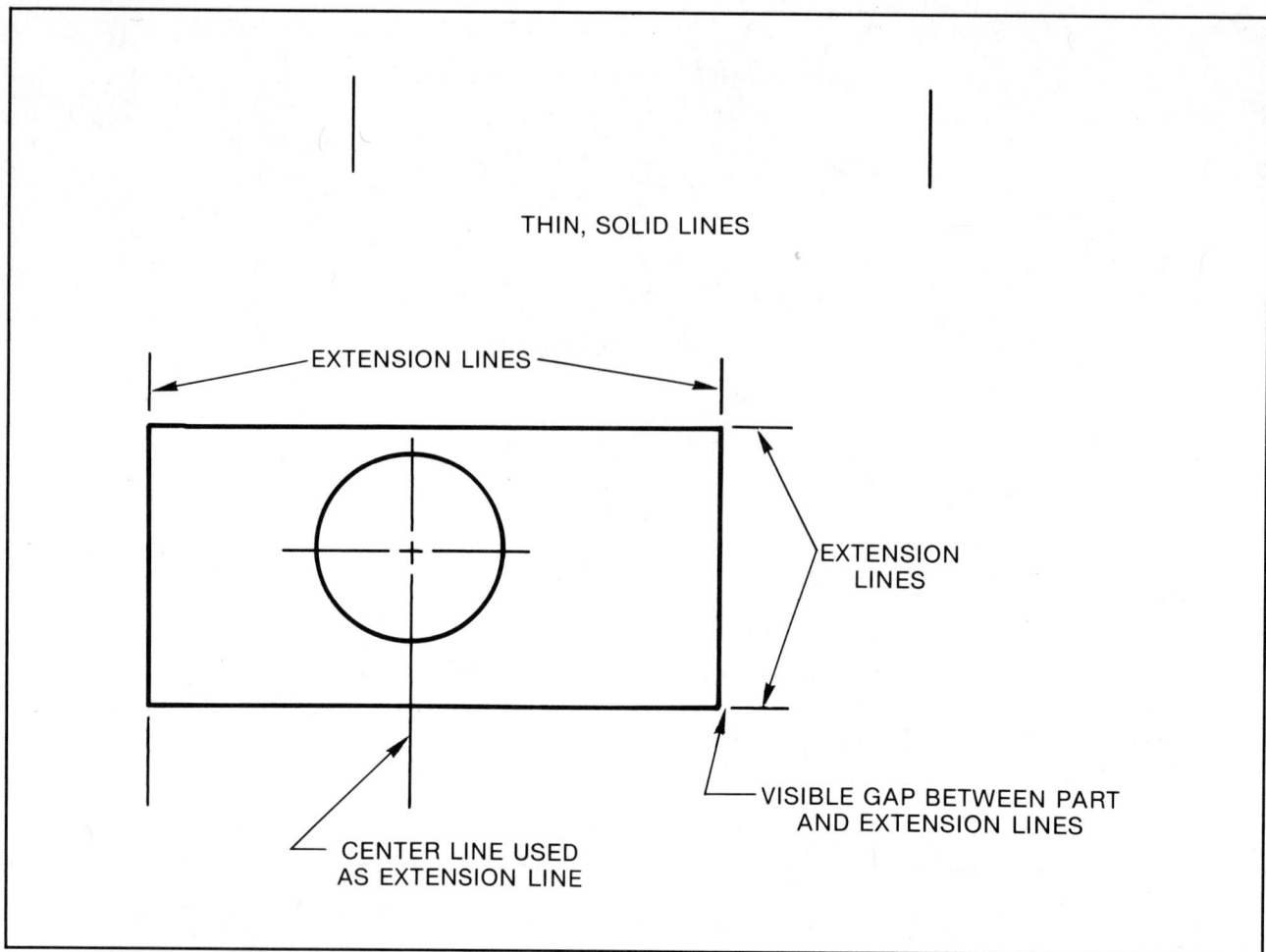

Figure 3-8. Extension lines

Leader Lines

Leader lines, or *leaders*, are single-end dimension lines. See Figure 3-9. Leaders are drawn as thin, solid lines with an arrowhead at one end and a "dogleg" bend at the other. Leaders indicate one-way dimensions, dimensional notes, and process or material specifications. Leaders, when referenced to a surface rather than to an edge or detail, end with a dot instead of an arrowhead.

Cutting Plane Lines

Cutting plane lines are used to indicate the position and path of an imaginary cut through the part. This cutting plane forms the surface shown in a sectional view of a part. See Figure 3-10. These cutting plane lines can be either *straight*, as in Figure 3-11A, or *offset*, as in Figure 3-11B. In either case, cutting plane lines are always drawn thick so they stand out clearly on the drawing. The arrowheads show the viewing direction of the section.

Letters are used with cutting plane lines to identify the specific sectional view indicated by these lines. The letters A-A are used to mark the cutting plane line and to identify the view: SECTION A-A. If more than one sectional view is used on a print, the cutting plane lines and sectional views are labeled consecutively as B-B, C-C, D-D, and so on.

Section Lines

Section lines are thin, solid lines used to show the portions of an object cut with a cutting plane line. See Figure 3-12.

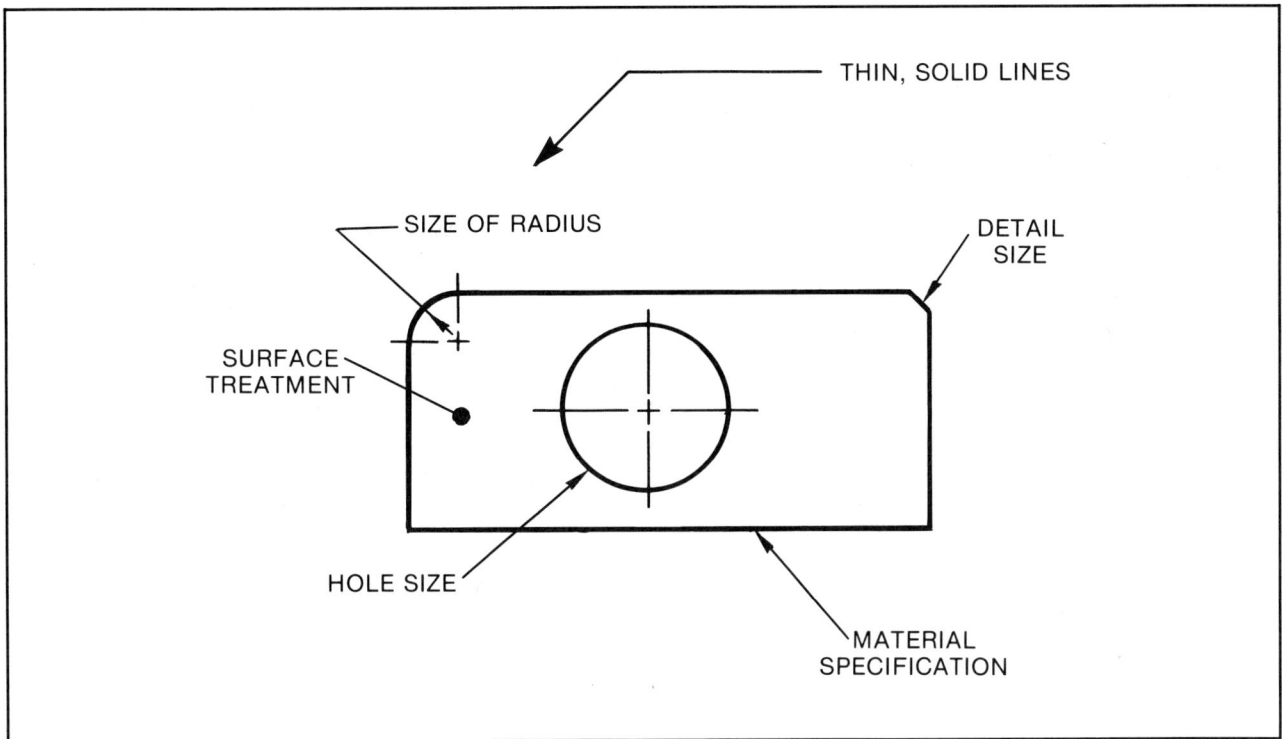

Figure 3-9. Leader lines

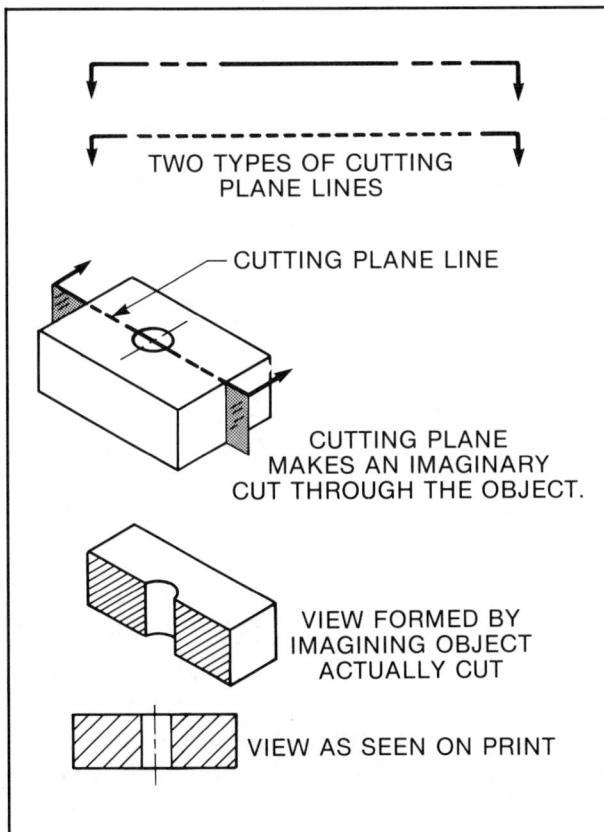

Figure 3-10. A cutting plane line locates a sectional view.

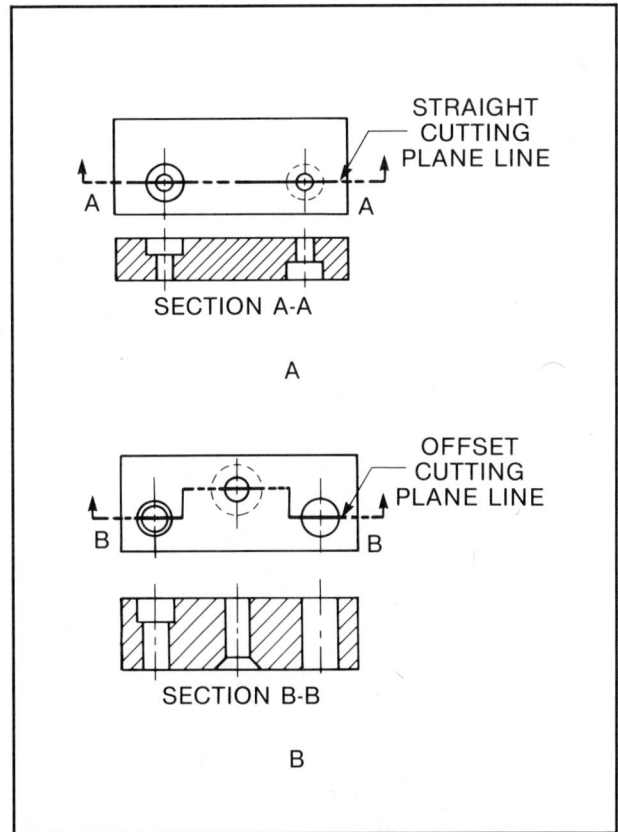

Figure 3-11. Cutting plane lines may be (A) straight or (B) offset.

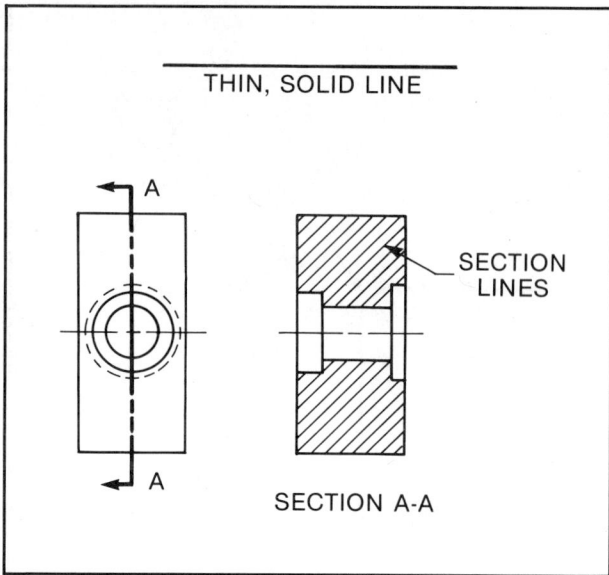

Figure 3-12. Section lines

Break Lines

Break lines are used to shorten objects that are too large to fit on the drawing sheet. When break lines are used, the section removed is assumed to be identical to the part of the object shown. Break lines are drawn as either long breaks or short breaks. See Figure 3-13. Long break lines are thin, solid lines with zigzags.

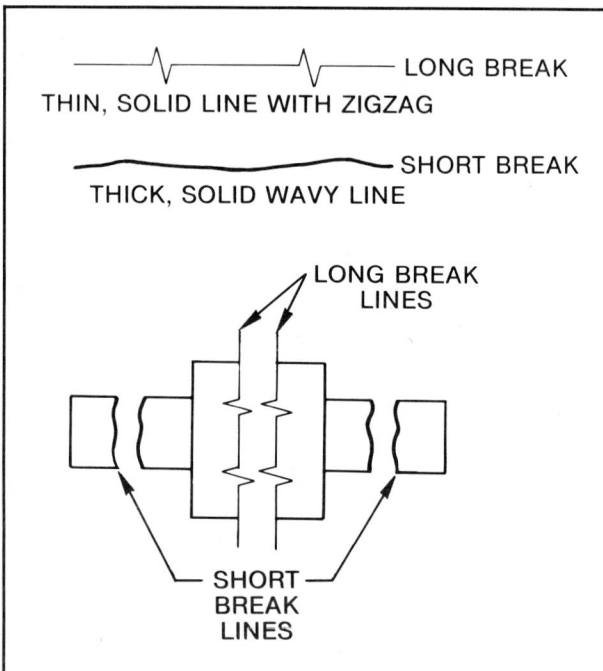

Figure 3-13. Break lines

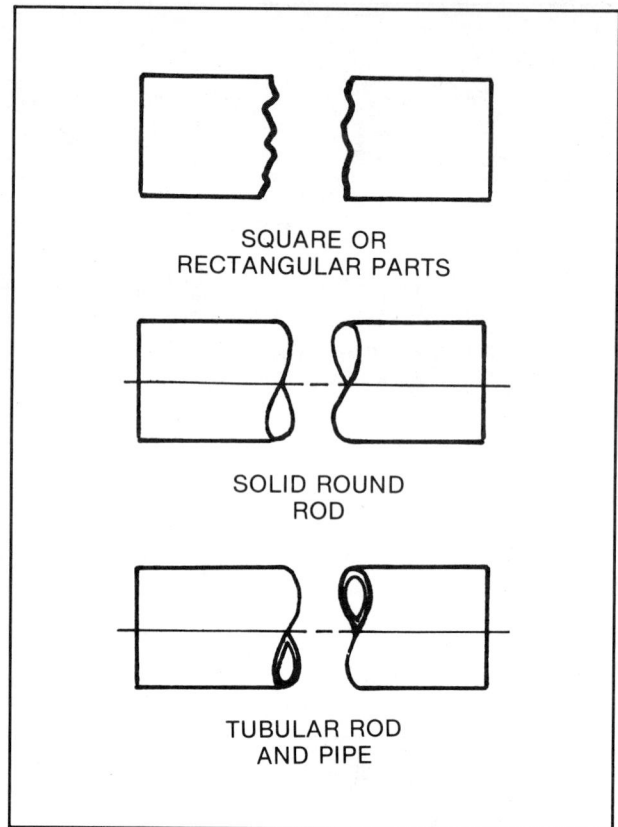

Figure 3-14. Other break line conventions

Short break lines are thick, freehand wavy lines. Other break line conventions are shown in Figure 3-14.

Chain Lines

Chain lines are thick alternating long and short dashes, the size of center lines. The chain line is used to indicate that a surface or surface zone is to receive additional treatment or consideration within the limits defined on the drawing.

MEANING AND PRECEDENCE OF LINES

The lines defined in this chapter have been discussed separately. Normally, however, several of these types of lines are used in an engineering drawing. To properly interpret the meaning of these lines and line combinations, drafters follow rules when making engineering drawings.

Meaning of Lines

Both a visible line and a hidden line can have two possible meanings:

1. The line, point, or edge where two surfaces meet as in Figure 3-15A

2. The edge of a contour or curved surface as in Figure 3-15B

To properly distinguish one meaning from the other, the object must be examined in each view. This can be seen in Figure 3-16A, which shows the front view of an object. By this view alone, it is impossible to determine what the lines are showing. However, by adding the additional views shown in Figure 3-16B, the meaning of the lines becomes clear.

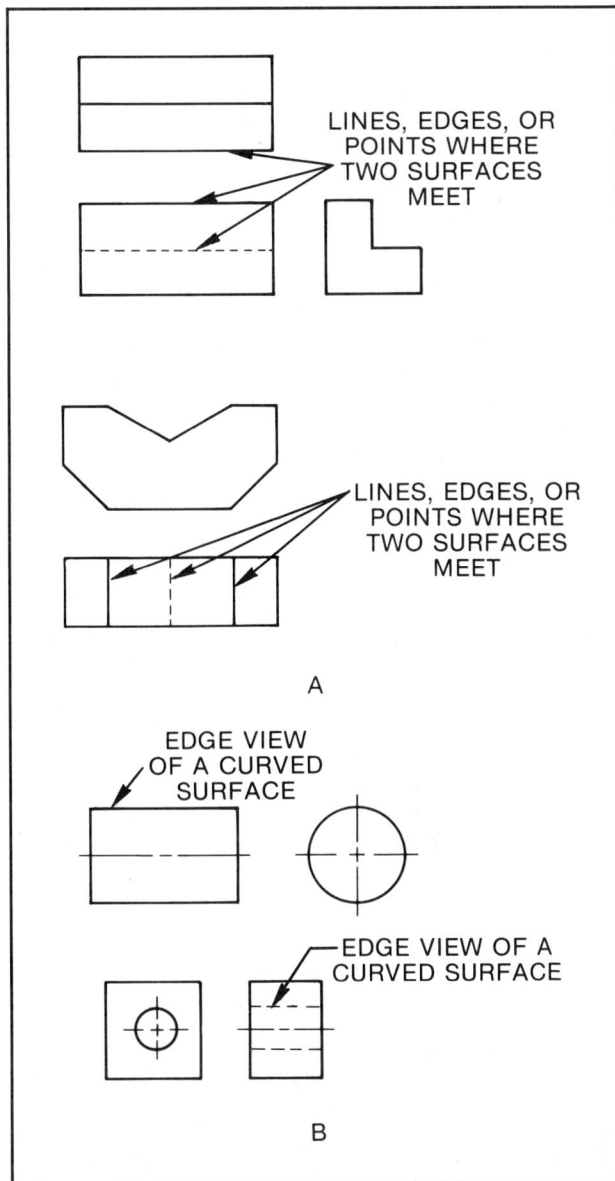

Figure 3-15. Visible and hidden lines can show (A) where two surfaces meet and (B) the edge view of a curved surface.

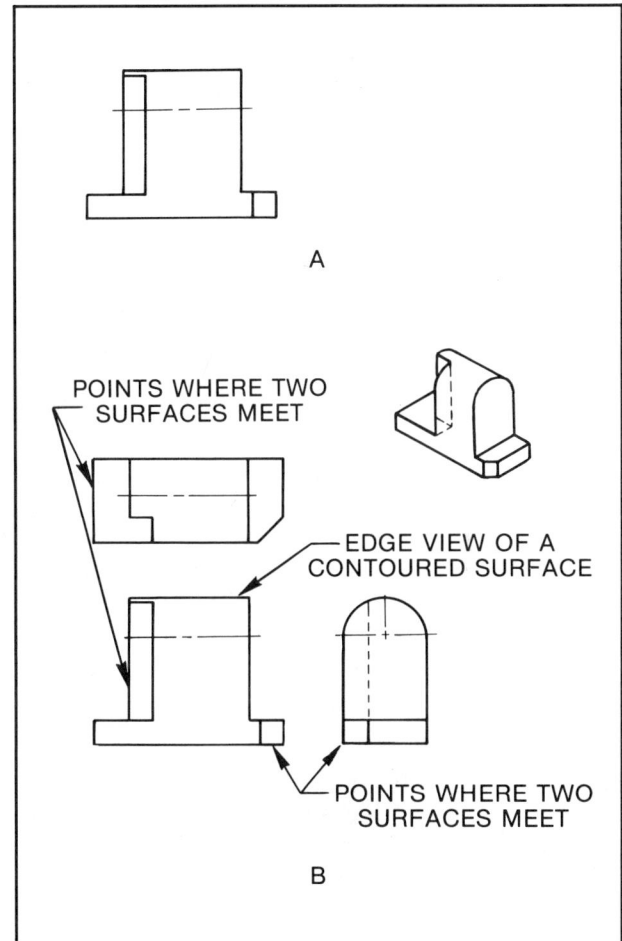

Figure 3-16. More than one view is usually needed to show the meanings of lines. The front view (A) does not show enough information. The additional views (B) clarify the meanings of the lines.

Precedence of Lines

Quite often on detailed prints, lines will "coincide," or occupy the same place. This occurs when one line in a view is aligned with another line in the same view as shown in Figure 3-17. When seen from another view, these lines will overlap each other. To properly determine the meaning, it is necessary to know the *precedence of lines*.

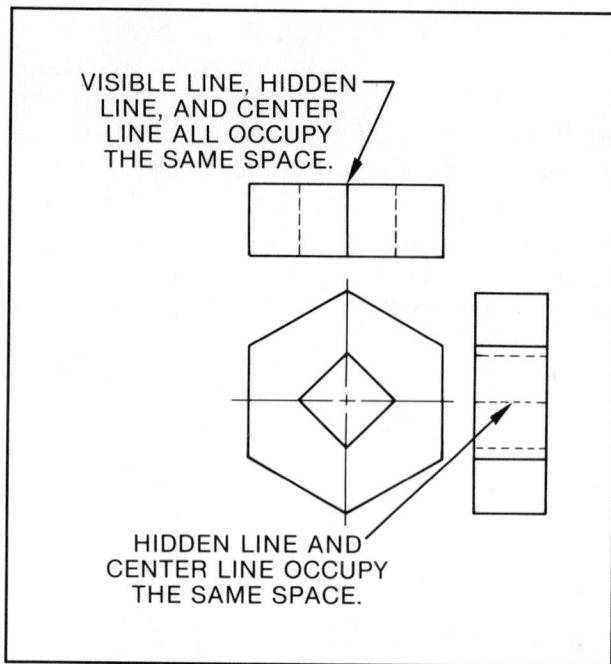

Figure 3-17. Examples of lines that overlap

Visible lines, hidden lines, center lines, and cutting plane lines occasionally overlap. Since only one line can be drawn where the lines overlap, one line must take precedence over the others. In these cases, drafters use the following order of line precedence when constructing drawings:

1. When a visible line, hidden line, and center line overlap, the visible line is shown. See Figure 3-18A.

2. When a hidden line and a center line overlap, the hidden line is shown. See Figure 3-18B.

3. When a cutting plane line and a center line overlap, the cutting plane line is shown. See Figure 3-18C.

4. In views where a center line is covered by either a visible line or a hidden line, the ends of the center line are shown detached from the part. See Figure 3-18D.

KEY TERMS

The following key terms were introduced in this chapter. Be sure you know the meaning of each term before proceeding to the review material.

Break Lines
Center Lines
Chain Lines
Conventions
Cutting Plane Lines
Dimension Lines
Extension Lines
Hidden Lines
Leader Lines
Phantom Lines
Precedence of Lines
Section Lines
Symmetry
Visible Lines

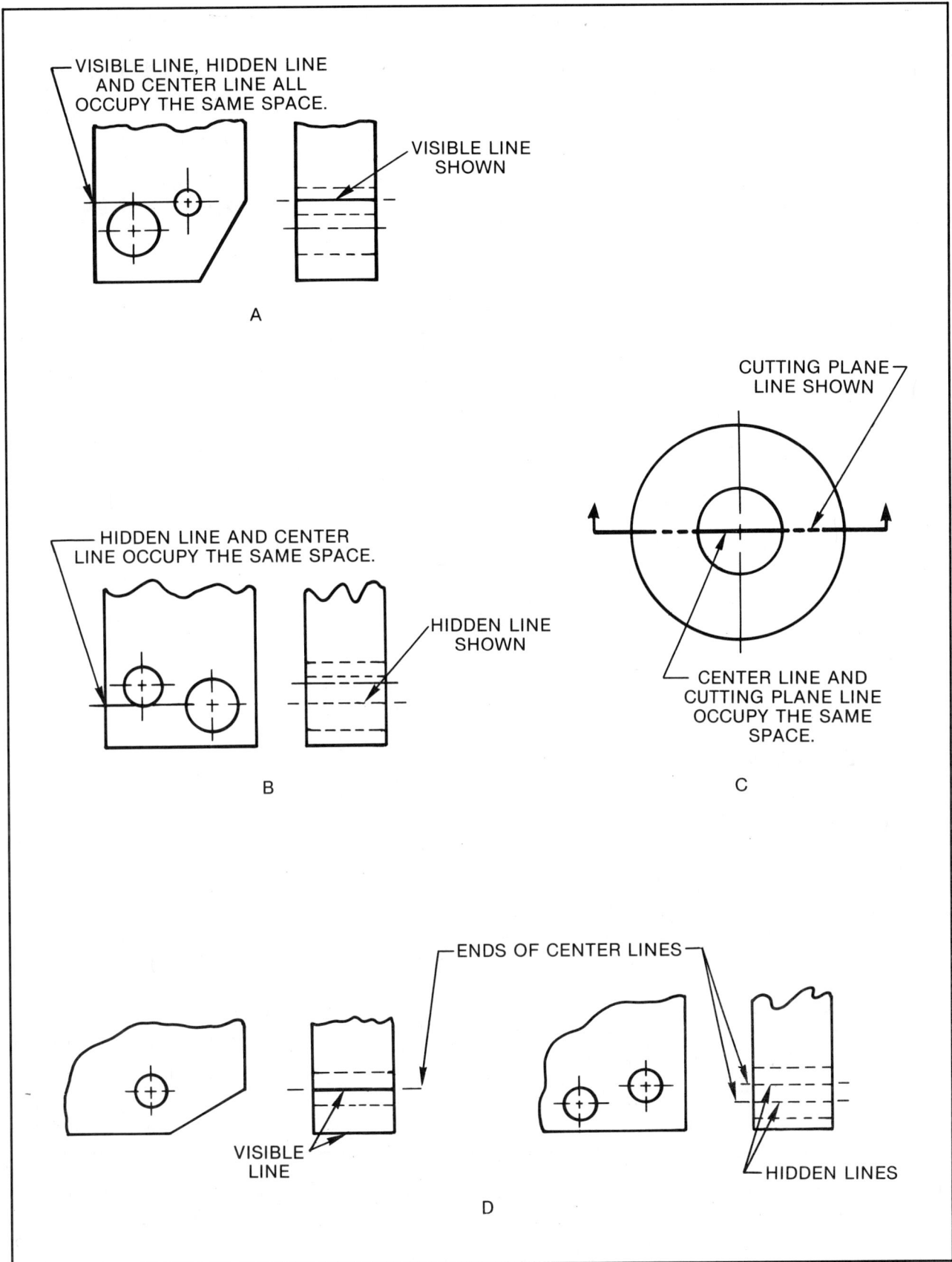

Figure 3-18. Precedence of lines

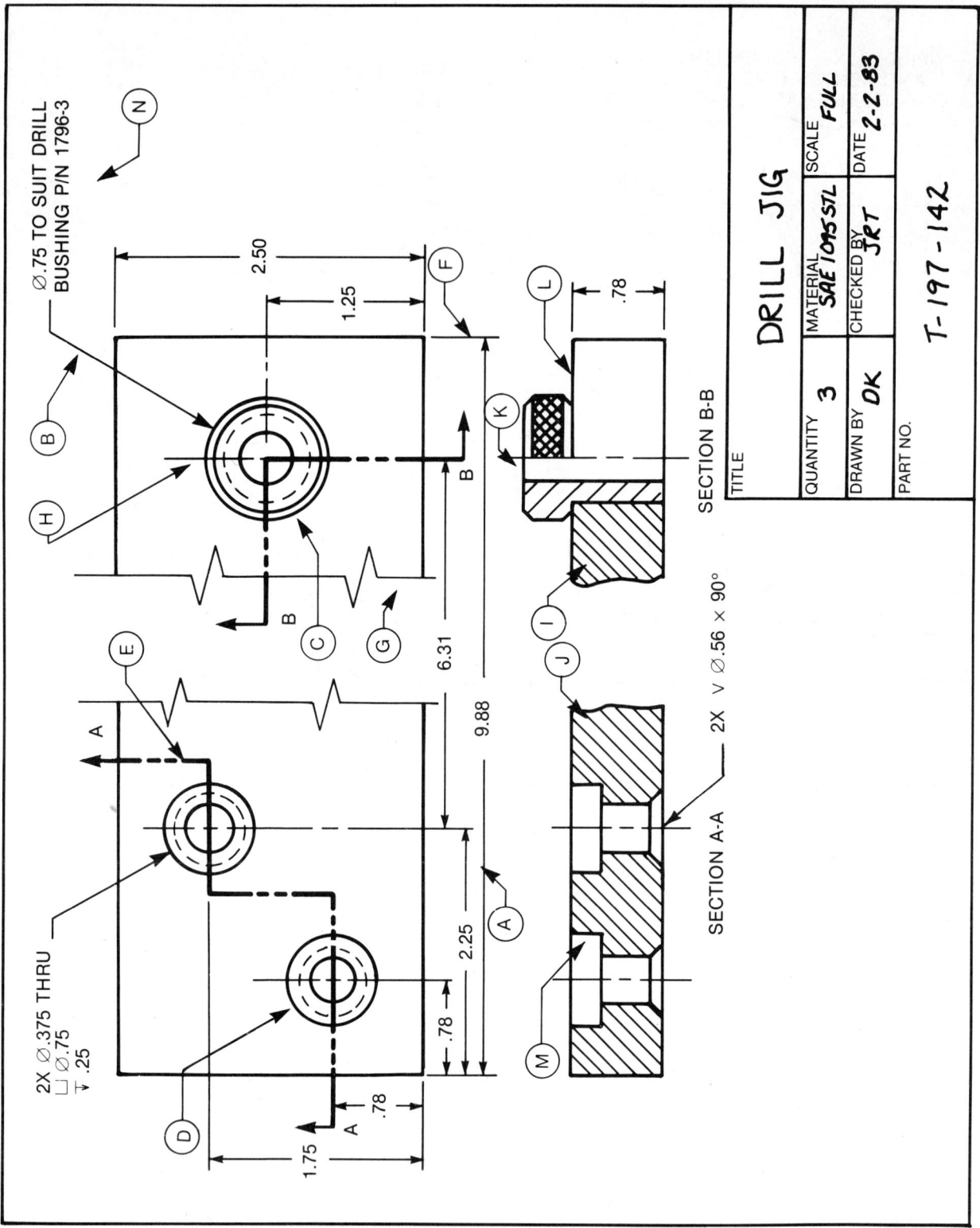

Figure R3-1.

Test your knowledge with this reinforcement study material. Write your answers to the questions in the spaces provided. Refer to Figure R3-1 on page 30 to answer the following questions.

1. What is the name of this part? _____

2. How many holes are in this part? _____

3. What is the part number on this print? _____

4. Describe note *N*. _____
_____

5. What is the purpose of the line shown at *B*? _____
_____

6. How large is this drawn object in relation to the actual part? _____
_____

7. What is shown by line *I*? _____
_____

8. What do the arrowheads at the ends of line *E* indicate? _____
_____

9. What does line *G* indicate? _____
_____

10. Identify the line convention for each line listed below.

   a. *A*: _____
   b. *B*: _____
   c. *C*: _____
   d. *D*: _____
   e. *E*: _____
   f. *F*: _____
   g. *G*: _____
   h. *H*: _____
   i. *I*: _____
   j. *J*: _____
   k. *K*: _____
   l. *L*: _____
   m. *M*: _____

Refer to Figure E3-1 on page 33 to answer the following questions.

1. What is the part number of this part? _____

2. What is indicated by line *B*? _____

   _____

   _____

3. What does line *A* show? _____

   _____

   _____

4. What does line *I* show? _____

   _____

   _____

5. What is indicated by line *F*? _____

   _____

   _____

6. Identify the line convention for each line listed below.

   a. *A*: _____

   b. *B*: _____

   c. *C*: _____

   d. *D*: _____

   e. *E*: _____

   f. *F*: _____

   g. *G*: _____

   h. *H*: _____

   i. *I*: _____

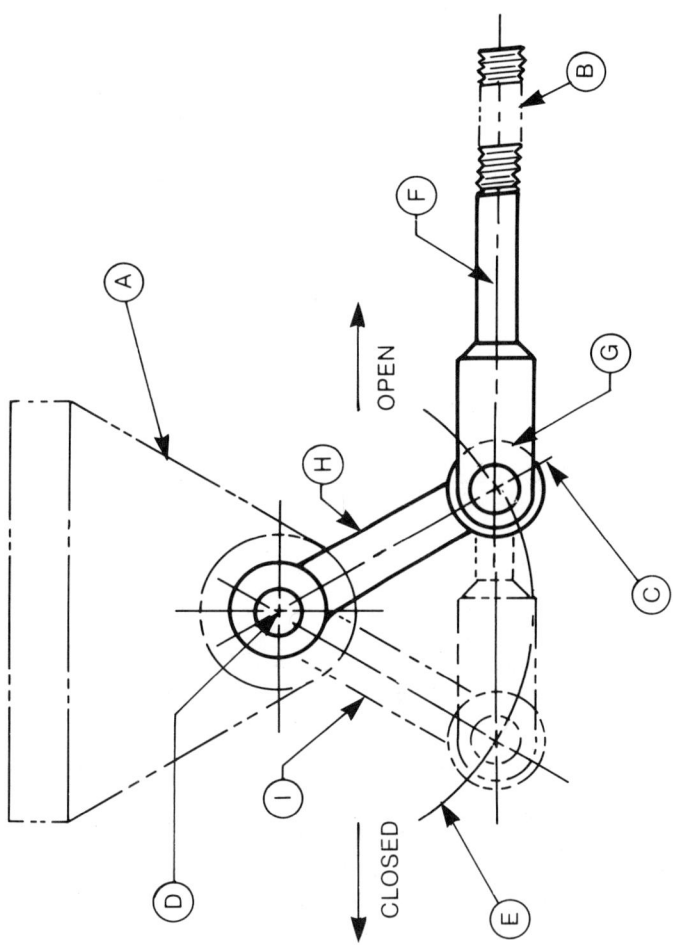

TITLE		SHIFT LINKAGE		SCALE	1/2 = 1"
QUANTITY	125	MATERIAL	SAE 1020 STL		
DRAWN BY	JPV	CHECKED BY	LR	DATE	12-16-82
PART NO.		5497-1			

OPEN

CLOSED

Figure E3-1.

Refer to Figure E3-2 on page 35 to answer the following questions.

1. What is the name of this part? _____

2. What material is specified for this part? _____

3. What type of line is shown at *A*? _____

   _____

4. What type of line is shown at *E*? _____

   _____

5. What is indicated by line *B*? _____

   _____

   _____

6. What does line *F* show? _____

   _____

   _____

7. How many parts are required? _____

8. What type of line is shown at *C*? _____

   _____

9. Identify the line convention for each line listed below.

   a. *F*: _____

   b. *G*: _____

   c. *H*: _____

   d. *I*: _____

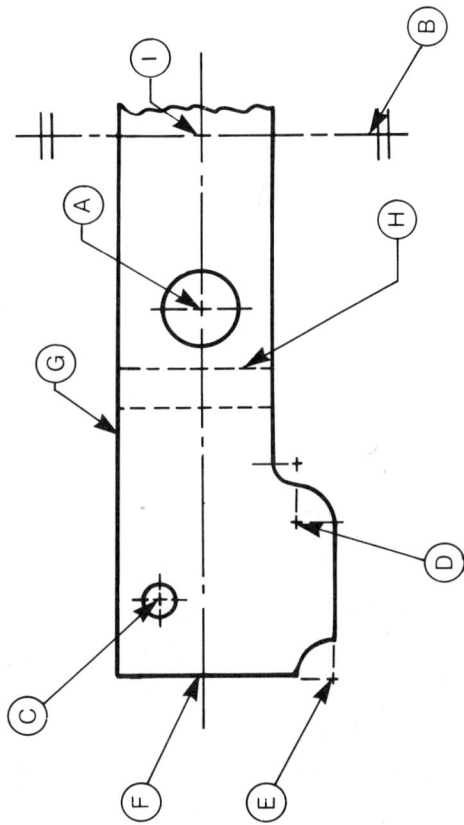

Figure E3-2.

TITLE	STIFFENER		SCALE FULL
QUANTITY 40	MATERIAL 7075 ALUM		DATE 1-13-83
DRAWN BY TC	CHECKED BY JK		
PART NO. 42-98447			

Refer to Figure E3-3 on page 37 to answer the following questions.

1. Which of the two *meanings of lines* do each of the following lines represent? Write *1* for the line, edge, or point where two surfaces meet; or *2* for the edge view of a contour or curved surface.

   a. *A*: _____

   b. *B*: _____

   c. *C*: _____

   d. *D*: _____

   e. *E*: _____

   f. *F*: _____

   g. *G*: _____

   h. *H*: _____

2. What do the lines extending from the part at *J*, *K*, and *L* indicate? _____

   _____

   _____

3. What do the notes used in this print describe? _____

   _____

   _____

4. What are the overall sizes of this part?

   a. Width = _____

   b. Depth = _____

   c. Height = _____

5. When a hidden line and a visible line occupy the same position, which line is shown?

   _____

6. List two specified lines that show precedence over another line.

   a. _____

   b. _____

7. List two specified lines that are not shown because of line precedence.

   a. _____

   b. _____

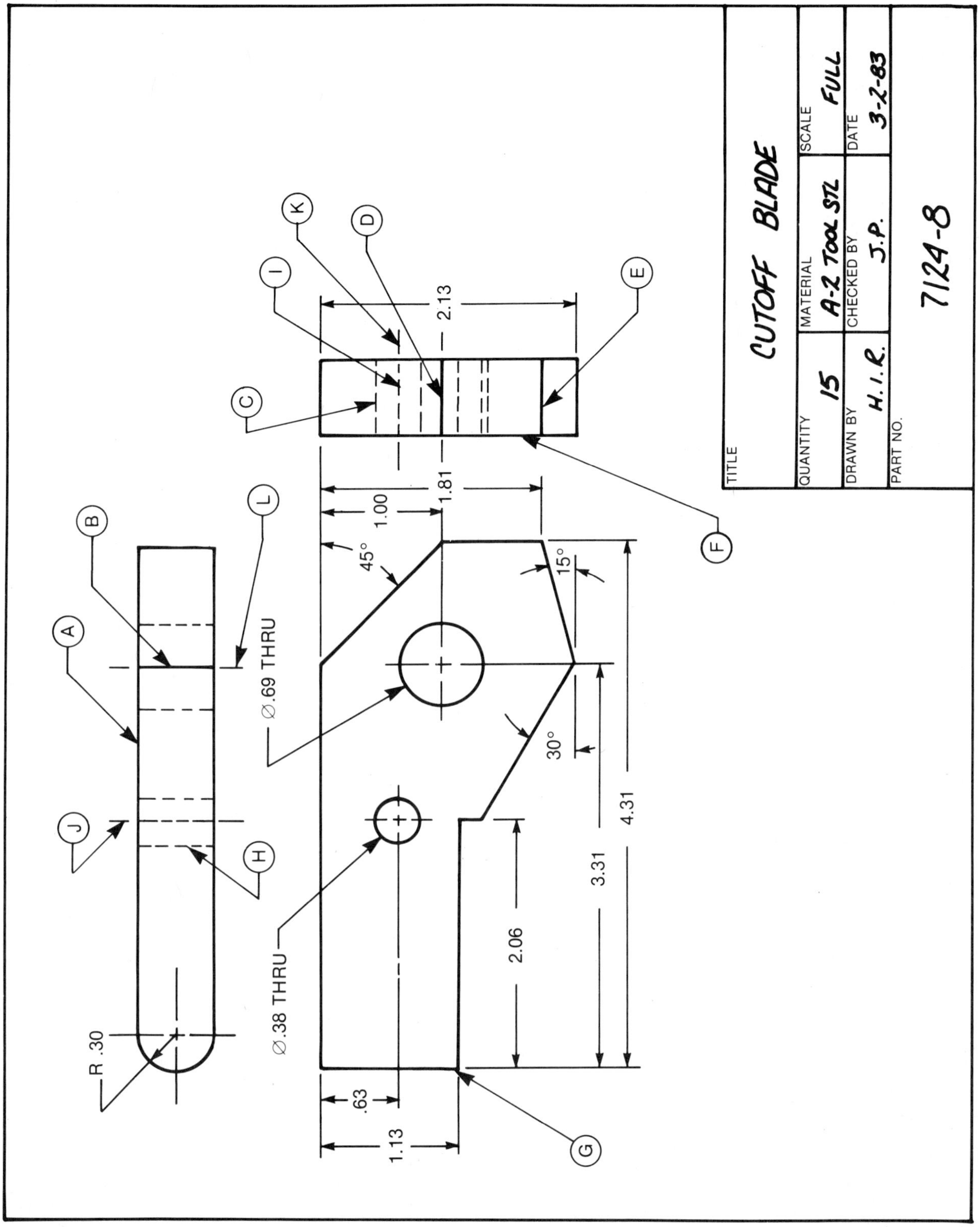
TITLE
CUTOFF BLADE
QUANTITY
15
MATERIAL
A-2 TOOL STL
SCALE
FULL
DRAWN BY
H.I.R.
CHECKED BY
J.P.
DATE
3-2-83
PART NO.
7124-8
A
B
C
D
E
F
G
H
I
J
K
L
R .30
∅.69 THRU
∅.38 THRU
2.13
1.81
1.00
45°
15°
30°
4.31
3.31
2.06
.63
1.13

Figure E3-3.

4

BASIC SHOP SKETCHING

OBJECTIVES

After studying this chapter, you will be able to:

■ Explain the purpose of shop sketching.

■ Identify the tools commonly used for shop sketching.

■ Sketch straight lines.

■ Sketch circles and arcs.

■ Sketch angular lines.

■ Describe the proper sketching techniques.

Shop sketching is a useful method of conveying technical information or ideas. Almost every part design has its start in the form of a sketch. By using sketches drafters can quickly and easily evaluate and modify technical ideas. In the shop sketches are used to clarify and simplify instructions, and to illustrate complicated ideas. Sketches also provide an inexpensive means to relay instructions for the manufacture of one-of-a-kind, custom parts.

The ability to make freehand sketches is an important asset to a machinist. Sketching is helpful not only in communicating ideas, but also in reading and interpreting shop prints.

SKETCHING TOOLS

Sketching does not require expensive drawing equipment and supplies. Shop sketching requires only three basic tools: a pencil, paper, and an eraser.

Pencils

The pencils used for shop sketching should be either medium (F) or soft (HB) lead. Pencils should be sharpened as shown in Figure 4-1. A pencil with a sharp point is used to mark the construction lines in the sketch. To make the part stand out, a pencil with a rounded, or blunt, point is used for thick lines and to darken the construction lines.

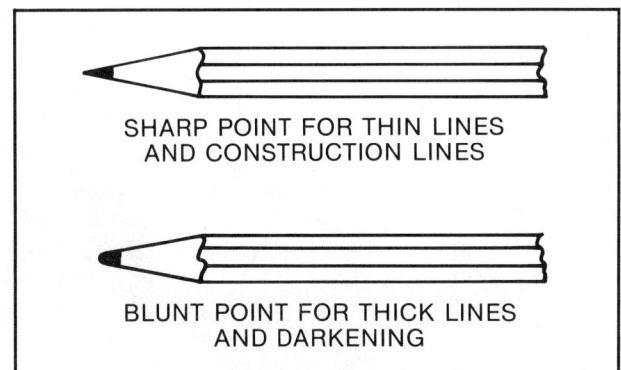

SHARP POINT FOR THIN LINES
AND CONSTRUCTION LINES

BLUNT POINT FOR THICK LINES
AND DARKENING

Figure 4-1. The proper pencil point shape is important in sketching.

Paper

Two types of paper are commonly used for sketching: *plain* and *lined*. Plain paper is usually the most available type and should be used while learning to sketch. To develop the proper technique and form, use plain, unlined paper while practicing the methods discussed in this chapter. After practicing the basic sketching techniques on plain paper, plan to use graph paper for most sketching problems. Graph paper is faster, easier, and more accurate to use.

Erasers

Erasers generally used for sketching are pink rubber or art gum types. While the eraser on the end of a pencil will work, better results are obtained by using one of these other types of erasers.

SHOP SKETCHING TECHNIQUES

The first step in mastering sketching techniques is learning how to hold a pencil properly. Hold the pencil in a comfortable position that allows for easy control without stiff movements. The point should extend approximately $\frac{3}{4}"$ to $1\frac{1}{2}"$ from the finger tips, and should be tilted as shown in Figure 4-2.

Sketching Straight Lines

A *straight line* is the basic element of any sketch. The three principal types of straight lines are *horizontal*, *vertical*, and *slanted*. The same method is used to sketch each of these lines.

To sketch straight lines, start by placing dots at the points where the line begins and ends. If the line is to be long, it may be helpful to use more dots along the path of the line. Next, tilt the pencil in the direction of travel, and using a smooth hand and arm motion, draw a series of dashes between the dots. During this process, keep your eye fixed on the last dot. This will help to maintain a relatively straight line.

The last step is to darken the line. When darkening construction lines, keep your eye just ahead of the pencil point. This will help to keep the pencil point on the construction line. The procedure for drawing straight lines is shown in Figure 4-3. A horizontal line is shown, but the procedure is the same for vertical and slanted lines.

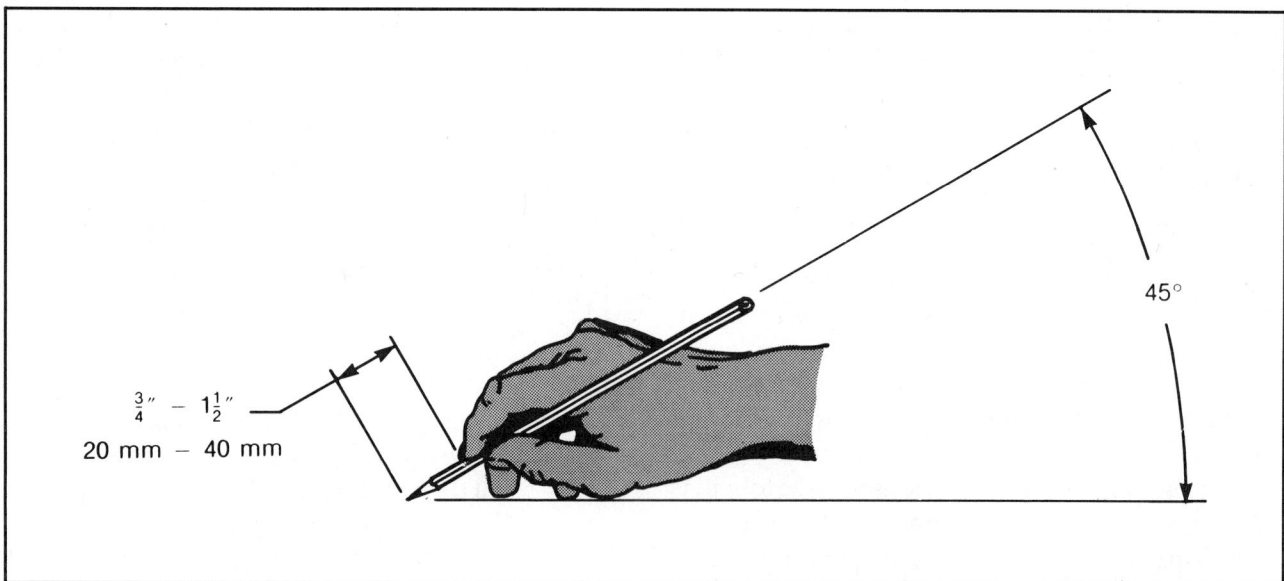

$\frac{3}{4}"$ — $1\frac{1}{2}"$

20 mm — 40 mm

45°

Figure 4-2. Holding the pencil

PLACE DOTS AT THE BEGINNING AND END OF THE LINE.

SKETCH SHORT DASHES BETWEEN THE DOTS.   KEEP YOUR EYE FIXED ON THIS POINT.

DARKEN THE LINE, KEEPING YOUR EYE FIXED SLIGHTLY AHEAD OF THE PENCIL POINT.

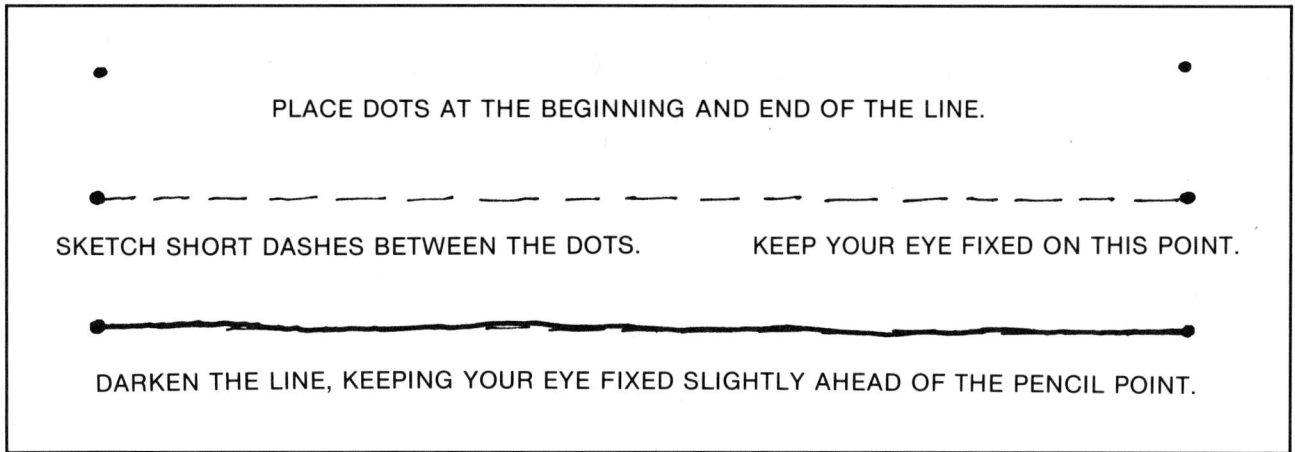

Figure 4-3. Procedure for sketching straight lines

When sketching straight lines, the pencil should always be pulled rather than pushed. This prevents tearing the paper and allows for greater control of the pencil. Try to sketch any straight line in a series of long dashes rather than in several overalpping strokes, as shown in Figure 4-4. This will prevent wide lines and will also reduce erasures.

In Practice Problems 4-1, 4-2, and 4-3, connect the dots with straight lines as indicated.

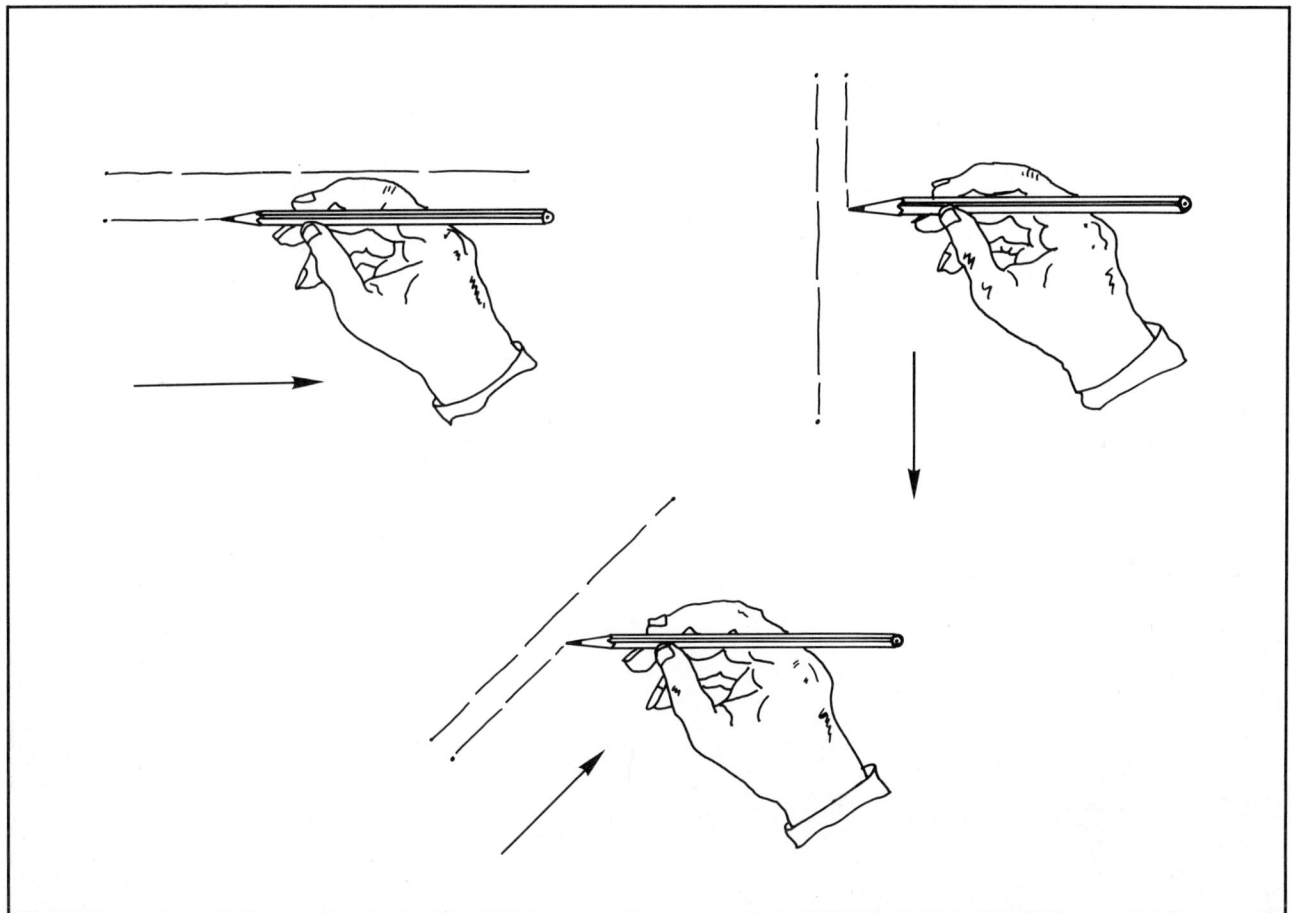

Figure 4-4. Sketching straight lines

PRACTICE PROBLEM 4-1

With a straight line, connect each lettered dot on the left to the dot with the same letter on the right.

A ●   ● A

B ●   ● B

C ●   ● C

D ●   ● D

E ●   ● E

F ●   ● F

G ●   ● G

H ●   ● H

PRACTICE PROBLEM 4-2

With a straight line, connect each lettered dot in the top row to the dot with the same letter in the bottom row.

H
●

G
●

F
●

E
●

D
●

C
●

B
●

A
●

●   ●   ●   ●   ●   ●   ●   ●
A   B   C   D   E   F   G   H

PRACTICE PROBLEM 4-3

With straight lines, connect the dots that have corresponding letters.

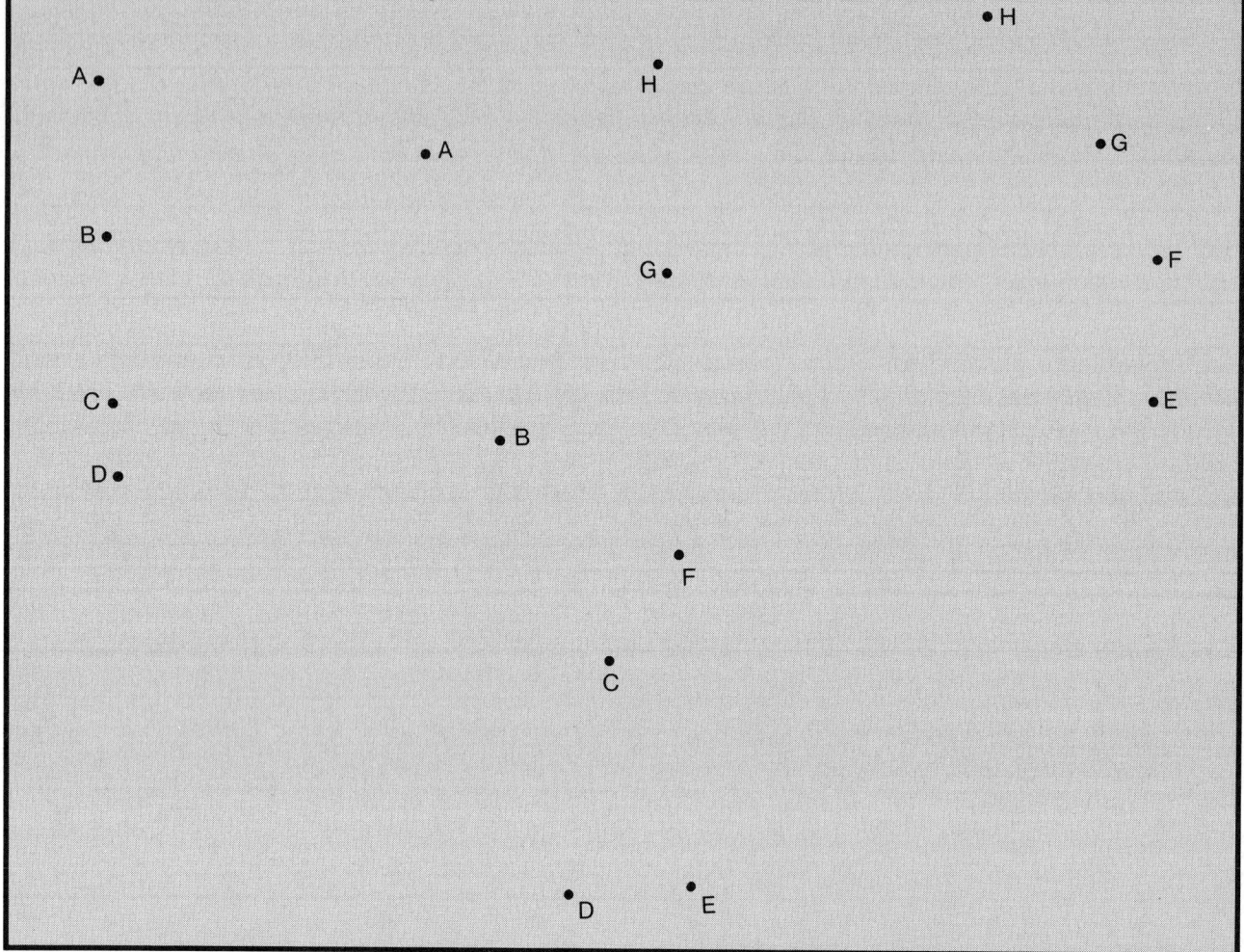

A •

•A

B •

C •

D •

•B

•H

H •

•G

G •

•F

•E

F •

C •

•D   •E

Sketching Circles and Arcs

When sketching, it is often necessary to draw *curved lines* as well as straight lines. The two principal types of curved lines are *circles* and *arcs* (parts of circles). Several methods can be used to draw these lines, but the simplest and most accurate is the *square and triangle method*. This simple six-step method for sketching circles is described below and shown in Figure 4-5.

STEP A. Locate the center of the circle and sketch the horizontal and vertical center lines.

STEP B. Indicate the size of the circle by marking half the diameter (the radius) on the horizontal and vertical center lines.

STEP C. Using the diameter size marks, sketch a square around the center lines.

STEP D. Draw diagonal (cross-corner) lines between the corners of each small square.

STEP E. Mark a dot in the center of each triangle to indicate the point through which the circle is to pass.

STEP F. Using the square frame and dots as a guide, sketch the circle. Erase all construction lines and darken the circle. The circle should appear as shown in G of Figure 4-5.

In the space provided in Practice Problem 4-4, sketch circles to the sizes indicated.

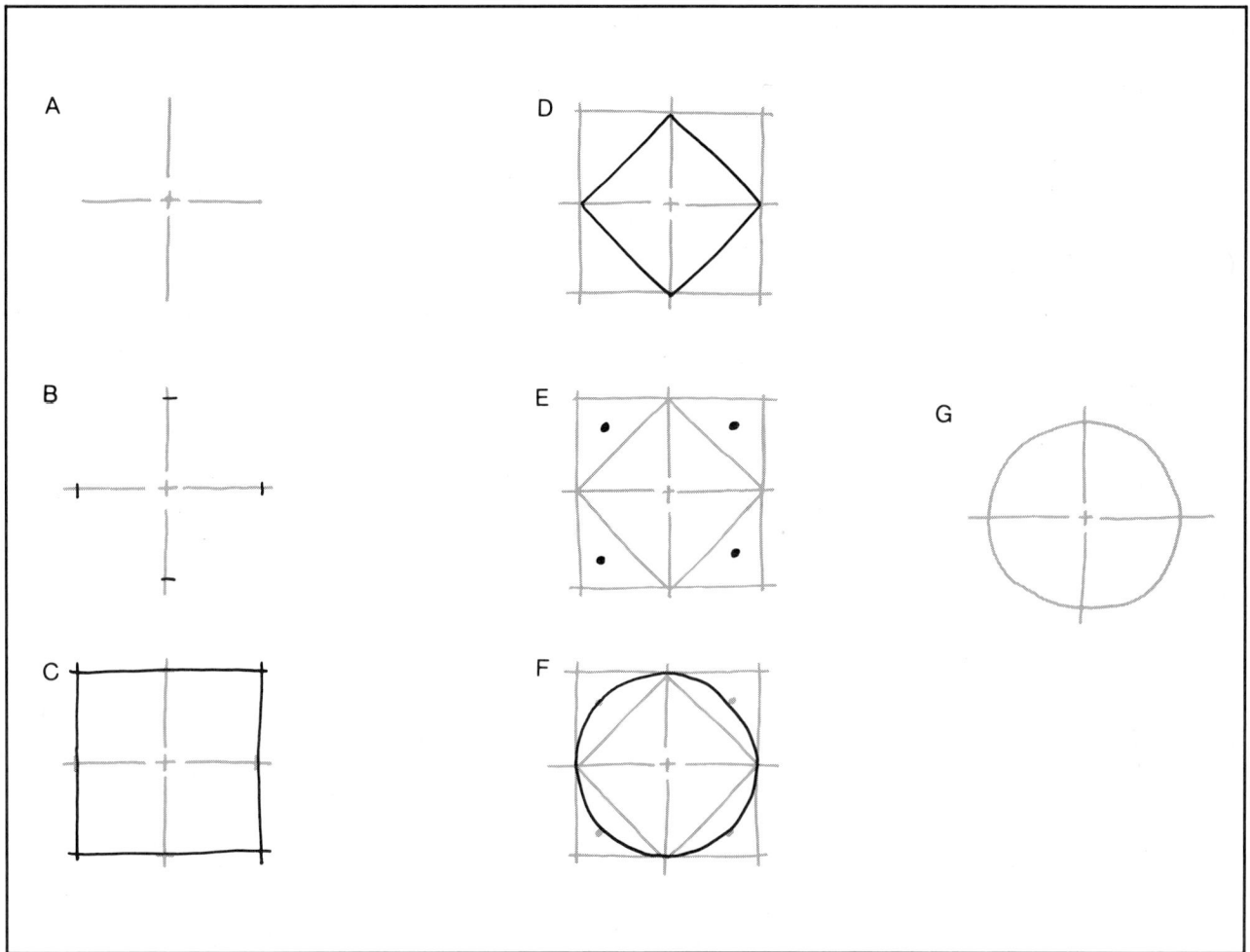

Figure 4-5. Procedure for sketching circles

PRACTICE PROBLEM 4-4

Sketch circles of the sizes indicated.

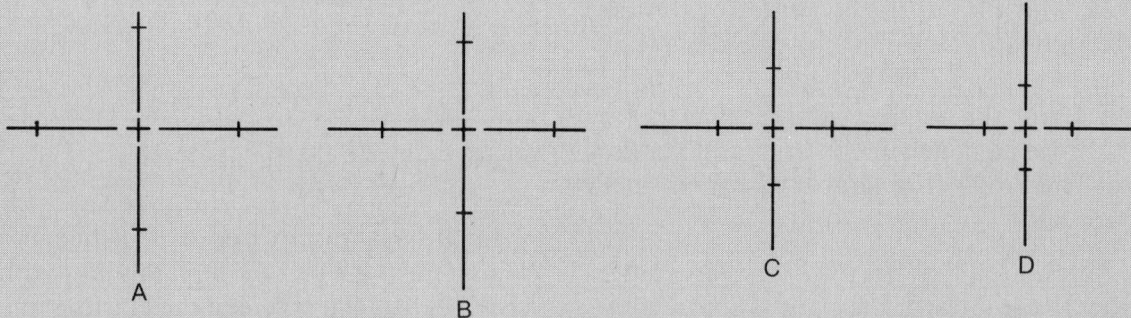

A B C D

Arcs are generally used to connect two other lines to form either a round or a fillet. A *round* is a curved external surface, and a *fillet* is a curved internal surface. See Figure 4-6. In some shops, the term *radius* is used to describe both of these details.

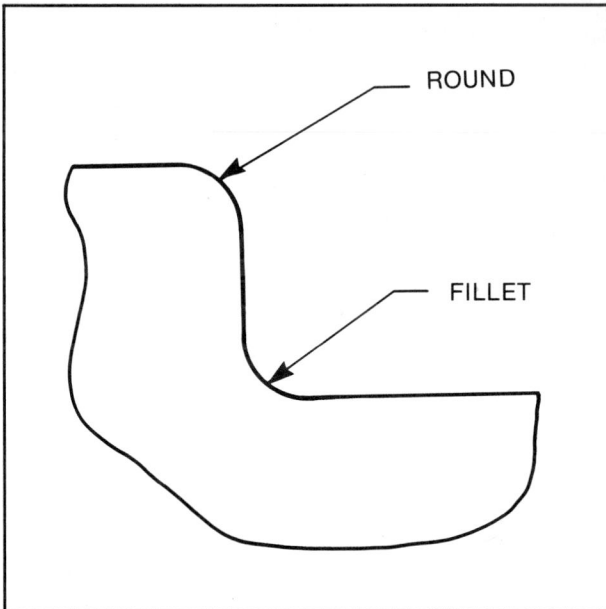

Figure 4-6. Round and fillet

The process used to sketch arcs is very similar to that used for circles. A five-step method that can be used to sketch arcs is described below and shown in Figure 4-7.

STEP A. Project the lines from the part until they meet.
STEP B. Mark the size of the arc on both intersecting lines.
STEP C. Connect the size marks with a straight line.
STEP D. Mark a dot in the center of the triangle to indicate the point through which the arc is to pass.
STEP E. Using the two lines and the dot, sketch the arc. Erase all construction lines and darken the arc. The arc should appear as shown in F of Figure 4-7.

To sketch arcs that are half-circles, simply construct the framework used to sketch a complete circle and draw only half of the circle. See Figure 4-8.

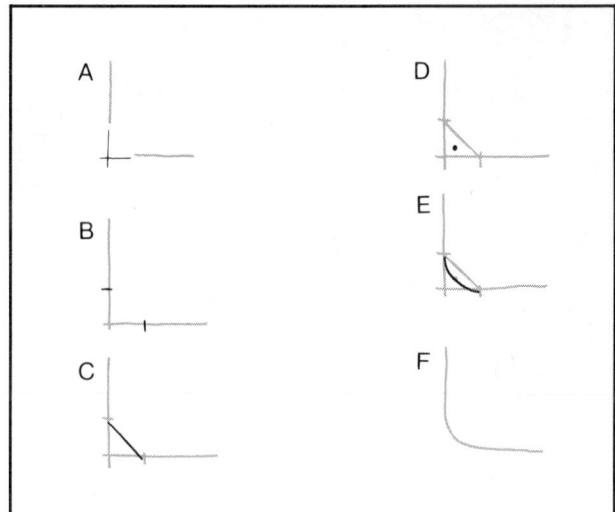

Figure 4-7. Procedure for sketching arcs

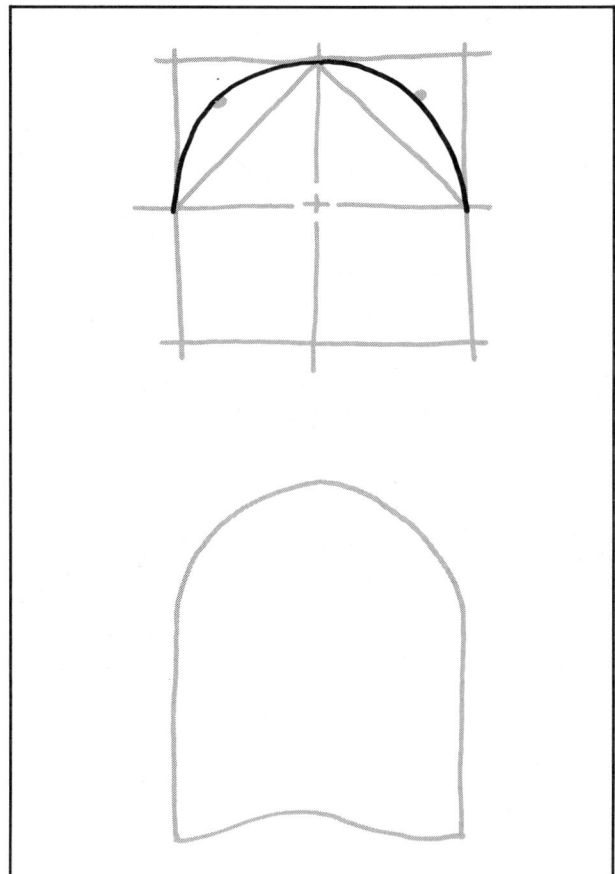

Figure 4-8. Sketching a half-circle

In the space provided in Practice Problem 4-5, connect the pairs of straight lines with arcs. Use the distance between the existing lines and the point where the lines would intersect as the size of each arc.

PRACTICAL SKETCHING TECHNIQUES

The size and shape of any part may be shown in a sketch by a combination of straight and curved lines. Now that the methods used to sketch have been discussed, these skills can be applied to practical sketching problems.

Starting the Sketch

Analyzing the part is the first step in sketching. Study the size, shape, and overall details of the part. Make mental notes of how each detail relates to the whole object, and how the part should appear in the sketch.

Every sketch must be an accurate description of the actual part. Regardless of the drawing scale, the sketch must be drawn proportional to the actual part. The *proportion* of a sketch is the size relationship between the drawn object and the actual object. To accurately sketch a part, each detail must be drawn proportionately to the whole object as well as to the other details in the sketch.

Estimating Proportion

When estimating proportions in sketching, remember that the sketch must be proportional to the dimensions of the part. Figure 4-9 shows two sketched objects. Block A is drawn proportional to the dimensions. Block B, however, does not relate to the dimensions shown. To be correct, this block should be drawn twice as wide as it is high.

The process of estimating proportion is simply making the size and shape of the sketched object appear the same as the actual object. This does not mean a sketch should be an exact drawing. However, the sketch should accurately show the correct proportional relationships.

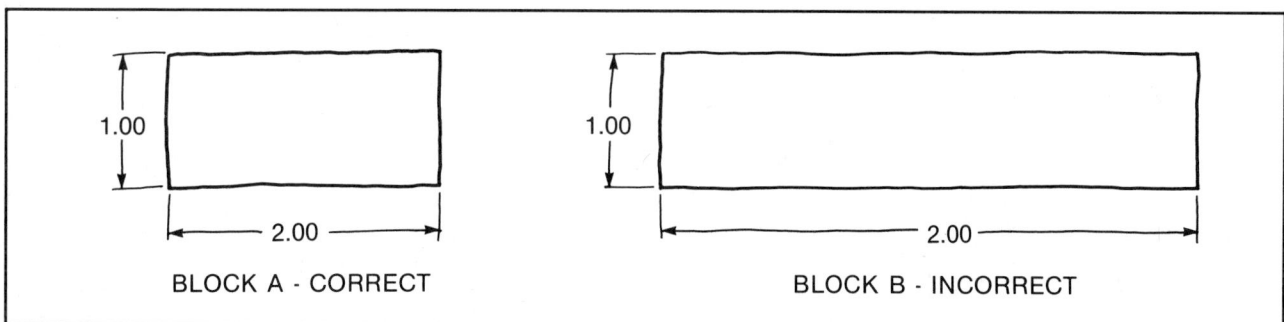

1.00	1.00
2.00	2.00
BLOCK A - CORRECT	BLOCK B - INCORRECT

Figure 4-9. Making a sketch proportional to the dimensions

Estimating Sizes

To estimate the overall size of a part, simply compare the height of the object to its width. In Figure 4-10, the height of the part is 3″ and the length is 4″. The height of the sketch should appear to be three-fourths of its length. The hole in the part is 1″ in diameter. Its center is located 1″ down from the top. It is centered with the length (L) of the part. Using the length as a guide, the hole should be positioned one-fourth the length from the top and one-half the length from either side. The hole diameter should be sketched as equal to one-fourth the length.

Using a common reference point for all estimates will help keep the sketch proportional. After a little practice, the process of estimating sizes will become almost automatic.

Estimating Angles

Inclined, or *slanted*, *lines* are generally sketched at a particular angle. To sketch these lines accurately, the angles must be estimated. The first step in estimating angles is drawing a right angle, Figure 4-11. A *right angle* is the 90-degree (90°) angle formed by a horizontal line and a

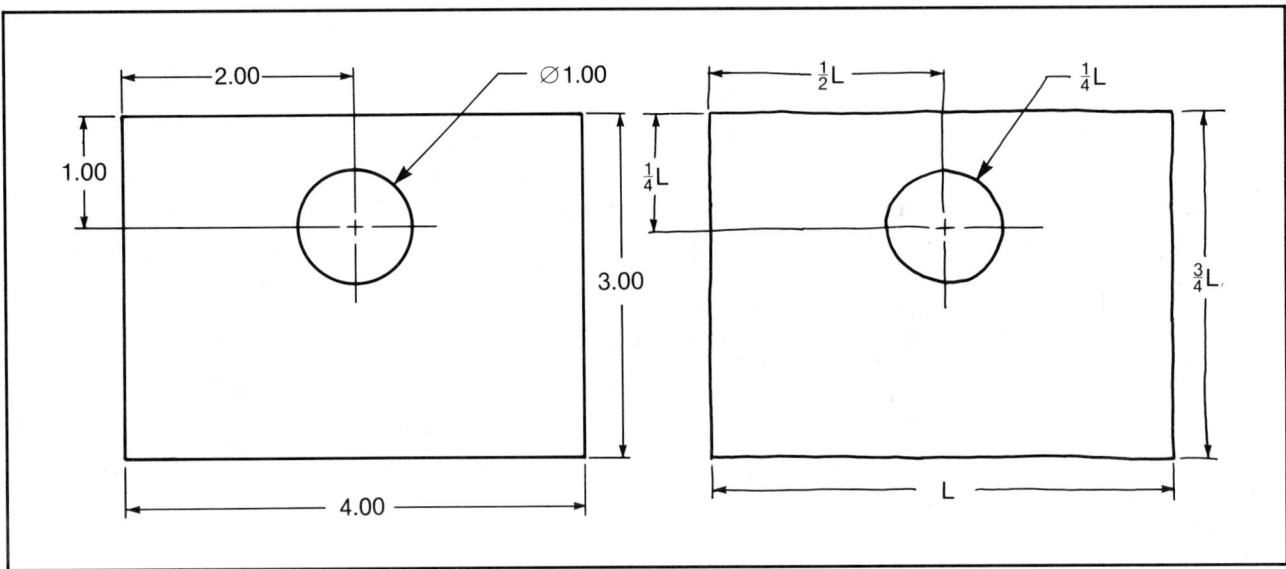

Figure 4-10. Estimating sizes

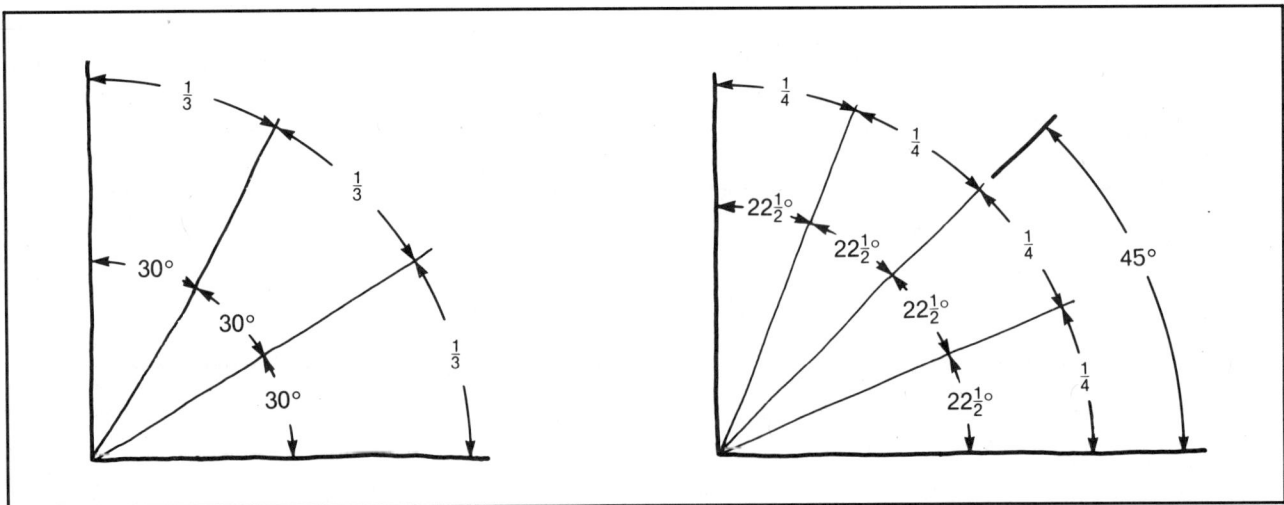

Figure 4-11. Estimating angles

vertical line. By dividing a right angle into equal parts, most of the more common angles can be sketched easily.

In the space provided in Practice Problem 4-6, sketch the angles indicated at each right angle.

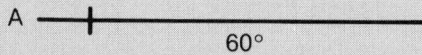

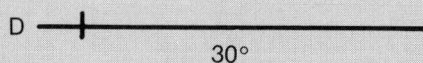

PRACTICE PROBLEM 4-6

Sketch the angles indicated at each right angle.

A ──┼──────────────────────
           60°

B ──┼──────────────────────
           45°

C ──┼──────────────────────
          15°

D ──┼──────────────────────
         30°

SKETCHING PROCEDURE

The easiest and fastest way to draw a sketch is called *blocking*. In this method, the basic outline of the part is sketched in a series of square and rectangular blocks. Once the part has been completely blocked, the individual details are sketched.

The blocking method is described below and shown in Figure 4-12. By using this method and a step-by-step approach, any object can be sketched quickly and easily.

STEP A. Draw two boxes to form the basic shape of the part. Remember to make the boxes proportional to the actual object. (The part to be sketched is shown at the top of Figure 4-12.)

STEP B. Mark the centers of the horizontal and vertical lines in the first box. Connect these points with center lines.

STEP C. Estimate the length of the base on the angular surface and mark the horizontal line. Next, draw a line from the top right corner of the first block to this point.

STEP D. Mark the diameter of the hole on the center lines and construct a square around the center position of the hole. Next, mark the size of the corner radius on the horizontal and vertical lines, and draw another square.

STEP E. Follow the procedures for circles and radii and sketch the hole and corner radius.

STEP F. Erase all construction lines and darken the visible lines.

While the method may vary from one part to another, the basic sketching procedure will remain the same. In general, first sketch the basic shape and form of the part and then add the details.

KEY TERMS

The following key terms were introduced in this chapter. Be sure you know the meaning of each term before proceeding to the review material.

Arc	*Diameter*
Blocking	*Proportion*
Circle	*Right Angle*
Diagonal	*Shop Sketching*

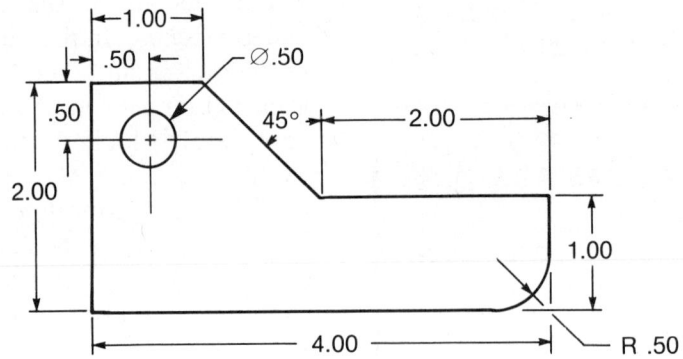

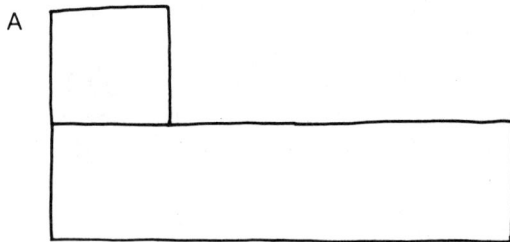

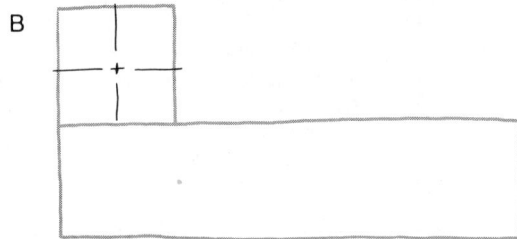

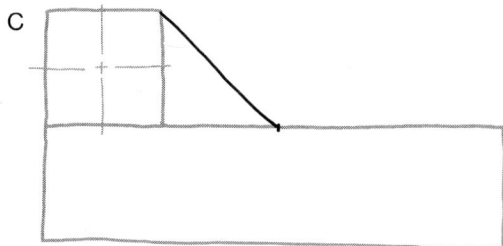

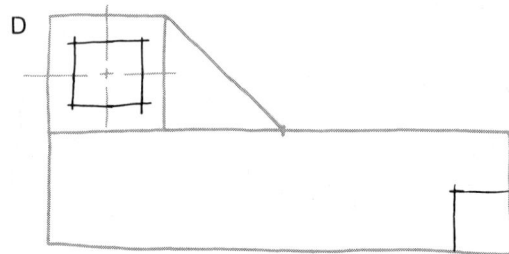

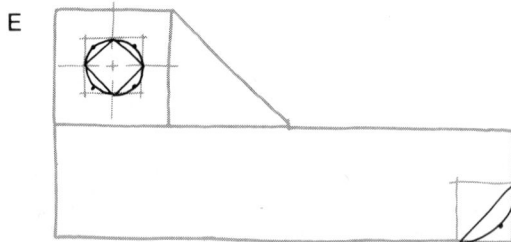

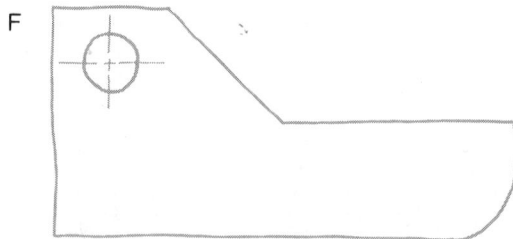

Figure 4-12. Using the blocking method

Test your knowledge with this reinforcement study material. Write your answers to the questions in the spaces provided.

1. What tools are normally required for sketching? _____
_____

2. Which style of pencil point should be used for construction lines? _____
_____

3. Which style of pencil point should be used for darkening a sketch? _____
_____

4. What are the three principal types of straight lines? _____
_____

5. When sketching a straight line, where should your eye be fixed? _____
_____

6. How should a long, straight line be sketched? _____
_____
_____

7. What is an arc? _____
_____

8. What is the most accurate method of sketching circles and arcs? _____
_____
_____

9. What is meant by making a sketch proportional? _____
_____
_____

10. What is the easiest way to estimate the size of a part? _____
_____
_____

11. How are the sizes of most angles determined in a sketch? _____
_____
_____

12. What is the first step in sketching any part? _____
_____
_____

13. What term is used to describe a line drawn between the corners of a square or rectangle?

_____

14. What is the purpose of marking two dots prior to sketching a straight line? _____

_____

_____

15. What method of constructing a sketch is the easiest and fastest? _____

_____

In Figure E4-1 sketch the part to the sizes indicated.

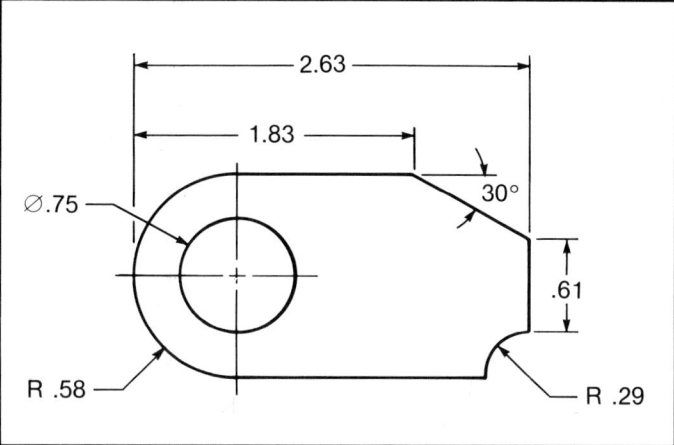

Figure E4-1.

In Figure E4-2 sketch the part to the sizes indicated.

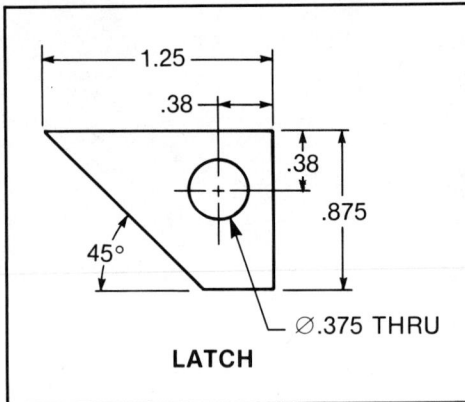

Figure E4-2.

In Figure E4-3 sketch the part to the sizes indicated.

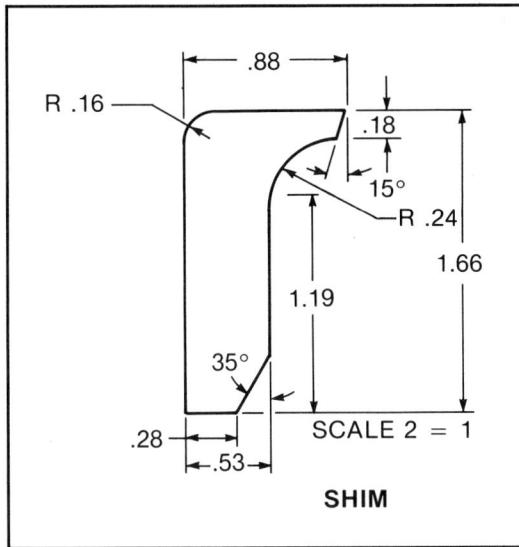

Figure E4-3.

In Figure E4-4 sketch the part to the sizes indicated.

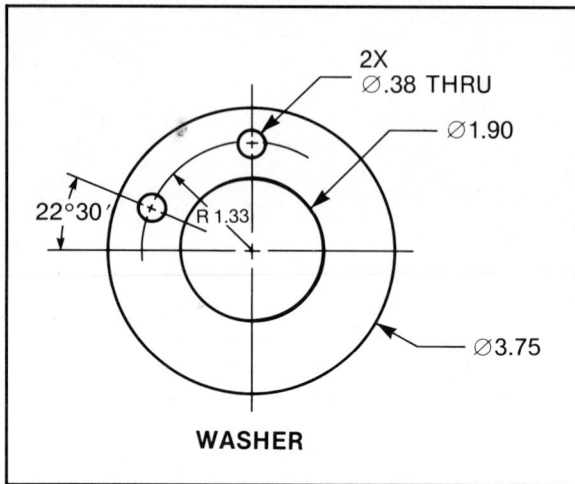

Figure E4-4.

5

MULTIVIEW PRINTS

OBJECTIVES

After studying this chapter, you will be able to:

■ Identify the projected views of an object.

■ Describe the standard view arrangement of a multiview print.

■ Identify and locate the standard three views used for multiview prints.

■ Identify and locate the alternate views that show the most detail.

■ List the steps used to visualize an object from a multiview print.

■ Explain the reasons for using auxiliary views.

■ Locate and identify auxiliary views.

Multiview prints are the most common type of prints used in industry today. These prints consist of a series of individual drawings, or views, used to completely describe the finished object.

Why is one view preferred over another? Why are the views positioned in a certain way? Why do the number of views vary from one print to another? To answer these and other questions, an understanding of the basic conventions (standard methods) used to construct multiview drawings is necessary.

PROJECTING VIEWS

Orthographic projection is the main method used by drafters to construct engineering drawings. Each view in a multiview print is a *projection* of one of the sides of the object. Projections are formed by extending imaginary lines of sight, called *projectors*, at right angles from each corner or edge of an object to the observer. This is shown in Figure 5-1. The views projected in this manner are called *orthographic views*, and are parallel to the surfaces they represent.

To understand how the views are projected to form a multiview print, imagine an object placed inside a transparent box. (See Figure 5-2.) Since this box has six sides, there are six possible views of the object. The views are developed by extending projectors from each side of the object to each side of the box. (See Figure 5-3.) When the box is opened and flattened, Figure 5-4 (page 60), the views unfold into six different drawings, or projections, of the object.

When a box is opened to form the six projections, each of the projections has a specific position in relation to the other views. (See Figure 5-4.) This placement represents the arrangement of the views in a multiview print. By using this standard view arrangement, both the drafter and the machinist will know exactly which area of the object, or workpiece, each view depicts.

Another way to visualize this standard arrangement of views is to imagine hinges placed on the box, as shown in Figure 5-5A. When the box is opened, the views automatically swing into their proper position, Figure 5-5B.

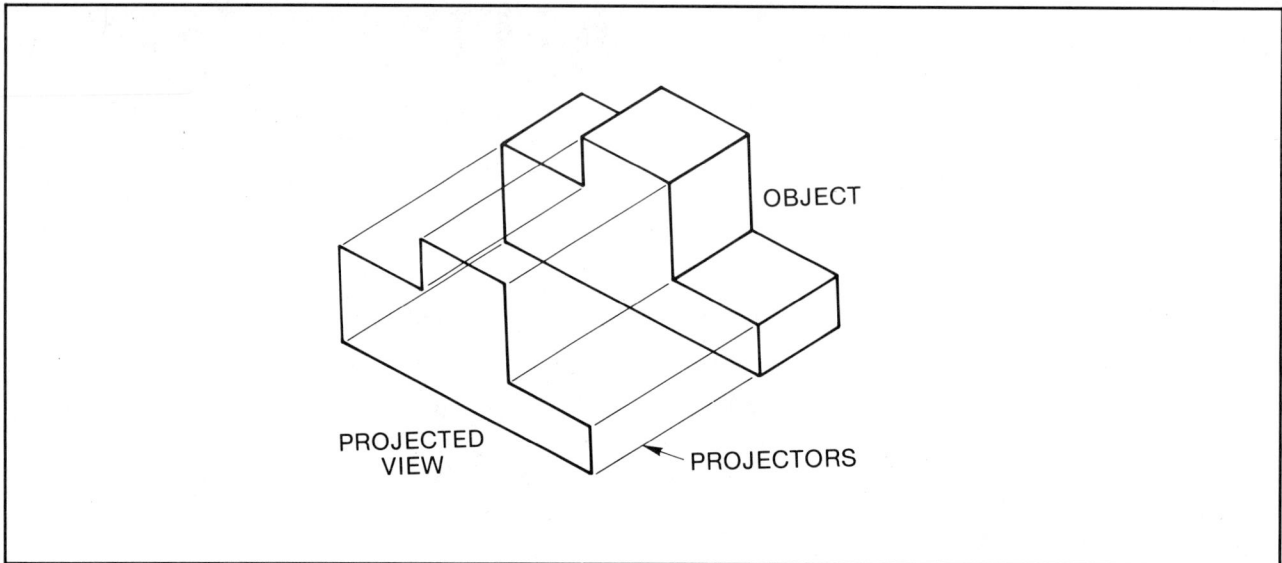

Figure 5-1. Projectors are used to develop an orthographic view of an object.

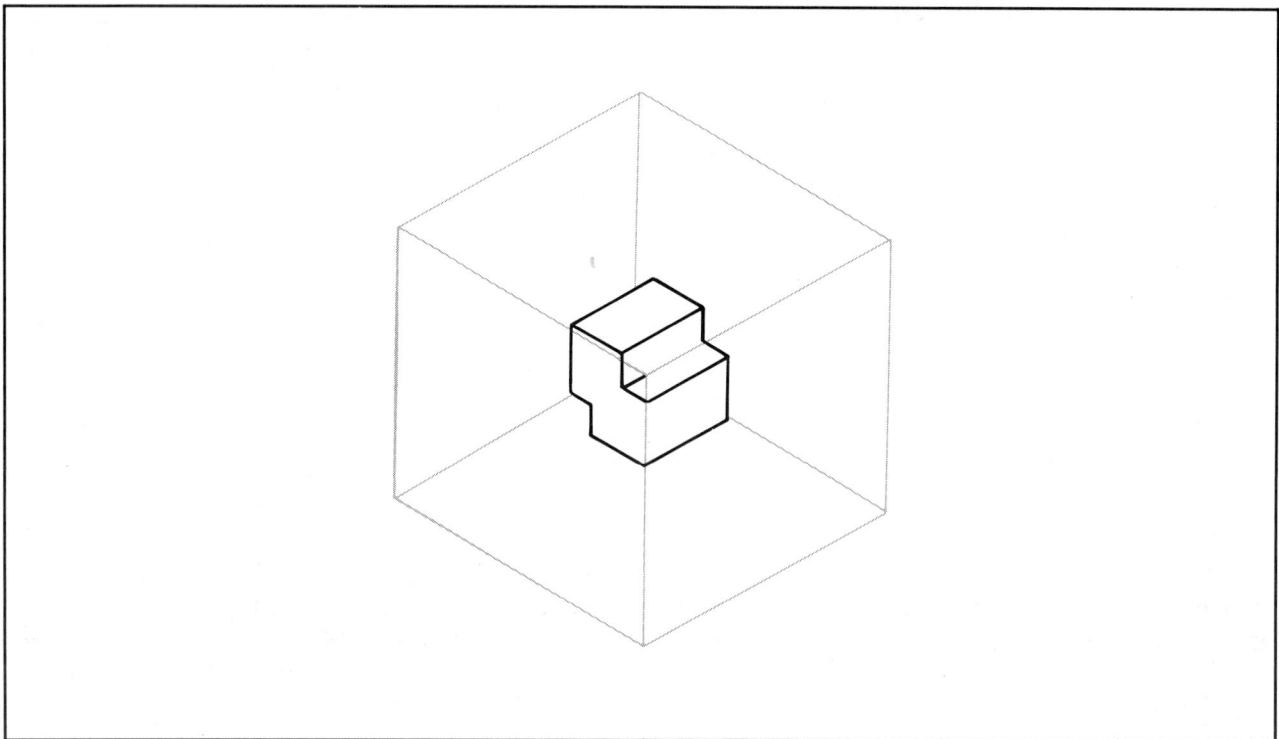

Figure 5-2. Enclosing an object in a transparent box

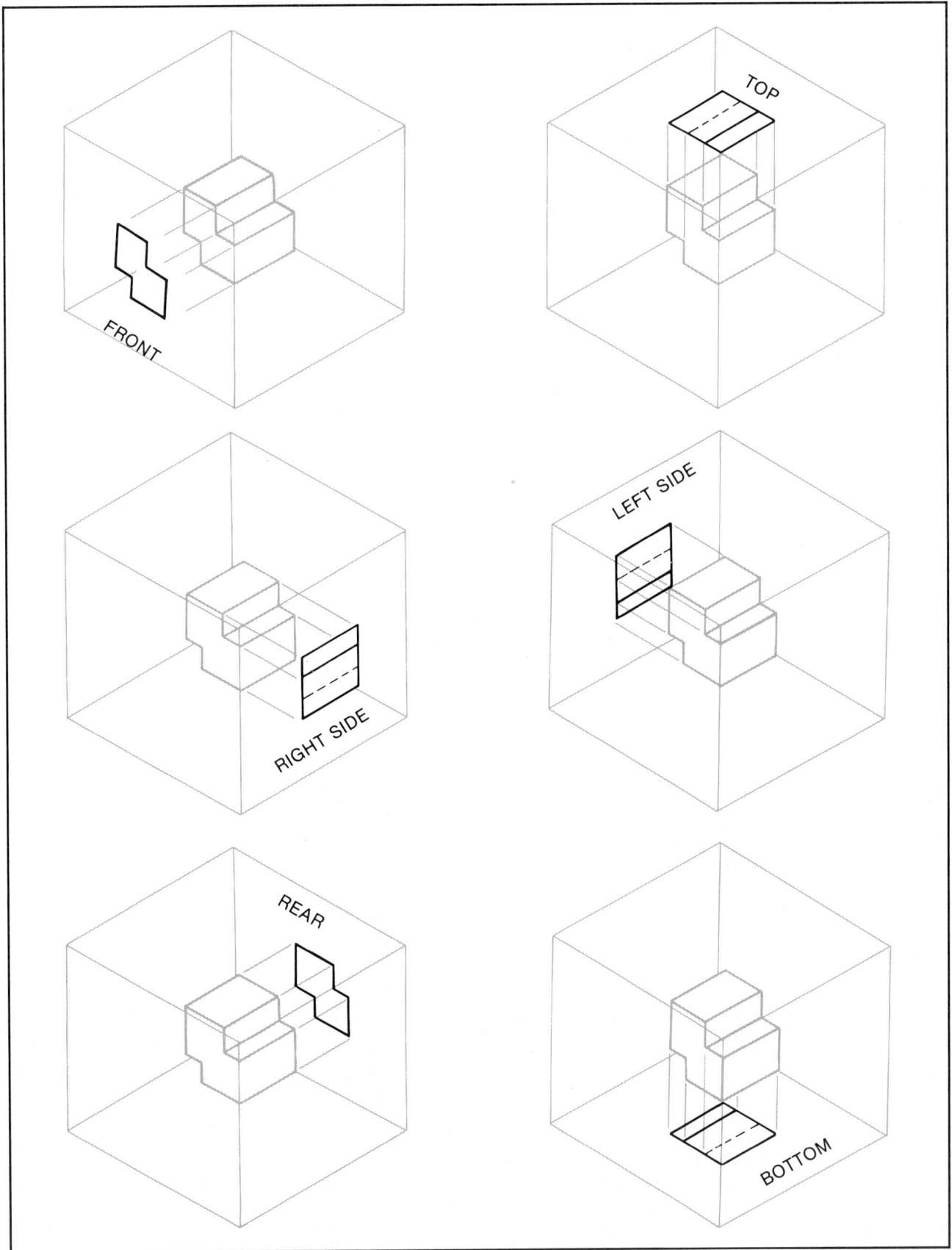

Figure 5-3. Projecting the views from the object to the box

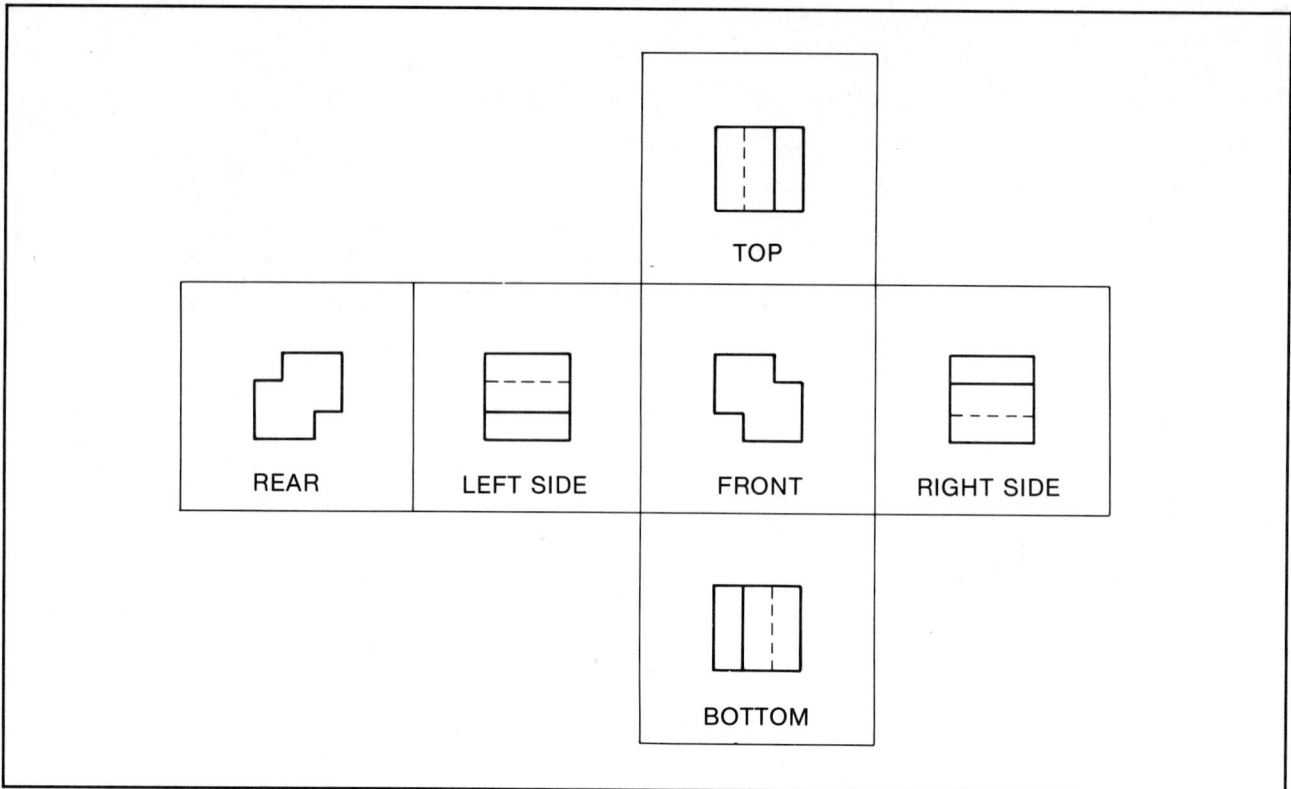

Figure 5-4. Opening the box to form six different views of the object

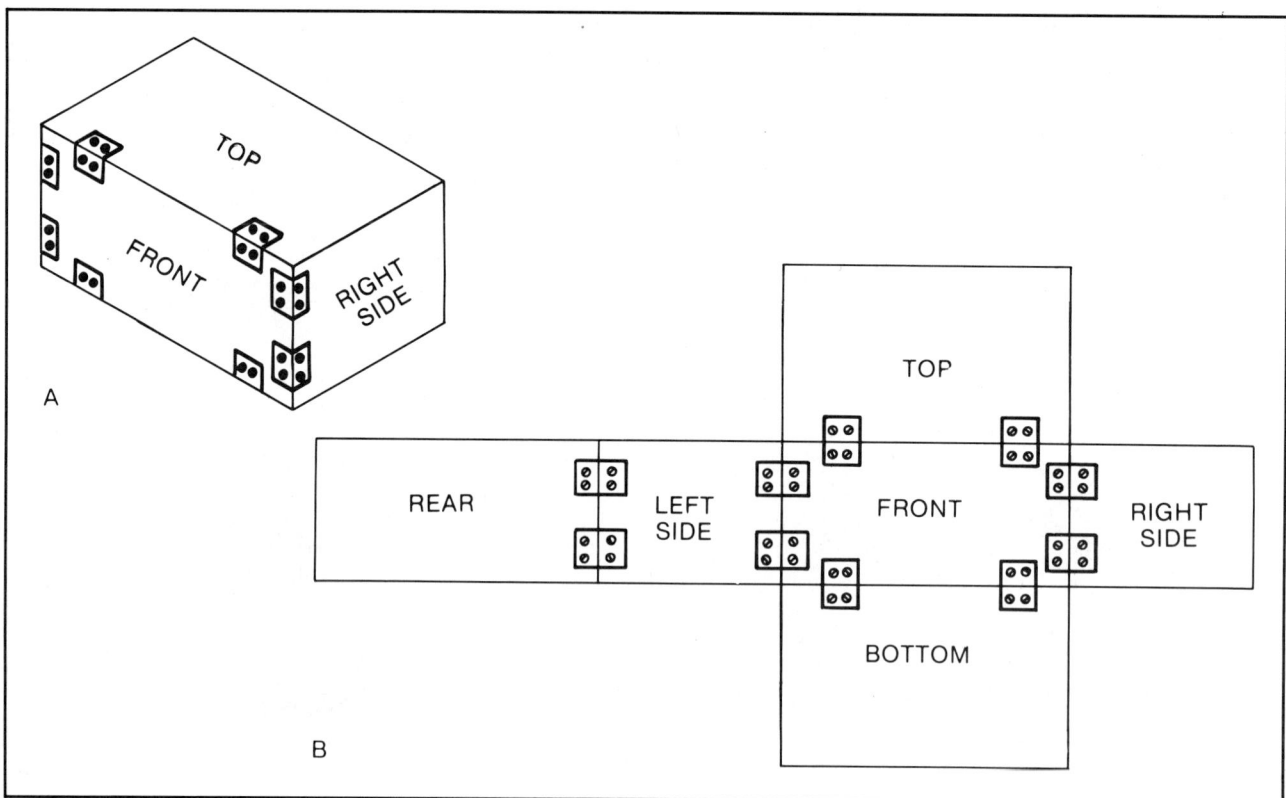

Figure 5-5. If hinges are hypothetically placed on the object (A), the views swing into their proper positions when the object is opened (B).

VIEW SELECTION

While there are six possible views of an object, it is unlikely that a print will show all six views. In most cases, the information contained in one view would simply be repeated in another view. To reduce such repetition, only those views that show the most information and detail are used in a drawing to describe a part.

The first rule of view selection is clarity. Drafters normally select and draw only those views that show the most detail and keep repeated details to a minimum. For example, Figure 5-6 shows six possible views of a part. Only three views (top, front, and right side) are needed to completely describe the part. The other views (rear, left side, and bottom) should be eliminated.

The standard views used for most multiview prints are the top, front, and right-side views. Whenever possible, the drafter will use these views to describe a part. There are times, however, when a different view will show more detail. In these cases, the views that show the most detail should be used in place of, or in addition to, the top, front, and right-side views. For example, in Figure 5-7, the left-side view is used rather than the right-side view. The left-side view is shown so that the print reader can see the visible lines of the sloped surface. In Figure 5-8, both the left-side and the right-side views are shown because both help to describe the part.

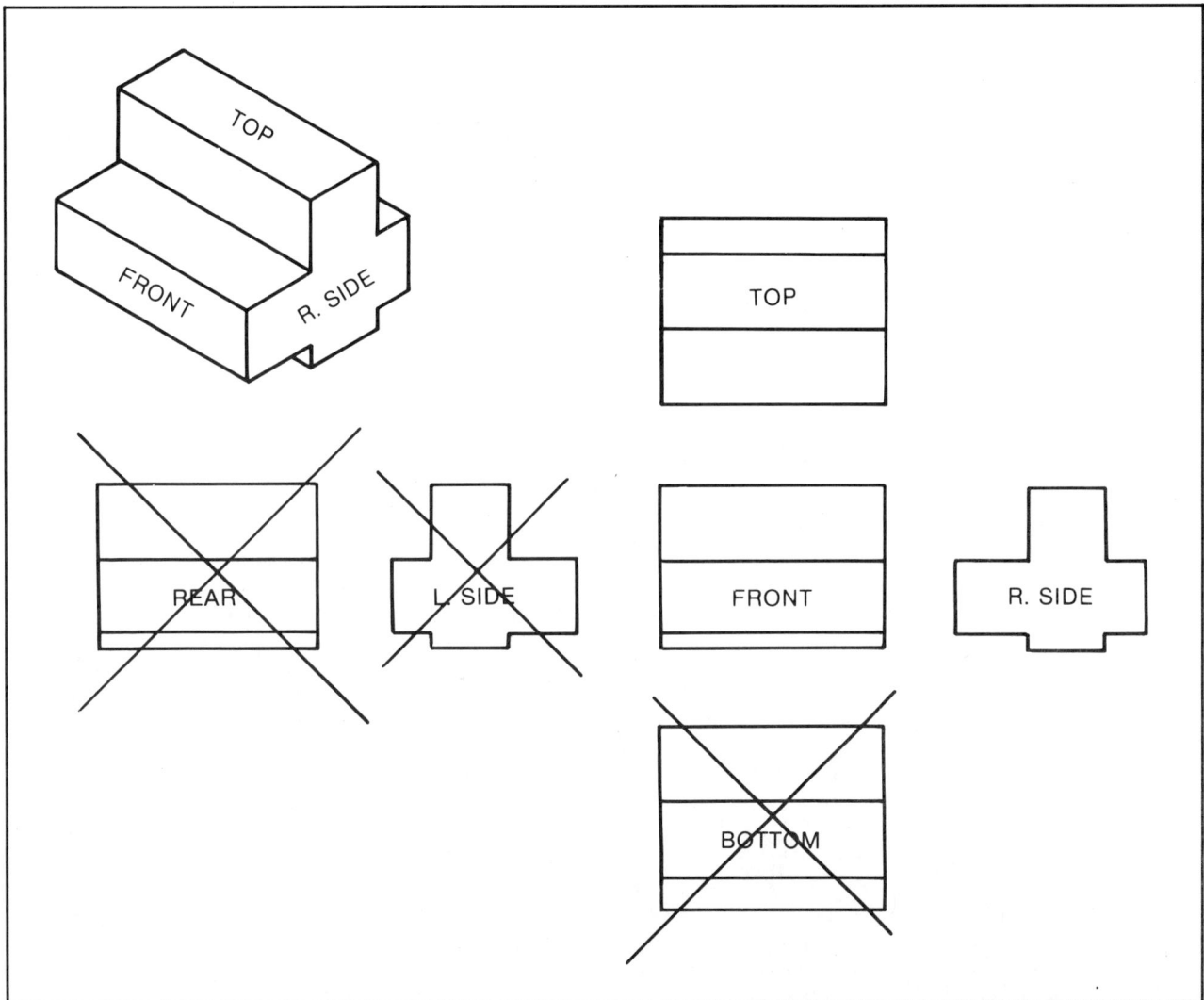

Figure 5-6. While there are six possible views, only three are needed to describe this object.

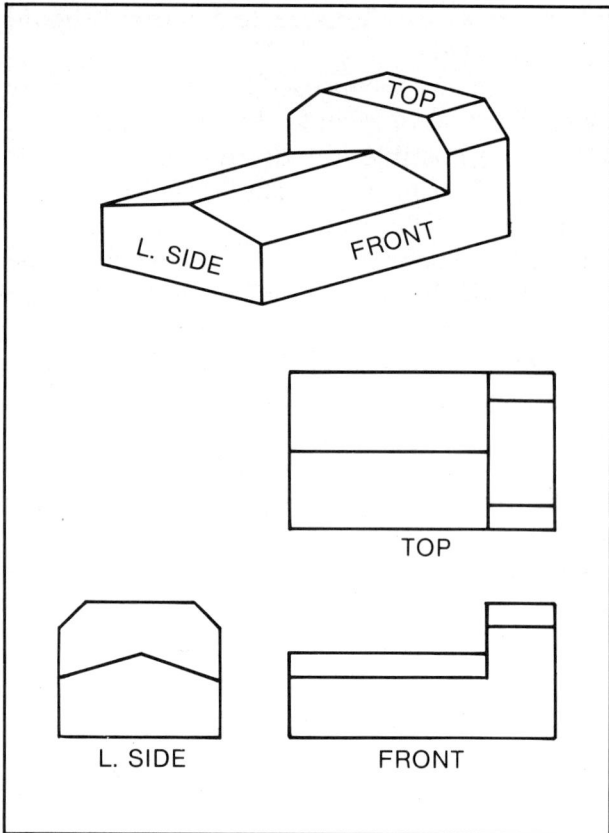

Figure 5-7. Using the left-side view to show important details

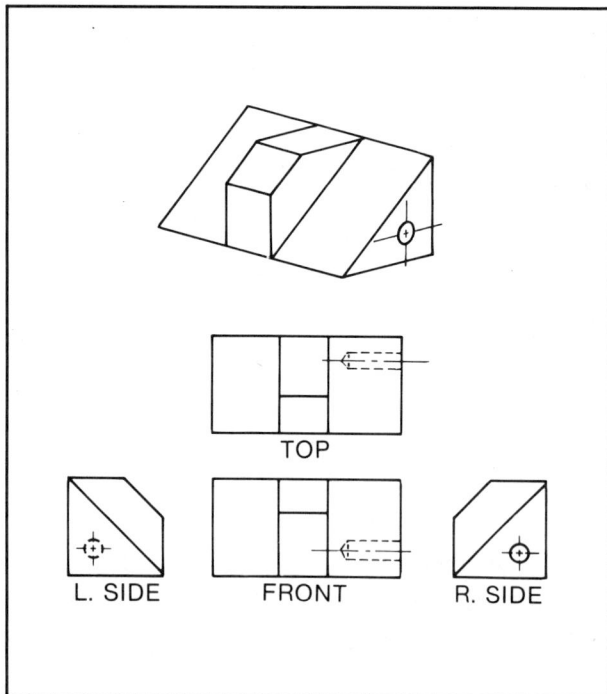

Figure 5-8. Using both the left-side and right-side views to show the details of a part

Occasionally, the standard arrangement of views will be changed to suit a special need. In Figure 5-9A, the arrangement is correct, but it causes crowding in one corner of the print. To solve this problem, the right-side view can be placed next to the top view rather than next to the front view, Figure 5-9B. Note the position of the right-side view in this new position. Rather than being hinged from the front view, this view is now hinged from the top view. Therefore, the right-side view is rotated 90° from its old position beside the front view.

The standard arrangement of views is generally followed without variation. However, as in the situation just described, the standard placement of the views is altered in some prints. In these cases, study the print carefully to make sure you understand the part structure fully before proceeding.

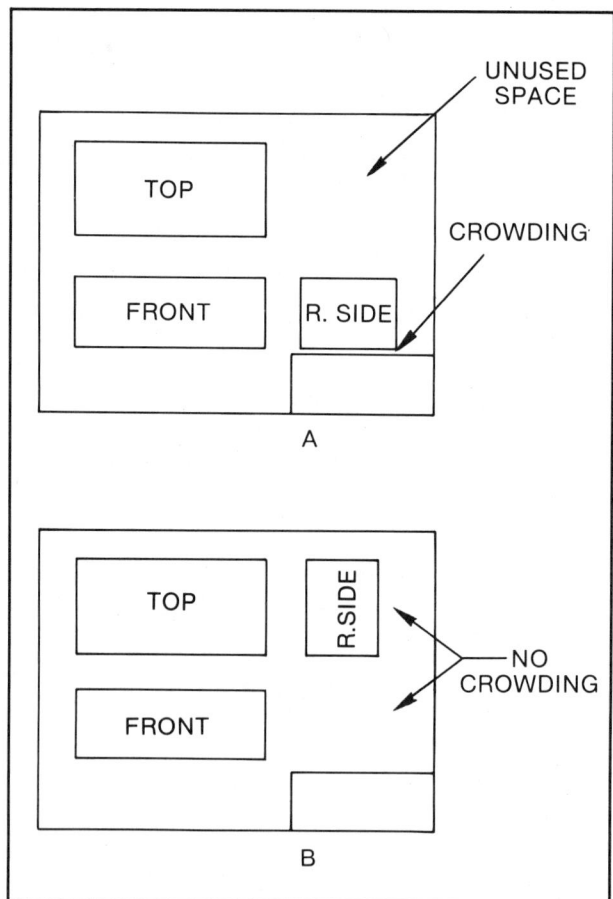

Figure 5-9. When the standard arrangement of views is used, the print may be crowded (A). The views can be varied to suit different situations (B).

While most multiview prints contain three views, there are times when three views are not necessary. In these situations, two-view or one-view drawings may be used. Two-view drawings are normally used for symmetrical parts such as shafts and collars, where a third view would simply repeat information already shown in the other two views. See Figure 5-10. One-view prints, Figure 5-11, are used when a second view would not add any important information. In such cases, a note can be used in place of a drawn view.

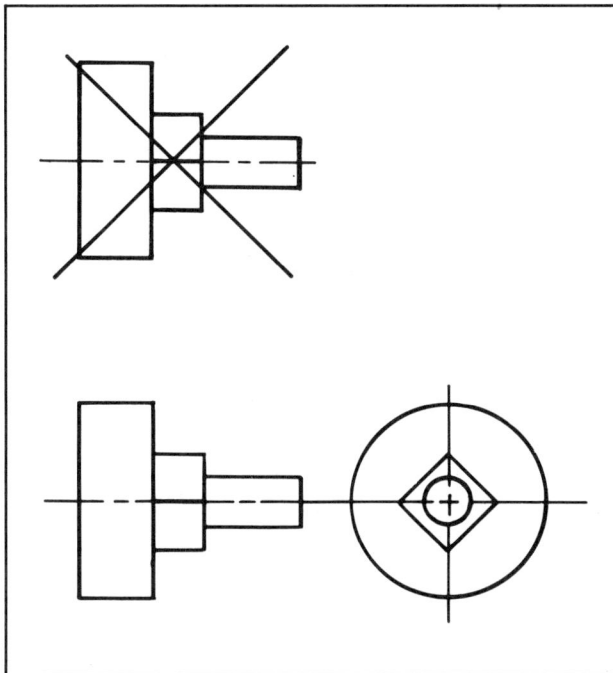

Figure 5-10. Two-view prints are often used when a third view would only repeat information already shown.

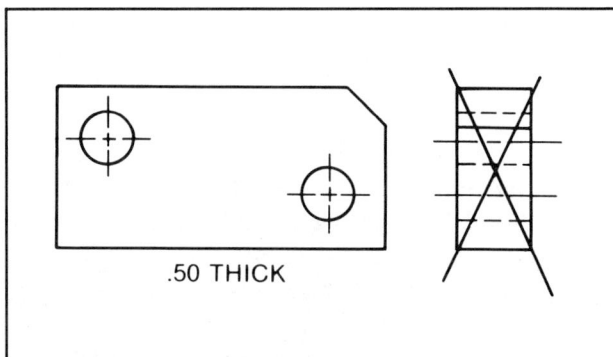

Figure 5-11. In one-view prints, a note is often used rather than a drawn view.

VISUALIZATION

Visualization is the process of studying the individual views of an object to form a mental picture of how the finished object should appear. By using a planned approach, even complicated prints can be easily transformed into the correct mental picture. The following procedures will help the print reader to quickly and accurately visualize any object in a multiview print.

1. Briefly look over the entire print to get an overall idea of what the views look like.
2. Study the front view to determine the height and width of the object. Look for any special details such as holes or slots.
3. Study the top view to find the depth of the object. Mentally add the front view to the top view to form a combined picture of the two views. Make note of any special details.
4. Study the right side view and add this to the mental image developed from the other views.
5. Finally, after determining the shape of the object, go back over the print. Look for any details you may have missed.

Visualizing a Two-View Print

When visualizing a two-view print, the first step is to look over the entire print, Figure 5-12A. Next study the front view, Figure 5-12B, to determine the height and width of the part. When viewed from the front, this part appears to have rectangular features. However, by observing the right-side view of Figure 5-12A, the true shape of the object, Figure 5-12C, can be seen. The final step is to recheck the print to make sure the mental picture is correct and that no important details have been missed. Try to visualize the multiview drawing as a picture of the object, as shown in Figure 5-12C.

Visualizing a Three-View Print

Visualizing a three-view print (Figure 5-13A) is basically the same as visualizing a two-view print. First, study the front view to determine

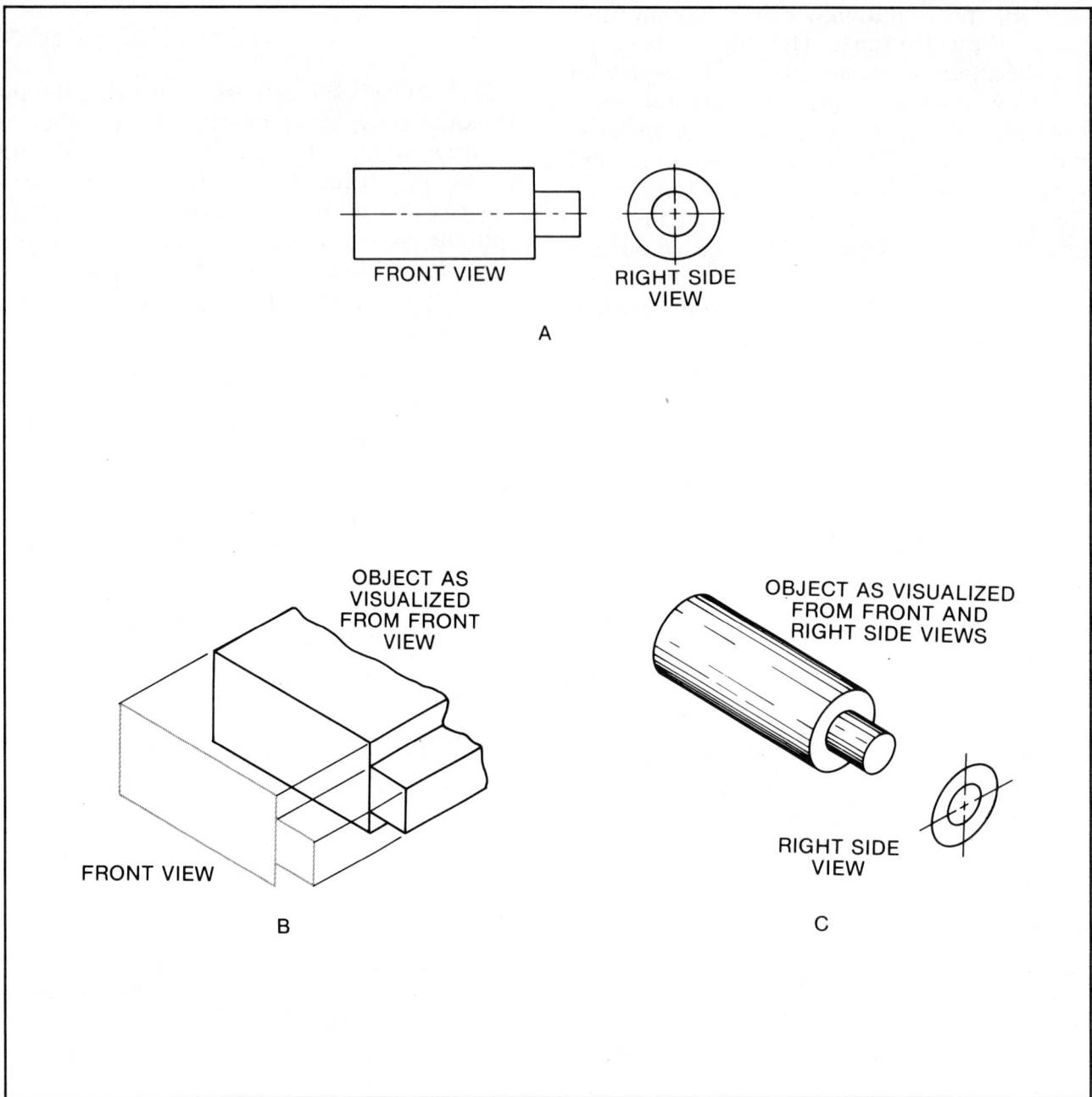

Figure 5-12. Reading a two-view print of a stepped cylinder (A) by only visualizing one view can be misleading (B). Visualizing the object from both views (C) will ensure interpretation of its true form.

the height and width of the part, Figure 5-13B. Here the front view is shown as two rectangles with a small square in one corner. When viewed from the front view, this is all that can be seen. Next, study the top view to find the depth of the part and to determine the meaning of the unknown lines in the front view. The part should now appear as shown in Figure 5-13C. Note that there are still two lines—one in the front view and one in the top view—that cannot be identified by these two views. By moving to the right-side view, Figure 5-13D, the meaning of these lines is easily determined. They represent the sloped surface on the short leg of the "L." The final step is to go back and recheck the views to make sure that no important details have been missed.

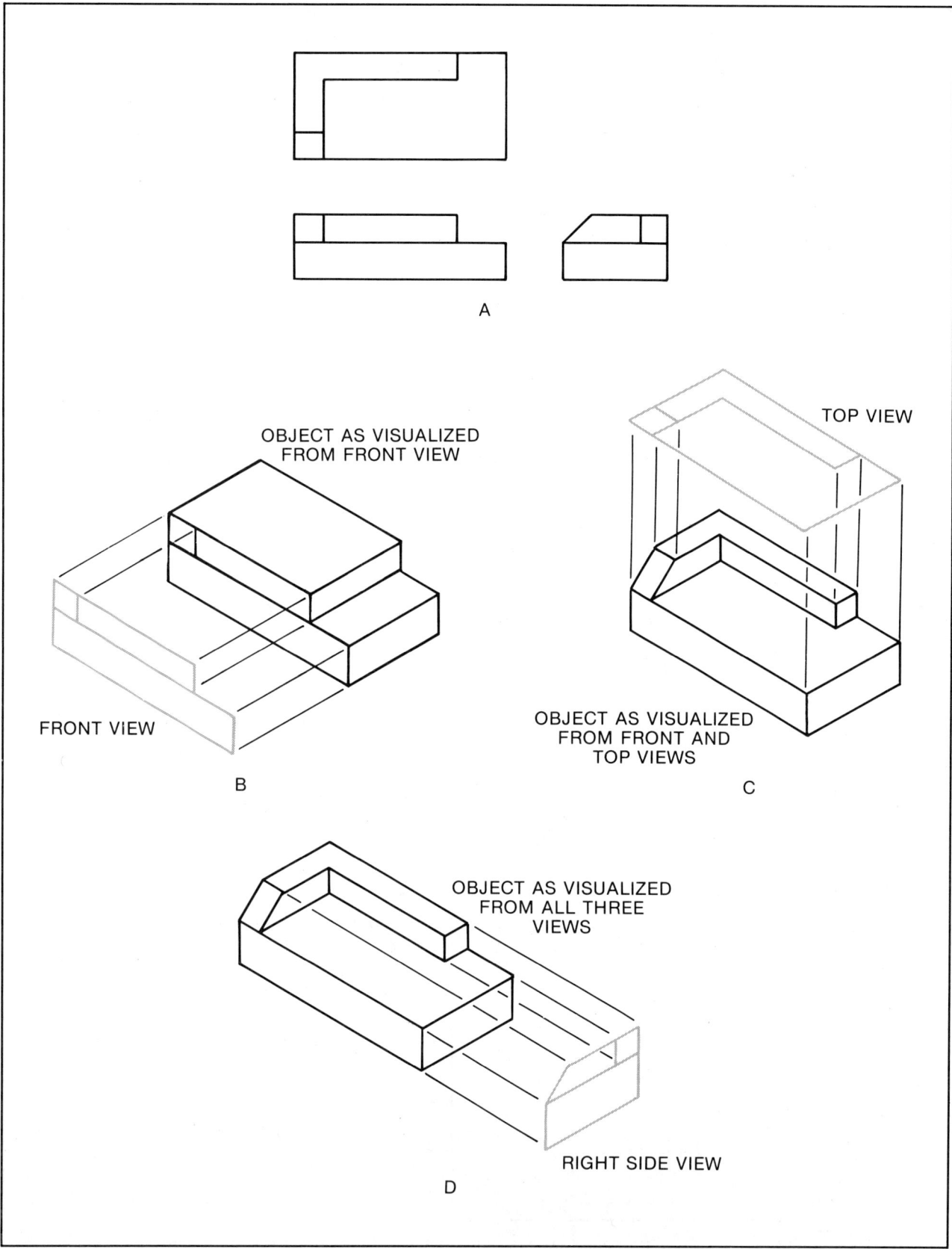

Figure 5-13. The object in a three-view print (A) should be visualized first from the front view (B), then from the front and top views (C), and finally from all three views (D).

AUXILIARY VIEWS

To provide a complete and accurate description of a part, all surfaces must be shown in their true shapes. However, this is not always possible with the standard views of a multiview print. If parts have surfaces that are not parallel to any normal viewing plane, they require auxiliary views to show their true shapes. *Auxiliary views* are typically used to show angles or inclined surfaces that cannot be accurately projected in a normal orthographic view. The use of auxiliary views can eliminate any confusion or misinterpretation in reading a print.

Parts that have inclined, or angled, surfaces normally appear distorted (not true size) when projected in any of the standard orthographic views. See Figure 5-14. This distortion is caused by the position of the inclined surface with respect to the other surfaces of the object. The angled lines appear shorter, and any round holes in the inclined surface are shown as *ellipses*, or oval-shaped holes, in each of the standard views. An auxiliary view, Figure 5-15, eliminates such distortion. In this example, the auxiliary view is used in addition to the standard

orthographic views. In some cases, such as that shown in Figure 5-16, an auxiliary view may also be used to replace a standard view.

Primary Auxiliary Views

Primary auxiliary views are shown adjacent to, or aligned with, the surface they represent in one of the orthographic views. These auxiliary views are identified by the principal view with which they are aligned. A *principal view* is considered

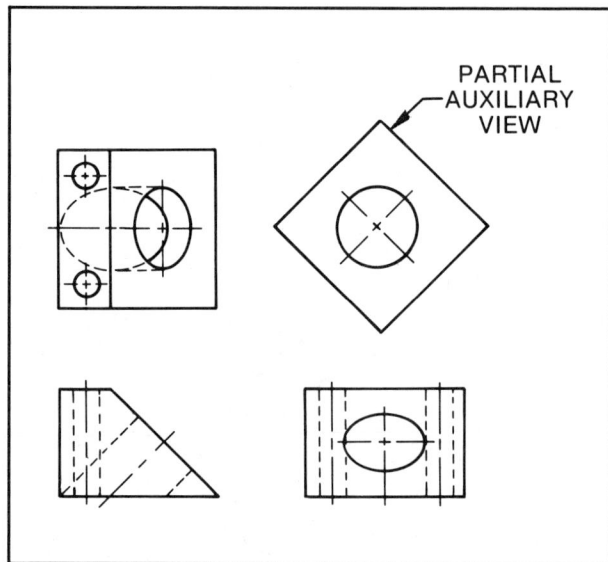

Figure 5-15. An auxiliary view shows the inclined surface in its true shape, without distortion.

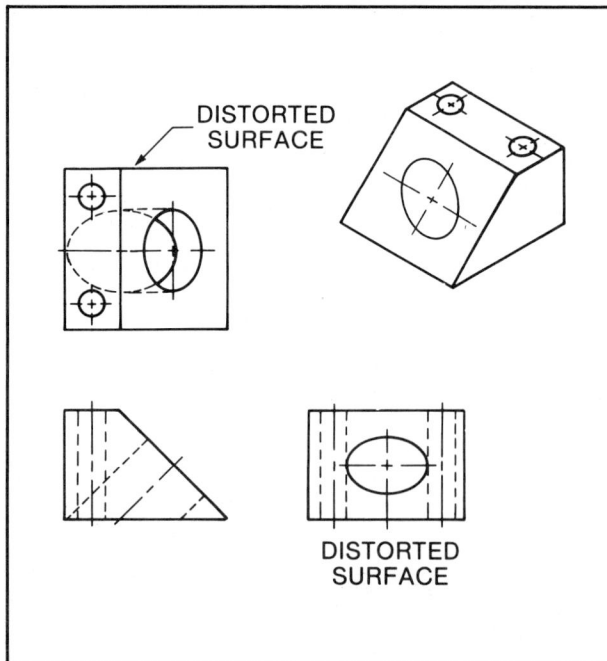

Figure 5-14. Angled, or inclined, surfaces appear distorted when shown in any of the principal views.

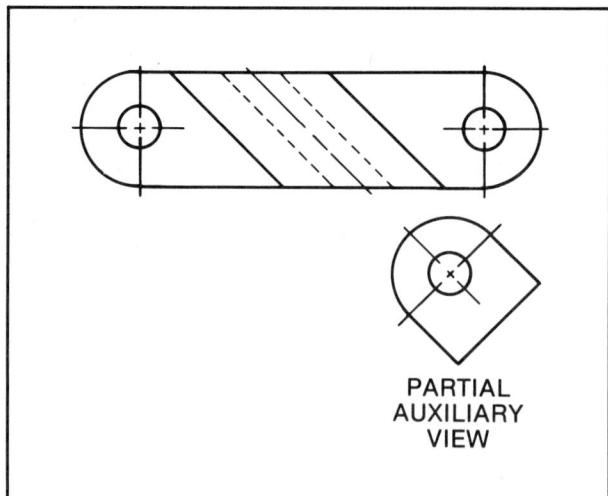

Figure 5-16. Occasionally an auxiliary view can be used in place of a principal view.

to be one of the standard orthographic views used to show the part in the print.

A *front auxiliary view*, Figure 5-17A, is shown perpendicular, or at a right angle, to the inclined surface in the front view. A *top auxiliary view*, Figure 5-17B, is shown perpendicular to the inclined surface in the top view. A *side auxiliary view* is drawn perpendicular to the inclined surface in the side view, Figure 5-17C.

One point to remember about the position and arrangement of auxiliary views is their relationship to the principal views. Each primary auxiliary view is directly related to only one of the principal views. Sometimes this can make the auxiliary view appear unrelated to the other views. However, auxiliary views are intended to clarify the details of the inclined surface only, not the entire part. For this reason only a partial auxiliary of the whole part is shown (the slanted surface).

Secondary Auxiliary Views

A *secondary auxiliary view* is shown perpendicular to a primary auxiliary view rather than to a principal view. Figure 5-18 shows how a secondary auxiliary view is shown on a drawing. Here a primary auxiliary view shows the true size of the round hole while the secondary auxiliary view shows the side view of the hole.

Secondary auxiliary views are shown in a print only when no other view can describe a specific detail. If a principal view or primary auxiliary view can clearly show the detail, a secondary auxiliary view will not be shown.

When either primary or secondary auxiliary views are shown on a print, only those details necessary to describe the inclined surface are shown. The remaining areas of the part are normally omitted.

The two main methods used to show both auxiliary and standard orthographic views are the complete view and the partial view as shown in Figure 5-19. A *complete view* is a view that shows every detail related to a part when viewed from a particular side. This is the type of view normally used for most prints. A *partial view* is shown when only a limited amount of information is required to be shown in the view. In most prints where a partial view is used,

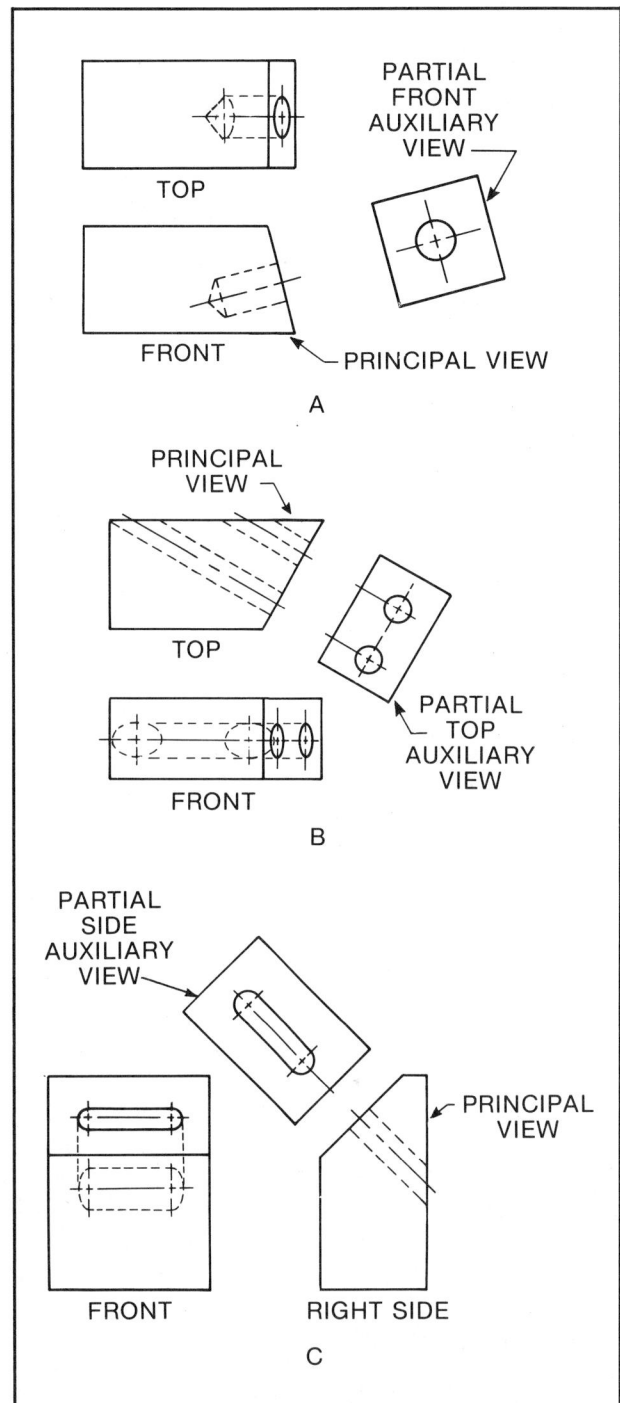

Figure 5-17. Primary auxiliary views are aligned with and identified by a principal view. A partial front auxiliary view is shown in (A), a partial top auxiliary view in (B), and a partial side auxiliary view in (C).

the information omitted is found in one of the other views. Information is omitted from the partial view to reduce confusion or duplication of information.

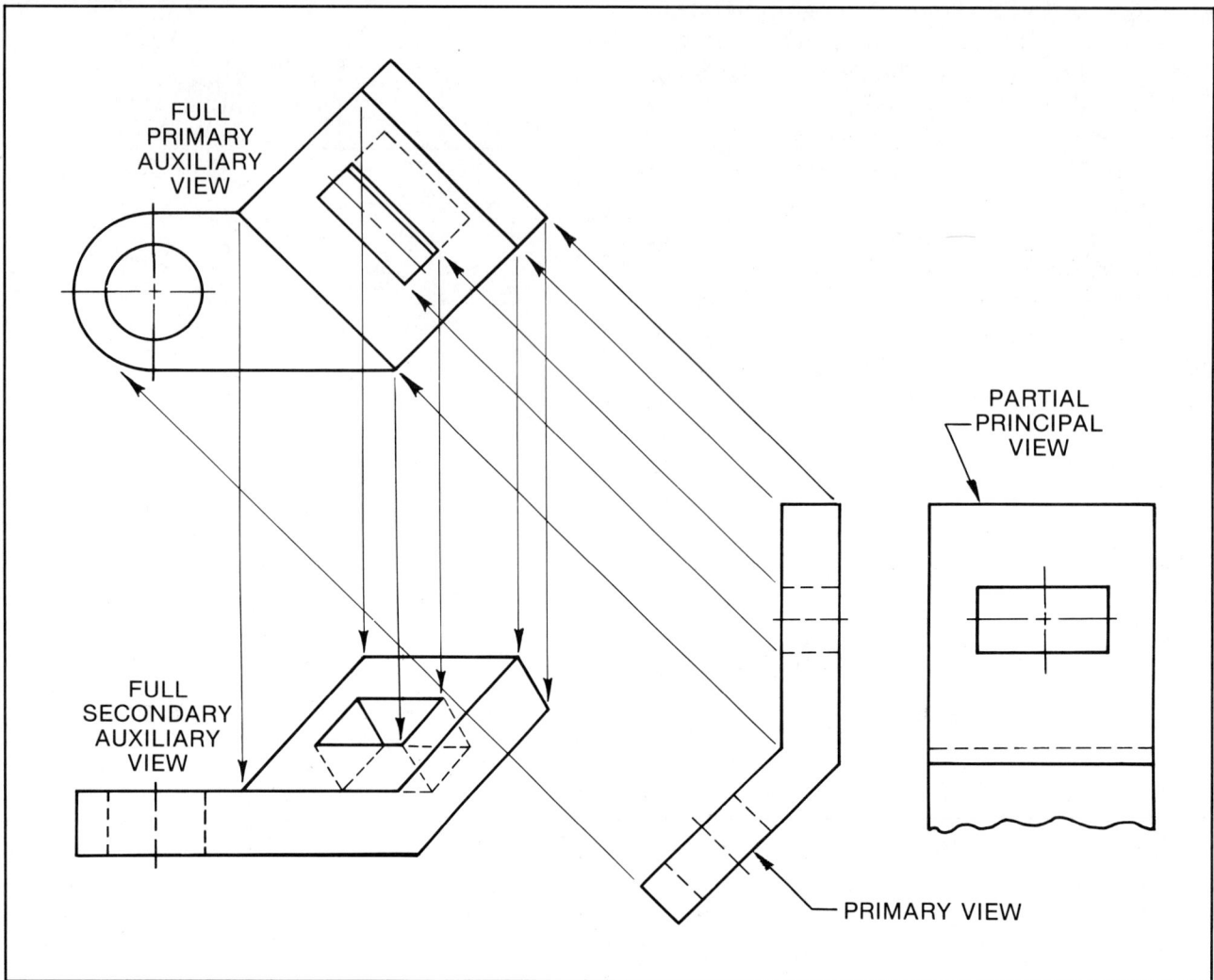

Figure 5-18. A secondary auxiliary view is aligned with a primary auxiliary view rather than with a principal view.

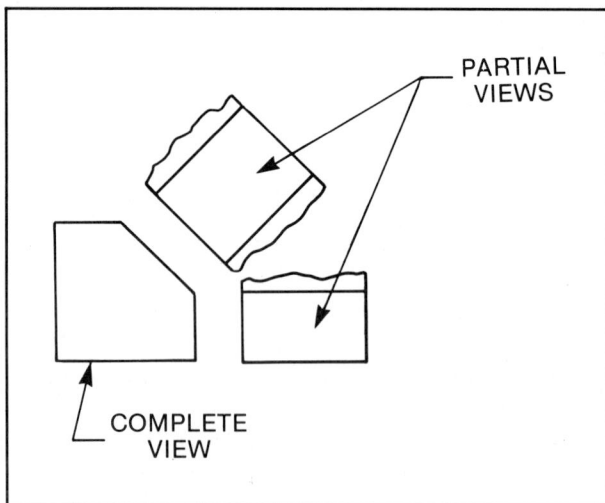

Figure 5-19. Using complete and partial views to describe a part

KEY TERMS

The following key terms were introduced in this chapter. Be sure you know the meaning of each term before proceeding to the review material.

Auxiliary View
Complete View
Orthographic Projection
Orthographic View
Partial View
Primary Auxiliary View
Principal View
Projectors
Secondary Auxiliary View
Visualization

Test your knowledge with this reinforcement study material. Write your answers to the questions in the spaces provided.

1. How are views of engineering drawings normally developed? _____
   _____

2. What type or types of views are normally shown in a multiview print? _____
   _____

3. How many possible views are there for most parts? _____

4. How many views are normally used to describe a part? _____

5. What views are normally shown in a multiview print? _____
   _____

6. What determines which views are used in a multiview print? _____
   _____

7. When would a print have only two views? _____
   _____

8. What can be used to replace a view in a one-view print? _____

9. What is the first step in the process of visualization? _____
   _____
   _____

10. When visualizing a part, what is the final step in the visualization process? _____
    _____
    _____

11. Which type of auxiliary view is shown perpendicular to another auxiliary view? _____
    _____

12. Which type of auxiliary view is shown aligned with a surface shown in an orthographic view?
    _____

13. How are auxiliary views identified? _____
    _____

14. To how many principal views is a single auxiliary view directly related? _____

15. What type of view is used to show specific information about a limited area of a part?
    _____

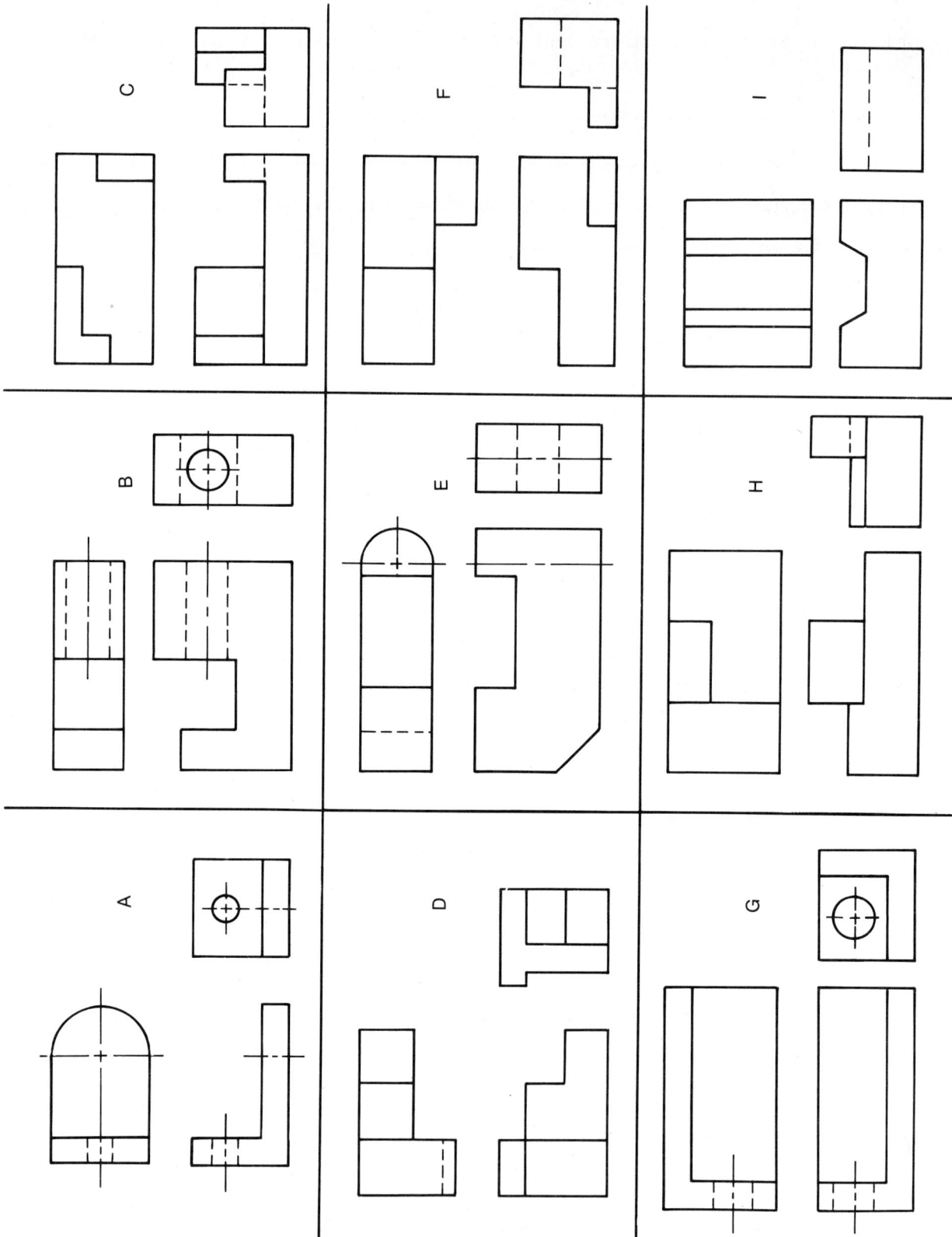

Figure E5-1A.

Match the multiview drawings in Figure E5-1A on page 70 with the part each represents in Figure E5-1B below. One answer is provided as an example.

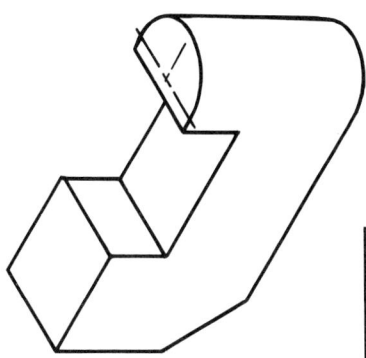

3. _____

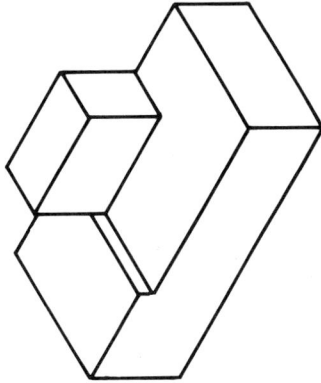

6. _____

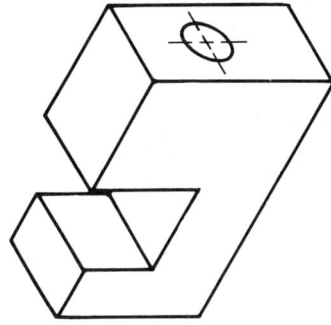

9. _____

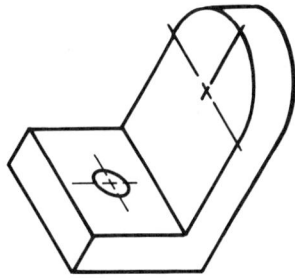

2. _____

A

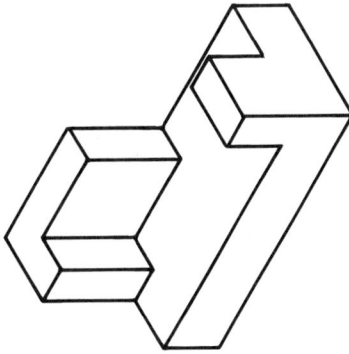

5. _____

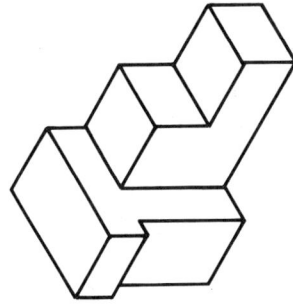

8. _____

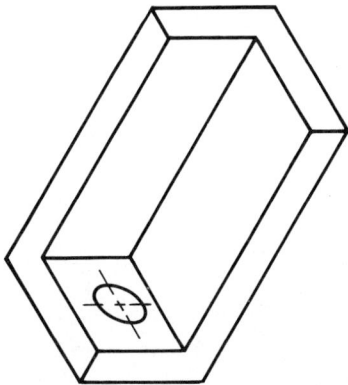

1. _____

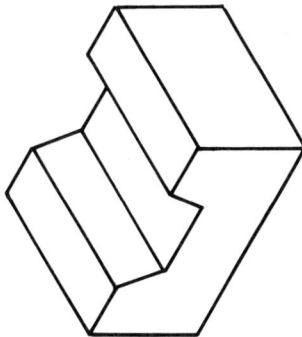

4. _____

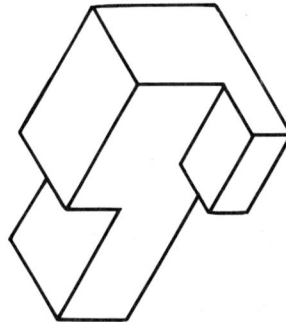

7. _____

Figure E5-1B.

Sketch the three views for the part shown in Figure E5-2.

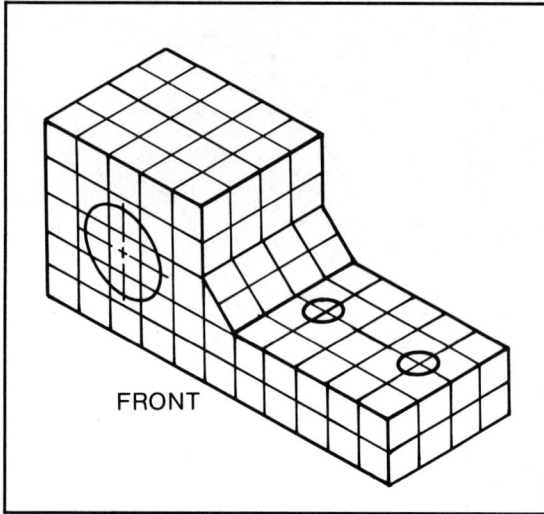

FRONT

Figure E5-2.

Sketch the three views for the part shown in Figure E5-3.

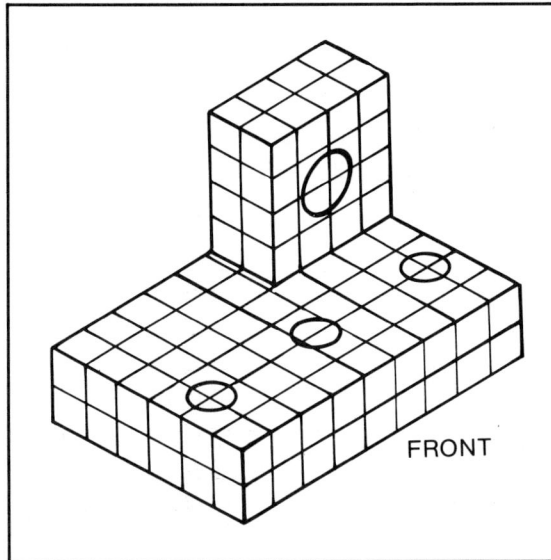

Figure E5-3.

Refer to Figure E5-4 on page 75 to answer the following questions.

1. How many views are used to describe this part? _____

2. Identify the following views.

    a. View *A*: _____

    b. View *B*: _____

    c. View *C*: _____

    d. View *D*: _____

    e. View *E*: _____

    f. View *F*: _____

3. Which views should be used to describe this part? _____

_____

4. List the three types of lines used to describe this part.

    a. _____

    b. _____

    c. _____

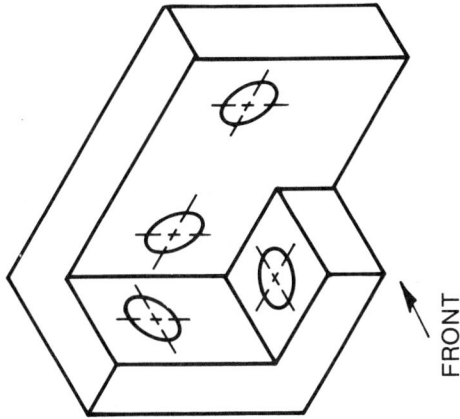

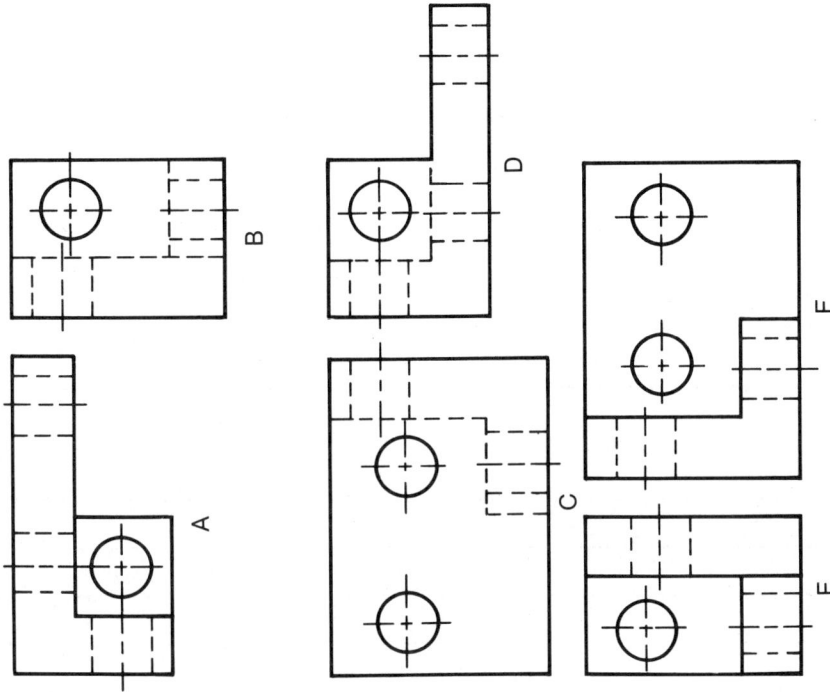

Figure E5-4.

Refer to Figure E5-5 on page 77 to answer the following questions.

1. How many views are used to describe this part? _____

2. Identify the following views.

   a. View *A*: _____

   b. View *B*: _____

   c. View *C*: _____

   d. View *D*: _____

3. What sizes are the holes in this part? _____
   _____

4. How many complete views are shown? _____

5. How many auxiliary views are shown? _____

6. Find the following dimensions.

   a. *E* = _____

   b. *F* = _____

   c. *G* = _____

7. Where are the general notes that concern this part located? _____
   _____

8. What are the specific instructions shown to the right called? _____
   _____

9. What five types of lines are shown in this print?

   a. _____

   b. _____

   c. _____

   d. _____

   e. _____

10. What term is used to describe the oval-shaped holes in the principal view shown at the top?
   _____

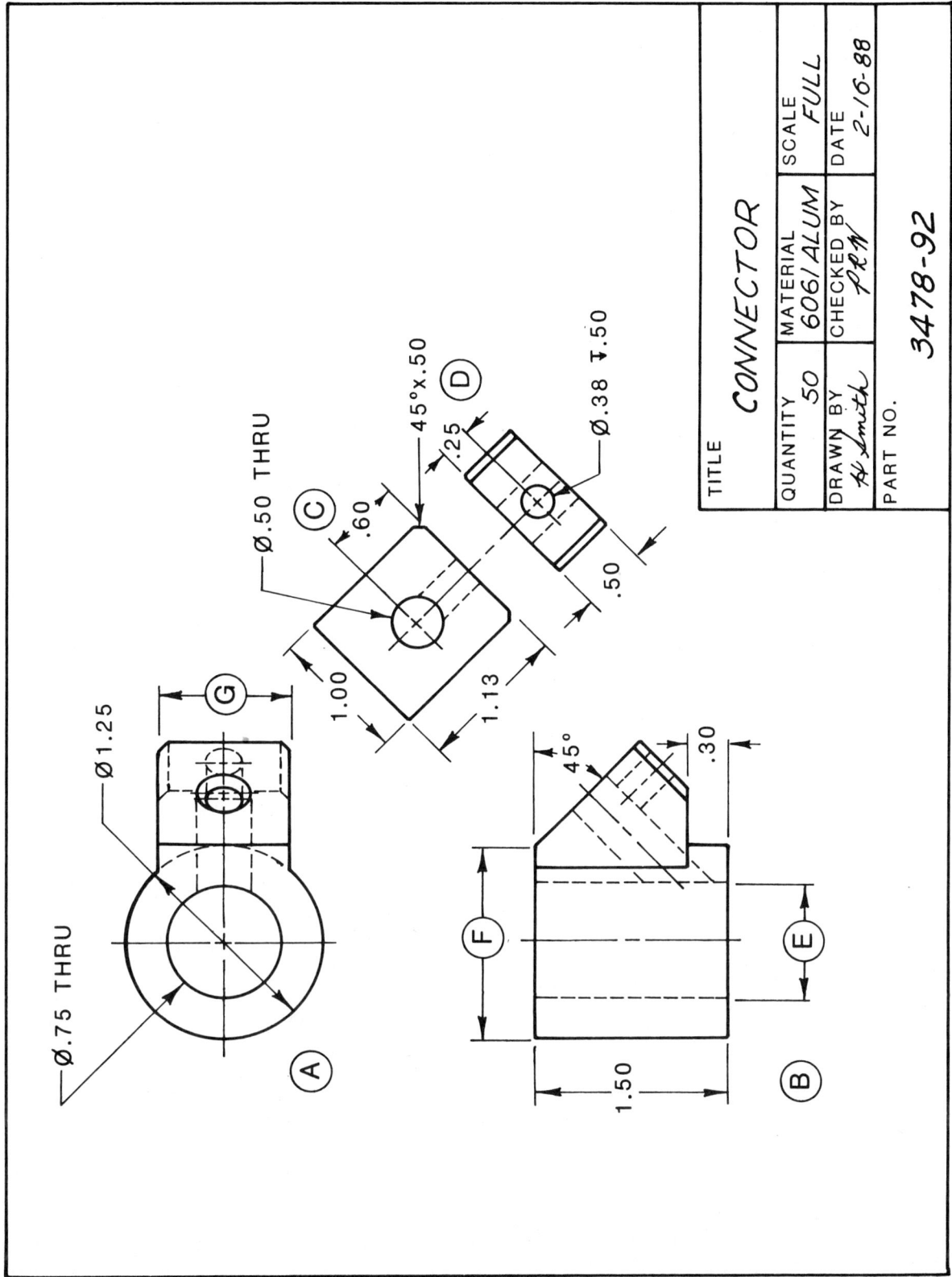

TITLE CONNECTOR

QUANTITY 50	MATERIAL 6061 ALUM	SCALE FULL
DRAWN BY A. Smith	CHECKED BY PRM	DATE 2-16-88
PART NO. 3478-92		

Figure E5-5.

6

SHOP MATHEMATICS

OBJECTIVES

After studying this chapter, you will be able to:

- Identify and explain common fractions.

- Perform basic calculations using common fractions.

- Identify and explain decimal fractions.

- Perform basic calculations using decimal fractions.

- Determine equivalent values between common fractions and decimal fractions.

Everyone who reads a print will occasionally need to make some calculations involving dimensions. Whether to find a dimensional value, to determine the amount of material required, or for a similar purpose, mathematics is often involved in reading prints. To make these calculations quickly and accurately, the print reader should be familiar with the procedures involved in working with common fractions and decimal fractions. This chapter provides a brief review of the steps necessary to make typical print calculations.

COMMON FRACTIONS

Common fractions are frequently used on prints to show thread sizes, rough stock sizes, and similar details. These fractions are used to denote parts of a whole unit. For example, one inch (1″) is considered a whole unit. Therefore, fractions such as $\frac{1}{32}$″, $\frac{1}{16}$″, and $\frac{1}{4}$″ are all considered to be parts of the 1″ unit.

Common fractions consist of two basic parts: the *denominator*, or bottom number, and the *numerator*, or top number. See Figure 6-1. The denominator shows into how many parts the whole unit has been divided. The numerator indicates the number of parts of the whole that are being considered. For example, if a fraction has a value of $\frac{5}{32}$, this means the whole unit has been divided into 32 equal parts, and 5 of these parts are to be considered in the value of the fraction. If a measurement is expressed as $\frac{1}{2}$″, it means the whole inch unit has been

$$\frac{\text{NUMERATOR}}{\text{DENOMINATOR}} \left(\frac{1}{2}\right)$$
or
NUMERATOR/DENOMINATOR (½)

Figure 6-1. Parts of a common fraction

divided into 2 equal parts and 1 of these parts is to be considered in the measurement.

Working with Common Fractions

The following are a few of the basic facts, rules, and terms concerning common fractions.

1. A *proper fraction* is a common fraction whose numerator is smaller than its denominator.

> **Examples:** $\frac{1}{4}$, $\frac{1}{2}$, $\frac{3}{4}$, $\frac{15}{16}$, $\frac{23}{32}$, and $\frac{47}{64}$ are all proper fractions.

2. An *improper fraction* is a common fraction whose numerator is greater than its denominator.

> **Examples:** $\frac{3}{2}$, $\frac{5}{4}$, $\frac{10}{8}$, $\frac{19}{16}$, $\frac{45}{32}$, and $\frac{84}{64}$ are all improper fractions.

3. A common fraction with the same numerator and denominator is equal to 1.

> **Examples:** $\frac{2}{2}$, $\frac{4}{4}$, $\frac{8}{8}$, $\frac{16}{16}$, $\frac{32}{32}$, and $\frac{64}{64}$ are all equal to 1.

4. A *reducible fraction* is a common fraction that can be reduced to lower terms.

> **Examples:** $\frac{2}{4}$, $\frac{6}{8}$, $\frac{12}{16}$, $\frac{28}{32}$, and $\frac{62}{64}$ are all reducible fractions: $\frac{2}{4} = \frac{1}{2}$; $\frac{6}{8} = \frac{3}{4}$; $\frac{12}{16} = \frac{3}{4}$; $\frac{28}{32} = \frac{7}{8}$; and $\frac{62}{64} = \frac{31}{32}$.

5. To reduce a common fraction to lowest terms, divide both the numerator and the denominator by the highest number by which both numbers can be divided.

> **Examples: (a)** $\frac{24}{32} = \frac{24 \div 8}{32 \div 8} = \frac{3}{8}$
>
> **(b)** $\frac{6}{8} = \frac{6 \div 2}{8 \div 2} = \frac{3}{4}$

6. A fraction may be changed to an equivalent fraction with a larger denominator by multiplying the numerator and denominator by the same number. This may be necessary before adding or subtracting fractions.

> **Examples: (a)** To change $\frac{1}{4}$ to 16ths:
>
> $\frac{1}{4} \times \frac{4}{4} = \frac{4}{16}$
>
> **(b)** To change $\frac{3}{8}$ to 24ths:
>
> $\frac{3}{8} \times \frac{3}{3} = \frac{9}{24}$

7. The *least common denominator* of two or more fractions is the smallest denominator value that is equally divisible by the denominator values of the fractions.

> **Example:** $\frac{1}{2} + \frac{1}{4} + \frac{3}{8}$; the least common denominator is 8.

8. A *mixed number* is a combination of a whole number and a common fraction.

> **Examples:** $2\frac{1}{2}$, $1\frac{7}{8}$, $3\frac{15}{16}$, and $1\frac{9}{32}$ are all mixed numbers.

9. To convert mixed numbers to improper fractions, multiply the whole number by the denominator of the fraction and add the numerator. The resulting number becomes the new numerator of the improper fraction. The denominator remains the same.

> **Examples: (a)** $2\frac{1}{2} = \frac{(2 \times 2) + 1}{2} = \frac{5}{2}$
>
> **(b)** $3\frac{7}{16} = \frac{(3 \times 16) + 7}{16} = \frac{55}{16}$

10. To convert a whole number to an improper fraction, divide the whole number by 1.

> **Examples: (a)** $4 = 4 \div 1 = \frac{4}{1}$
>
> **(b)** $3 = 3 \div 1 = \frac{3}{1}$

11. To change a whole number to a common fraction with a specific denominator value, convert the whole number to a fraction. Then multiply the numerator and denominator by the desired denominator value.

Examples: (a) To change 4 to 16ths:

$$\frac{4}{1} \times \frac{16}{16} = \frac{4 \times 16}{1 \times 16} = \frac{64}{16}$$

(b) To change 3 to 32nds:

$$\frac{3}{1} \times \frac{32}{32} = \frac{3 \times 32}{1 \times 32} = \frac{96}{32}$$

---

12. To convert an improper fraction to a mixed number, divide the numerator by the denominator, and reduce the fraction to its lowest terms.

Examples: (a) $\frac{17}{8} = 17 \div 8 = 2\frac{1}{8}$

(b) $\frac{26}{16} = 26 \div 16 = 1\frac{10}{16} = 1\frac{5}{8}$

Adding Fractions and Mixed Numbers

Follow these steps to add common fractions:

• Find the least common denominator for the fractions in the problem.

• Change the fractions to equivalent fractions with all denominators the same.

• Add the numerators.

• Convert the answer to a mixed number, if necessary.

• Reduce the fraction to lowest terms.

Example: $\frac{1}{4} + \frac{3}{16} + \frac{7}{8}$:

$$\frac{1}{4} = \frac{4}{16}$$
$$\frac{3}{16} = \frac{3}{16}$$
$$+ \frac{7}{8} = \frac{14}{16}$$
$$\overline{\frac{21}{16} = 1\frac{5}{16}}$$

Follow these steps to add mixed numbers:

• Find the least common denominator for the fractions in the problem.

• Change the fractions to equivalent fractions with all denominators the same.

• Add the numerators.

• Add the whole numbers.

• Reduce the answer to lowest terms.

Example: $2\frac{1}{2} + 4\frac{1}{4} + 1\frac{15}{32}$:

$$2\frac{1}{2} = 2\frac{16}{32}$$
$$4\frac{1}{4} = 4\frac{8}{32}$$
$$+ 1\frac{15}{32} = 1\frac{15}{32}$$
$$\overline{7\frac{39}{32} = 8\frac{7}{32}}$$

PRACTICE PROBLEM 6-1

Add the following common fractions and mixed numbers.

(a) $\frac{3}{8} + \frac{7}{32} = $ _____

(b) $\frac{1}{8} + \frac{15}{16} + \frac{1}{2} = $ _____

(c) $\frac{17}{32} + \frac{9}{64} = $ _____

(d) $\frac{13}{16} + 2\frac{7}{8} + \frac{5}{64} = $ _____

(e) $1\frac{1}{2} + \frac{3}{32} + \frac{7}{8} = $ _____

(f) $3\frac{1}{64} + 2\frac{7}{16} = $ _____

(g) $\frac{13}{16} + \frac{7}{8} + 1\frac{3}{4} = $ _____

(h) $2\frac{5}{64} + 7\frac{1}{32} + \frac{1}{2} = $ _____

(i) $\frac{1}{4} + \frac{3}{8} + 3\frac{19}{64} = $ _____

(j) $2\frac{3}{4} + 1\frac{7}{16} + \frac{1}{8} = $ _____

(k) $\frac{3}{4} + \frac{9}{64} = $ _____

(l) $\frac{7}{64} + 1\frac{1}{8} = $ _____

(m) $\frac{1}{8} + \frac{7}{32} = $ _____

(n) $\frac{19}{32} + \frac{7}{16} = $ _____

(o) $\frac{11}{16} + \frac{1}{2} = $ _____

Subtracting Fractions and Mixed Numbers

Follow these steps to subtract common fractions:

• Find the least common denominator for the fractions in the problem.

• Change the fractions to equivalent fractions with the same denominator.

• Subtract the numerators.

• Reduce the answer to lowest terms, if necessary.

Follow these steps to subtract mixed numbers:

• Find the least common denominator for the fractions in the problem.

• Change the fractions to equivalent fractions with the same denominator.

• Subtract the numerators.

• Subtract the whole numbers.

• Reduce the answer to lowest terms, if necessary.

Example: $\frac{15}{16} - \frac{7}{32}$:

$$\frac{15}{16} = \frac{30}{32}$$
$$- \frac{7}{32} = \frac{7}{32}$$
$$\frac{23}{32}$$

Example: $2\frac{3}{8} - 1\frac{1}{16}$:

$$2\frac{3}{8} = 2\frac{6}{16}$$
$$- 1\frac{1}{16} = 1\frac{1}{16}$$
$$1\frac{5}{16}$$

PRACTICE PROBLEM 6-2

Subtract the following common fractions and mixed numbers.

(a) $\frac{7}{8} - \frac{23}{64} =$ _____

(b) $\frac{15}{16} - \frac{7}{8} =$ _____

(c) $1\frac{1}{8} - \frac{23}{64} =$ _____

(d) $3\frac{3}{16} - 1\frac{1}{8} =$ _____

(e) $4\frac{7}{8} - 3\frac{61}{64} =$ _____

(f) $\frac{1}{2} - \frac{1}{32} =$ _____

(g) $1\frac{13}{16} - \frac{19}{32} =$ _____

(h) $\frac{9}{16} - \frac{9}{32} =$ _____

(i) $2\frac{23}{32} - 1\frac{7}{8} =$ _____

(j) $1\frac{1}{4} - \frac{7}{16} =$ _____

(k) $\frac{23}{64} - \frac{1}{8} =$ _____

(l) $\frac{17}{64} - \frac{7}{32} =$ _____

(m) $\frac{3}{8} - \frac{5}{64} =$ _____

(n) $\frac{63}{64} - \frac{3}{16} =$ _____

(o) $2\frac{1}{2} - \frac{9}{32} =$ _____

Multiplying Fractions and Mixed Numbers

Follow these steps to multiply common fractions:

- Multiply the numerators.
- Multiply the denominators.
- Convert improper fractions to mixed numbers, if necessary.

> **Example:** $\frac{3}{4} \times \frac{7}{16} = \frac{3 \times 7}{4 \times 16} = \frac{21}{64}$

Follow these steps to multiply mixed numbers:

- Convert the mixed numbers to improper fractions.
- Multiply the numerators.
- Multiply the denominators.
- Convert improper fractions to mixed numbers, if necessary.

> **Example:** $2\frac{1}{4} \times 3\frac{1}{2} = \frac{9}{4} \times \frac{7}{2} = \frac{9 \times 7}{4 \times 2} = \frac{63}{8}$
> $= 7\frac{7}{8}$

PRACTICE PROBLEM 6-3

Multiply the following common fractions and mixed numbers.

(a) $\frac{3}{16} \times \frac{7}{8} = $ _____

(b) $\frac{7}{32} \times 1\frac{1}{2} = $ _____

(c) $\frac{1}{2} \times \frac{1}{4} = $ _____

(d) $\frac{17}{64} \times \frac{3}{8} = $ _____

(e) $\frac{1}{8} \times \frac{7}{64} = $ _____

(f) $\frac{3}{8} \times \frac{21}{32} = $ _____

(g) $2\frac{5}{16} \times 1\frac{15}{32} = $ _____

(h) $\frac{1}{16} \times \frac{7}{8} = $ _____

(i) $1\frac{1}{64} \times 2\frac{3}{16} = $ _____

(j) $6\frac{17}{32} \times 7\frac{23}{64} = $ _____

(k) $1\frac{5}{8} \times 2\frac{7}{16} = $ _____

(l) $4\frac{1}{8} \times \frac{31}{32} = $ _____

(m) $2\frac{23}{32} \times \frac{5}{64} = $ _____

(n) $6\frac{1}{2} \times 4\frac{1}{8} = $ _____

(o) $1\frac{1}{2} \times 1\frac{1}{8} = $ _____

Dividing Fractions and Mixed Numbers

Follow these steps to divide common fractions:

- Write the fractions to be divided.
- Invert (switch) the numerator and denominator in the dividing fraction.
- Multiply the numerators and denominators.
- Convert improper fractions to mixed numbers, if necessary.

> **Example:** $\frac{3}{4} \div \frac{1}{2} = \frac{3}{4} \times \frac{2}{1} = \frac{3 \times 2}{4 \times 1} = \frac{6}{4} = 1\frac{1}{2}$

Follow these steps to divide mixed numbers:

- Convert the mixed numbers to improper fractions.
- Write the improper fractions to be divided.
- Invert (switch) the numerator and denominator in the dividing fraction.
- Convert improper fractions to mixed numbers, if necessary.

> **Example:** $2\frac{1}{2} \div 1\frac{7}{8} = \frac{5}{2} \div \frac{15}{8} = \frac{5}{2} \times \frac{8}{15}$
> $= \frac{5 \times 8}{2 \times 15} = \frac{40}{30} = 1\frac{1}{3}$

PRACTICE PROBLEM 6-4

Divide the following common fractions and mixed numbers.

(a) $\frac{1}{4} \div \frac{3}{16} =$ _____

(b) $\frac{5}{16} \div 1\frac{1}{2} =$ _____

(c) $3\frac{3}{4} \div 2\frac{1}{16} =$ _____

(d) $1\frac{15}{16} \div 2\frac{3}{16} =$ _____

(e) $\frac{15}{32} \div 1\frac{1}{4} =$ _____

(f) $1\frac{7}{8} \div 2\frac{3}{4} =$ _____

(g) $\frac{3}{8} \div 2\frac{1}{4} =$ _____

(h) $\frac{63}{64} \div \frac{1}{2} =$ _____

(i) $4\frac{5}{32} \div 1\frac{1}{4} =$ _____

(j) $1\frac{7}{16} \div 2\frac{1}{64} =$ _____

(k) $\frac{3}{32} \div \frac{1}{8} =$ _____

(l) $1\frac{5}{8} \div \frac{7}{16} =$ _____

(m) $3\frac{1}{4} \div \frac{3}{16} =$ _____

(n) $2\frac{1}{8} \div \frac{7}{16} =$ _____

(o) $\frac{31}{32} \div \frac{1}{8} =$ _____

DECIMAL FRACTIONS

Decimal fractions, like common fractions, are used to denote parts of a whole unit. However, decimal fractions do not have a denominator value written into the fraction. Their denominators are implied, rather than stated. Decimal fractions all have implied denominators that are multiples of ten. Therefore, a decimal fraction of .1 has a value of one-tenth $(\frac{1}{10})$. A decimal fraction of .01 has a value of $\frac{1}{100}$. A decimal fraction of .001 has a value of $\frac{1}{1,000}$.

These examples show that as the number of decimal place values increases, the value of the decimal number changes by a multiple of ten. A single number placed to the right of a decimal point has a value expressed in tenths. Two numbers to the right of a decimal point have a value expressed in hundredths. Three numbers to the right have a value expressed in thousandths, and four numbers are expressed in ten-thousandths. See Figure 6-2. Since the denominator is implied, the number of decimal places in the numerator indicates the value of the decimal fraction.

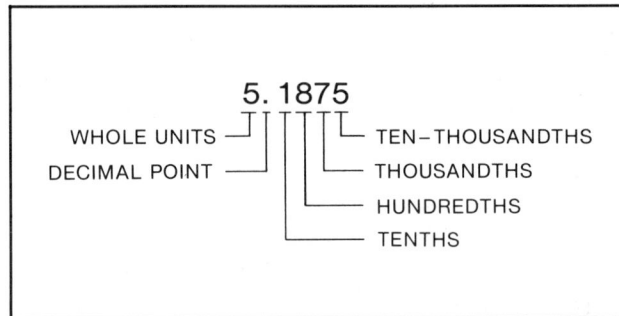

Figure 6-2. Place values of a decimal fraction

A decimal fraction expressed as .125 means the whole unit has been divided into 1,000 parts, and 125 of these parts are considered in the value of the decimal fraction. Similarly, if a measurement is expressed as .250″, it means the whole inch unit is divided into 1,000 parts, and 250 of these parts are considered in the measurement.

Note: In industry, most decimal fractions are expressed in terms of thousandths rather than in tenths or hundredths. Thus, a decimal fraction of .2 is usually written .200 and read 200 thousandths. A value of .75 is usually written .750 and read 750 thousandths. However, in the case of decimals having more than three places, the values are not expressed in terms of thousandths. Therefore, a value of .1875 is read as 1 thousand 8 hundred 75 ten-thousandths.

Working with Decimal Fractions

The following are a few of the basic facts, rules, and terms concerning decimal fractions.

1. Whole units are shown to the left of a decimal point. Fractional parts of a whole unit are shown to the right.

Example: 10.125

2. Zeros are not normally used to the left of a decimal point on prints when indicating values less than 1.

Example: .875, not 0.875

3. Zeros may be used to the right of a decimal point to show place values in certain dimensions.

Example: .500 not .5 .492 .492

4. Most decimal dimensions shown on prints are written as two-place decimals. The exception to this is where added precision is required. Then three- or four-place decimal fractions may be used.

Examples: (a) .50, not .500 nor .5000 (b) When added precision is required: .1875, not .188 nor .19

5. In most cases, the calculations involving decimal fractions are handled in much the same way as the calculations used for whole numbers. The only difference is in the way the decimal point is handled.

Adding Decimal Fractions

Follow these steps to add decimal fractions:

- Write the problem with all decimal points aligned vertically.
- Add the numbers as whole number values.
- Insert the decimal point in the answer in the same vertical column as in the problem.

Examples: (a)		(b)	
	.125		1.750
	1.0625		.875
	2.50		.125
+	.1875	+	2.0005
	3.8750		4.7505

PRACTICE PROBLEM 6-5

Add the following decimal fractions.

(a) .75 + .0625 = _____

(b) .125 + .539 = _____

(c) .8125 + .50 = _____

(d) .9875 + .365 = _____

(e) 2.350 + .896 = _____

(f) .875 + .375 = _____

(g) 3.569 + 1.2087 = _____

(h) 1.8676 + 2.227 = _____

(i) 4.0002 + .0003 = _____

(j) .6283 + .5503 = _____

(k) 1.558 + 2.674 = _____

(l) .0065 + 1.330 = _____

(m) .0006 + 1.1800 = _____

(n) .8776 + 1.0083 = _____

(o) .0202 + 1.0022 = _____

Subtracting Decimal Fractions

Follow these steps to subtract decimal fractions:

• Write the problem with all decimal points aligned vertically.
• Subtract the numbers as whole number values.
• Insert the decimal point in the answer in the same vertical column as in the problem.

Examples:	(a)	1.750	(b)	2.625
		− .250		− 1.125
		1.500		1.500

PRACTICE PROBLEM 6-6

Subtract the following decimal fractions.

(a) .500 − .245 = _____

(b) .9999 − .1119 = _____

(c) .50 − .4751 = _____

(d) 1.500 − .0025 = _____

(e) 1.00 − .7508 = _____

(f) .789 − .0034 = _____

(g) .30 − .0007 = _____

(h) 2.592 − .01 = _____

(i) 4.375 − .3997 = _____

(j) 3.75 − 2.89 = _____

(k) 2.65 − .9007 = _____

(l) 3.890 − .0008 = _____

(m) 1.020 − .9073 = _____

(n) 2.006 − 1.097 = _____

(o) .010 − .0007 = _____

Multiplying Decimal Fractions

Follow these steps to multiply decimal fractions:

• Write the problem with the decimal points aligned.
• Multiply the values as whole numbers.

• Count the number of decimal places in both multiplied values.
• Counting from right to left in the answer, insert the decimal point so the number of decimal places in the answer equals the total number of decimal places in the numbers multiplied. (This may require the insertion of zeros to the left of the answer, as in Example (b) below.)

Examples:	(a)	.75	(Four decimal
		× .25	places)
		.1875	(Four decimal places in the answer)
	(b)	1.625	(Six decimal
		× .033	places)
		.053625	(Six decimal places in the answer)

PRACTICE PROBLEM 6-7

Multiply the following decimal fractions. Round off each answer to four decimal places or less.

(a) .375 × .50 = _____

(b) .5023 × .25 = _____

(c) 2.0075 × .1007 = _____

(d) 1.125 × .875 = _____

(e) 3.089 × .278 = _____

(f) 1.897 × .006 = _____

(g) 3.008 × 6.45 = _____

(h) 3.4375 × 2.1002 = _____

(i) 4.562 × .0097 = _____

(j) 4.875 × 1.509 = _____

(k) 6.4523 × .1007 = _____

(l) 1.250 × 1.50 = _____

(m) .550 × 1.7505 = _____

(n) 2.9375 × .1006 = _____

(o) 2.50 × 1.0081 = _____

Dividing Decimal Fractions

Follow these steps to divide decimal fractions:

- Write the problem as whole number division.

- Move the decimal point of the dividing number to the extreme right. (This will make the decimal fraction appear as a whole number.) Note the number of places the decimal point was moved.

- Move the decimal point in the divided number to the right the same number of places.

- Locate the decimal point on the horizontal line directly above the new position in the divided number.

- Divide as a whole number problem.

- Continue the division until the desired number of decimal places in the answer is reached. Add zeros, if necessary.

Examples: (a) 2.75 ÷ .625

```
          4.4
625.)2750.0
     2500
      250 0
      250 0
          0
```

(b) 5.6682 ÷ 2.557

```
           2.216738
2557.)5668.200000
      5114
       554 2
       511 4
        42 80
        25 57
        17 230
        15 342
         1 8880
         1 7899
            9810
            7671
           21390
           20456
             934
```

CONVERTING COMMON AND DECIMAL FRACTIONS

Common fractions and decimal fractions must sometimes be converted to their equivalent values. This may be accomplished by using a decimal equivalent chart, Table 6-1, or by calculating the equivalent values.

Table 6-1. DECIMAL EQUIVALENT CHART

FRACTION	DECIMAL	FRACTION	DECIMAL	FRACTION	DECIMAL	FRACTION	DECIMAL
$\frac{1}{64}$	.01562	$\frac{17}{64}$	.26562	$\frac{33}{64}$	.51562	$\frac{49}{64}$	.76562
$\frac{1}{32}$	.03125	$\frac{9}{32}$	.28125	$\frac{17}{32}$	.53125	$\frac{25}{32}$	.78125
$\frac{3}{64}$	.04687	$\frac{19}{64}$	.29687	$\frac{35}{64}$	.54687	$\frac{51}{64}$	.79687
$\frac{1}{16}$	.0625	$\frac{5}{16}$	.3125	$\frac{9}{16}$	.5625	$\frac{13}{16}$	.8125
$\frac{5}{64}$	.07812	$\frac{21}{64}$	.32812	$\frac{37}{64}$	.57812	$\frac{53}{64}$	.82812
$\frac{3}{32}$	.09375	$\frac{11}{32}$	.34375	$\frac{19}{32}$	.59375	$\frac{27}{32}$	.84375
$\frac{7}{64}$	.10937	$\frac{23}{64}$	.35937	$\frac{39}{64}$	.60937	$\frac{55}{64}$	.85937
$\frac{1}{8}$	.125	$\frac{3}{8}$	.375	$\frac{5}{8}$	.625	$\frac{7}{8}$	.875
$\frac{9}{64}$	.14062	$\frac{25}{64}$	.39062	$\frac{41}{64}$	.64062	$\frac{57}{64}$	.89062
$\frac{5}{32}$	.15625	$\frac{13}{32}$	.40625	$\frac{21}{32}$	.65625	$\frac{29}{32}$	.90625
$\frac{11}{64}$	.17187	$\frac{27}{64}$	.42187	$\frac{43}{64}$	.67187	$\frac{59}{64}$	.92187
$\frac{3}{16}$	.1875	$\frac{7}{16}$	.4375	$\frac{11}{16}$	.6875	$\frac{15}{16}$	.9375
$\frac{13}{64}$	.20312	$\frac{29}{64}$	.45312	$\frac{45}{64}$	.70312	$\frac{61}{64}$	.95312
$\frac{7}{32}$	.21875	$\frac{15}{32}$	.46875	$\frac{23}{32}$	.71875	$\frac{31}{32}$	.96875
$\frac{15}{64}$	.23437	$\frac{31}{64}$	.48437	$\frac{47}{64}$	.73437	$\frac{63}{64}$	.98437
$\frac{1}{4}$	.250	$\frac{1}{2}$	.500	$\frac{3}{4}$	.750	1	1.000

Converting Common Fractions to Decimal Fractions

Follow these steps to convert common fractions to decimal fractions:

• Divide the numerator of the common fraction by its denominator.

• Continue the division until there is no remainder, or until the desired number of decimal places in the answer is reached. (Normally divide to five places and round the answer to four places.)

Examples: (a) Convert $\frac{1}{2}$ to a decimal fraction.

$$
\begin{array}{r}
.5 \\
2\overline{)1.0} \\
\underline{1\ 0} \\
0
\end{array}
$$

(b) Convert $\frac{1}{8}$ to a decimal fraction.

$$
\begin{array}{r}
.125 \\
8\overline{)1.000} \\
\underline{8} \\
20 \\
\underline{16} \\
40 \\
\underline{40} \\
0
\end{array}
$$

(c) Convert $\frac{3}{16}$ to a decimal fraction.

$$
\begin{array}{r}
.187 = .19 \text{ rounded} \\
16\overline{)3.000}\quad \text{to two deci-} \\
\underline{1\ 6}\quad\ \ \text{mal points} \\
1\ 40 \\
\underline{1\ 28} \\
120 \\
\underline{112} \\
8
\end{array}
$$

PRACTICE PROBLEM 6-9

Convert the following common fractions to decimal numbers.

(a) $\frac{7}{8}$ = _____

(b) $\frac{1}{4}$ = _____

(c) $\frac{17}{32}$ = _____

(d) $\frac{7}{16}$ = _____

(e) $\frac{15}{16}$ = _____

(f) $\frac{31}{32}$ = _____

(g) $\frac{3}{8}$ = _____

(h) $\frac{3}{4}$ = _____

(i) $\frac{31}{64}$ = _____

(j) $\frac{11}{32}$ = _____

(k) $\frac{27}{64}$ = _____

(l) $\frac{23}{64}$ = _____

(m) $\frac{5}{8}$ = _____

(n) $\frac{3}{16}$ = _____

(o) $\frac{3}{64}$ = _____

PRACTICE PROBLEM 6-10

Convert the following decimal fractions to common fractions.

(a) .0781 = _____

(b) .0313 = _____

(c) .6875 = _____

(d) .0625 = _____

(e) .1406 = _____

(f) .4688 = _____

(g) .2344 = _____

(h) .0156 = _____

(i) .9844 = _____

(j) .1562 = _____

(k) .8125 = _____

(l) .2031 = _____

(m) .3906 = _____

(n) .1719 = _____

(o) .2188 = _____

Converting Decimal Fractions to Common Fractions

Follow these steps to convert decimal fractions to common fractions:

• Write the decimal fraction as a common fraction. Remember to use the proper number of zeros in the denominator.

• Reduce this common fraction to lowest terms.

Examples: (a) $.5 = \frac{5}{10} = \frac{1}{2}$

(b) $.125 = \frac{125}{1000} = \frac{1}{8}$

KEY TERMS

The following key terms were introduced in this chapter. Be sure you know the meaning of each term before proceeding to the review material.

Common Fraction
Decimal Fraction
Denominator
Improper Fraction
Least Common Denominator
Mixed Number
Numerator
Proper Fraction
Reducible Fraction

Test your knowledge with this reinforcement study material. Write your answers to the questions in the spaces provided.

1. What is the top number in a common fraction called? _____

2. What is the bottom number in a common fraction called? _____

3. What is the term used to describe a common fraction whose bottom number is larger than its top number? _____

4. What is the term used to describe a common fraction whose top number is larger than its bottom number? _____

5. What is the combination of a common fraction and a whole number called? _____

_____

6. What is the value of a common fraction with the same top and bottom numbers? _____

7. How many decimal places do most decimal fractions shown on prints have? _____

8. How are two-place and three-place decimal fractions normally expressed? _____

_____

9. What type of chart can be used to convert decimal fractions and common fractions?

_____

10. What do all common fractions and decimal fractions show? _____

_____

11. Add these common fractions and mixed numbers.

  a. $2\frac{1}{2} + \frac{3}{4} + \frac{9}{16} =$ _____

  b. $\frac{17}{32} + \frac{31}{64} + 1\frac{1}{16} =$ _____

  c. $4\frac{7}{16} + \frac{1}{32} + \frac{7}{8} =$ _____

  d. $\frac{23}{32} + \frac{5}{8} + \frac{15}{64} =$ _____

12. Subtract these common fractions and mixed numbers.

  a. $\frac{3}{4} - \frac{5}{64} =$ _____

  b. $1\frac{1}{2} - 1\frac{1}{64} =$ _____

  c. $\frac{23}{64} - \frac{3}{32} =$ _____

  d. $2\frac{1}{4} - \frac{61}{64} =$ _____

13. Multiply these common fractions and mixed numbers.

  a. $\frac{3}{8} \times \frac{17}{32} =$ _____

  b. $1\frac{7}{8} \times 3\frac{15}{64} =$ _____

  c. $5\frac{1}{2} \times 2\frac{3}{4} =$ _____

  d. $2\frac{9}{16} \times \frac{5}{8} =$ _____

14. Divide these common fractions and mixed numbers.

    a. $1\frac{17}{64} \div \frac{5}{16} =$ _____

    c. $3\frac{5}{8} \div \frac{1}{8} =$ _____

    b. $2\frac{3}{4} \div \frac{7}{8} =$ _____

    d. $3\frac{19}{32} \div 1\frac{3}{16} =$ _____

15. Add these decimal fractions.

    a. .625 + .375 + .1875 = _____

    c. .75 + .125 + 2.1882 = _____

    b. 1.50 + .4375 + .563 = _____

    d. 1.998 + 3.0006 + 2.24 = _____

16. Subtract these decimal fractions.

    a. 1.368 − .9987 = _____

    c. .7601 − .2003 = _____

    b. 3.7506 − 1.9003 = _____

    d. 3.87 − 1.0008 = _____

17. Multiply these decimal fractions.

    a. .25 × .1875 = _____

    c. 2.0917 × 4.68 = _____

    b. 1.388 × 2.9808 = _____

    d. 3.652 × .0007 = _____

18. Divide these decimal fractions.

    a. 5.689 ÷ 2.67 = _____

    c. .895 ÷ 3.509 = _____

    b. 2.667 ÷ .8652 = _____

    d. 5.1429 ÷ 4.449 = _____

19. Convert these common fractions to decimal fractions.

    a. $\frac{13}{64} =$ _____

    c. $\frac{59}{64} =$ _____

    b. $\frac{17}{32} =$ _____

    d. $\frac{9}{16} =$ _____

20. Convert these decimal fractions to common fractions.

    a. .2031 = _____

    c. .0496 = _____

    c. .7812 = _____

    d. .3594 = _____

EXERCISE 6-1

Refer to Figure E6-1 to answer the questions below.

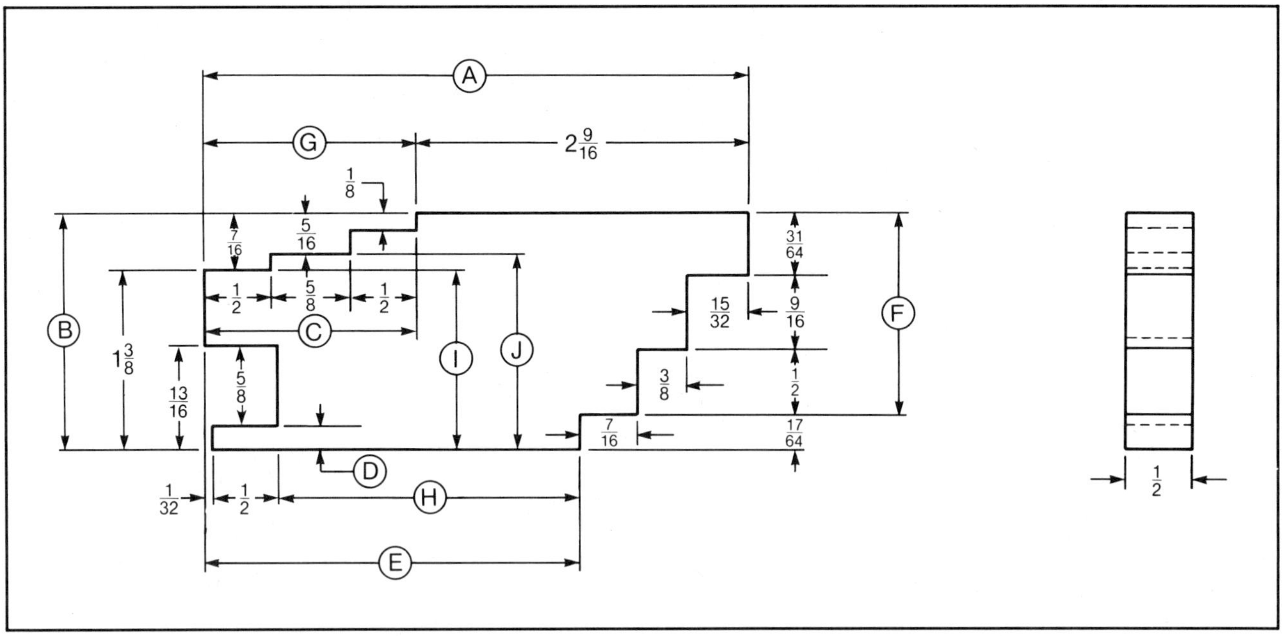

Figure E6-1.

1. What type or types of views are shown on this print? ______________________

____

2. How many views are shown? ______________________

3. What is the "reading direction" of all dimensional values? ______________________

____

4. Identify the four types of lines shown on this print.

a. ______________________

b. ______________________

c. ______________________

d. ______________________

5. Find the following dimensions.

a. $A =$ ______________ **f.** $F =$ ______________

b. $B =$ ______________ **g.** $G =$ ______________

c. $C =$ ______________ **h.** $H =$ ______________

d. $D =$ ______________ **i.** $I =$ ______________

e. $E =$ ______________ **j.** $J =$ ______________

EXERCISE 6-2

Refer to Figure E6-2 to answer the questions below.

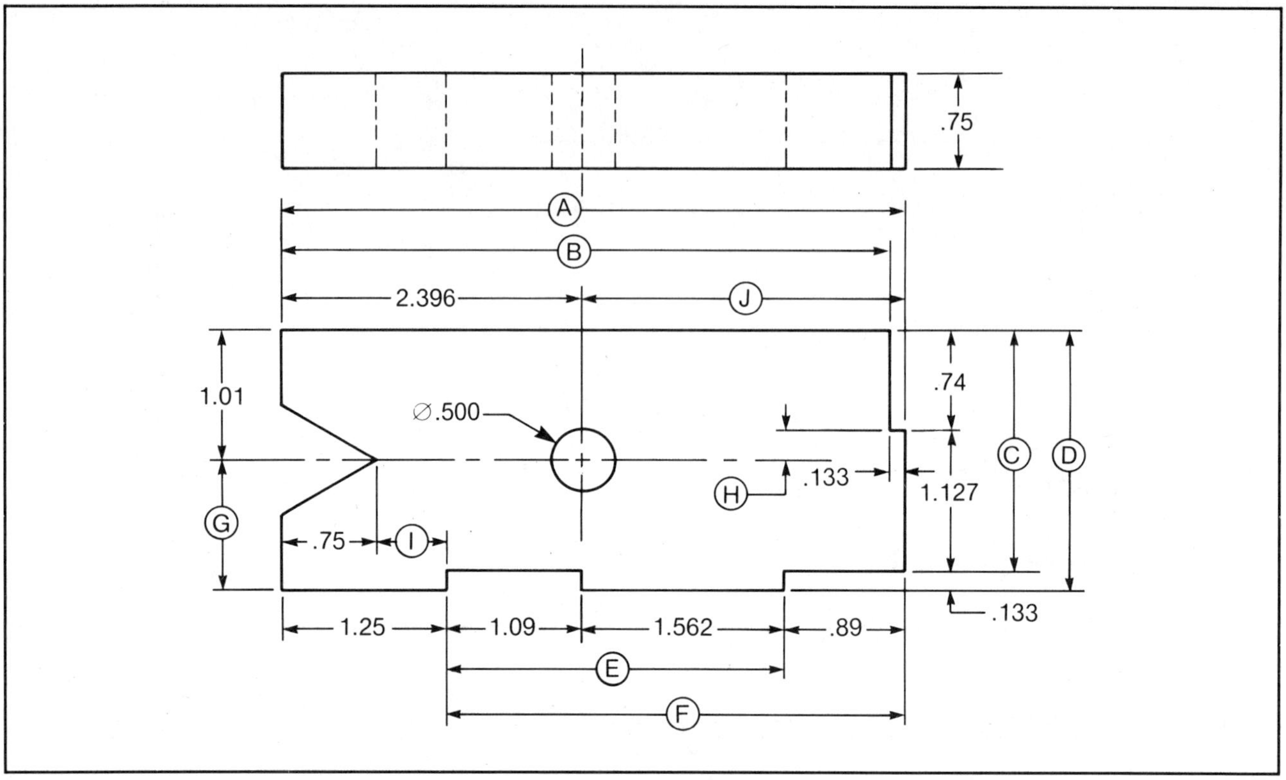

Figure E6-2.

1. What views are shown on this print? ______

2. What is the size of the hole? ______

3. What type of line is used to show the hole size? ______

4. What are the overall dimensions of this part?

a. Height = ______

b. Width = ______

c. Depth = ______

5. Find the following dimensions.

a. A = ______

b. B = ______

c. C = ______

d. D = ______

e. E = ______

f. F = ______

g. G = ______

h. H = ______

i. I = ______

j. J = ______

7

USING SHOP MEASURING TOOLS

OBJECTIVES

After studying this chapter, you will be able to:

■ Identify common shop measuring tools.

■ Describe the procedures used in making measurements.

■ Determine the values of steel rule measurements.

■ Determine the values of micrometer measurements.

■ Determine the values of vernier caliper measurements.

■ Determine the values of angular measurements.

Every dimension on a print represents a specific measurement value that is required for the part. Familiarity with the methods and tools used in the shop to make measurements is necessary to accurately interpret dimensions.

Most measurements made in the shop are classified as either linear or angular. *Linear measurements* are measurements of length, and are normally expressed in inches (in.) or millimeters (mm). *Angular measurements* are measurements of angles, and are expressed in degrees (°) and minutes ('). These measurement units, as well as the tools and methods normally used to measure print dimensions, are discussed in this chapter.

STEEL RULES

The *steel rule* is the most basic measuring tool used in the shop. This tool is a flat, thin strip of steel with evenly spaced marks along each edge, Figure 7-1. These marks are called *graduations*. Each series of marks, or lines, along an edge of the rule is called a *scale*. Most steel rules have four scales, one on each edge, on both sides. The scales most commonly used on steel rules have 8, 16, 32, and 64 graduations per inch.

The spacing of the graduations determines the discrimination of the scale. The *discrimination* is the finest, or smallest, division of a scale that can be read reliably. A scale that has 16 divisions in each inch is said to have a discrimination of $\frac{1}{16}$ inch. A scale with 100 divisions per inch has a discrimination of $\frac{1}{100}$ inch.

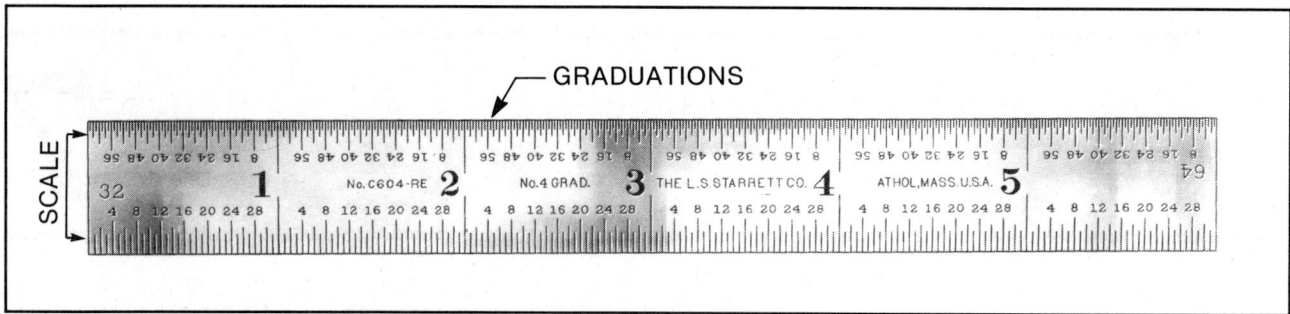

GRADUATIONS

SCALE

Figure 7-1. Steel rule (Courtesy of L. S. Starrett Co.)

Two variations of the basic steel rule are the decimal inch rule and the millimeter rule. *Decimal inch rules* have 10, 50, and 100 graduations per inch. *Millimeter rules* are graduated in millimeters and half-millimeters.

The scale used to make a measurement on a part is determined by the dimensional size stated on the print. If, for example, a dimension is shown as $\frac{3}{4}''$, any of the common scales can be used to make the measurement. However, the 8th scale would be the easiest to read and would normally be used for this size. If the dimension is $\frac{23}{64}''$, however, only the 64th scale can be used to accurately measure this size. As a rule, use the scale that best suits the dimension to be measured and is easiest to read. For example, while a 64th scale could be used to measure the $\frac{3}{4}''$ size, the 8th scale would be easier to read and thus should be used.

Reading a Steel Rule

The first step in reading a steel rule is selecting the proper scale. The scale that is easiest to read should always be used to reduce the chance of error. After selecting the proper scale, position the rule on the workpiece, with the end aligned with the starting point of the measurement. This is called the *reference point* and can be either the end of the rule or one of the full-inch graduation lines, Figure 7-2. Next, read the graduation line that is aligned with the end of the measurement. This is called the *measured point*, Figure 7-3.

If the end of the rule is used as the reference point, the measured point should be read directly. If one of the full-inch graduation lines is used as the reference point, however, the value of this graduation must be subtracted from the measured point to determine the proper size of the part.

To determine the value of the measurement, simply treat the measured value as a common fraction, with the number of graduations between the reference point and the measured point as the numerator, and the scale discrimination as the denominator. For example, if the measurement is 9 lines on the 16th scale, the value would be expressed as $\frac{9}{16}''$, Figure 7-4. Whenever possible, measurements expressed as

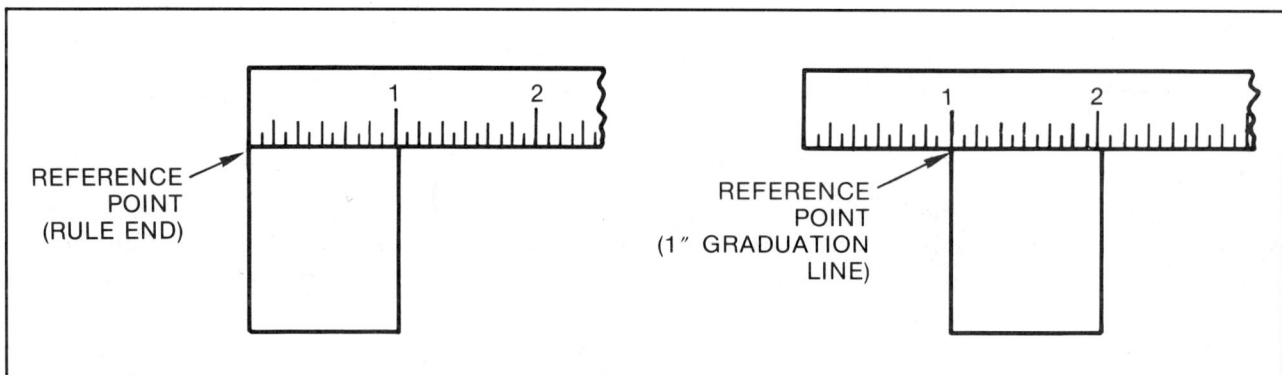

REFERENCE POINT (RULE END)

REFERENCE POINT (1″ GRADUATION LINE)

Figure 7-2. Locating the reference point on a rule

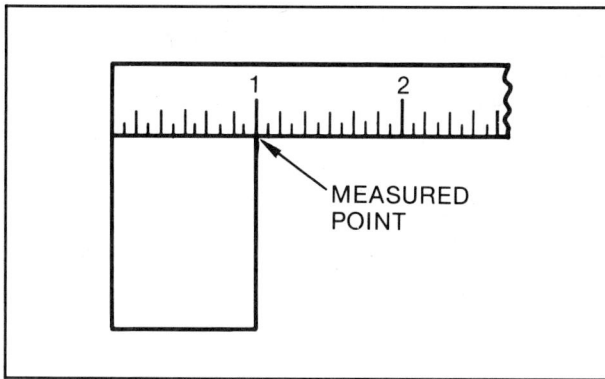

Figure 7-3. Reading the measured point on a rule

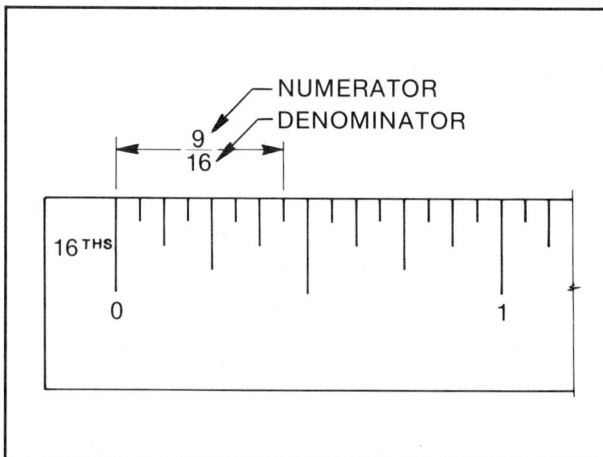

Figure 7-4. Expressing measurements as common fractions

common fractions should be reduced to their lowest terms. If the measurement is 4 lines on the 8th scale, it should be expressed as $\frac{1}{2}$", not as $\frac{4}{8}$". When making measurements that are larger than 1", always remember to add the value of the full-inch units to the measurement.

When reading a *decimal inch rule*, Figure 7-5, the measured value is expresed in tenths, hundredths, or thousandths of an inch. A reading of 5 lines on the 10th scale may be expressed as 500 thousandths, 50 hundredths, or 5 tenths. The same is true when reading the 100th scale. A reading of 75 lines on the 100th scale is expressed as 750 thousandths, 75 hundredths, or 7.5 tenths.

The millimeter is the standard unit of measurement for metric drawings used for industrial purposes. Measurements made with *millimeter rules*, Figure 7-6, are always expressed in millimeters, not in centimeters, decimeters, or meters. Therefore, a measurement of 50 lines on a millimeter scale is expressed as 50 millimeters, not as 5 centimeters.

A conversion chart for converting between inch and millimeter values is included in the Appendix of this book.

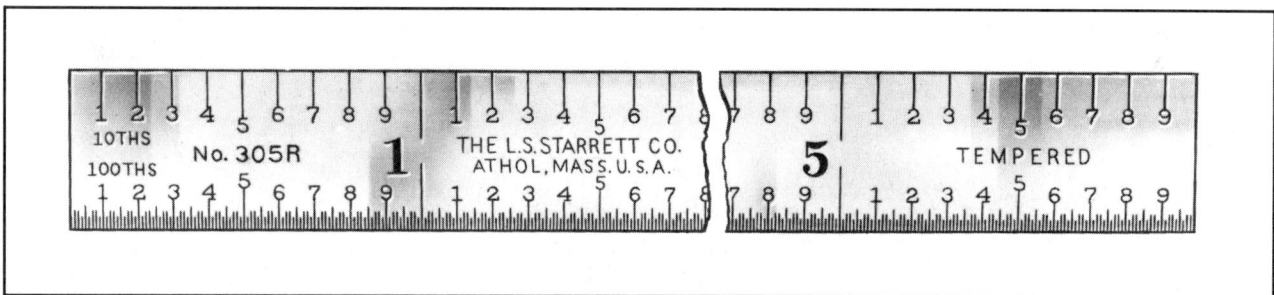

Figure 7-5. Decimal inch rule (Courtesy of L. S. Starrett Co.)

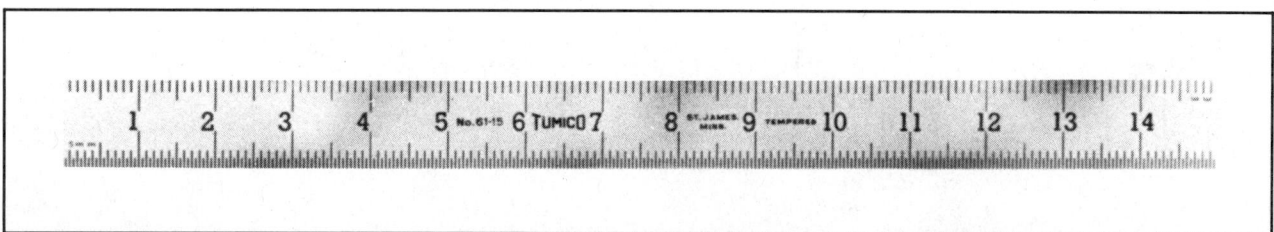

Figure 7-6. Millimeter rule (Courtesy of Rank Scherr-Tumico, Inc.)

PRACTICE PROBLEM 7-1

Make readings *A* through *J*. Record your answers in the spaces provided. Reduce the answers where possible.

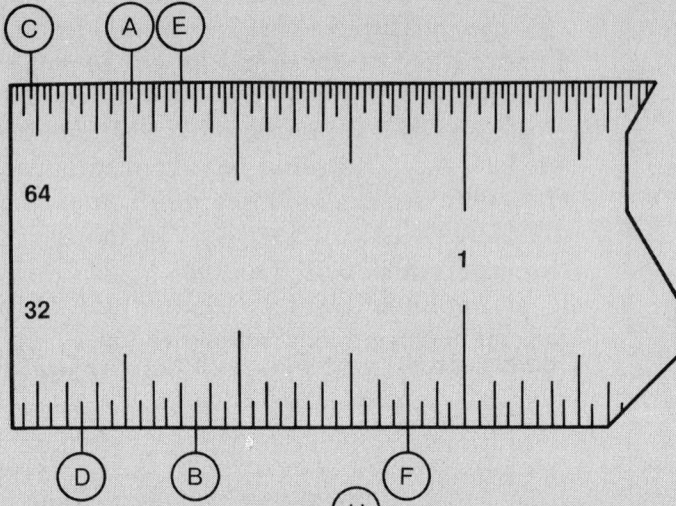

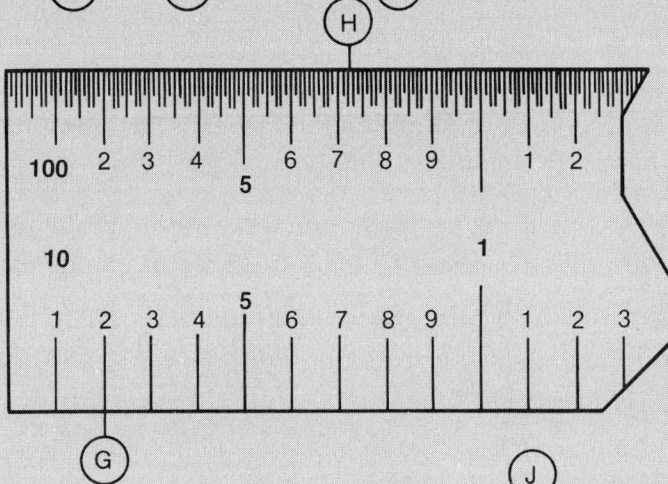

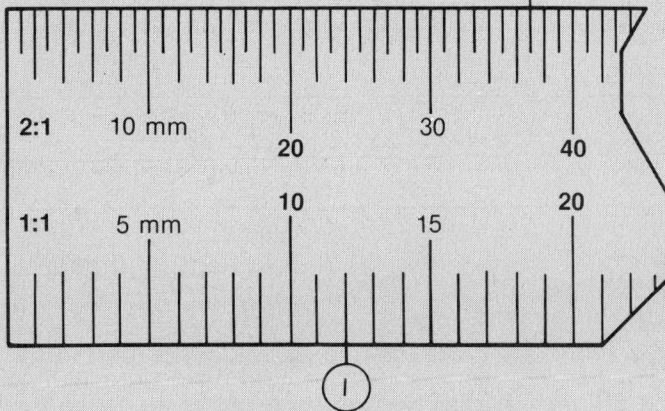

A = _____

B = _____

C = _____

D = _____

E = _____

F = _____

G = _____

H = _____

I = _____

J = _____

MICROMETERS

Micrometers are linear measuring tools. They are used to make measurements to accuracies of .001 inch or .0001 inch. The metric measurements have accuracies of .01 mm or .002 mm.

The most common type of micrometer is the *micrometer caliper*. As shown in Figure 7-7, this instrument consists of several parts. The frame is the main unit of the micrometer caliper. The frame maintains the alignment of the contact points on the ends of the spindle and anvil. The sleeve is also attached to the frame. On *inch micrometers*, the sleeve has graduations that divide the travel of the spindle into 40 equal parts. The thimble, which rotates around the sleeve, is also graduated. These graduations divide the rotation of the thimble into 25 equal parts. Together, the sleeve and thimble graduations of the inch micrometer caliper divide an inch into 1,000 (40 × 25 = 1,000) parts. The ratchet stop, located at the end of the thimble, is used to limit the pressure between the contact points. When making measurements with a micrometer, using the ratchet stop will assure the correct measuring pressure.

Some inch micrometer calipers also have a graduated scale on the top side of the sleeve, Figure 7-8. This scale is called the *vernier scale* and is used to make measurements to an accuracy of .0001″.

Millimeter micrometers, Figure 7-9, have the same basic construction as inch micrometers. The only difference is in the graduations. The sleeve on a millimeter micrometer is graduated in half-millimeter units. The thimble is divided into 50 equal parts. Together, the sleeve and thimble graduations divide a millimeter into 100 parts. The millimeter micrometer may also have a vernier scale on the top side of the sleeve. A vernier scale further divides the micrometer reading and permits readings to an accuracy of .002 mm.

Reading an Inch Micrometer

Each graduation on the sleeve of an inch micrometer is equal to $\frac{1}{40}″$. Since inch micrometers are read in decimal inches, this value is expressed as .025″ (1 ÷ 40). Every fourth line is numbered to aid in reading the micrometer. Each of these numbered lines is equal to .100″, Figure 7-10.

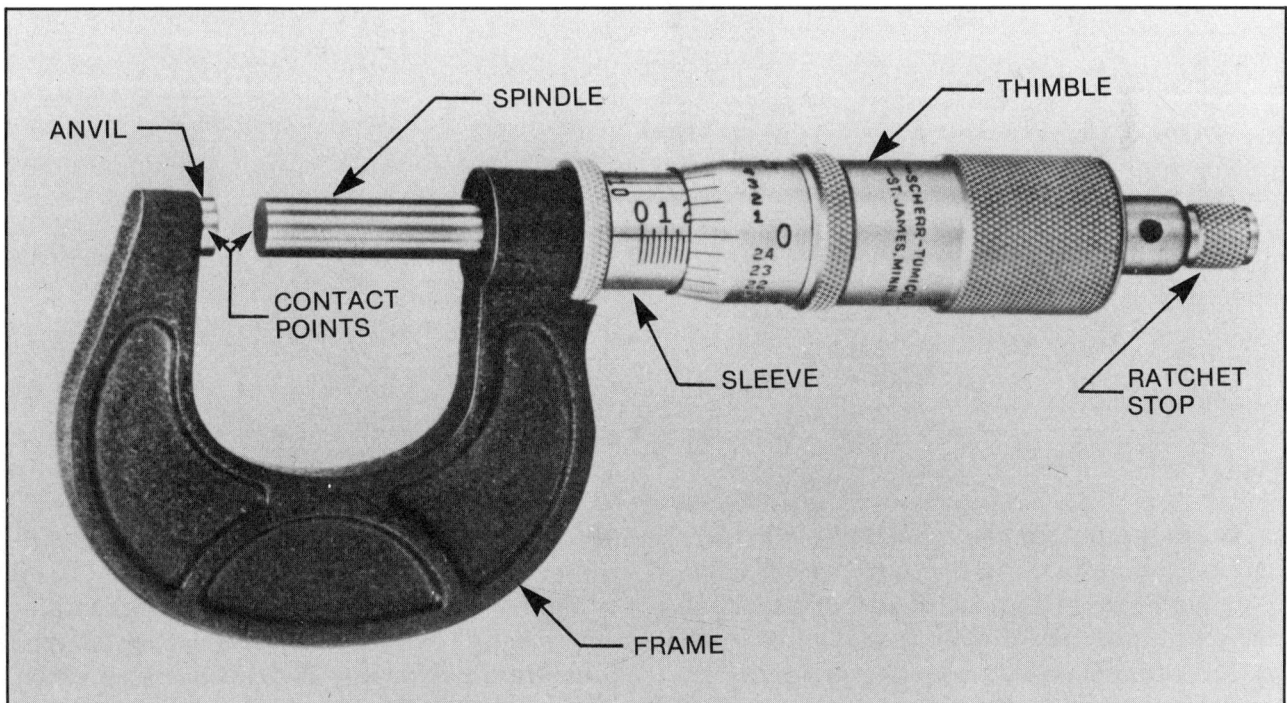

Figure 7-7. Micrometer caliper (Courtesy of Rank Scherr-Tumico, Inc.)

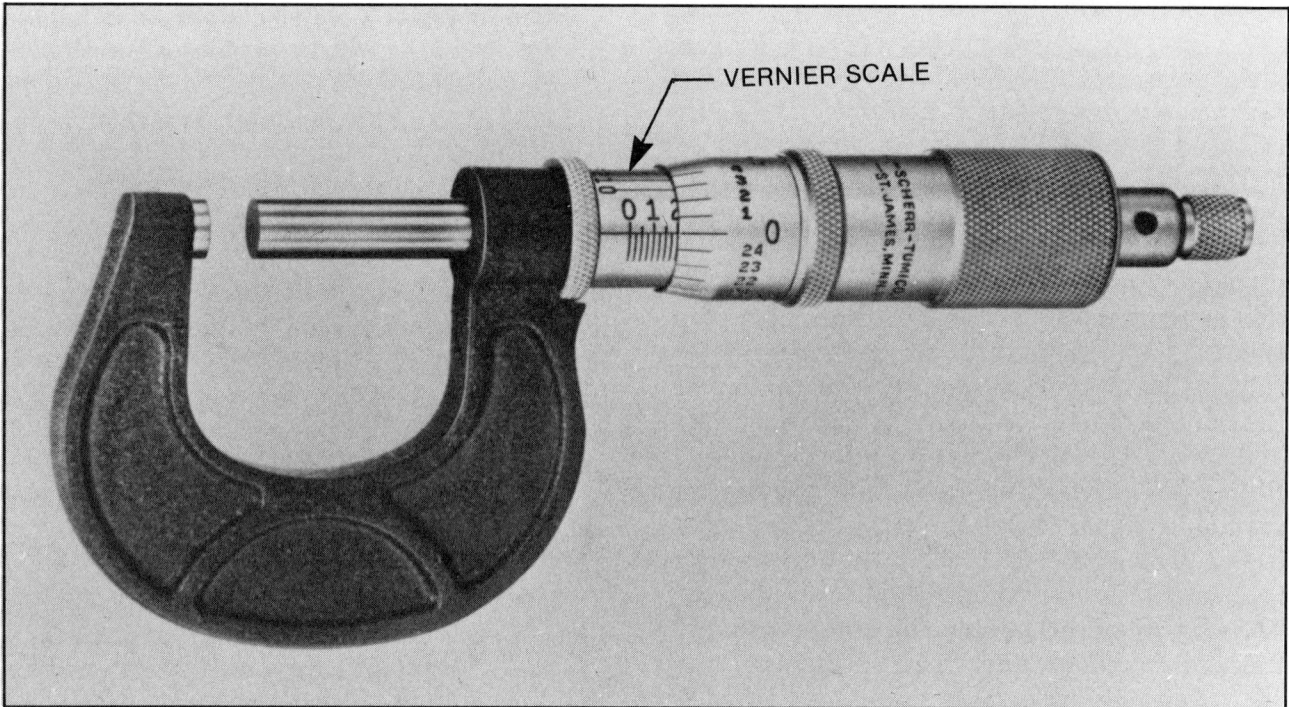

Figure 7-8. Vernier scale on an inch micrometer (Courtesy of Rank Scherr-Tumico, Inc.)

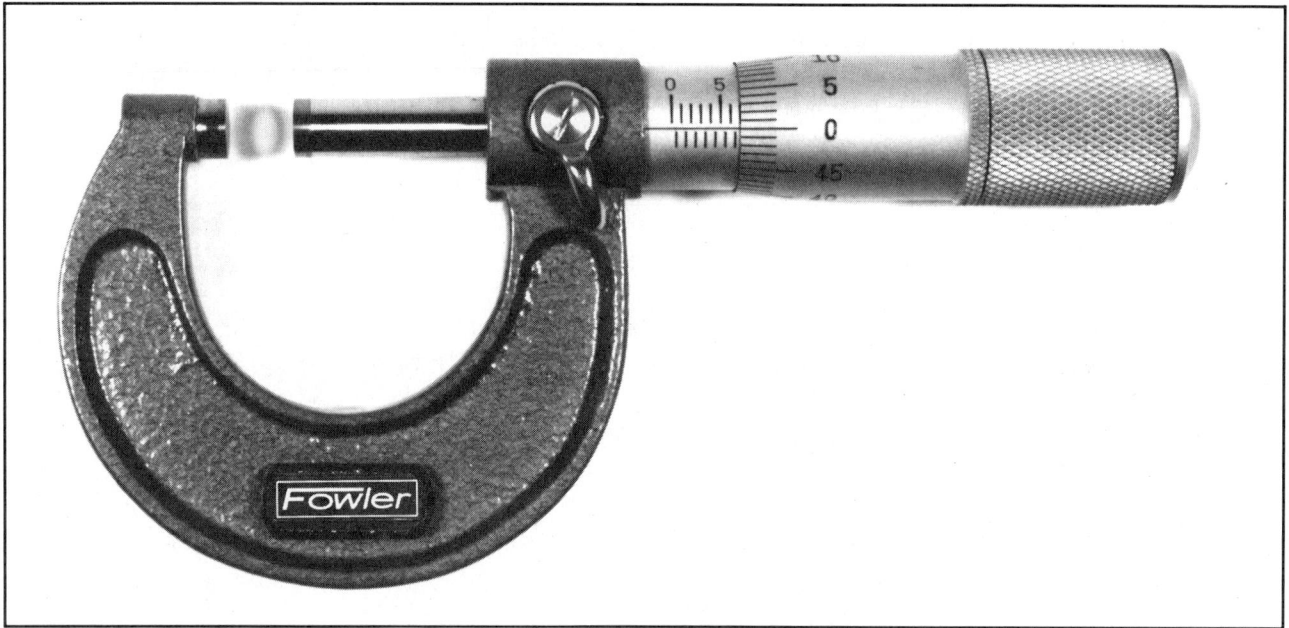

Figure 7-9. Millimeter micrometer (Courtesy of Fred V. Fowler Co.)

To determine the value of the sleeve graduations, simply count the number of graduations visible to the left of the thimble. Do not count any lines that are partially covered by the thimble. Next, note the value of the thimble graduation aligned with the reading line on the sleeve, and add this amount to the value of the sleeve graduations. The micrometer reading shown in Figure 7-11 has a sleeve value of .250″ (250 thousandths, 10 lines). The thimble graduation aligned with the reading line is 16 (16 thousandths). Adding these together gives the total reading of .266″.

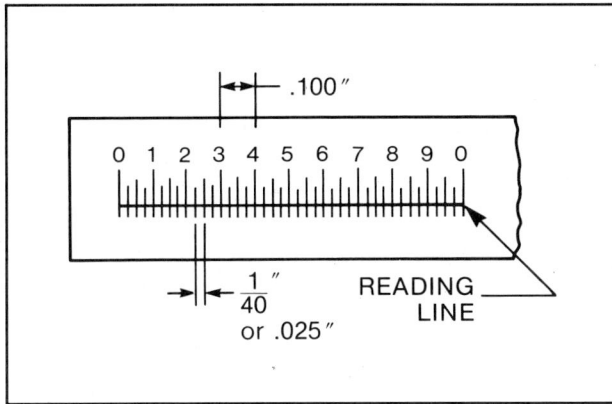

Figure 7-10. Each sleeve graduation on an inch micrometer is equal to .025".

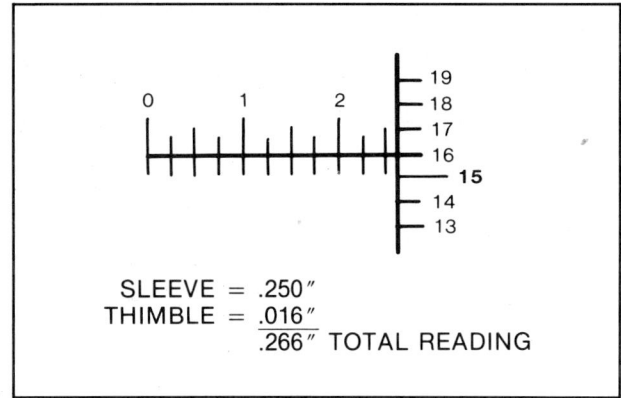

SLEEVE = .250"
THIMBLE = .016"
       .266" TOTAL READING

Figure 7-11. Reading a micrometer to .001"

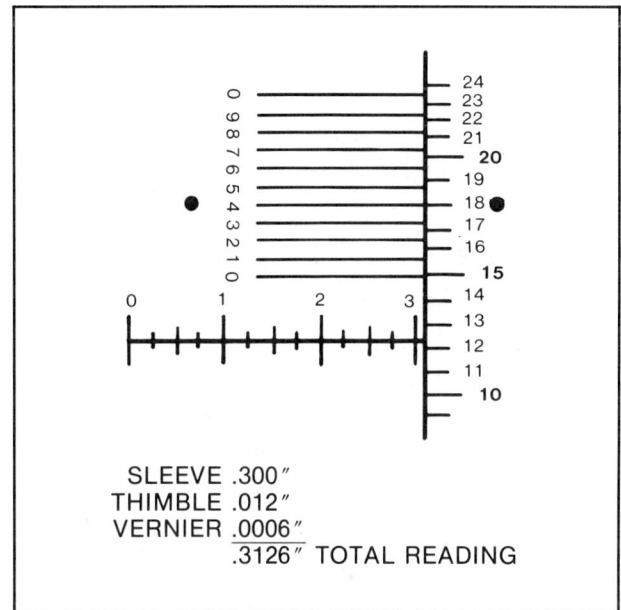

SLEEVE .300"
THIMBLE .012"
VERNIER .0006"
     .3126" TOTAL READING

Figure 7-12. Reading a micrometer to .0001"

To read an *inch vernier micrometer*, simply read the micrometer as just described, and add the value of the vernier scale. To read the vernier scale, first turn the micrometer so that the complete scale is visible. Next, find the line on the vernier scale that is aligned with one of the thimble lines. There will only be one line perfectly aligned at any single time. Remember, use the value of the vernier scale, not the thimble graduation. The purpose of a vernier scale on an inch micrometer is to divide the space between the thimble graduations into 10 equal parts. Therefore, when reading this micrometer, the thimble graduations may not be exactly aligned with the reading line on the sleeve. In such cases, always use the lesser value. The micrometer reading shown in Figure 7-12 depicts how this type of micrometer should be read. The sleeve graduations visible to the left of the thimble equal .300". The thimble is between the numbers 12 and 13, and the vernier is aligned on the number 6. Added together, the reading is .3126". This value should be read as "three thousand one hundred twenty-six ten-thousandths of an inch."

Reading a Millimeter Micrometer

The process of reading a millimeter micrometer is similar to that used for the inch micrometer. First, find the value of the sleeve graduations visible to the left of the thimble. On millimeter micrometers, each graduation is equal to .5 mm. Most millimeter micrometers only number every 10 millimeters on the sleeve, so pay close attention when counting these lines. After noting the value of the sleeve graduations, note the value of the thimble graduation aligned with the reading line, and add the two values. In the example shown in Figure 7-13, the sleeve graduations equal 7.5 mm and the thimble is aligned on the number 35. Added together, this reading equals 7.85 mm.

When reading a *millimeter vernier micrometer*, first determine the value of the measurement to an accuracy of .01 mm. Next, find the line on the vernier scale that is aligned with a thimble graduation, and add this amount to the first reading. In the example shown in Figure 7-14, the sleeve graduations equal 5 mm. The thimble is between the numbers 18 and 19 and the vernier is aligned on the number 6. Added together, this reading equals 5.186 mm. It should be read as "five and one hundred eighty-six thousandths millimeters."

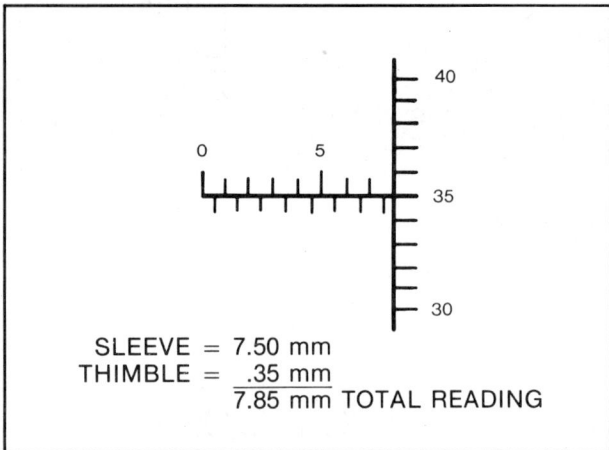

SLEEVE = 7.50 mm
THIMBLE = .35 mm
7.85 mm TOTAL READING

Figure 7-13. Reading a micrometer to .01 mm

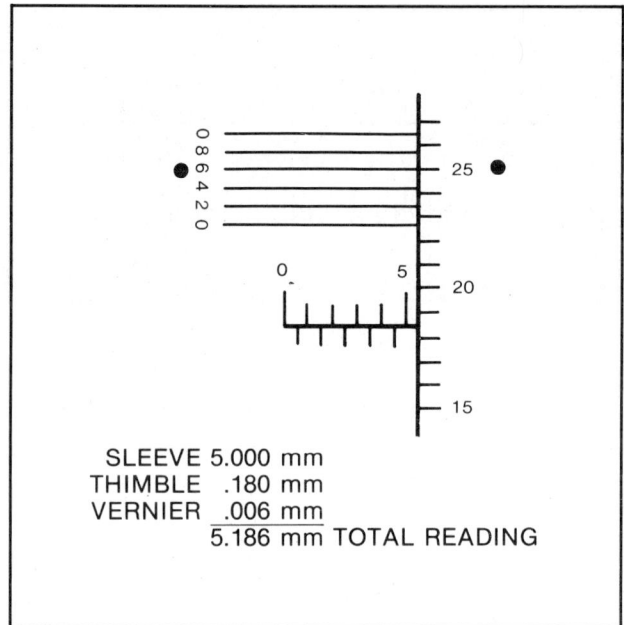

SLEEVE 5.000 mm
THIMBLE .180 mm
VERNIER .006 mm
5.186 mm TOTAL READING

Figure 7-14. Reading a micrometer to .002 mm

PRACTICE PROBLEM 7-2

Make the micrometer readings below. Record your answers in the spaces provided.

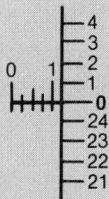

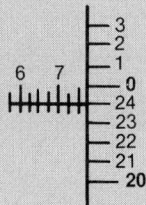

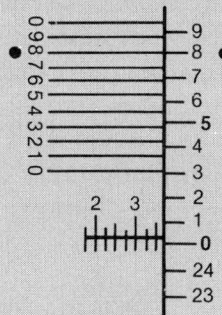

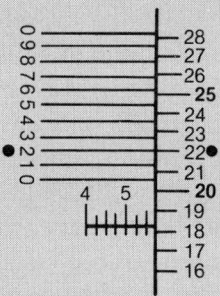

(a) _____

(b) _____

(c) _____

(d) _____

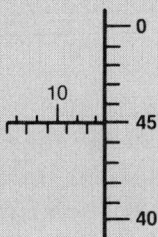

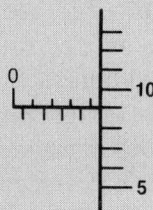

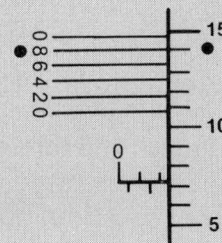

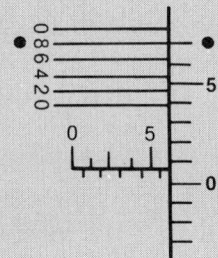

(e) _____

(f) _____

(g) _____

(h) _____

VERNIERS

Verniers are another form of linear measuring tool frequently used for making measurements in the shop. The most popular type of vernier instrument is the *vernier caliper*, Figure 7-15.

The vernier caliper consists of a beam containing a solid jaw and main scale graduations. A movable jaw is mounted on the beam. It can be moved and clamped in any desired position along the length of the beam with the clamp screws. A vernier scale is contained within the movable jaw and permits the vernier caliper to measure to within .001" or .02 mm. Some types of vernier calipers also have a fine adjustment slide to aid in making accurate measurements.

The major variations of the vernier caliper are the 25-division vernier, the 50-division vernier, the millimeter vernier, and the dial caliper.

Reading a 25-Division Vernier

The beam of the *25-division vernier* has a main scale graduated in increments of $\frac{1}{40}''$. Each graduation on the main scale is equal to .025", like the sleeve on a micrometer. The vernier plate on this vernier, like the micrometer thimble, has 25 divisions. It is used to further divide the main scale graduations to .001".

To read this vernier, first locate the left zero on the vernier plate with reference to the main scale. Include all graduations to the left of the zero in the reading. Like the micrometer sleeve, the main scale graduations are numbered at each .100" value. The larger numbers on the main scale represent full-inch graduations. Always remember to include any full-inch values in measurements with a vernier. Once the zero is located and the value is noted, find the line on the vernier scale that is aligned with a line on the main scale, and add this value to the total reading. Here again, remember to use the value of the vernier graduation and not the main scale number when determining this value. To read the vernier shown in Figure 7-16, first find the main scale value. In this case the value is 1.525". Next, find the vernier scale line that is aligned with a line on the main scale. In this example, the number 10 is aligned. Added together, the reading equals 1.535".

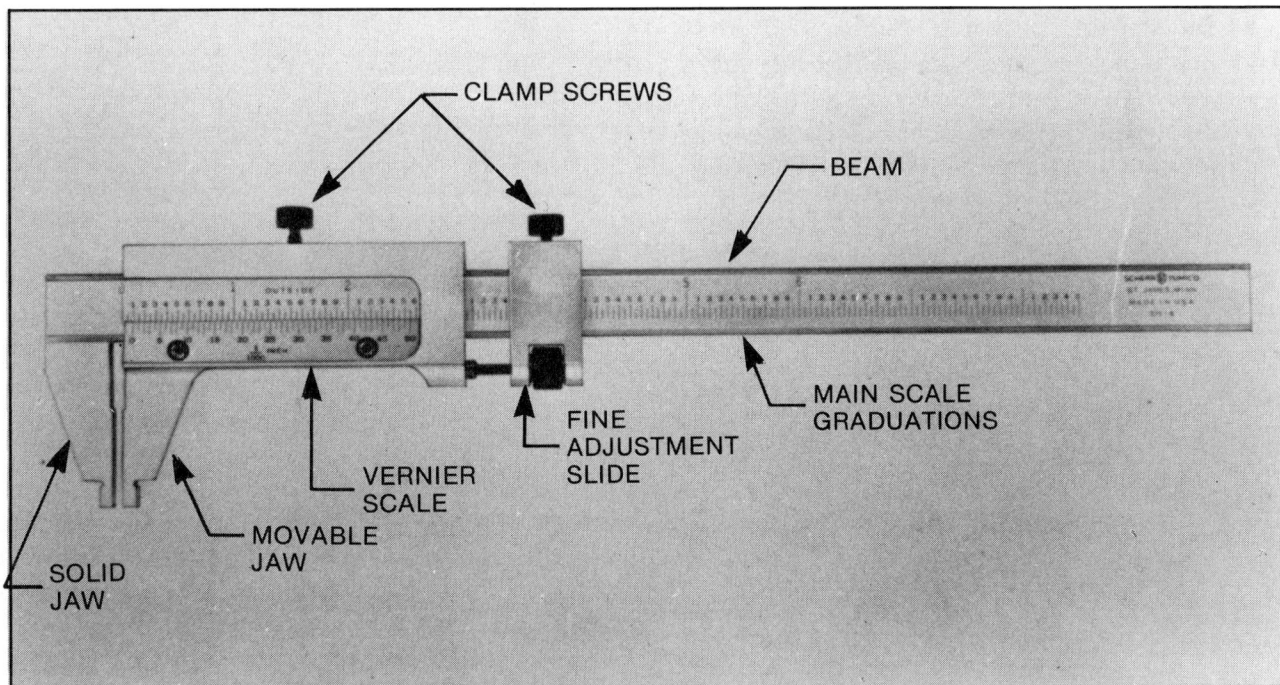

Figure 7-15. Vernier caliper (Courtesy of Rank Scherr-Tumico, Inc.)

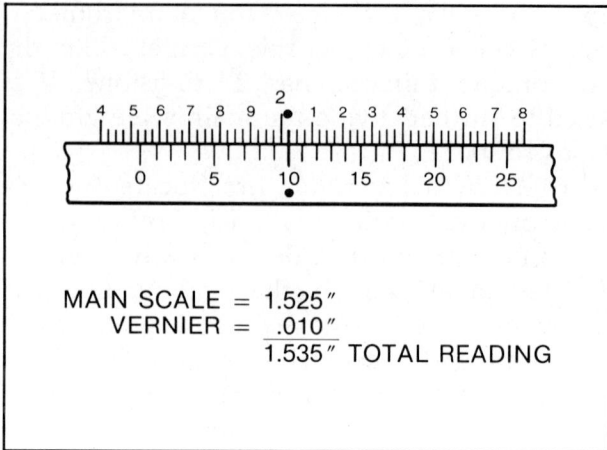

Figure 7-16. Reading a 25-division vernier

MAIN SCALE = 1.525″
VERNIER = .010″
1.535″ TOTAL READING

Reading a 50-Division Vernier

The *50-division vernier* is read in much the same way as the 25-division vernier. The only difference is in the main scale and vernier scale graduations. The graduations on the main scale of this vernier are equal to .050″, and the vernier scale has 50, rather than 25 divisions. To read this vernier, first find the value of the main scale graduations to the left of the zero. Then find the vernier scale graduation that is aligned with one of the lines on the main scale. Then add the values together. To read the vernier shown in Figure 7-17, first find the main scale value. In this case the value is 2.550″. Next, find the vernier scale value. On this vernier the number 22 is the only number that is aligned. Added together, this reading equals 2.572″.

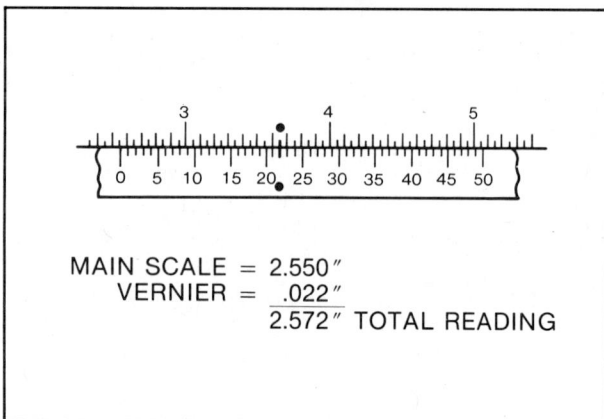

MAIN SCALE = 2.550″
VERNIER = .022″
2.572″ TOTAL READING

Figure 7-17. Reading a 50-division vernier

Reading a Millimeter Vernier

Millimeter verniers are read in much the same way as inch verniers. The main scale graduations on the millimeter vernier are graduated in full-millimeter units, and the vernier scale is divided into 50 parts, each equal to .02 mm. To read this vernier, first locate the position of the zero and note the value of the reading to the left of the zero. Next, find the line on the vernier scale that is aligned with a main scale graduation, and add the two values. In the example shown in Figure 7-18, the main scale reading is 35 mm, and the vernier scale is aligned at the at the number .48. Added together, this reading equals 35.48 mm.

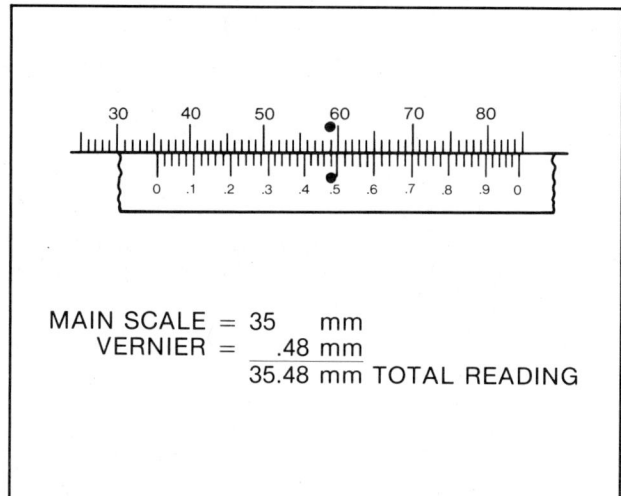

MAIN SCALE = 35 mm
VERNIER = .48 mm
35.48 mm TOTAL READING

Figure 7-18. Reading a millimeter vernier

Dial Calipers

Another variation of the vernier caliper is the *dial caliper*, Figure 7-19. Dial calipers are read by finding the value of the main scale graduations visible to the left of the reference edge and adding the dial reading. With these calipers, the dial replaces the vernier scale. This instrument is generally easier and faster to read than the standard vernier caliper. *Inch dial calipers* are accurate to .001″. *Millimeter dial calipers* are accurate to .02 mm.

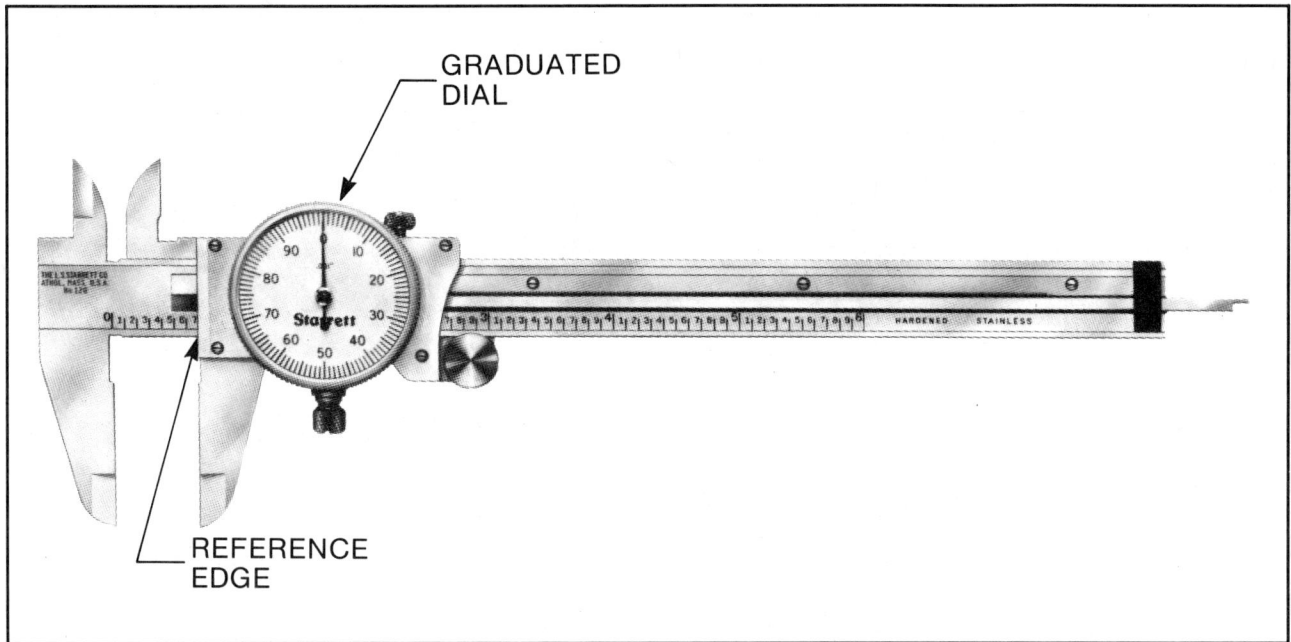

Figure 7-19. Dial caliper (Courtesy of L. S. Starrett Co.)

PRACTICE PROBLEM 7-3

Make the vernier caliper readings below and on the next page. Record your answers in the spaces provided.

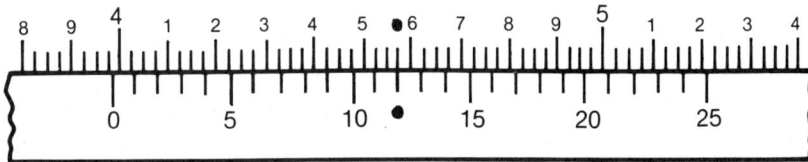

(a) _____

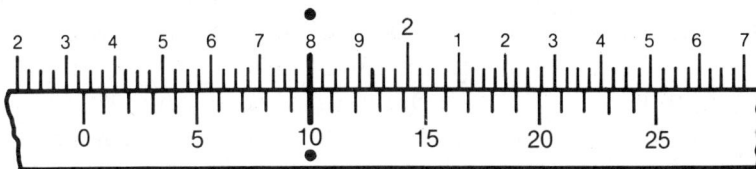

(b) _____

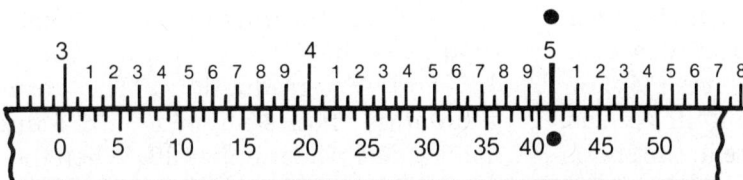

(c) _____

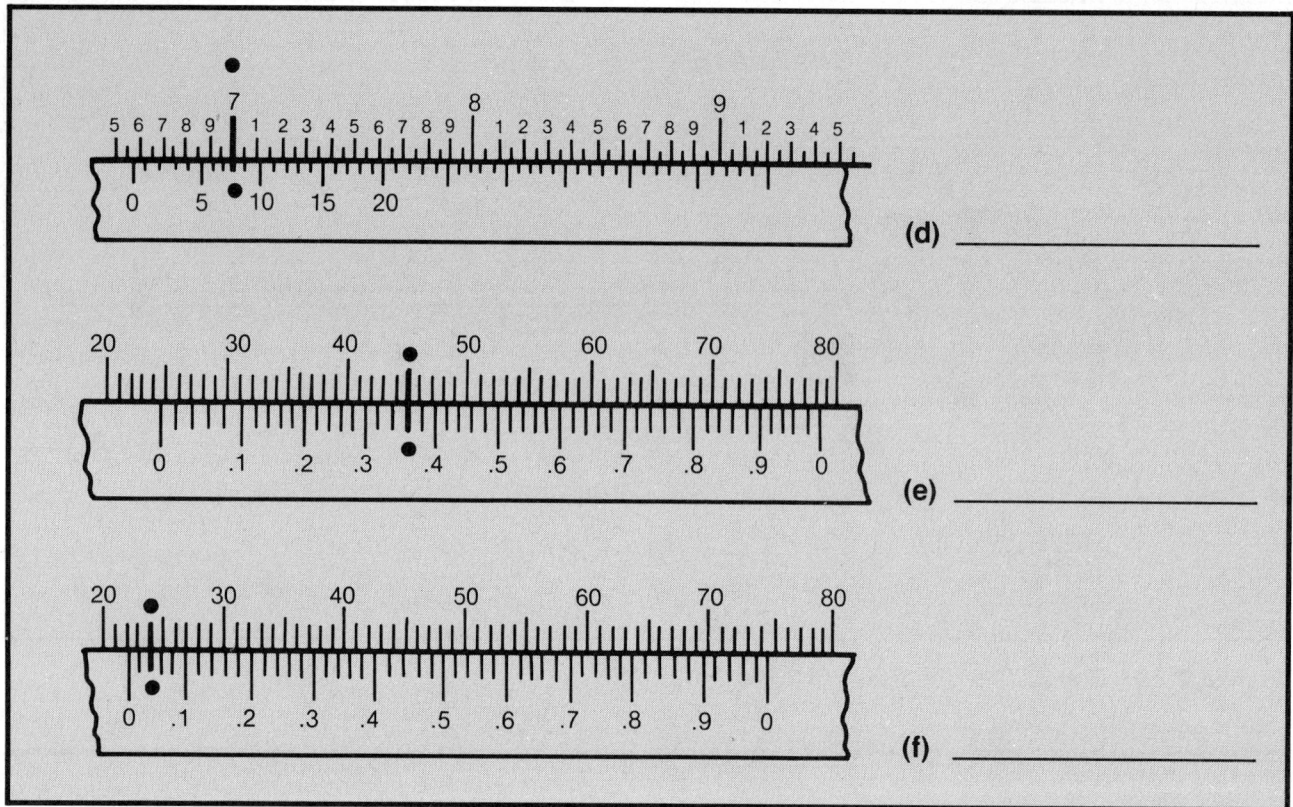

(d) _____

(e) _____

(f) _____

PROTRACTORS

Protractors are tools used to measure angles. The two basic types of protractors used in the shop are the *steel protractor* (Figure 7-20A) and the *vernier bevel protractor* (Figure 7-20B). The basic difference between these two protractors is accuracy. The steel protractor and its variations, such as the protractor head on a combination square, are accurate to 1 degree (1°). The vernier bevel protractor is accurate to 5 minutes (5′).

The first step in measuring angles is understanding how angles are expressed. When an angle is specified on a print, it is normally noted in units of degrees and minutes or decimal parts of a degree. Every circle contains 360°. Each degree contains 60′. Therefore, an angle specified as 15°30′ is the same as 15.5 degrees. Minutes are also divided into seconds (″). For all practical purposes, however, minutes are as close as you will be able to measure in the shop.

When reading a steel protractor, position the protractor against the angle to be measured. Then remove the protractor and note the position of the reference line against the protractor scale. The reading in Figure 7-21 is 82°. Steel protractors are read directly. The value shown on the protractor scale is the total value of the angle within 1°.

When angles must be measured to an accuracy closer than 1°, a vernier bevel protractor may be used. To read a vernier bevel protractor, note the position of the zero reference line on the vernier scale. The vernier scale on this tool may be read in either direction, depending on the direction of rotation of the main scale. If the main scale is rotated to the right, the vernier scale to the right of the zero is used for the measurement. If the main scale is rotated to the left, then the scale to the left of the zero is read. In the example shown in Figure 7-22, the main scale is rotated to the right, and the zero is positioned between the 42° and 43° graduations. The vernier scale graduation that is aligned with a line on the main scale is the number 30. When these values are added together, the reading is expressed as 42°30′.

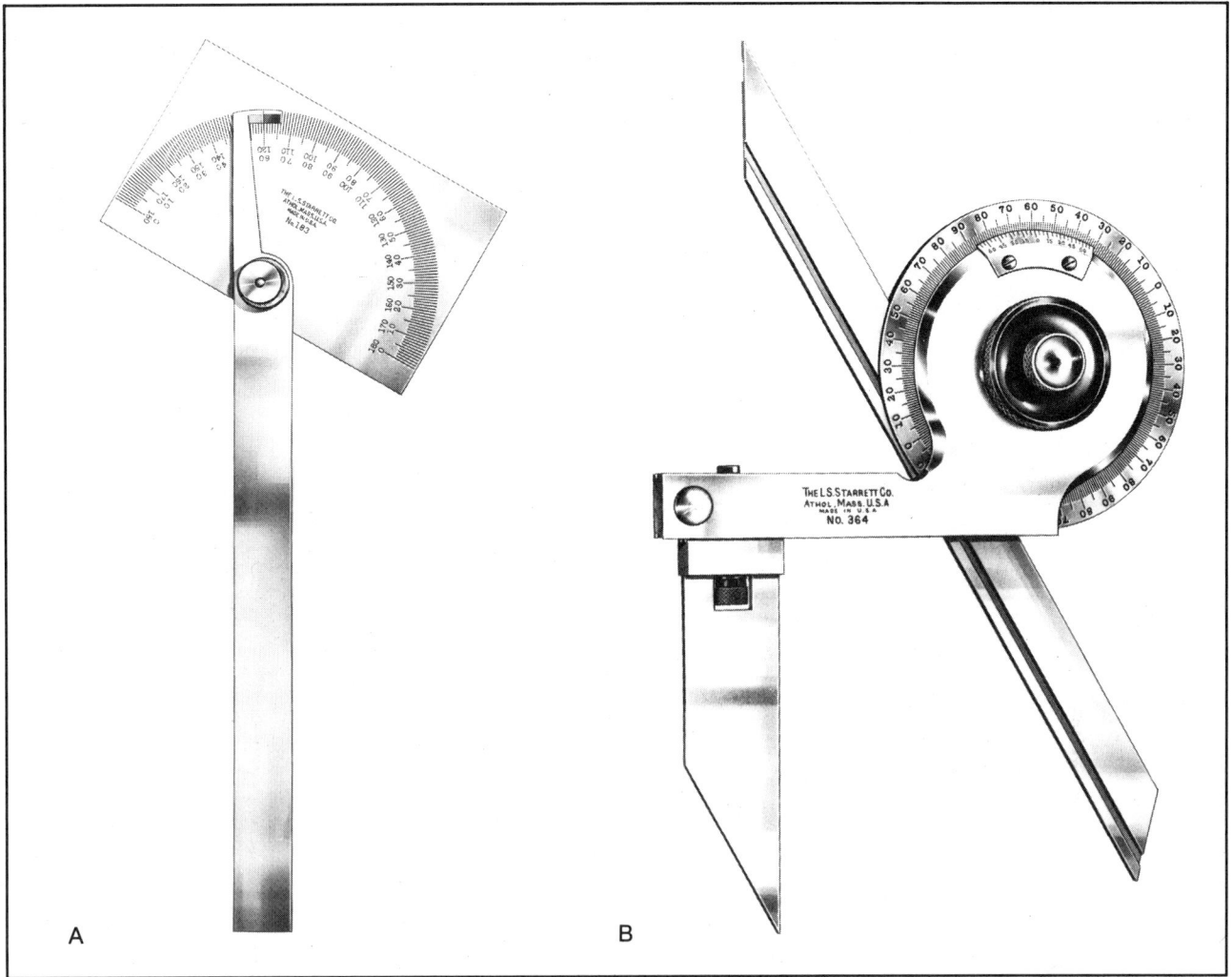

Figure 7-20. Shops use (A) steel protractors (Courtesy of L. S. Starrett Co.) and (B) vernier bevel protractors (Courtesy of L. S. Starrett Co.).

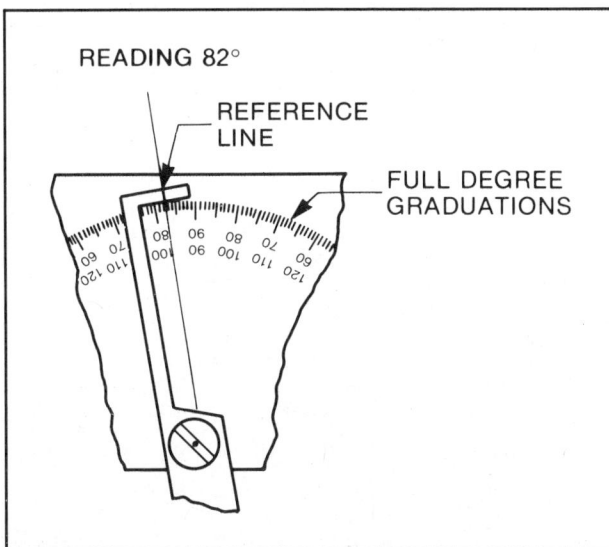

Figure 7-21. Reading a protractor to 1 minute

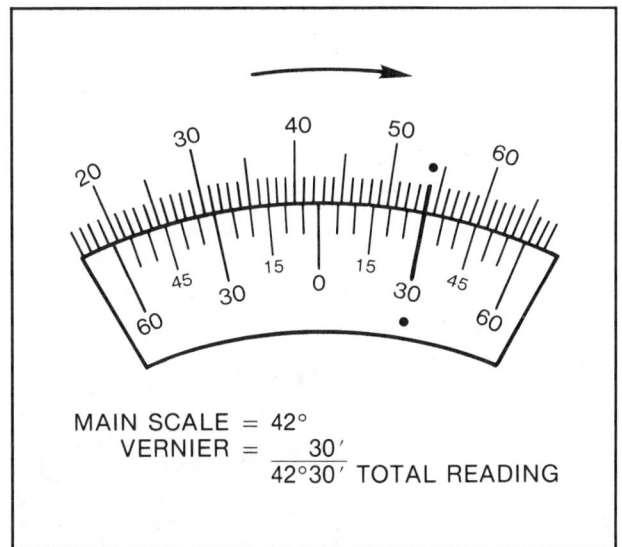

MAIN SCALE = 42°

VERNIER = $\dfrac{30'}{42°30'}$ TOTAL READING

Figure 7-22. Reading a protractor to 5 minutes

PRACTICE PROBLEM 7-4

Make the following protractor readings. Record your answers in the spaces provided.

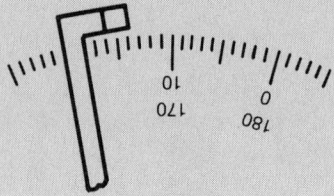

(a) _____

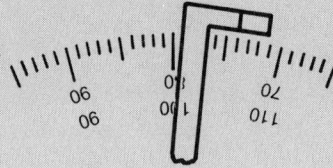

(b) _____

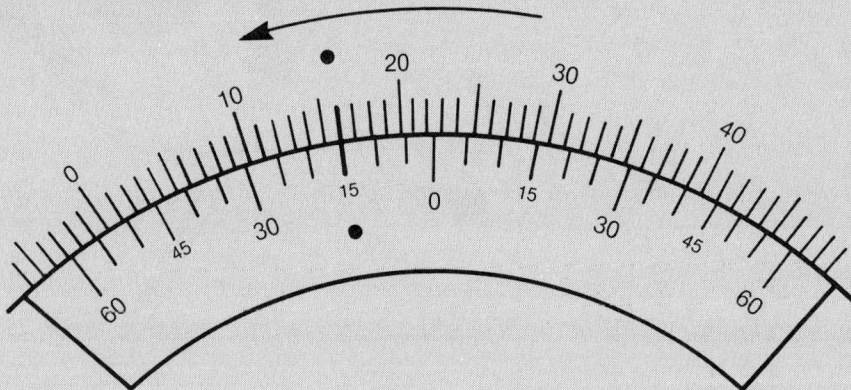

(c) _____

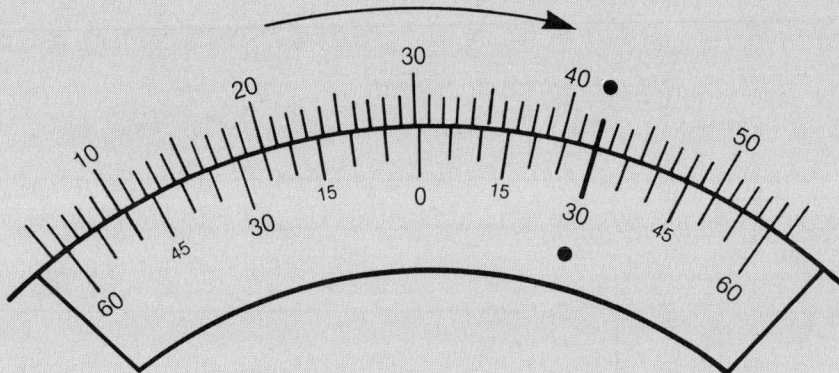

(d) _____

KEY TERMS

The following key terms were introduced in this chapter. Be sure you know the meaning of each term before proceeding to the review material.

Angular Measurement
Degree
Dial Caliper
Discrimination
Graduations
Linear Measurement

Measured Point
Micrometer
Micrometer Caliper
Minute
Reference Point
Scale
Second
Steel Protractor
Steel Rule
Vernier Bevel Protractor
Vernier Caliper

Test your knowledge with this reinforcement study material. Write your answers to the questions in the spaces provided.

1. What are the two classifications for most measurements made in the shop? _____

_____

2. Which type of measurement is expressed in degrees and minutes? _____

_____

3. Which type of measurement is normally expressed in inches or millimeters? _____

_____

4. What term is used to describe the finest division of a scale that can be read reliably?

_____

5. How many scales are on most steel rules? _____

_____

6. How are most steel rules graduated? _____

_____

7. How are decimal inch rules graduated? _____

_____

8. How are millimeter rules graduated? _____

_____

9. What is the term used to describe the starting point of a measurement? _____

_____

10. What is the end point of a measurement called? _____

_____

11. What is the standard unit of measurement for metric prints? _____

_____

12. How accurate are the two types of inch micrometers? _____

_____

13. How accurate are the two types of millimeter micrometers? _____

_____

14. What are the four variations of the vernier caliper? How accurate is each type? _____

_____

_____

_____

_____

15. What two tools are used for measuring angles? How accurate is each? _____

_____

_____

Find the values of the steel rule measurements a through j.

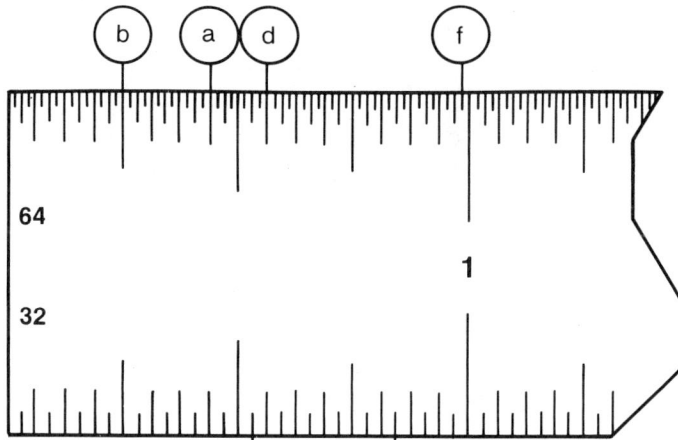

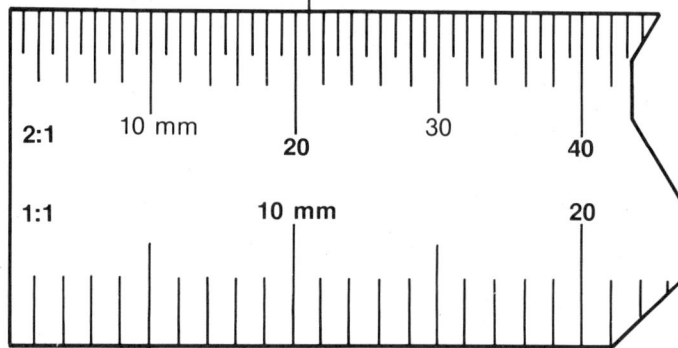

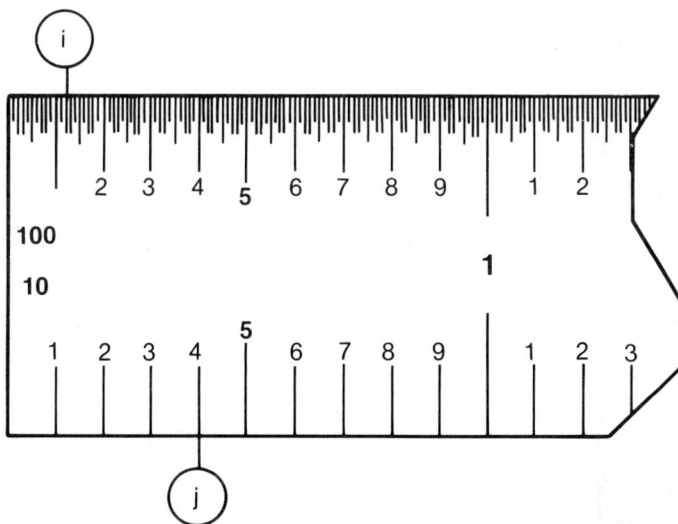

a. _____

b. _____

c. _____

d. _____

e. _____

f. _____

g. _____

h. _____

i. _____

j. _____

Find the values of the micrometer settings a through j.

a

f

b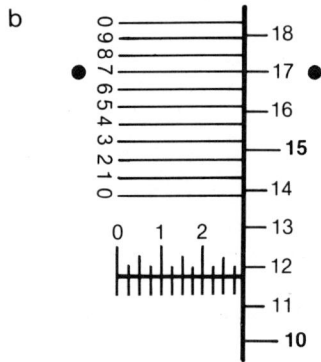

g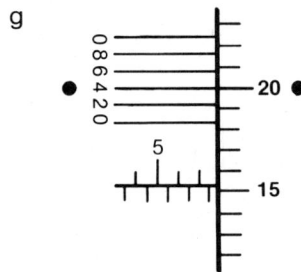

h

c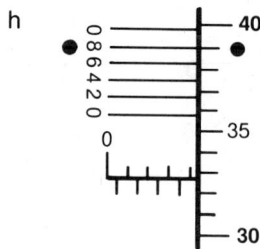

d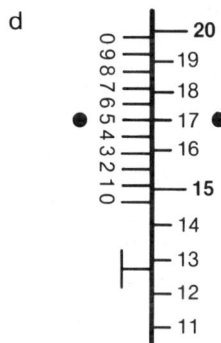

i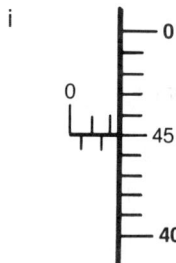

e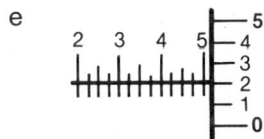

j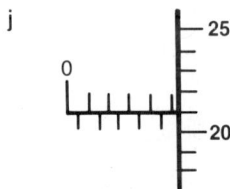

a. _____

b. _____

c. _____

d. _____

e. _____

f. _____

g. _____

h. _____

i. _____

j. _____

Find the values of the vernier caliper settings a through j.

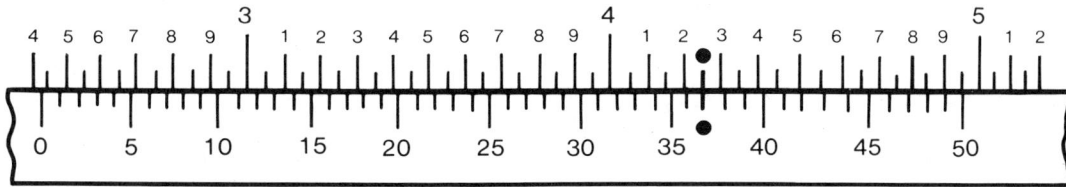

a. _____

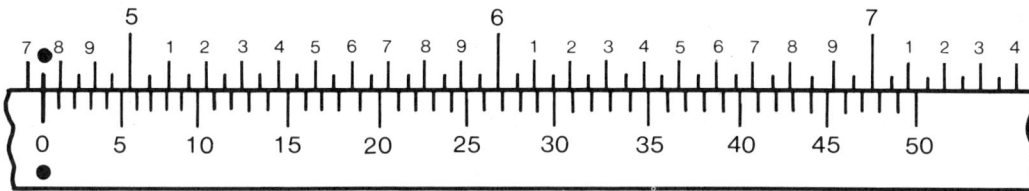

b. _____

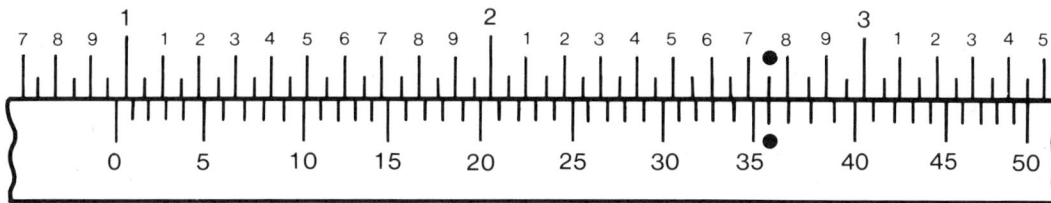

c. _____

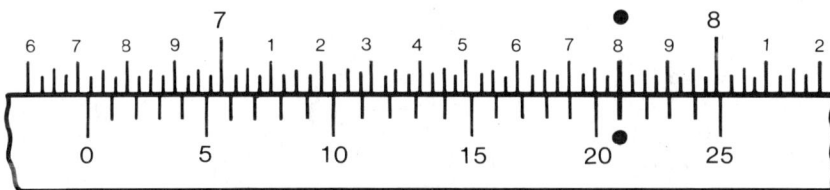

d. _____

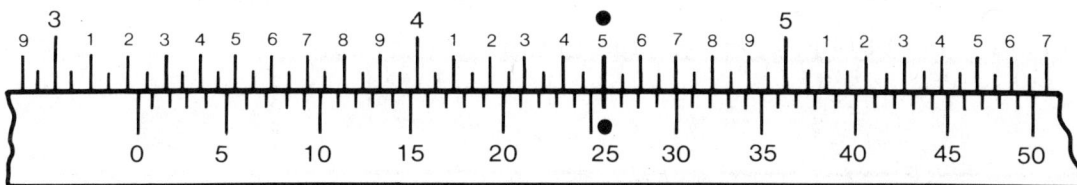

e. _____

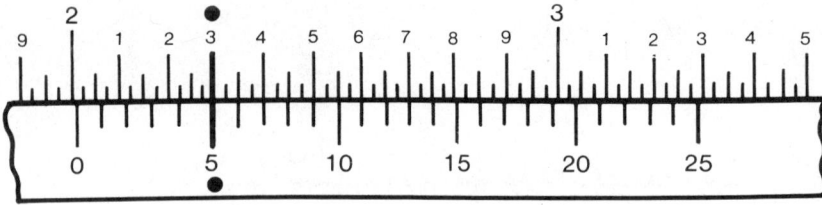

f. _____

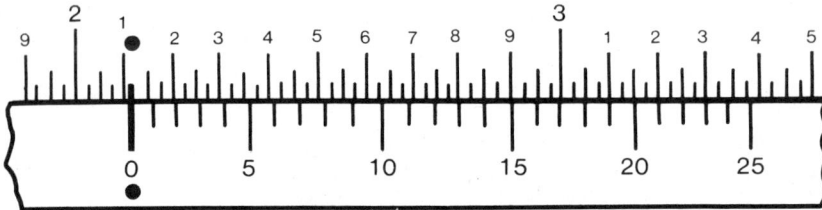

g. _____

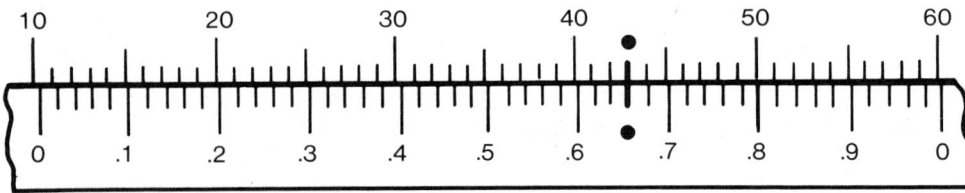

h. _____

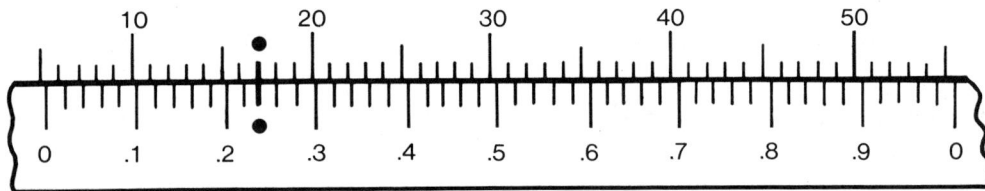

i. _____

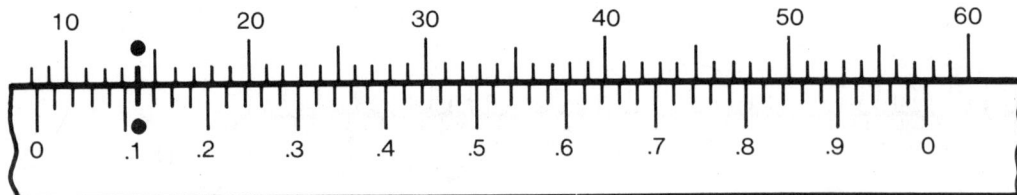

j. _____

Find the values of the angular measurements a through d.

a. _____

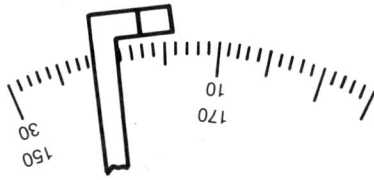

b. _____

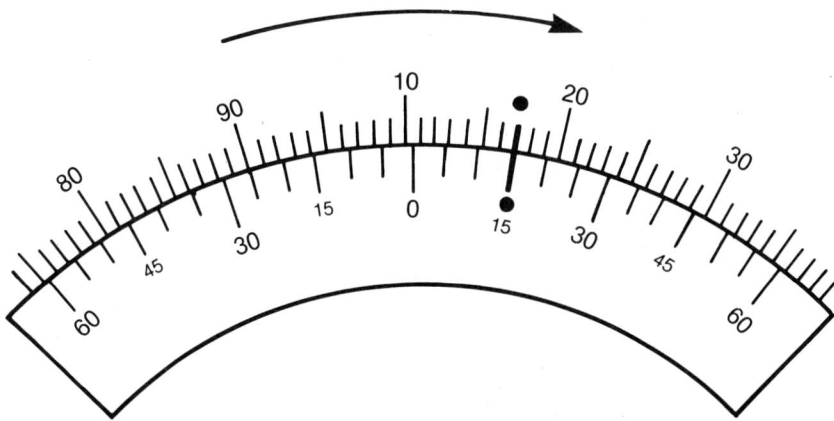

c. _____

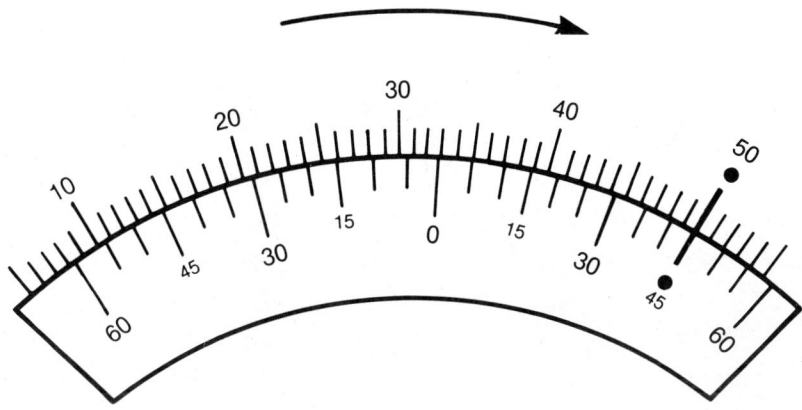

d. _____

INTERPRETING PRINT DIMENSIONS

OBJECTIVES

After studying this chapter, you will be able to:

■ Explain the basic rules for reading print dimensions.

■ Identify and describe the five common dimensional forms.

■ Identify and describe the three primary methods of placing dimensions.

■ Describe the standard methods used to dimension internal part features.

■ Describe the standard methods used to dimension external part features.

■ List and define some of the common abbreviations found on prints.

■ Define tolerance, allowance, and related terms.

■ Calculate tolerance and allowance values.

The basic shape and detail of a part are clearly shown by the views in a multiview drawing. However, to show the exact size of the part and the size and location of part features, dimensions must be specified. *Dimensions* are numbers that are used to accurately describe the size and form of the part.

Dimensions are mainly used to show either size or location, Figure 8-1. *Size dimensions* show the overall size of the part and the size of part details, such as holes, slots, and radii. *Location dimensions* show the exact position of part details, such as the location of holes or slots.

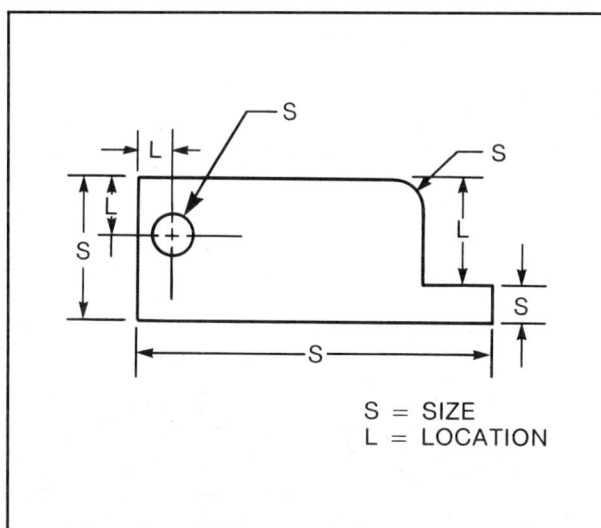

Figure 8-1. Dimensions show either size or location of part details.

BASIC RULES FOR READING PRINT DIMENSIONS

Every engineering drawing is drawn and dimensioned in accordance with some type of standard. The accepted standard by all American drafting firms is the American National Standards Institute (ANSI). All drafters should follow these standards to unify all dimensioning on drawings. To correctly read a print, it is necessary to know how these drafting methods relate to reading dimensions. The basic rules for reading print dimensions are summarized here.

1. Never measure a print. Drafters are not responsible for the scale accuracy of the drawn object. In addition, when prints are made, there is sometimes a slight amount of shrinkage or stretching. This makes the print slightly different than the original. Therefore, always use the dimensions noted on the print. If a dimension is not shown in one view, look in the other views.

Sometimes it may be necessary to calculate dimensions, as shown in Figure 8-2. In these cases, add or subtract the dimensions shown to find the required size. If the dimension cannot be found or calculated, check with a supervisor. Remember, never measure the print.

2. As a rule, dimensions are placed in the view that best describes the detail. For example, the part shown in Figure 8-3 has two holes. These holes are shown in all three views, but only

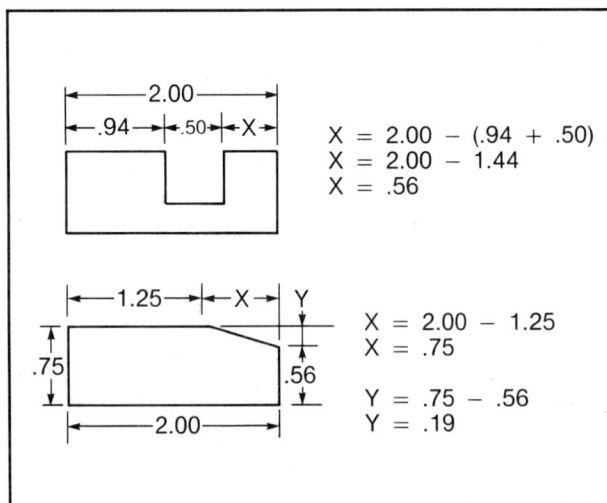

Figure 8-3. Dimensions are generally placed in the view that best describes the detail.

the front view shows the true profile of the holes. Thus, only the front view shows the dimension.

3. Dimensions are shown on a drawing by using either dimension lines or leader lines. *Dimension lines* show the direction of the dimension. *Extension lines* mark the limits of the dimension. *Leader lines* are used to show where a dimension or note applies on the drawing, such as the diameter or a hole or the size of a radius. See Figure 8-4.

4. Dimensions may be considered as being either working dimensions or reference dimensions. *Working dimensions* are those that control the size of the part. *Reference dimensions* are

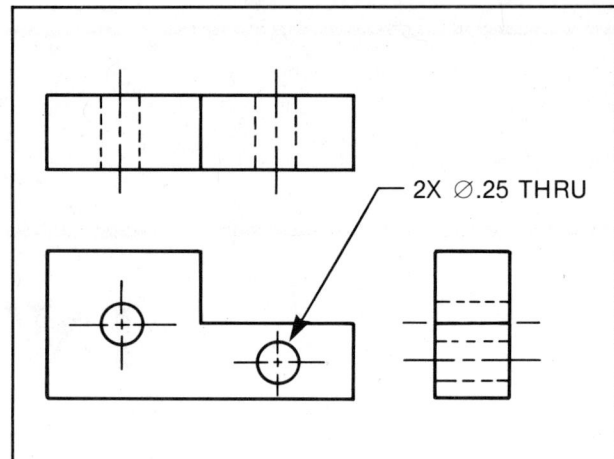

Figure 8-2. It is sometimes necessary to calculate dimensions.

$$X = 2.00 - (.94 + .50)$$
$$X = 2.00 - 1.44$$
$$X = .56$$

$$X = 2.00 - 1.25$$
$$X = .75$$

$$Y = .75 - .56$$
$$Y = .19$$

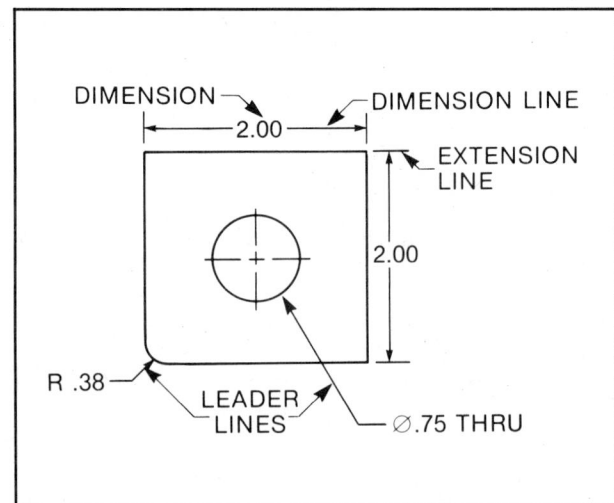

Figure 8-4. Use of dimension lines, extension lines, and leader lines

shown for convenience and are derived from the other dimensions. Reference dimensions are shown by using parentheses ( ), Figure 8-5. Reference dimensions are intended to give general information rather than specific instructions. In the example in Figure 8-5, the reference dimension is used to show the center distance from the other side of the part. Only working dimensions should be used when producing a part.

5. Sometimes a print will show *in-process dimensions*. These dimensions are intended to show the size of a part after a specific machining process, such as milling or turning. In such cases, the fact that the dimensions are in-process sizes and not final sizes is normally noted on the print.

6. The scale of the drawing does not affect the dimensions. Parts are always dimensioned to show the finished size. The scale only refers to the size relationship between the drawn object and the actual part, Figure 8-6. A scale may be noted in one of the following manners: (a) 1:2 (ratio), (b) $\frac{1}{2}'' = 1''$ (ratio), or (c) HALF SIZE (spelled out).

7. Every dimension has a tolerance. The tolerance may be shown directly on the dimensions or in a general dimensional note in the title block, Figure 8-7. The tolerance is the total amount a dimension is allowed to vary. In Figure 8-8, the dimension is ⌀.50″ and the tolerance is plus or minus (±) .03″. This means the dimension can vary either up or down (plus or

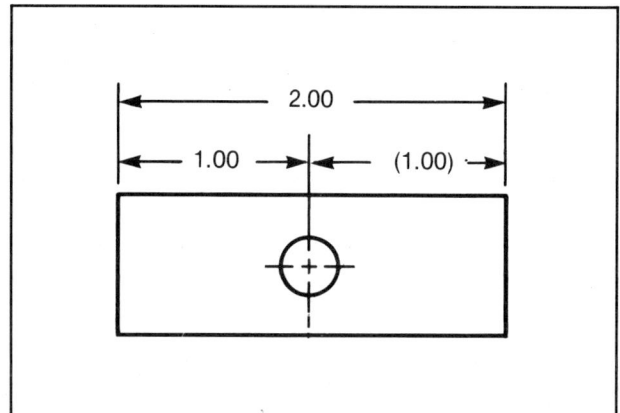

Figure 8-5. Reference dimensions are only used for convenience and are shown in parentheses.

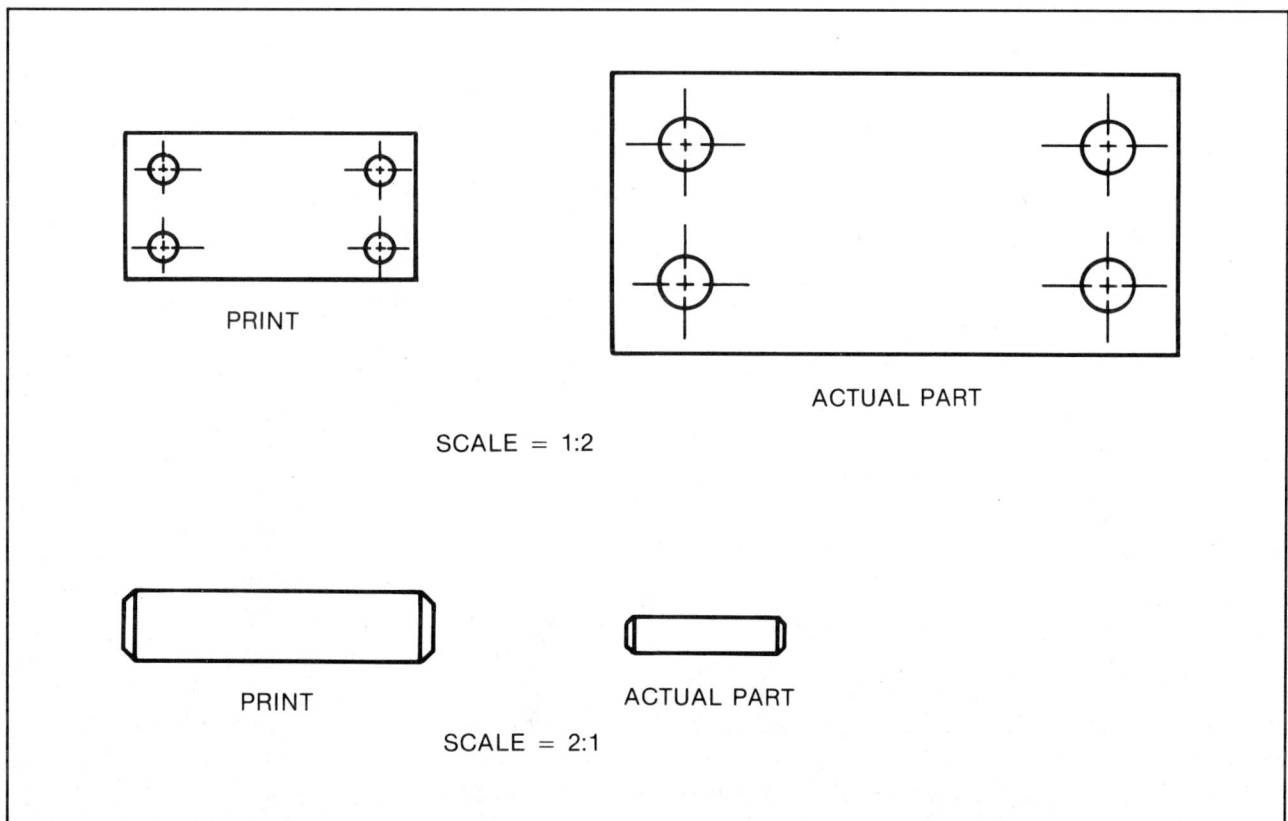

Figure 8-6. Scale is the size relationship between the actual part and the view in the print.

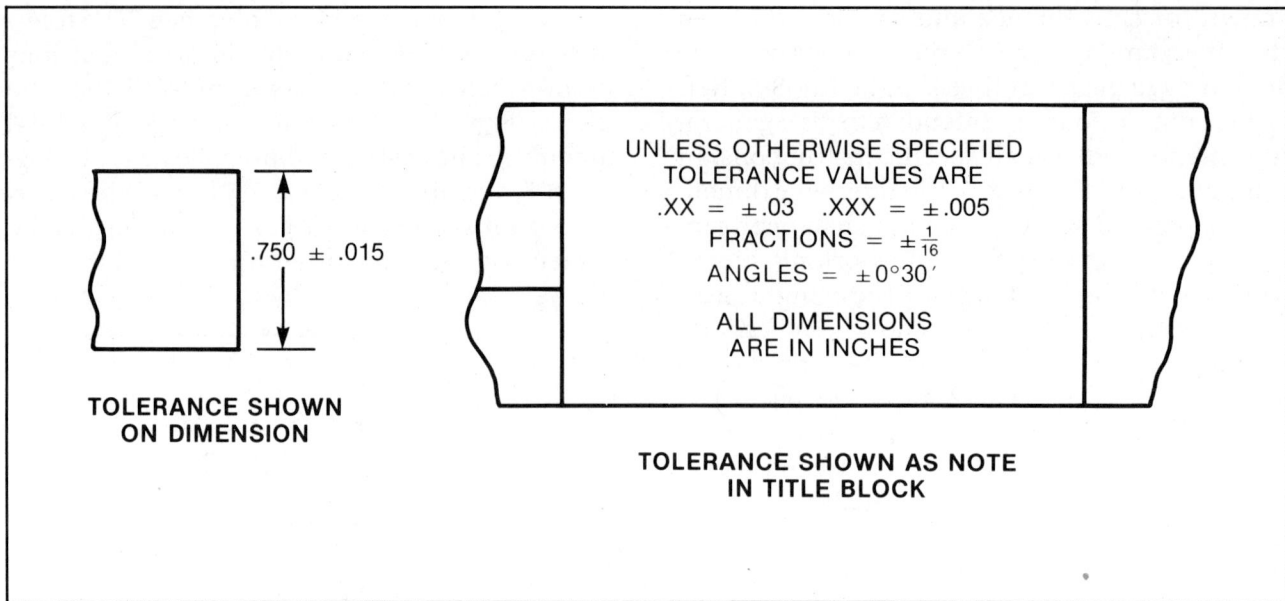

Figure 8-7. Tolerances are shown either on the dimension or in the title block.

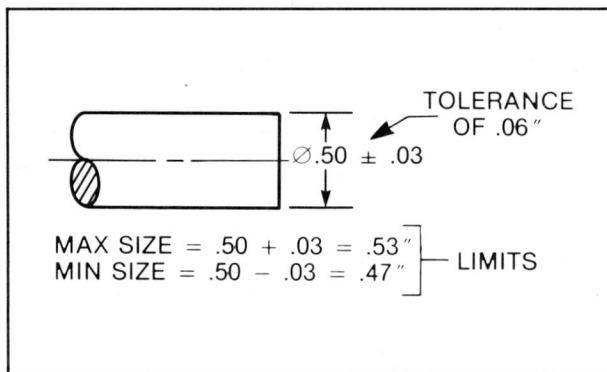

Figure 8-8. Tolerance and limits

minus) .03″. Thus, the finished part can be between .47″ and .53″ in diameter and still be correct. The upper and lower sizes of this dimension are called the limits.

BASIC RULES OF DIMENSIONING

Drafters follow some general rules when dimensioning in decimal inch units. Decimal inch units are usually used now to show all dimensional values on prints. Fractional inch units are rarely used, except for noting such things as thread sizes or sizes of rough stock. Knowing

the following rules will help the print reader to read dimensions quickly and accurately.

1. Dimensions are generally specified with two-place decimal values, Figure 8-9. However, when greater precision is required, three- or four-place decimal values may also be used.

2. The dimension and tolerance should have the same number of decimal places, Figure 8-10. If necessary, zeros should be added to make them equal.

3. The tolerance should never be assumed by the number of decimal places. *Note:* Always check the title block if no tolerance is shown on a dimension.

4. For values less than one inch, zeros are not used before the decimal point, Figure 8-11.

5. Decimal and fractional dimensions should not be mixed on a drawing. The only exception to this rule is when common hardware sizes are specified.

6. If a fractional size is required, it should be converted to a decimal value, except when used as noted in rule #5 above.

7. Commas are *not* used to show values greater than one thousand using inch dimensions.

8. The abbreviation (in.) or symbol (″) for inch is not included in a dimension. All dimensions are assumed to be in inches unless otherwise noted.

```
2.00 NOT 2.
1.50 NOT 1.5
1.06 NOT 1.0625
```

Figure 8-9. Dimensions are normally shown as two-place decimals.

```
.75 ± .01 NOT .75 ± .010
1.500 ± .010 NOT 1.500 ± .01
5.63 ± .03 NOT 5.63 ± .030
```

Figure 8-10. A tolerance value should have the same number of decimal places as the dimension.

```
.56 NOT 0.56
.750 ± .002 NOT 0.750 ± 0.002
```

Figure 8-11. Zeros are not used before the decimal point in dimensions less than one inch.

DIMENSIONAL FORMS

There are five basic dimensional forms commonly used to dimension most parts. These forms are linear, angular, radial, coordinate, and tabular dimensions.

Linear Dimensions

Linear dimensions are used to show length along a straight line. A linear dimension is always placed parallel to the surface it defines, as shown in Figure 8-12A. Generally, the dimension is placed in the middle of the dimension line, using either the aligned or unidirectional methods, Figure 8-12B. However, in cases where several dimensions are close together or where space is limited, dimensions are staggered or positioned as shown in Figure 8-12C.

Angular Dimensions

Angular dimensions are used to show the sizes of angular part features. These dimensions are shown as either linear or angular values. When reading an angular dimension, pay close attention to the dimension line and dimensional units. As shown in Figure 8-13, angular dimensions are specified by either the angle of the arc, the length of the arc, or the length of the chord.

The *angle of the arc* is the size of the angle, shown in angular units. These units are stated in degrees (°), minutes ('), and seconds ("), Figure 8-14. A complete circle contains 360°, a degree contains 60', and a minute contains 60", Figure 8-15. In most cases the prints seen in the shop are only dimensioned in degrees or in degrees and minutes. Seconds are very difficult to measure without sophisticated inspection equipment. Angular measurements may be expressed in terms of *decimal degree units*. A dimension of 15.5° is the same as 15°30'. Angular dimensions that show the angle of the arc are shown with a curved dimension line.

The *length of the arc*, Figure 8-16, is generally shown in decimal inch units. These dimensions are normally used only for informational purposes, since they cannot be measured readily with "normal" shop measuring instruments. The length of the arc dimension, like the angle of the arc dimension, is shown with a curved dimension line. The major difference between the length of the arc and the angle of the arc dimensions is the small arc shown above the length of the arc dimension.

The *length of the chord*, Figure 8-17, is the length of the straight line that joins two points on a curve. This dimension is usually stated in decimal inch units. It is always shown with a straight dimension line.

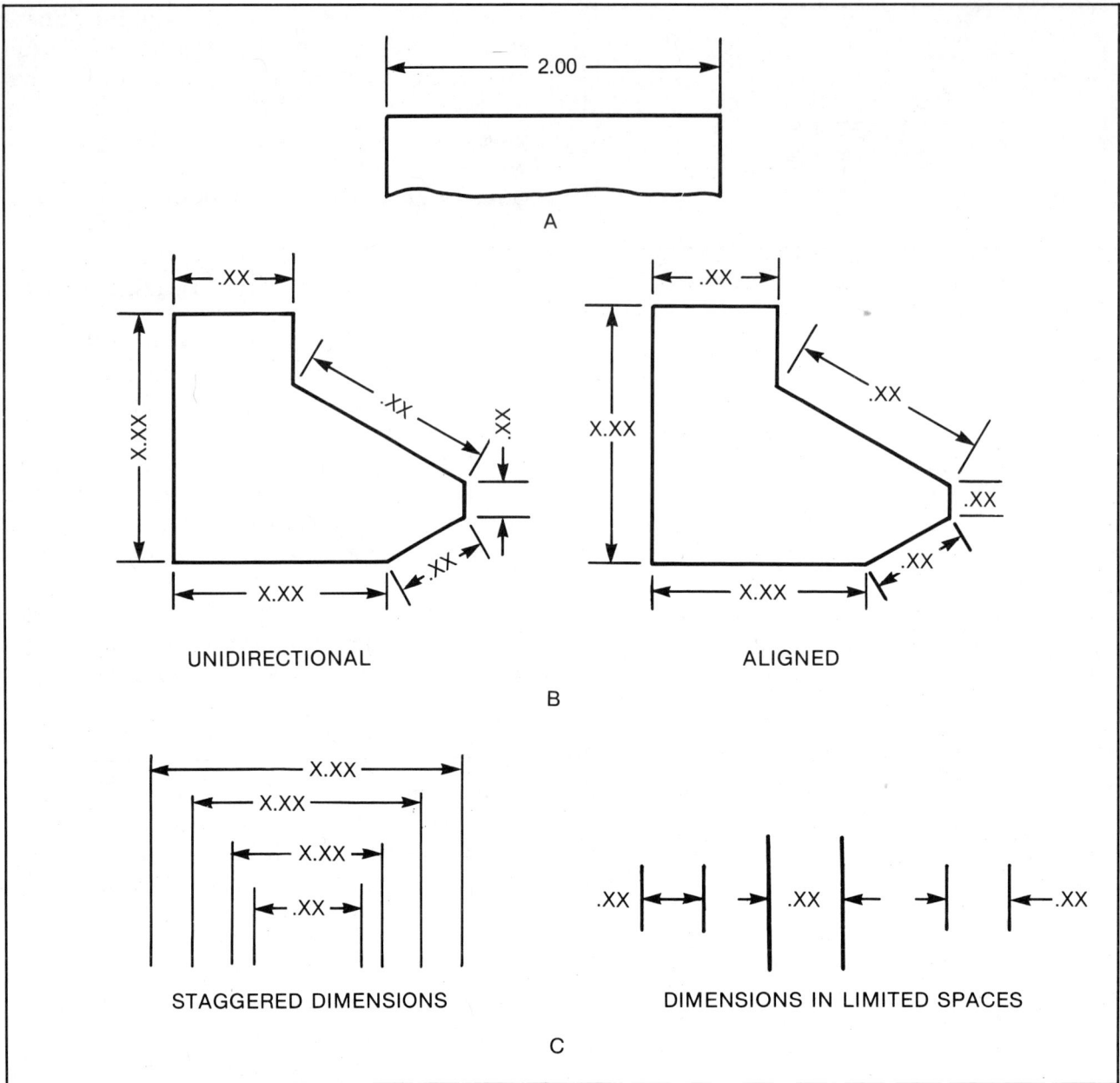

Figure 8-12. Linear dimensions are always placed parallel to the surfaces they define (A). They may be placed by the aligned method or the unidirectional method (B). If space is limited, it may be necessary to change the standard placement of linear dimensions (C).

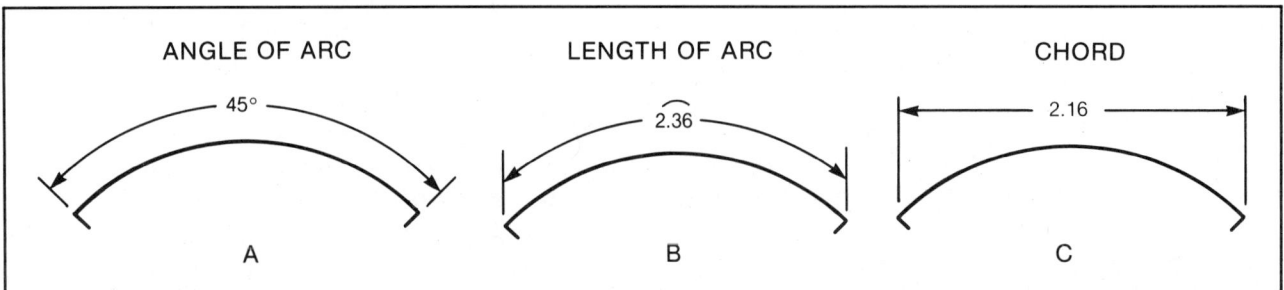

Figure 8-13. Dimensioning (A) angles, (B) arcs, and (C) chords

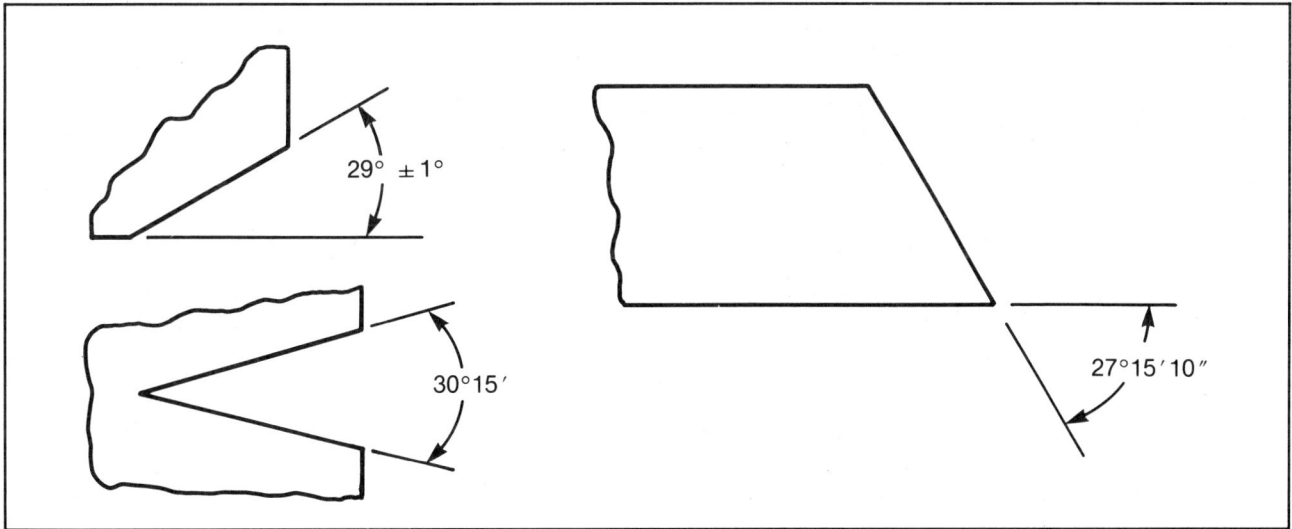

Figure 8-14. Angular dimensional units

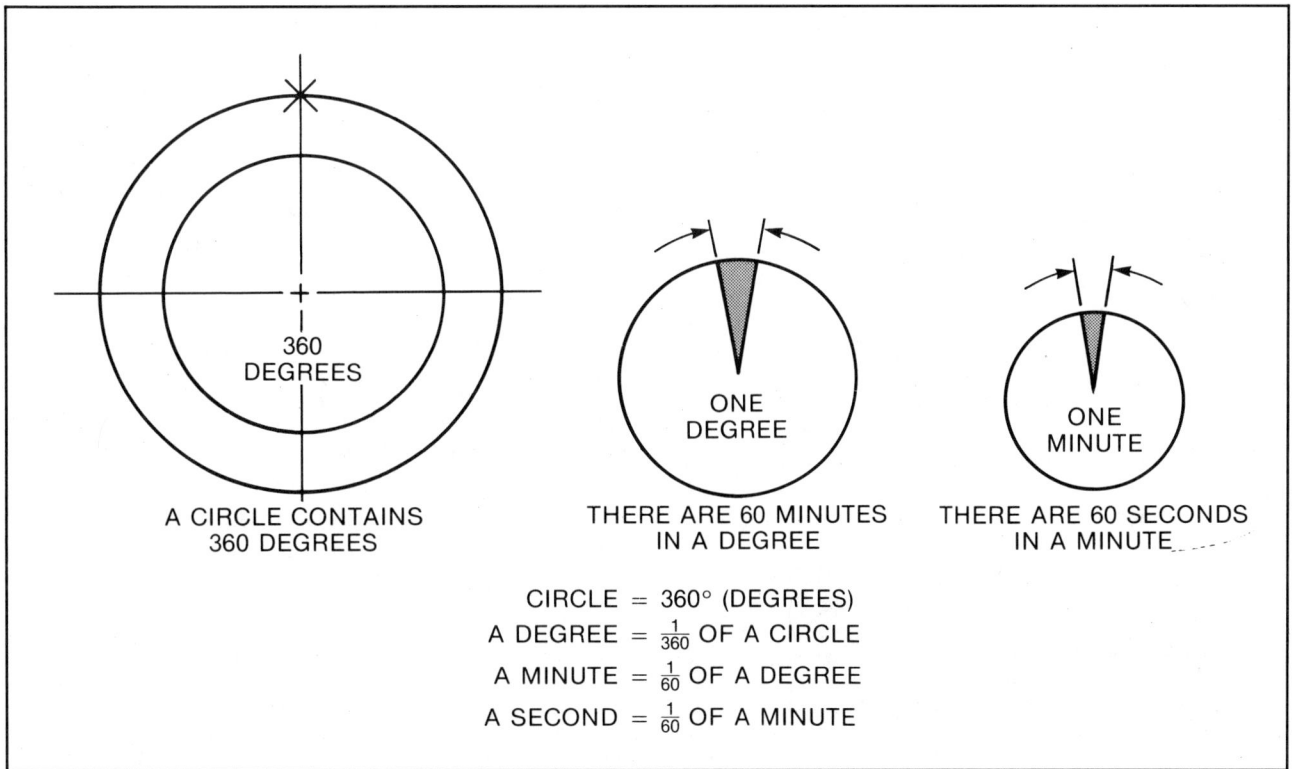

A CIRCLE CONTAINS
360 DEGREES

THERE ARE 60 MINUTES
IN A DEGREE

THERE ARE 60 SECONDS
IN A MINUTE

CIRCLE = 360° (DEGREES)

A DEGREE = $\frac{1}{360}$ OF A CIRCLE

A MINUTE = $\frac{1}{60}$ OF A DEGREE

A SECOND = $\frac{1}{60}$ OF A MINUTE

Figure 8-15. Relationship of degrees, minutes, and seconds

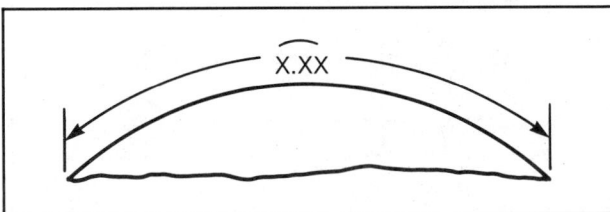

Figure 8-16. Length of arc dimension

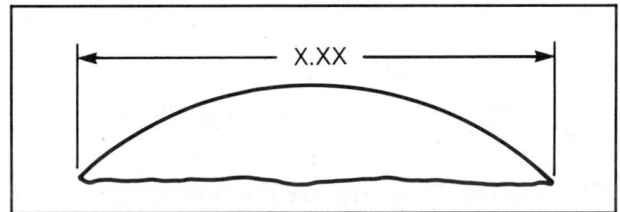

Figure 8-17. Length of chord dimension

Radial Dimensions

Radial dimensions are used to show the size of radii, Figure 8-18A. (*Note:* The term *radii* is the plural of *radius*.) These dimensions are given in decimal inch units and are shown with leader lines. A few variations of radial dimensions are shown in Figure 8-18B. These variations include shortened radii, true radii, and spherical radii dimensions.

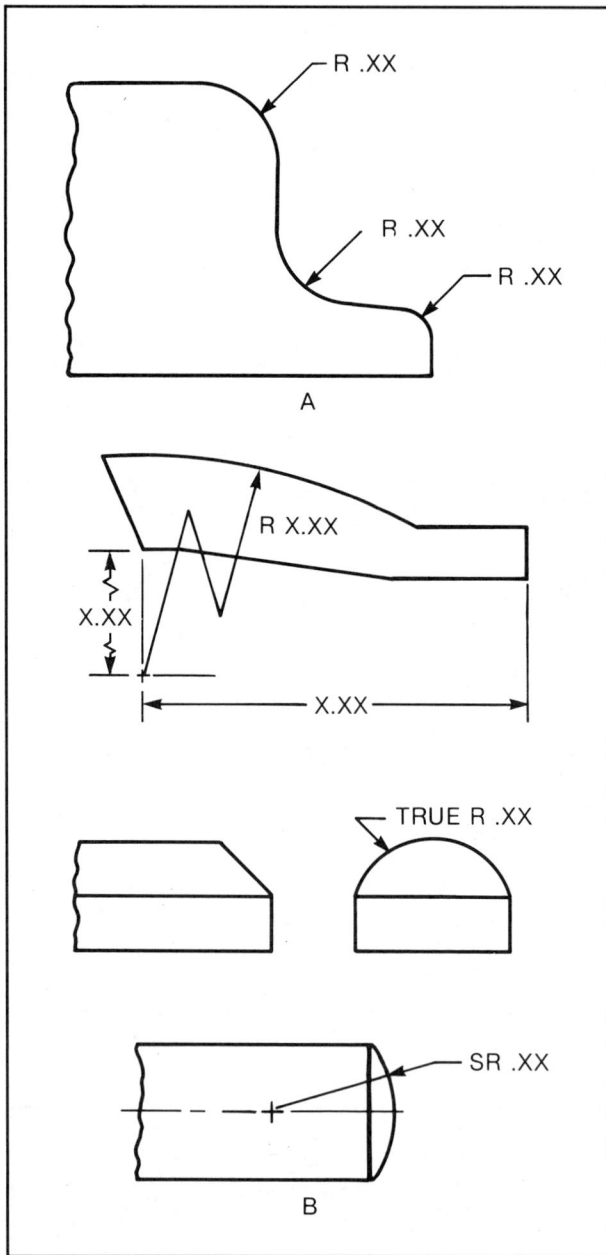

Figure 8-18. Radii are usually dimensioned as in (A). Some variations of radial dimensioning are shown in (B).

A *shortened radius dimension* is used when the center position of the radius is located outside the drawing, or when it would interfere with another view. *True radius* and *spherical radius dimensions* are indicated with a radial dimension and a note. The note *TRUE R* (true radius) means the radius dimension shown is the actual size of the radius, regardless of how it appears in a particular view. The method used to show a spherical radius is the abbreviation *SR* preceeding the size value. The *SR* means *spherical radius*. When the spherical form specified is a diameter, the abbreviation *S∅* is used to note the feature.

Coordinate Dimensions

Coordinate dimensions are used for dimensioning parts from a single, or common, reference point. These reference points are called *datums*. Datums are specific lines, points, planes, surfaces, or features used as a reference for dimensions. In Figure 8-19, the datum surfaces are the top and left edges of the part. The top edge, or datum, controls all vertical dimensions. The left edge, or datum, controls all horizontal dimensions.

The two main types of coordinate dimensions are rectangular and polar. See Figure 8-20. *Rectangular coordinate dimensions* can be shown with or without dimensions lines, and can be used to dimension both size and location. *Polar coordinate dimensions* all start at a single center position, or datum, and are generally dimensioned as angular units.

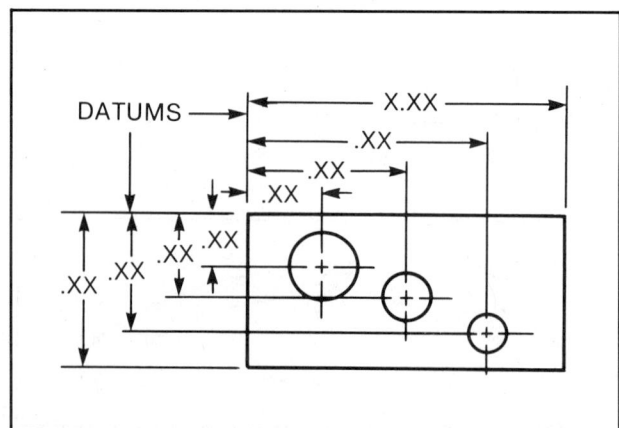

Figure 8-19. Coordinate dimensioning

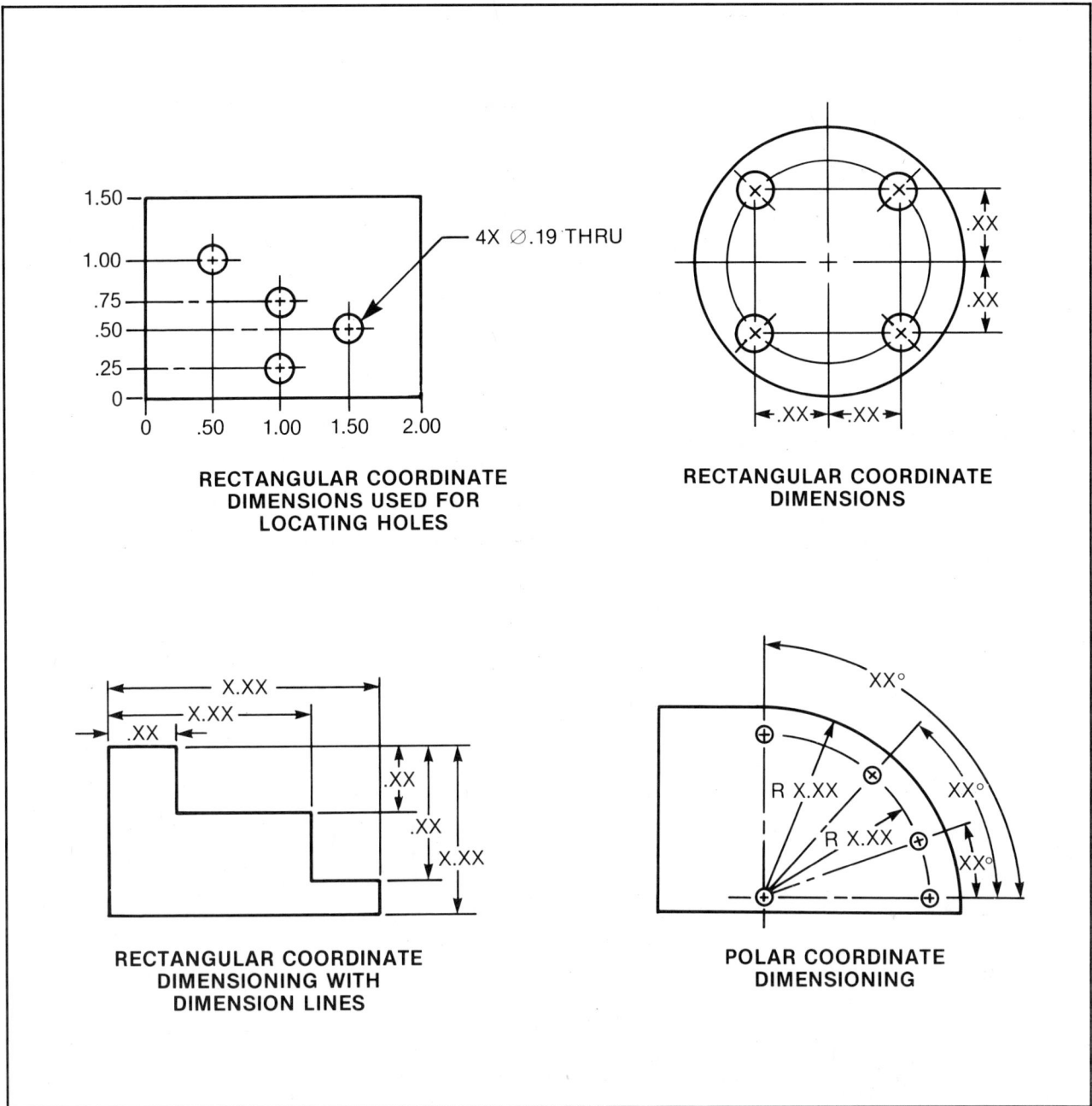

Figure 8-20. Rectangular and polar coordinate dimensioning

Tabular Dimensions

Tabular dimensions are mainly used for two purposes. First, they show the sizes and locations of part features without cluttering the print with excessive dimension lines and extension lines. See Figure 8-21. Here the size and location of each hole is specified in the chart. The datum surfaces, marked *X* and *Y*, are the left and bottom surfaces of the part. The second use of tabular dimensions is to show the different sizes of parts having the same proportional shape. As shown in Figure 8-22, the part is dimensioned with letter values that refer to specific sizes noted in the chart. With this application, the same drawn object can be used to represent many different parts having the same shape but different sizes.

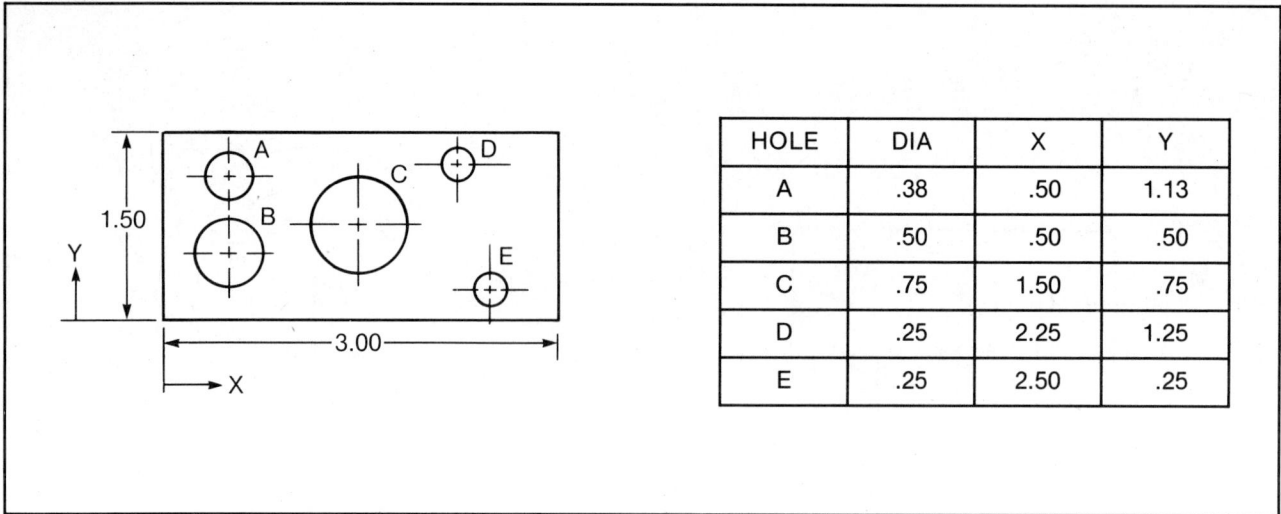

HOLE	DIA	X	Y
A	.38	.50	1.13
B	.50	.50	.50
C	.75	1.50	.75
D	.25	2.25	1.25
E	.25	2.50	.25

Figure 8-21. Tabular dimensions used to show the size and location of holes

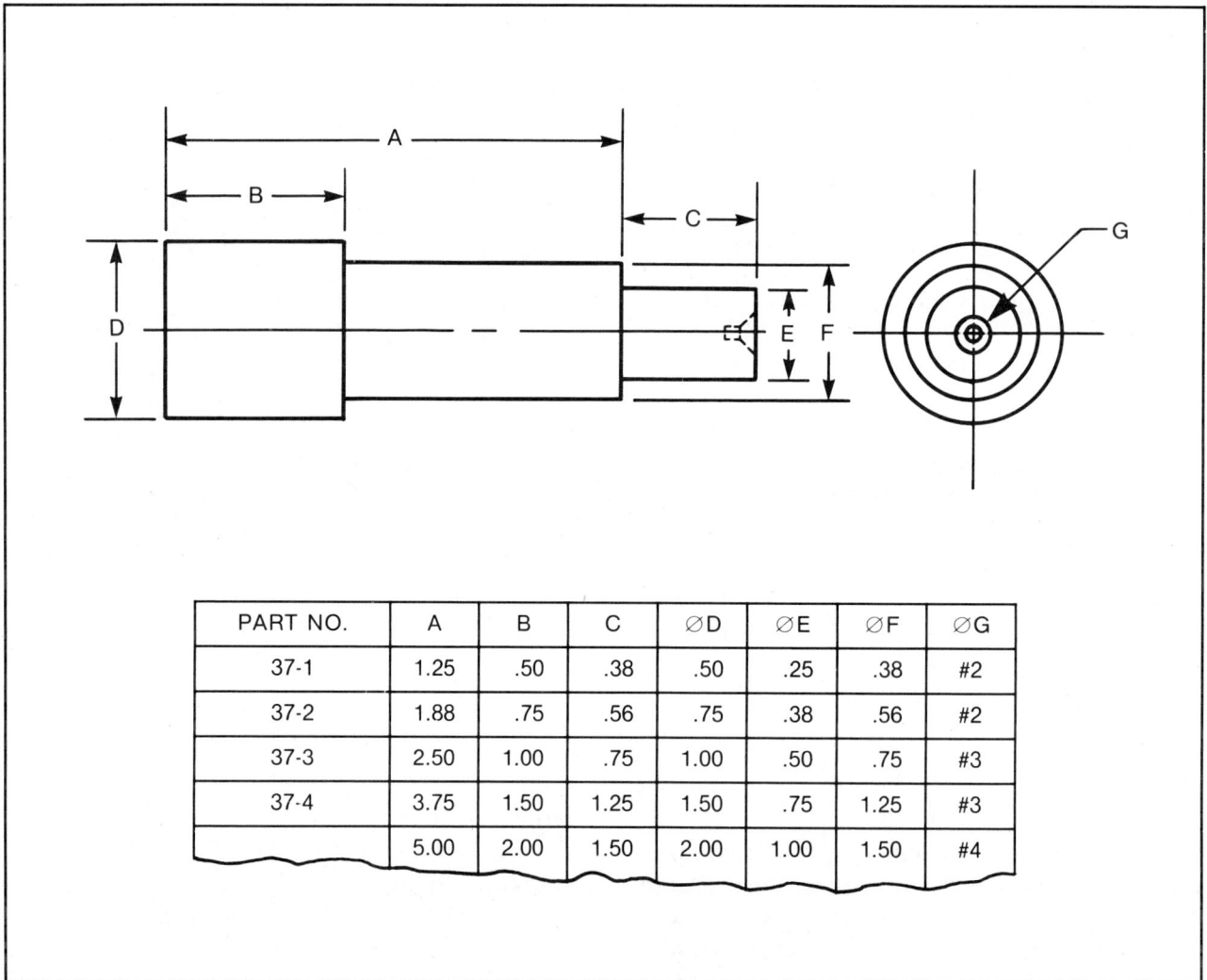

PART NO.	A	B	C	ØD	ØE	ØF	ØG
37-1	1.25	.50	.38	.50	.25	.38	#2
37-2	1.88	.75	.56	.75	.38	.56	#2
37-3	2.50	1.00	.75	1.00	.50	.75	#3
37-4	3.75	1.50	1.25	1.50	.75	1.25	#3
	5.00	2.00	1.50	2.00	1.00	1.50	#4

Figure 8-22. Tabular dimensions used to show the different sizes of parts with the same proportional shape

METHODS OF PLACING DIMENSIONS

The proper placement of dimensions on a drawing is important. This is especially true when specific relationships between features must be maintained. The three primary methods of placing dimensions are chain dimensioning, datum dimensioning, and direct dimensioning.

Chain Dimensioning

Chain dimensions are used to show a relationship between details in a series. See Figure 8-23A. The location of each hole is shown as it relates to the other holes in the series. In this case, the relationship of the holes to one another is more important than the relationship of the holes to the part. Chain dimensioning is also referred to as *incremental dimensioning.*

Datum Dimensioning

Datum dimensions are used to show the relationship of the part details to a common reference point. See Figure 8-23B. In this case, the datums are the top and left edges of the part. By dimensioning a part in this way, the location of each hole in relation to the datums is the most important factor. Datum dimensioning is also referred to as *base line* or *absolute dimensioning.*

Direct Dimensioning

Direct dimensions are used to control the relationship of specific details, regardless of the rest of the part. As an example, in Figure 8-23C the most important relationship is between hole *A* and hole *C*. In this case, the direct dimension shows the distance between these two holes. The dimensions on the top of the part are only intended to locate the position of the center hole in relation to hole *A* and the right edge of the part. In those cases where the origin of a dimension must be shown, the symbol *O* is used at one end of the dimension line. This symbol denotes the point of origin of the dimension. See Figure 8-23D.

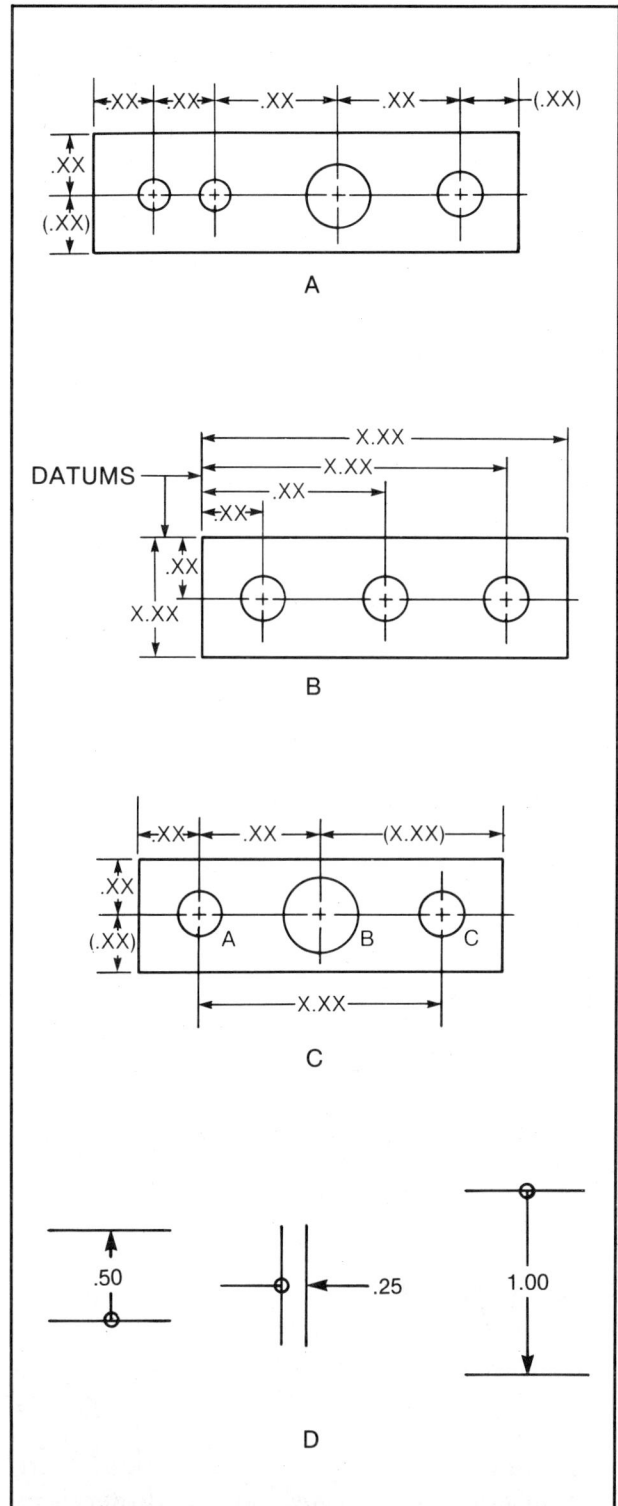

Figure 8-23. The placement of dimensions shows specific relations between details of an object. Chain dimensioning (A), datum dimensioning (B), and direct dimensioning (C) are the methods most commonly used to show specific relations. If the origin of a dimension must be shown, it is indicated as in (D).

DIMENSIONING MACHINED FEATURES

Besides showing the overall size and shape of a part, prints must also identify and describe all machined features. To properly interpret a print, the print reader must understand these machining operations and know the standard methods used by drafters to specify machined features. Due to the changes in the ANSI standards covering dimensioning and tolerancing (*Y14.5*), it is necessary to be familiar with both the old and new methods used to dimension these machined features. This text will use the latest ANSI standards.

Holes

Round holes are the most common type of machined feature. These features are normally specified by their diameter and depth, Figure 8-24A. The symbol ∅ precedes the diameter size to indicate the measurement refers to a diameter.

The operation to be used to produce the hole is normally not specified on the print. The exception to this rule is when engineering control must be exercised in manufacturing. In such cases, the machining operation is noted as well as the size. See Figure 8-24B. Remember not to note the machining operation unless it is absolutely necessary for the clarity of the part.

Hole Patterns

Holes are often grouped into *hole patterns*. The two primary types of hole patterns are circular hole patterns and linear hole patterns. Hole patterns are normally dimensioned by the hole size, depth, and spacing.

Circular hole patterns, Figure 8-25A, are dimensioned with either angular dimensions or a dimensional note. In either method, the diameter of the bolt circle is also shown. The *bolt circle* is the imaginary circle along which all the hole centers are located.

Linear hole patterns, Figure 8-25B, are used for repetitive features and dimensions. When a linear hole pattern is shown, the location of

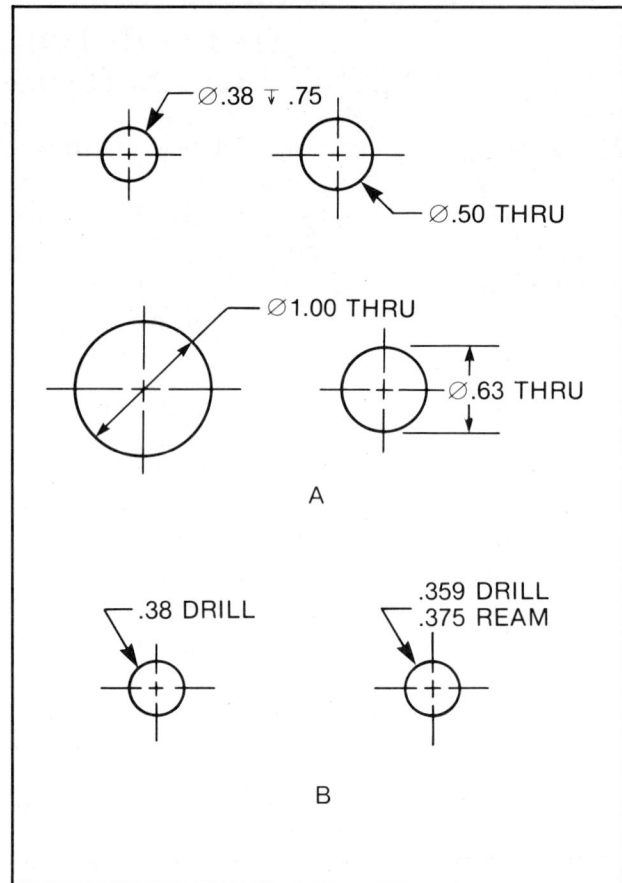

Figure 8-24. Holes are dimensioned with their diameter and depth (A). When necessary, the machining operation may also be shown (B).

the hole centers can be shown either by dimensioning each hole or with a note as shown. When the holes are specified in a note, either the number of holes or the spacing is specified. In newer prints, the number of holes normally precedes the size, and is shown with an X to note the number of places.

Slotted Holes

Slotted holes are specified on a print in several ways. When a part has several slots, the location of each slot may be dimensioned as shown in Figure 8-26A. Here the location of each slot is specified from a datum surface, and the size and number of slots are specified in a note. When the overall size of the slot is the most important factor, the slot may be dimensioned as shown in Figure 8-26B. When the center-to-center distance is important, the method

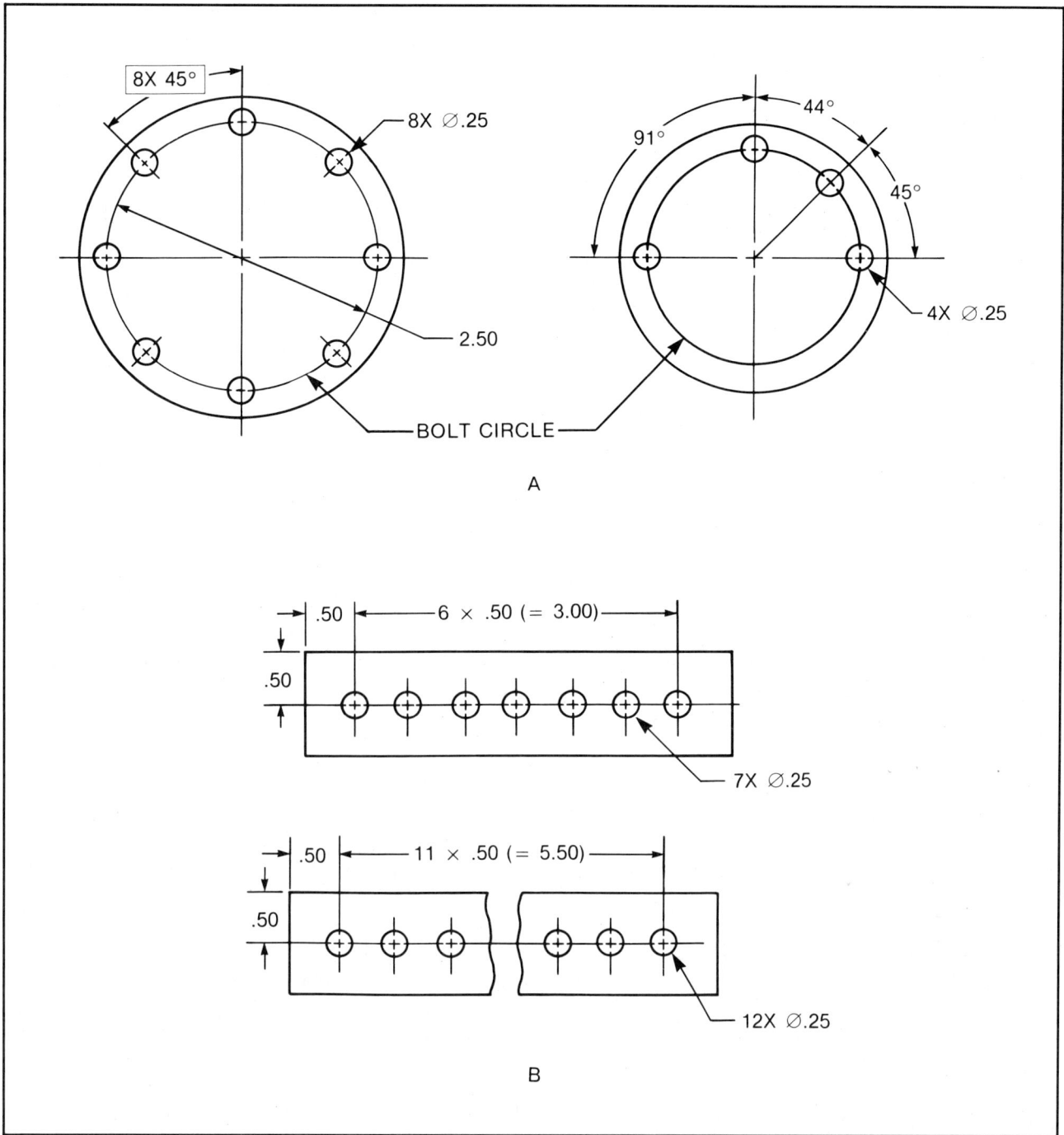

Figure 8-25. Hole patterns may be (A) circular or (B) linear. Hole patterns are dimensioned by the size, depth, and spacing of the holes.

shown in Figure 8-26C may be used. When a slot has rounded ends, the radius is noted without a specific size, Figures 8-26A, 8-26B, and 8-26C. However, if a slot has a partially rounded end radius or a radius not equal to half the slot width, the radius is dimensioned as shown in Figure 8-26D.

Reamed Holes

Reaming is a machining process used to enlarge holes to a specific size. A reamed hole is smoother, straighter, and closer to size than a drilled or punched hole. Reamed holes are specified by either a close tolerance or by a

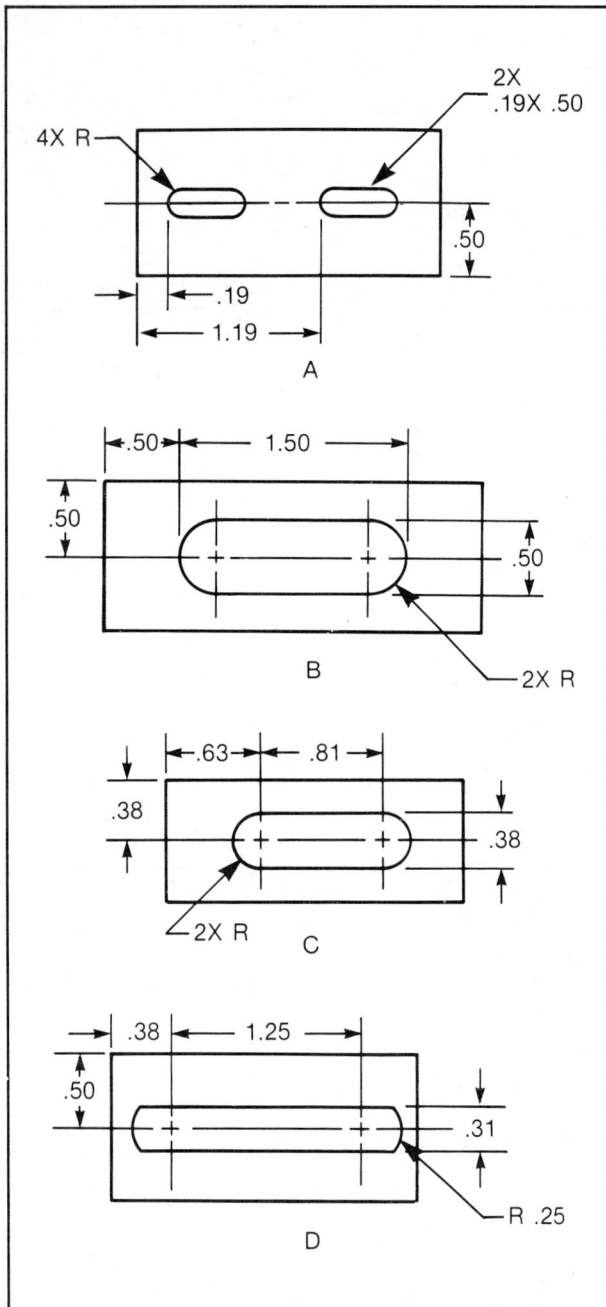

Figure 8-26. Dimensioning slotted holes

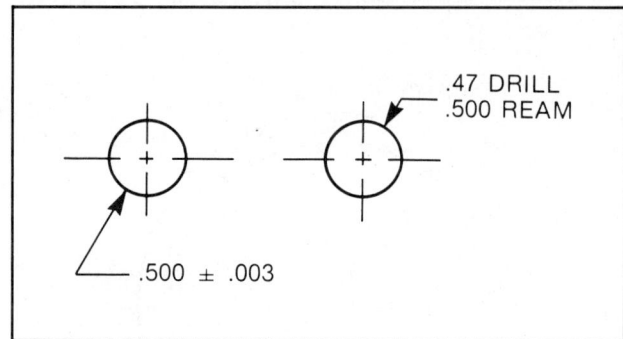

Figure 8-27. Specifying reamed holes

Bored Holes

Boring is another operation used to enlarge holes. However, unlike reamed holes, bored holes are not limited to a fixed size. Boring is generally performed on holes that are larger than the normal drill sizes, or on holes that are between standard drill or reamer sizes. There are also cases when boring is specified for holes because they must be made perfectly round and straight prior to other machining. Boring, like reaming, requires a starting hole slightly smaller than the finished size. Bored holes are noted on a print as illustrated in Figure 8-28.

Countersinking

Countersinking is an operation that forms a conical edge on a hole. Countersunk holes are primarily used for flathead screws, Figure 8-29A. The standard angles of countersunk holes are 60°, 82°, 90°, and 100°. When countersunk

direct note, Figure 8-27. Reaming can either be done by hand or machine. In either case, a hole slightly smaller than the reamer size must be drilled prior to reaming. The exact size of the drilled hole may be specified on the print, or the drilled size may be left to the judgment of the machinist rather than the drafter. This is why the ANSI standards recommends machining operations to be left off the working drawing.

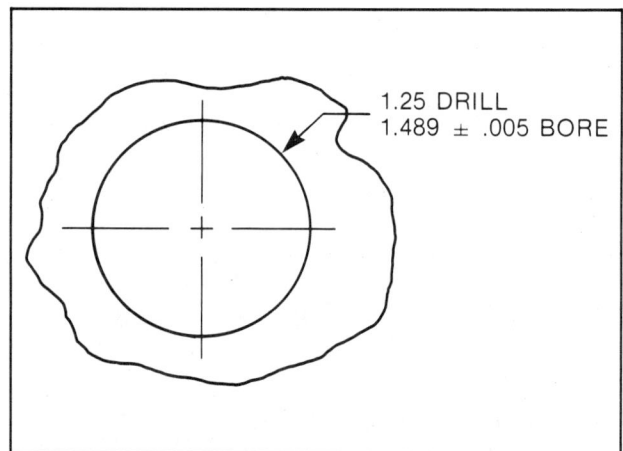

Figure 8-28. Noting a bored hole

holes are specified on a print, the information given is normally the diameter at the top of the hole and the angle. See Figure 8-29B. Newer prints show the size of a countersunk hole with the symbol ∨.

When a countersunk hole is specified on a curved surface, such as the surface of a shaft, the dimension given is the small diameter of the countersunk edge, Figure 8-29C. A simple note, Figure 8-29D, is generally used for noncritical applications. The abbreviation CSK may be used to indicate a countersunk hole.

Counterboring

Counterboring is an operation that enlarges the top of a hole, forming a recessed area for a bolt head. Counterbored holes are specified by their diameter and depth, Figure 8-30A. When necessary, the corner radius is specified as shown. In cases where the thickness of the remaining material is critical, the counterbore may be dimensioned as shown in Figure 8-30B. In less critical applications, a simple note is used, Figure 8-30C. In newer prints, the counterbore

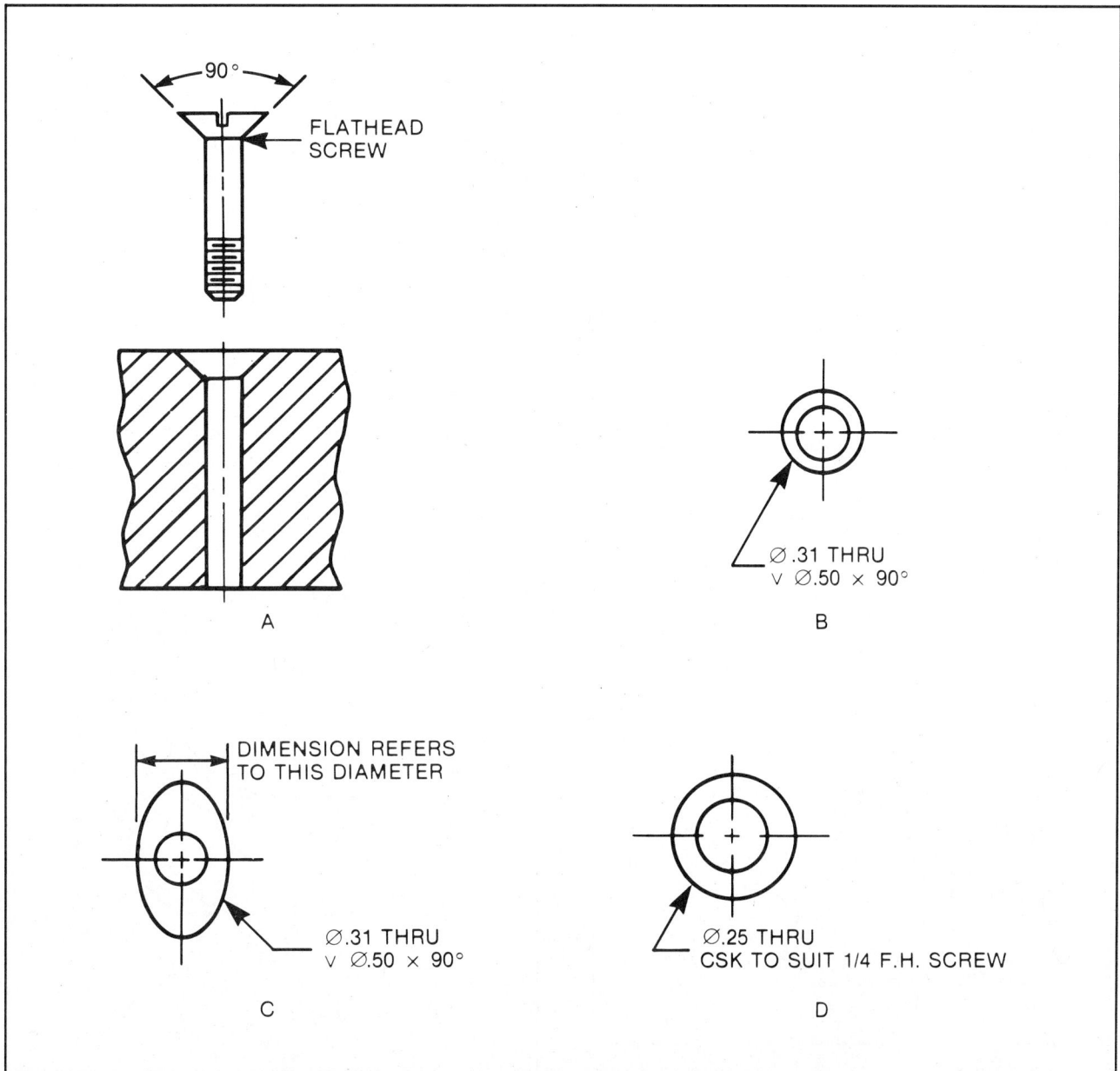

Figure 8-29. Dimensioning countersunk holes, such as for flathead screws

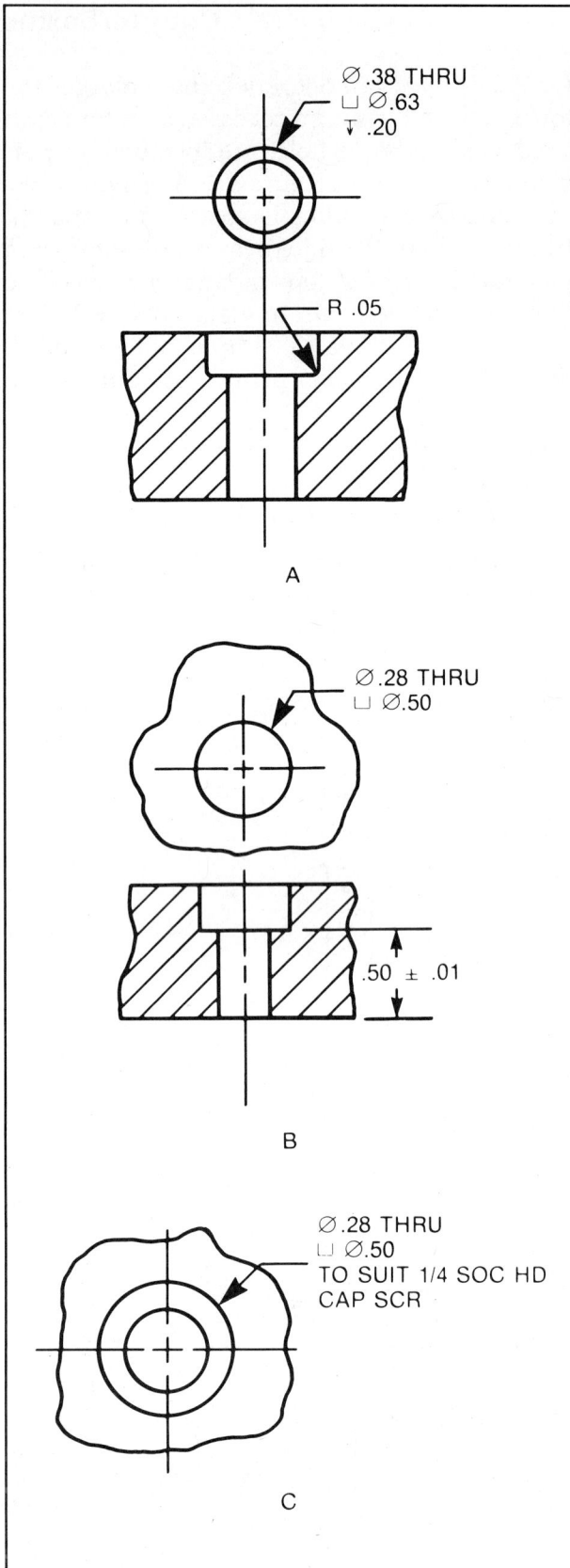

Figure 8-30. Dimensioning counterbored holes, such as for bolt heads

symbol ⊔ is used to denote a counterbore, as shown. The depth symbol ↧ is used to note the depth of the feature.

Spotfacing

Spotfacing is an operation that forms a smooth, flat surface around the top edge of a hole, Figure 8-31A. Spotfaced holes are normally used to provide flat seats for bolts and nuts in irregular parts such as castings. When specified on a print, only the diameter of the spotface is usually given. The depth is left to the judgment of the machinist. As a rule, a spotface is only machined deep enough to provide a good bolt seat. In cases where the thickness of the remaining material is critical, the spotface is dimensioned as shown in Figure 8-31B. The symbol ⊔ is used to indicate a spotfaced hole, as shown. Note that this is the same symbol used for counterboring.

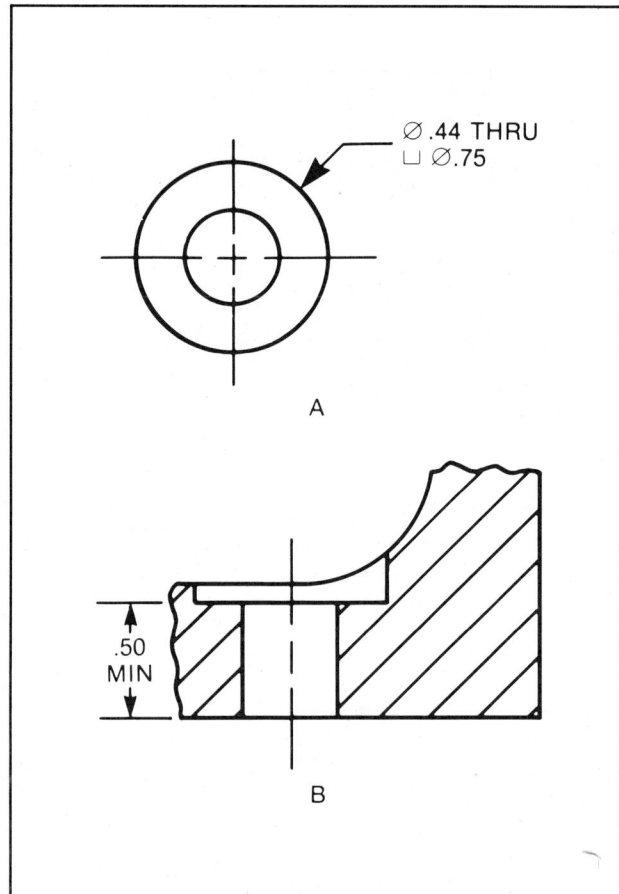

Figure 8-31. Dimensioning spotfaced holes

Internal Chamfering

Internal chamfering is a process of removing the sharp edge of a hole. Chamfers are usually cut at 45° and are specified on the print by a note that describes the angle and depth, Figure 8-32A. In cases where the size is critical, the diameter, rather than the depth, is specified as shown in Figure 8-32B.

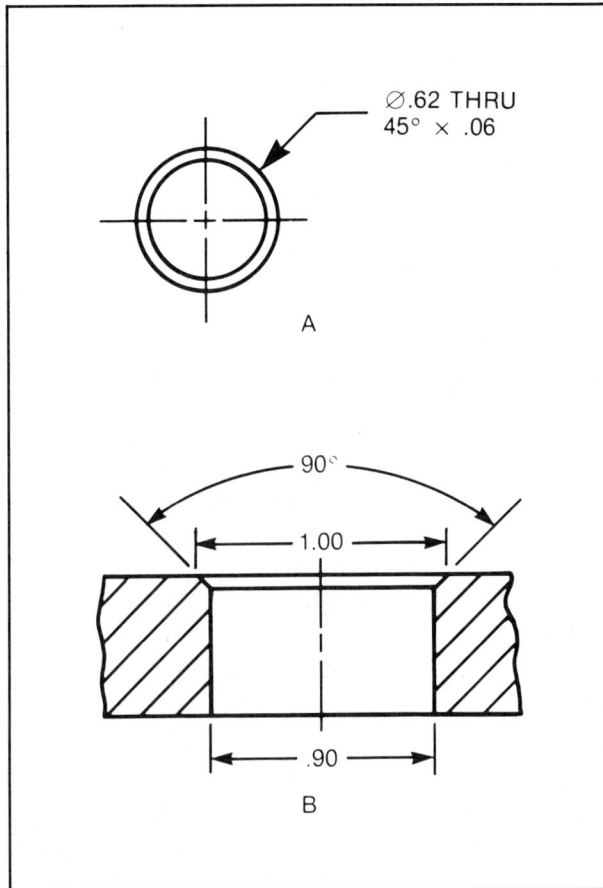

Figure 8-32. Dimensioning internal chamfers

Undercutting

Undercutting, or internal grooving, is an operation performed inside a hole to provide tool runout, thread clearance, or seats for O-rings or lock rings. When specified on a print, an undercut is dimensioned as shown in Figure 8-33. Only the width and the depth of one side are normally shown. When the end view is shown, it will generally include the diameter of the undercut. Here again, the depth symbol ⊽ may also be used.

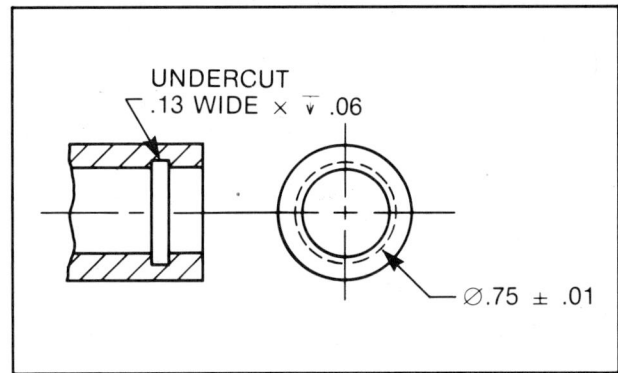

Figure 8-33. Dimensioning an undercut (internal groove)

Necking

Necking, or external grooving, is an operation performed on the outside surface of cylindrical parts. These external grooves are used to provide square shoulders for bearing seats, lock-ring or O-ring seats, or thread runout.

Several methods are used to dimension these details. The most common methods are shown in Figure 8-34A.

When reading a neck dimension, always look for exactly what the dimension specifies. These dimensions will generally specify the width, and either the depth of the groove or the diameter of the part at the bottom of the groove. Necks that have a rounded bottom are dimensioned with a radius rather than a width, Figure 8-34B. The notes *THD. RELIEF* and *GROOVE* may or may not be shown, depending on the print.

Flats

Flats are used to provide a flat surface, or seat, on round shafts. This permits either the shaft or a pulley to be held in a fixed position with a set screw. Flats are usually dimensioned by the length of the full flat and the distance from the flat to the opposite side of the shaft, Figure 8-35.

External Chamfers

External chamfers are generally dimensioned in one of two ways. The size can be shown by either the angle and length or by the length of both sides, as shown in Figure 8-36.

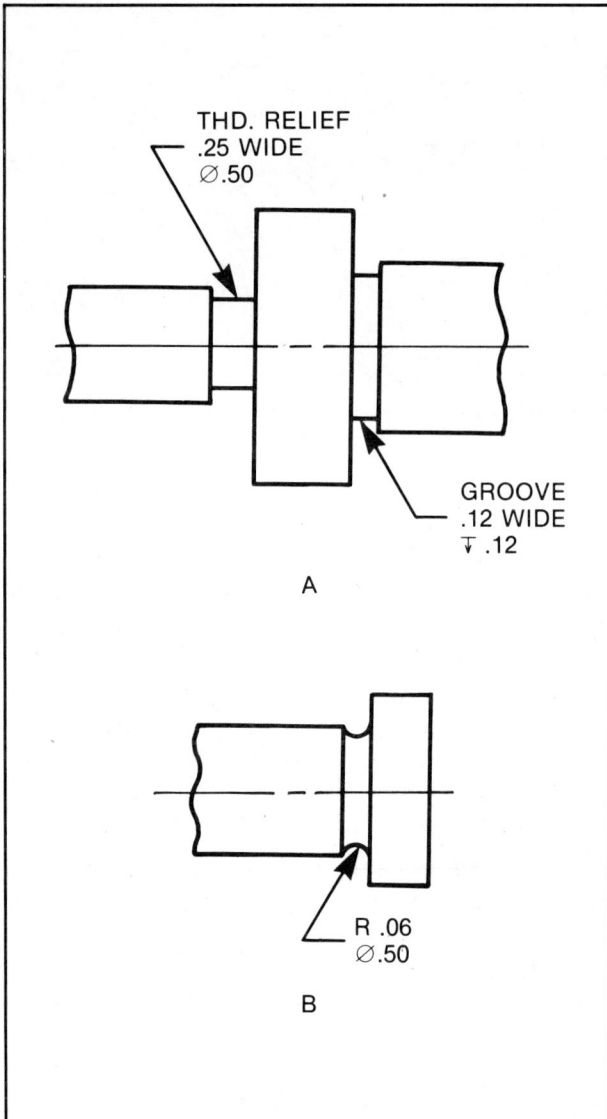

Figure 8-34. Dimensioning a neck (external groove)

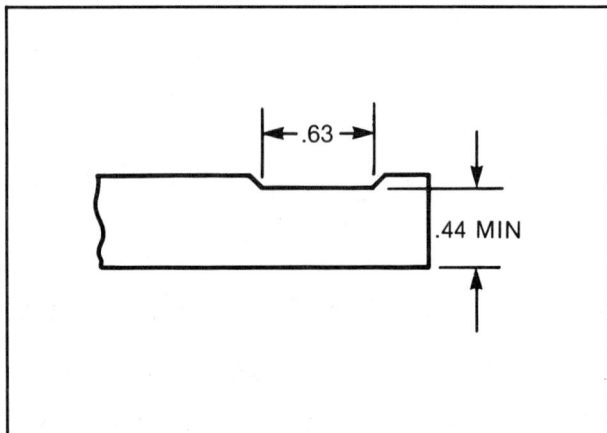

Figure 8-35. Dimensioning flats

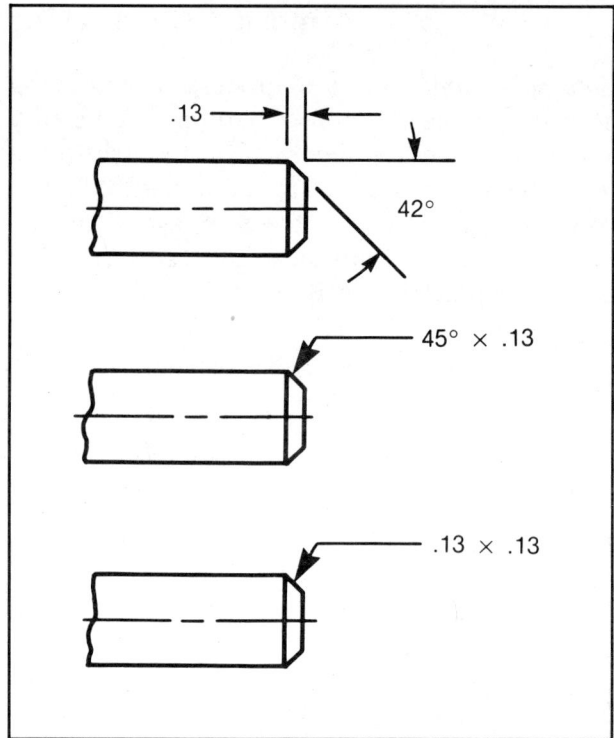

Figure 8-36. Dimensioning external chamfers

Fillets, Rounds, and Radii

Fillets and *rounds*, Figure 8-37A, are normally found on cast or welded parts. A fillet is the rounded area on an inside corner. A round is the rounded outside edge, or corner. When shown on a print, fillets and rounds are dimensioned by their radii, Figure 8-37B.

Radii are the machined equivalents of fillets and rounds. When a part has rounded inside or outside machined corners, the corners are referred to as radii. Radii are also dimensioned by their radius. When a part has several radii, Figure 8-37C, the size and number of radii are shown in a note.

Knurling

Knurling is an operation that puts a raised, roughened surface, called a *knurl*, on a shaft to provide a better hand grip. It can also enlarge a shaft for a press fit into a mating hole. The two most common forms of knurls are diamond and straight, Figure 8-38A. *Diamond knurls* are generally used for handles with a nonslip grasp. *Straight knurls* are usually used for press fits.

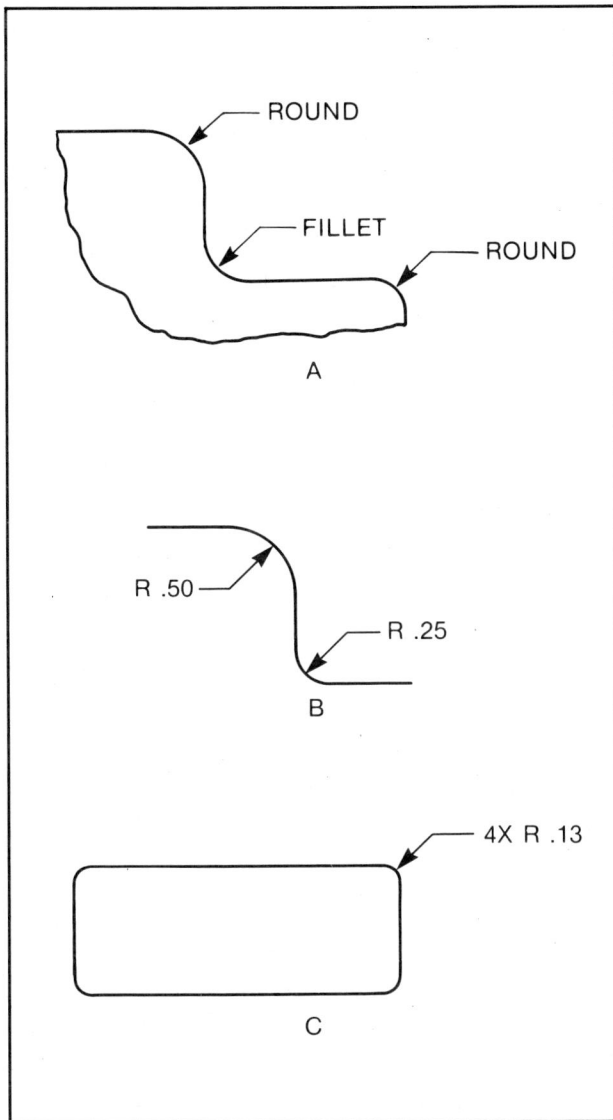

Figure 8-37. Fillets and rounds (A) are dimensioned as shown in (B). Radii (the machined equivalents of fillets and rounds) are dimensioned as shown in (C).

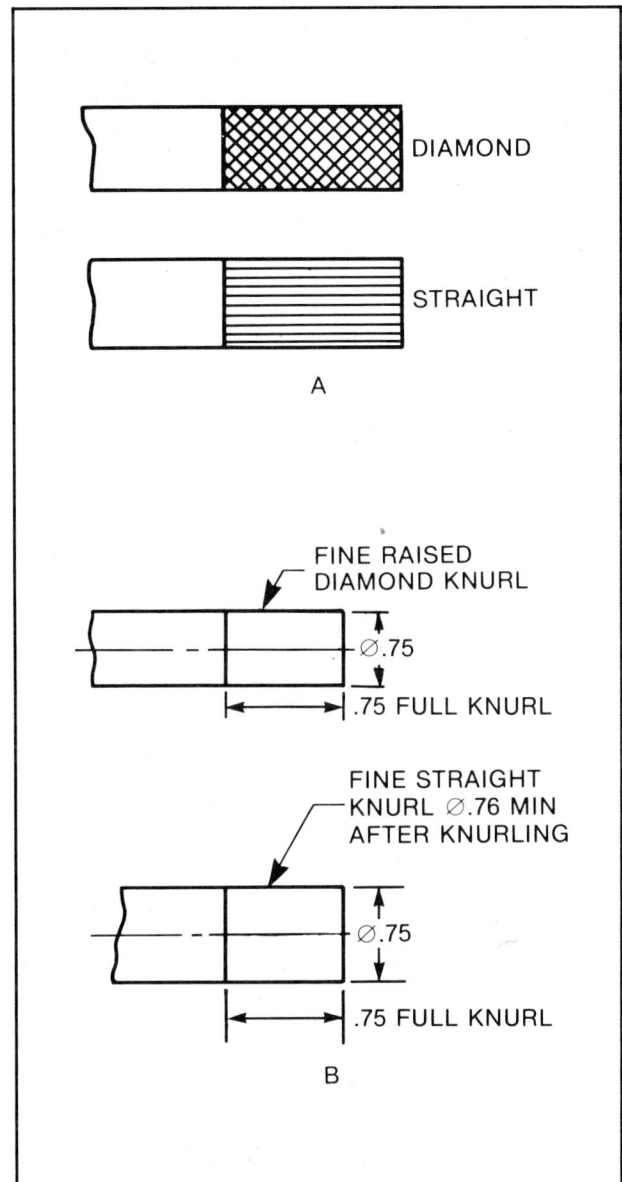

Figure 8-38. Diamond and straight knurls (A) are dimensioned as shown in (B).

The method used to dimension a knurl depends on the intended use. For example, a knurl used for a hand grip is specified by the length and pitch of the knurl. The *pitch* of a knurl refers to the number of points per inch of surface. While some prints may specify the pitch as a numerical value, many prints express the pitch in terms of fine, medium, and coarse. Knurling used for press fits is also specified by length and the pitch of the knurl, but here the finished diameter is also dimensioned, Figure 8-38B.

Dimensional and Process Notes

Dimensional and process notes are generally intended to show information that cannot be shown or dimensioned on the part. They are also intended to save space on a print. For example, if a part has 15 holes, all .50″ in diameter, a note could read *15× ∅.50*. This saves the drafter's time since each hole does not have to be dimensioned. It also prevents the drawing from being cluttered with unnecessary lines.

To save space, usually dimensional and process notes are made in abbreviations or symbols rather than in complete words. A list of these abbreviations and their meanings is included in Appendix B of this book. The abbreviations will become easy to understand with increased experience in reading prints.

Limited Length or Area Indication

Sometimes additional processing or treatment, such as grinding, polishing, or similar operations, is required on a specific limited area of a part. In such cases, the *limited length or area indication symbol* may be used. This symbol is a chain line resembling a thick center line, drawn around the area requiring the special treatment.

When the part feature is cylindrical, the line is only shown on one side, Figure 8-39A. When the feature requires dimensions, Figure 8-39B, the dimensions of the area are also shown. If the chain line clearly shows the location and extent of the area intended for special treatment, the dimensions are not usually shown, Figure 8-39C.

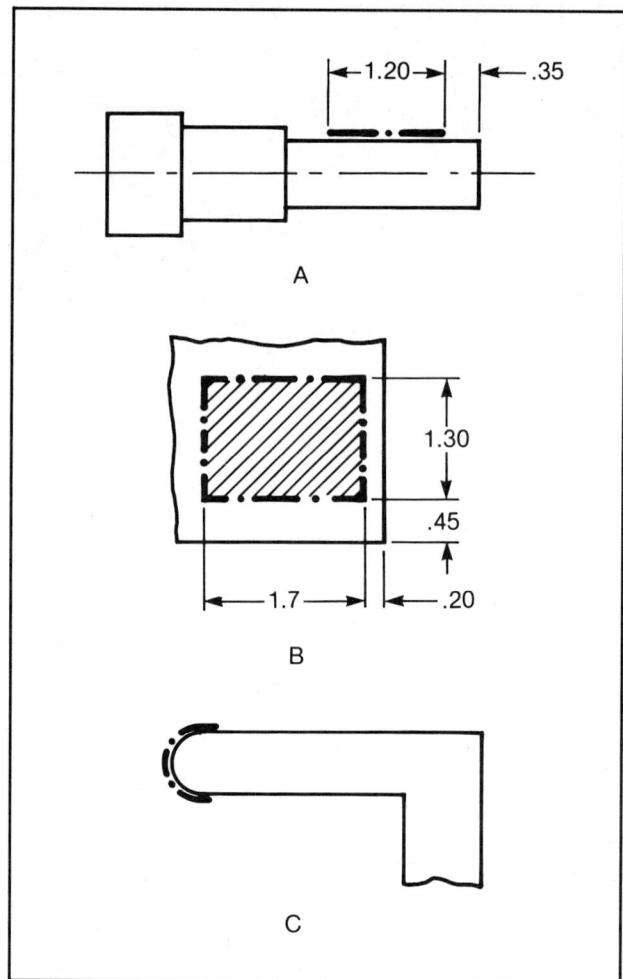

Figure 8-39. Dimensioning using chain lines to indicate limited lengths or areas

TOLERANCE AND ALLOWANCE

In any machining operation, the parts made must conform to the dimensions shown on the print. However, since a "perfect" part is impossible to make, an allowance, or margin for error, must be made when designing a part. The engineer must specify which dimensions are critical and which are noncritical. To indicate the critical (most important) dimensions, an engineer uses close tolerances to control the size of the critical feature. Those features that are not critical normally have a greater tolerance value, since their size is not as important to the overall function of the part. The machinist must know how to interpret these tolerance values to properly make the part shown in the print.

TOLERANCING TERMS

The terms explained in this section are often used in describing tolerancing values.

Nominal Size

The *nominal size* of a part is the size used for general identification. For example, a steel bolt with a nominal size of 1″ actually measures between .976″ and .998″ in diameter when measured across the threads. See Figure 8-40A. Lumber and pipe also have different nominal sizes and measured sizes, Figure 8-40B. The nominal size is used only for identification; it frequently does not reflect the true size of the object.

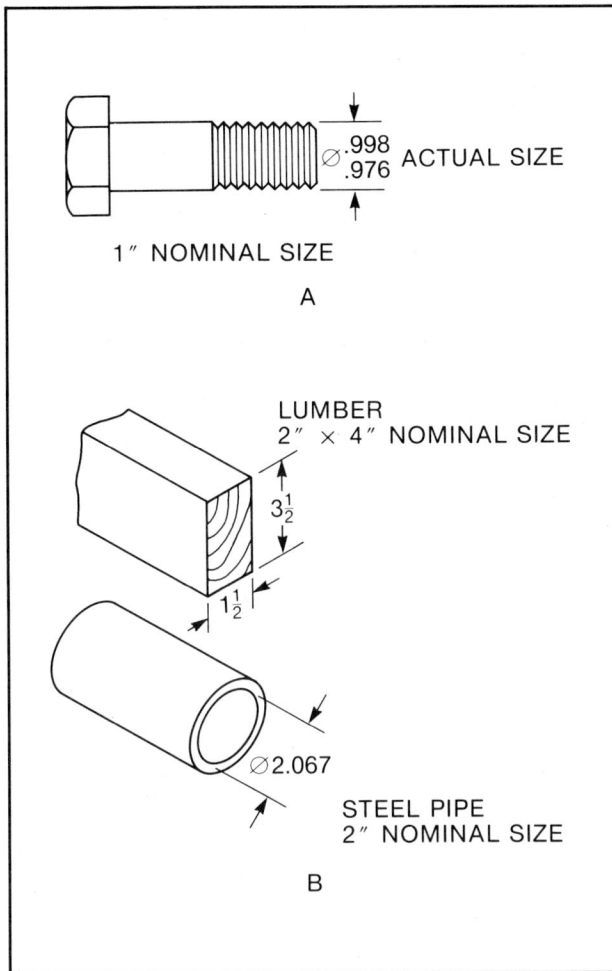

Figure 8-40. The nominal size of an object is used for identification and may not indicate its true size.

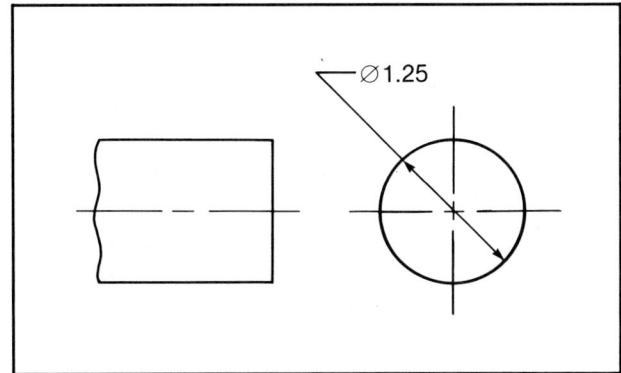

Figure 8-41. A basic size is the dimension to which the tolerance is applied.

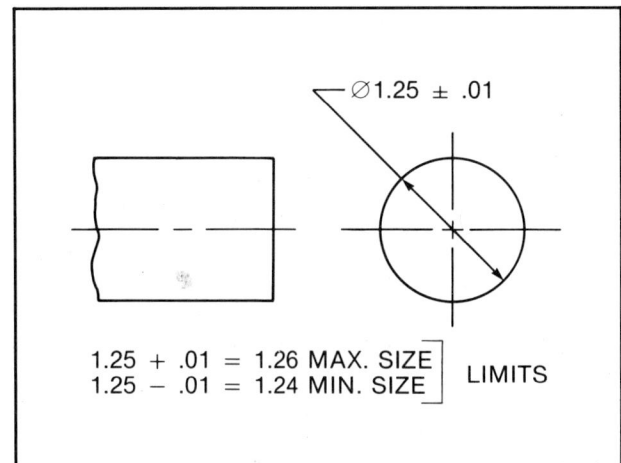

1.25 + .01 = 1.26 MAX. SIZE ⎤
1.25 − .01 = 1.24 MIN. SIZE ⎦ LIMITS

Figure 8-42. The limits of size are the maximum and minimum sizes allowed by a tolerance.

Actual Size

The *actual size* is the measured size of the part.

Basic Size

The *basic size* is the base dimension of the part. In practice, the basic size is the dimension to which the tolerance is applied. For example, the part shown in Figure 8-41 has a base size of 1.25″.

Limits of Size

The *limits of size* are the maximum and minimum sizes of the part allowed by the tolerance. See Figure 8-42.

Tolerance

Tolerance is the total allowable variation in the actual size of the part. Remember, every dimension on a print has a tolerance. The tolerance may be applied directly to the dimension or shown as a general note in the title block (refer to Figure 8-7).

Often the tolerance has a direct effect on how the part is made. Study the two parts shown in Figure 8-43. Part *A*, which has a tolerance on the hole size of plus or minus (±) .03″, could be drilled since the accuracy of a drilled hole is well within the tolerance allowed. Part *B*, however, has a tolerance of ±.005″. In order to maintain this tolerance, the hole must be drilled and reamed to ensure the proper size. The machinist's judgment in these situations is very important.

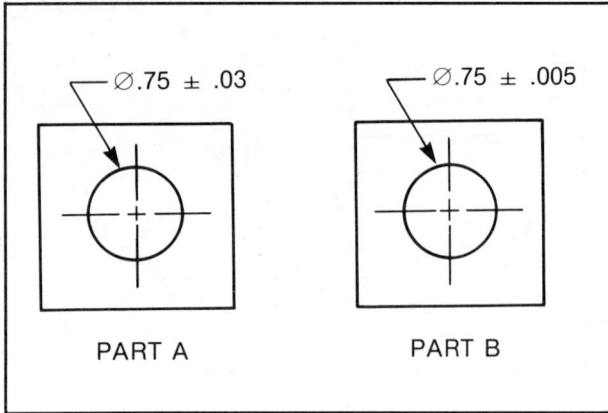

Figure 8-43. Tolerances often determine machining methods.

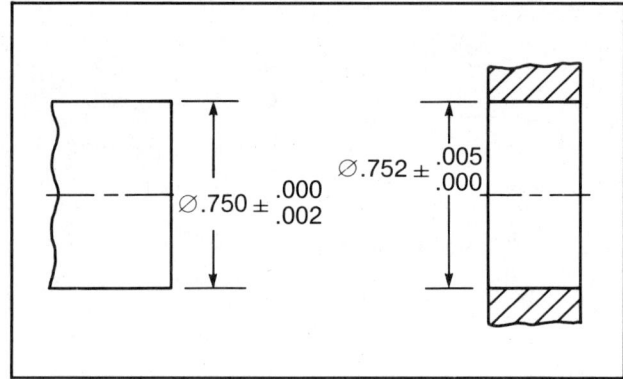

Figure 8-44. Unilateral (one-way) tolerances

Tolerance also has a direct effect on the cost of a part. In almost all cases, close tolerances result in added time to ensure accuracy. This means a higher cost per part. The machinist must pay close attention to the tolerance. Every part should be made within the specified tolerance, but there is no need to be overly accurate. Trying to make a part to within ±.002″ when the tolerance only calls for ±.030″ takes too much time and adds to the cost of the part.

The primary methods used to show tolerances on a print are plus and minus dimensioning and limit dimensioning. In *plus and minus dimensioning*, the basic size is applied to the part, and the amount of allowable variation is stated with plus and minus values that must be added to or subtracted from the basic size. In *limit dimensioning*, the upper and lower sizes of the feature are stated directly and no further calculations are required. The two principal forms of plus and minus dimensioning are unilateral and bilateral.

Unilateral tolerance is a tolerance that varies from the specified basic size in only one direction. As shown in Figure 8-44, the .750″ diameter shaft may be smaller (by .002″) but not larger than .750″. Similarly, the .752″ hole may be larger (by .005″) but not smaller than .752″. In these examples, the tolerance is only allowed to move in one direction. The shaft can be made to the basic size or up to .002″ smaller, while the hole can be machined to the basic size or up to .005″ larger. In this example these sizes must be tightly controlled to ensure that the parts will fit together.

Bilateral tolerance is a tolerance in which variation is allowed in both directions from the basic size. Bilateral tolerances can be either equal or unequal. As shown in Figure 8-45A, *equal bilateral tolerances* allow the .750″ basic dimension to be either larger or smaller by .005″. Therefore, any part made within .755″ (.750″ + .005″) and .745″ (.750″ − .005″) is correct. *Unequal bilateral tolerances* also permit movement in both directions, but, as shown in Figure 8-45B, the amounts of movement

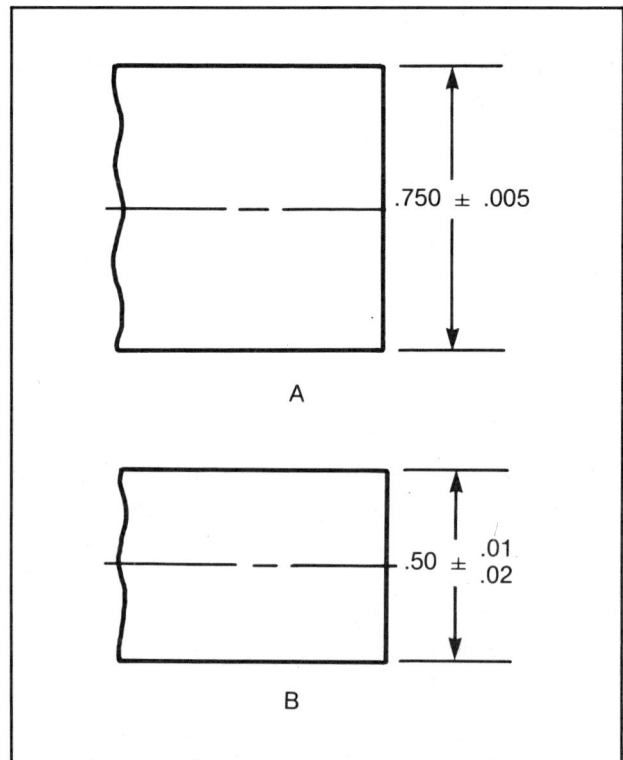

Figure 8-45. Bilateral (two-way) tolerances may be (A) equal or (B) unequal.

are not equal. Here the .50″ dimension is allowed to be smaller by .02″, but only larger by .01″. Therefore, any part made within .51″ (.50″ + .01″) and .48″ (.50″ − .02″) is correct.

Limit dimensioning specifies the maximum and minimum sizes of the feature directly. The dimensions are always positioned with the larger size over the smaller, or, in cases where the dimension is shown horizontally, the smaller size is shown to the left of the larger. When limit dimensioning is used, any part within the larger and smaller sizes is correct.

Figure 8-46 shows several methods that are commonly used to show limit tolerances and plus and minus tolerances.

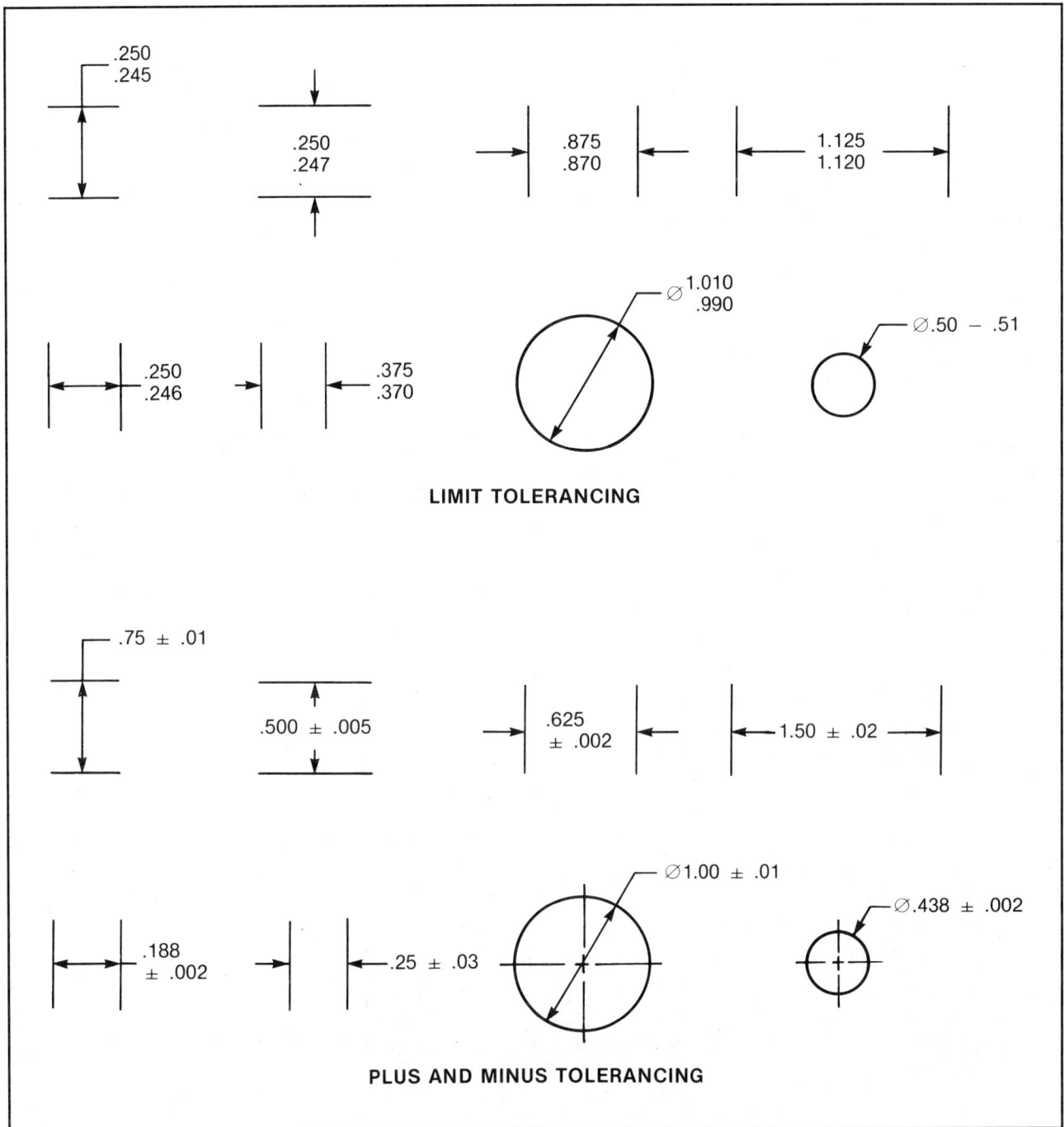

Figure 8-46. Methods of showing tolerances

Allowance

Allowance is the intentional difference in size between two mating parts. In practice, allowance is referred to as the difference in size between the largest shaft size and the smallest hole size. By varying the basic sizes and tolerances, a specific fit can be achieved. The *fit* is the general term used to describe the degree of tightness between mating parts. If a shaft is smaller than the hole, a positive, or clearance, fit results. However, if the shaft is larger than the hole, a negative, or interference, fit results. A *positive fit* is a running or sliding fit and a *negative fit* is a force or shrink fit. In addition to these two categroies of fits, there are *transition fits*. These are the range of fits between clearance and interference fits. For more information about fits and the specific tolerances and allowances required for each category, refer to a machining handbook.

The parts shown in Figure 8-47 illustrate how tolerance and allowance work together. The shaft size is $1.500'' \pm {.000 \atop .002}$ in diameter. The hole in the mating part is shown as $1.502'' \pm {.002 \atop .000}$ in diameter. The allowance can be calculated by comparing the sizes. The allowance is equal to the difference between the smallest hole size and the largest shaft size (Allowance = Smallest Hole Size − Largest Shaft Size). In this example, the smallest hole size is 1.502″ and the largest shaft size is 1.500″. Therefore, 1.502″ minus 1.500″ equals .002″ allowance. The difference between the largest hole size and the smallest shaft size is called the *maximum clearance*. To calculate the maximum clearance, simply subtract the smallest shaft size from the largest hole size (Maximum Clearance = Largest Hole Size − Smallest Shaft Size). In this example, the maximum clearance is 1.504″ minus 1.498″ equals .006″ maximum clearance.

In Figure 8-47, the fit is a positive, or clearance, fit, since the shaft is smaller than the hole. A running or sliding fit will result from this assembly. If, however, a force or shrink fit were required, the shaft would have to be made larger than the mating hole.

KEY TERMS

The following key terms were introduced in this chapter. Be sure you know the meaning of each term before proceeding to the review material.

Actual Size
Allowance
Angular Dimensions
Basic Size
Bilateral Tolerance
Bolt Circle
Boring
Chain Dimensioning
Chamfering
Coordinate Dimensioning
Counterboring
Countersinking
Datum
Datum Dimensioning
Dimensional and Process Notes
Dimensions
Direct Dimensioning
Fillet
Fit
Flats
Hole Pattern
Knurling
Limit Dimensioning
Limits
Linear Dimensions
Maximum Clearance
Necking
Nominal Size
Plus and Minus Tolerancing
Radial Dimensions
Radius
Radii
Reaming
Reference Dimensions
Round
Spotfacing
Tabular Dimensions
Tolerance
Undercutting
Unilateral Tolerance
Working Dimensions

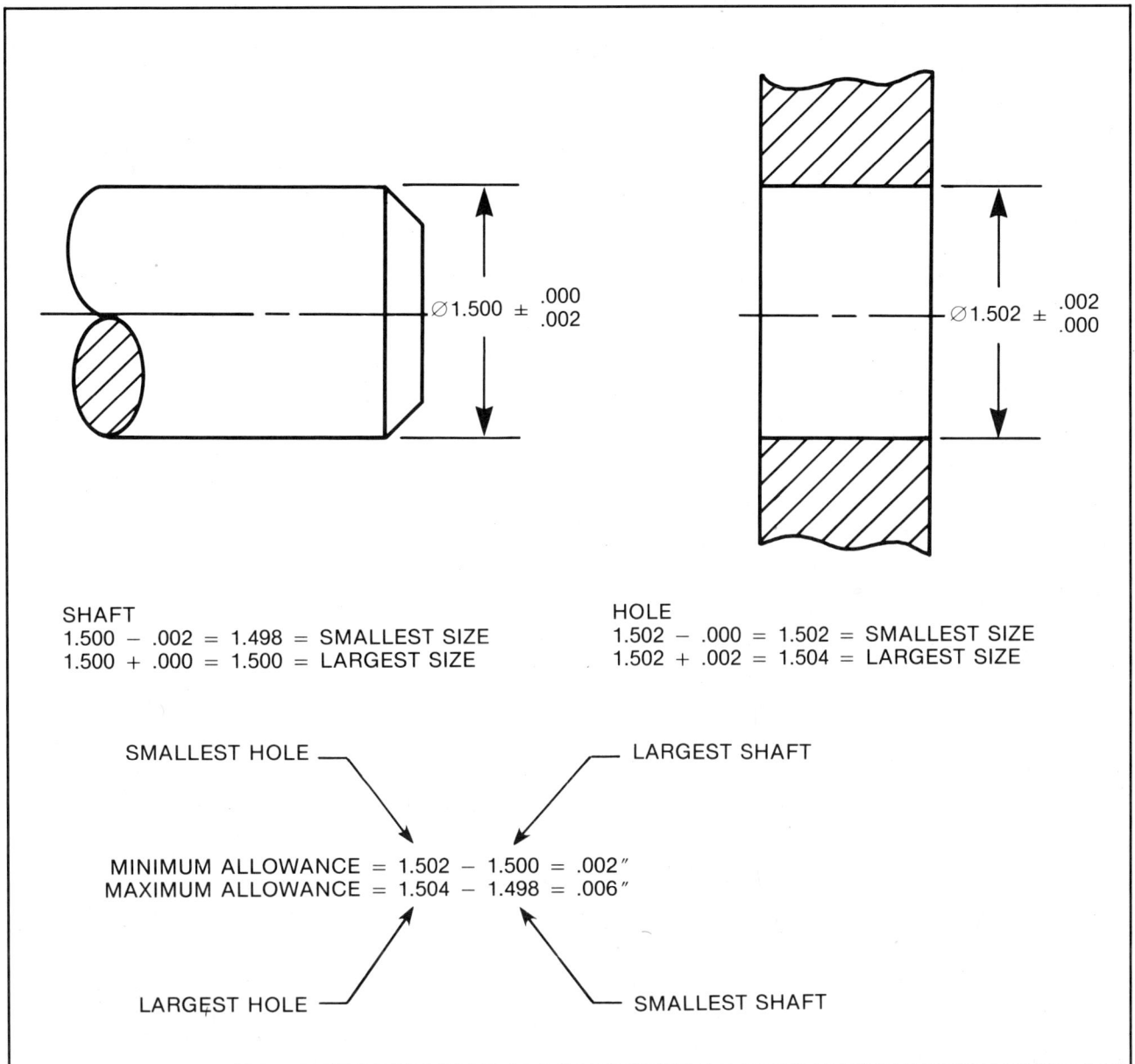

SHAFT
1.500 − .002 = 1.498 = SMALLEST SIZE
1.500 + .000 = 1.500 = LARGEST SIZE

HOLE
1.502 − .000 = 1.502 = SMALLEST SIZE
1.502 + .002 = 1.504 = LARGEST SIZE

SMALLEST HOLE ⎯⎯⎯⎯⎯⎯⎯⎯⎯ LARGEST SHAFT

MINIMUM ALLOWANCE = 1.502 − 1.500 = .002″
MAXIMUM ALLOWANCE = 1.504 − 1.498 = .006″

LARGEST HOLE ⎯⎯⎯⎯⎯⎯⎯⎯⎯ SMALLEST SHAFT

Figure 8-47. Calculating allowance and maximum clearance

Test your knowledge with this reinforcement study material. Write your answers to the questions in the spaces provided.

1. What type of dimension is used only for convenience? _____

   _____

2. How does the scale of a print affect the dimensions shown? _____

   _____

3. Of the following, which shows a proper dimension?

   a. 1,417.50 _____ c. 1.50″ _____

   b. 4.93750 _____ d. 3.75 _____

4. Which of the following dimensions is properly toleranced?

   a. 1.50 ± .010 _____ c. 1.5000 ± .05 _____

   b. 1.500 ± .005 _____ d. 1.50 ± .0005 _____

5. List the five basic dimensional forms.

   a. _____

   b. _____

   c. _____

   d. _____

   e. _____

6. Which dimensional form uses a chart to show the dimensions? _____

   _____

7. What is the term used to describe the surface from which dimensions are referenced?

   _____

8. List the three methods of placing dimensions on a print.

   a. _____

   b. _____

   c. _____

9. How are holes dimensioned on a print? _____

   _____

   _____

   _____

10. What are the two most common forms of hole patterns? _____

    _____

11. What is another term used to describe an internal grooving operation? _____

_____

12. What is meant by the term *necking*? _____

_____

_____

13. What is a nominal size? _____

_____

14. To what type of dimension is the tolerance applied? _____

_____

15. What are the maximum and minimum sizes allowed by the tolerance called? _____

_____

16. What is the measured size of a feature called? _____

_____

17. What is tolerance? _____

_____

_____

_____

18. What is allowance? _____

_____

_____

_____

19. What are the two primary methods used to show tolerances on a part?

    a. _____

    b. _____

20. What term is used to describe the difference between the largest hole size and the smallest shaft size? _____

21. Identify the following dimensioning symbols.

    a. ⊽ _____   f. ∨ _____

    b. ∅ _____   g. ( ) _____

    c. SR _____   h. ∅—.XX—⊣ _____

    d. ⊔ _____   i. — · — _____

    e. S∅ _____

Match the dimensional forms with the terms. Some terms are shown more than once; some are not shown at all.

a. Linear
b. Angular
c. Radial
d. Coordinate
e. Tabular
f. Spherical radius
g. True radius
h. Length of arc
i. Chordal
j. Polar coordinate

1.

2.

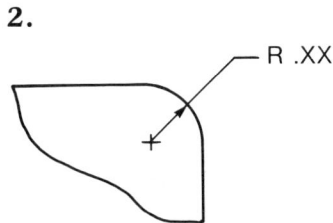

R .XX

3.

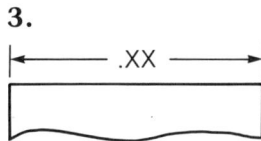

.XX

4.

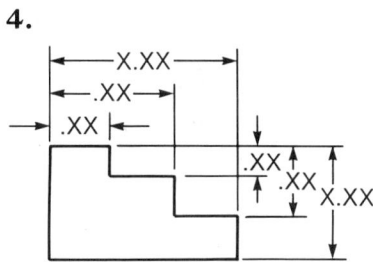

X.XX
.XX
.XX
.XX
.XX X.XX

5.

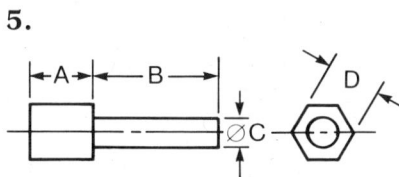

A B D
∅C

P/N	A	B	C	D
139-2	.50	.88	.25	.50
137-8	.50	1.00	.25	.50
18-4	.75	1.50	.38	.63

6.

X.XX

7.

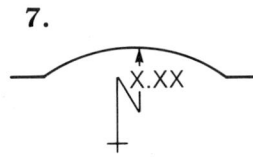

X.XX

8.

SR .XX

9.

XX°

10.

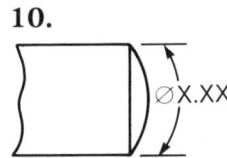

∅X.XX

11.

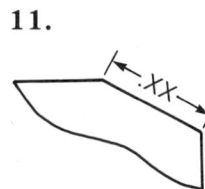

.XX

12.

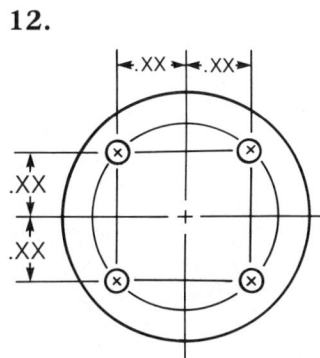

.XX .XX
.XX
.XX

Answers

1. _____
2. _____
3. _____
4. _____
5. _____
6. _____
7. _____
8. _____
9. _____
10. _____
11. _____
12. _____

Match the dimensional examples below and on page 147 with the operation or feature they describe.

a. Drill and chamfer
b. Linear hole pattern
c. Drill
d. Drill and ream
e. Drill and counterbore
f. Radius
g. Circular hole pattern
h. Drill and bore
i. Undercut
j. Drill and spotface
k. Drill and countersink
l. Multiple radii
m. Press fit knurl
n. Radius groove
o. Slotted hole
p. Groove for thread runout
q. Flat
r. Chamfer

1.

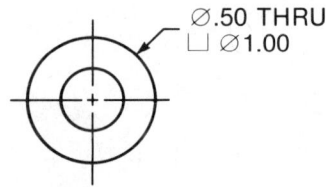

⌀.50 THRU
⌴ ⌀1.00

2.

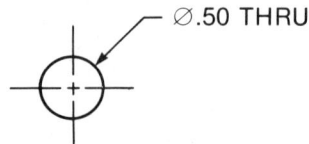

⌀.50 THRU

3.

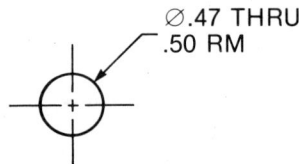

⌀.47 THRU
.50 RM

4.

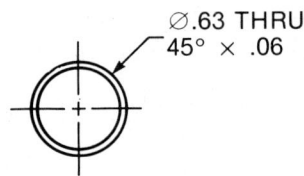

⌀.63 THRU
45° × .06

5.

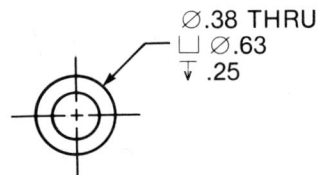

⌀.38 THRU
⌴ ⌀.63
▼ .25

6.

⌀1.13 THRU
1.25 BORE THRU

Answers

1. _____

2. _____

3. _____

4. _____

5. _____

6. _____

7. _____

8. _____

9. _____

10. _____

11. _____

12. _____

13. _____

14. _____

15. _____

16. _____

17. _____

18. _____

7.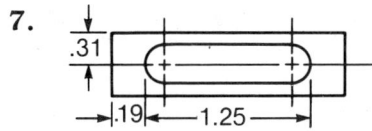

13.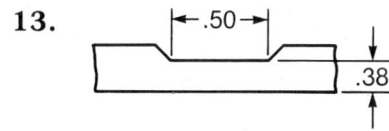

8.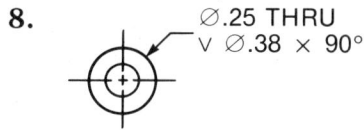

∅.25 THRU
∨ ∅.38 × 90°

14.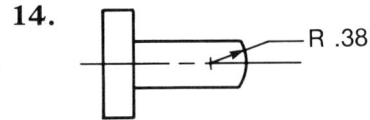

R .38

9.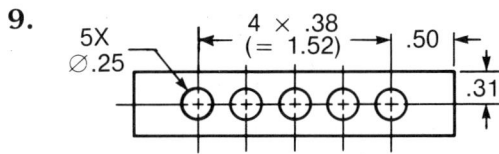

5X
∅.25

4 × .38
(= 1.52)

.50

.31

15.

4X R .19

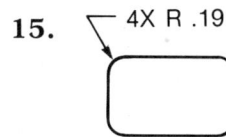

10.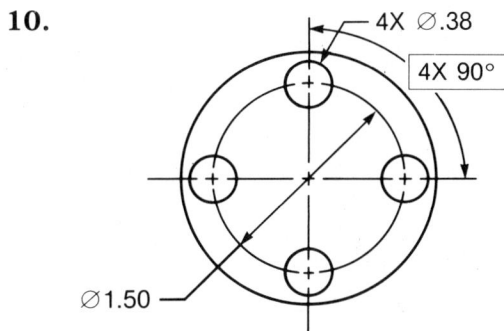

4X ∅.38

4X 90°

∅1.50

16.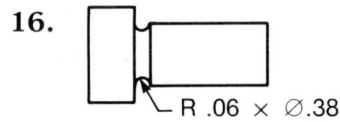

R .06 × ∅.38

11.

96 DP STR KNURL
∅.51 MIN AFTER
KNURLING

.50

.75 FULL KNURL

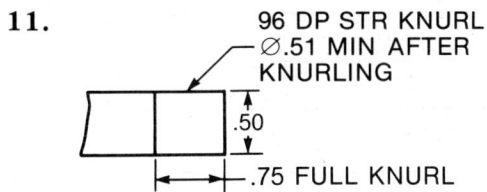

17.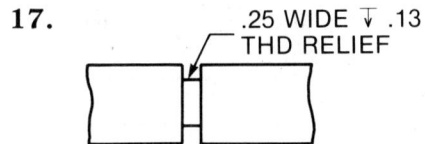

.25 WIDE ⊤ .13
THD RELIEF

18.

UNDERCUT
.13 WIDE
⊤ .13

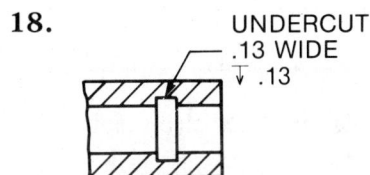

12.

.06 × .06

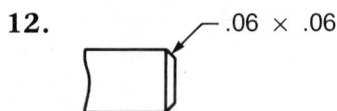

Refer to Figure E8-1 on page 149 to answer the following questions.

1. What is the name of this part? _____

2. What material is specified for the part? _____

3. What views are shown? _____

   _____

4. Identify the following lines.

   a. Line *E:* _____

   b. Line *I:* _____

   c. Line *J:* _____

   d. Line *K:* _____

5. What is the diameter at the bottom of groove *D?* _____

6. What does line *M* show? _____

   _____

7. What type of dimensions are shown at the following?

   a. *B:* _____   e. *G:* _____

   b. *C:* _____   f. *H:* _____

   c. *E:* _____   g. *L:* _____

   d. *F:* _____

8. What methods of placing dimensions are shown at the following?

   a. *A:* _____

   b. *L:* _____

9. What are the tolerances of the following dimensions?

   a. *C:* _____

   b. *H:* _____

   c. *N:* _____

10. What form of tolerancing is used on this print? _____

    _____

11. What size of chamfer is specified on the ends of the part? _____

    _____

12. What diameter is the bolt circle for the five holes? _____

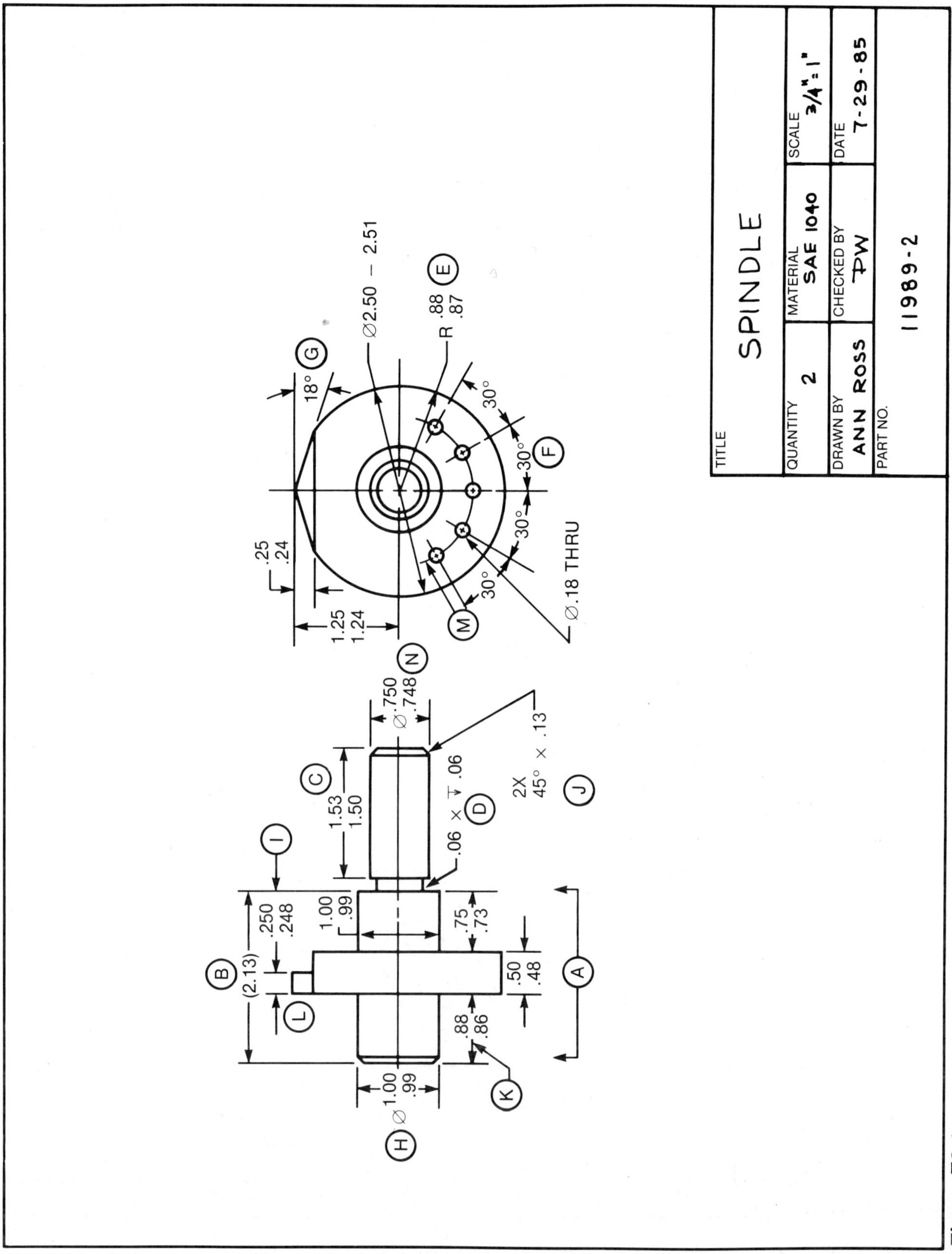
TITLE
SPINDLE
QUANTITY
2
MATERIAL
SAE 1040
SCALE
3/4" = 1"
DRAWN BY
ANN ROSS
CHECKED BY
PW
DATE
7-29-85
PART NO.
11989-2
Ø2.50 – 2.51
R .88 .87
18°
.25 .24
1.25 1.24
30°
Ø.18 THRU
Ø.750 .748
2X 45° × .13
.06 × ⊽ .06
1.53 1.50
.250 .248
(2.13)
1.00 .99
.75 .73
.50 .48
.88 .86
Ø 1.00 .99

Figure E8-1.

Refer to Figure E8-2 on page 151 to answer the following questions.

1. What is the part number? _____

2. What is the scale of this print? _____

3. What method of dimensional placement is used at *N*? _____

   _____

4. Identify the following lines.

   a. *E*: _____

   b. *F*: _____

   c. *H*: _____

5. Find the following dimensions. Calculate the basic size only.

   a. *G* = _____

   b. *I* = _____

   c. *K* = _____

   d. *M* = _____

6. What are the tolerances of the following dimensions?

   a. *C*: _____

   b. *J*: _____

   c. *L*: _____

7. What types of tolerance are shown at the following?

   a. *A*: _____

   b. *B*: _____

   c. *D*: _____

8. What form of tolerancing is used on this print? _____

   _____

9. What are the limits of size on the following dimensions?

   a. *D*: _____

   b. *J*: _____

   c. *B*: _____

10. What are the overal sizes of this part? Consider the basic sizes only.

    a. Height = _____

    b. Width = _____

    c. Depth = _____

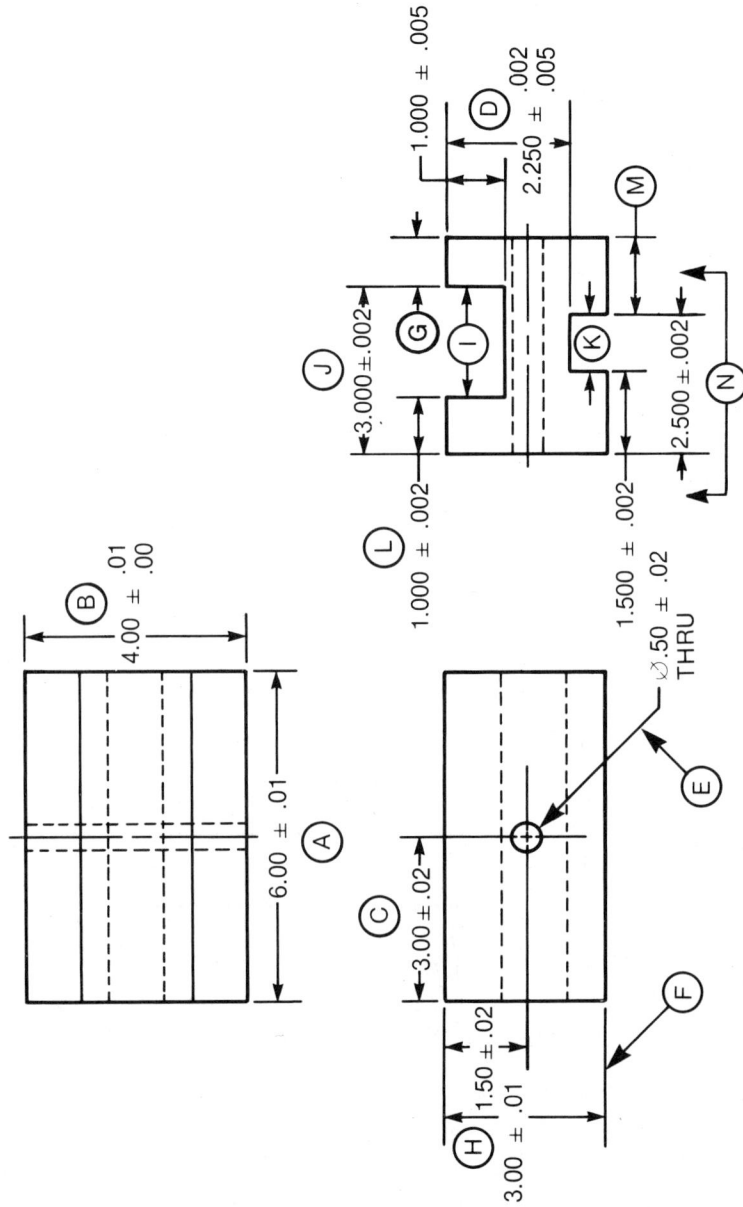

Figure E8-2.

Refer to Figure E8-3 below to answer the following questions.

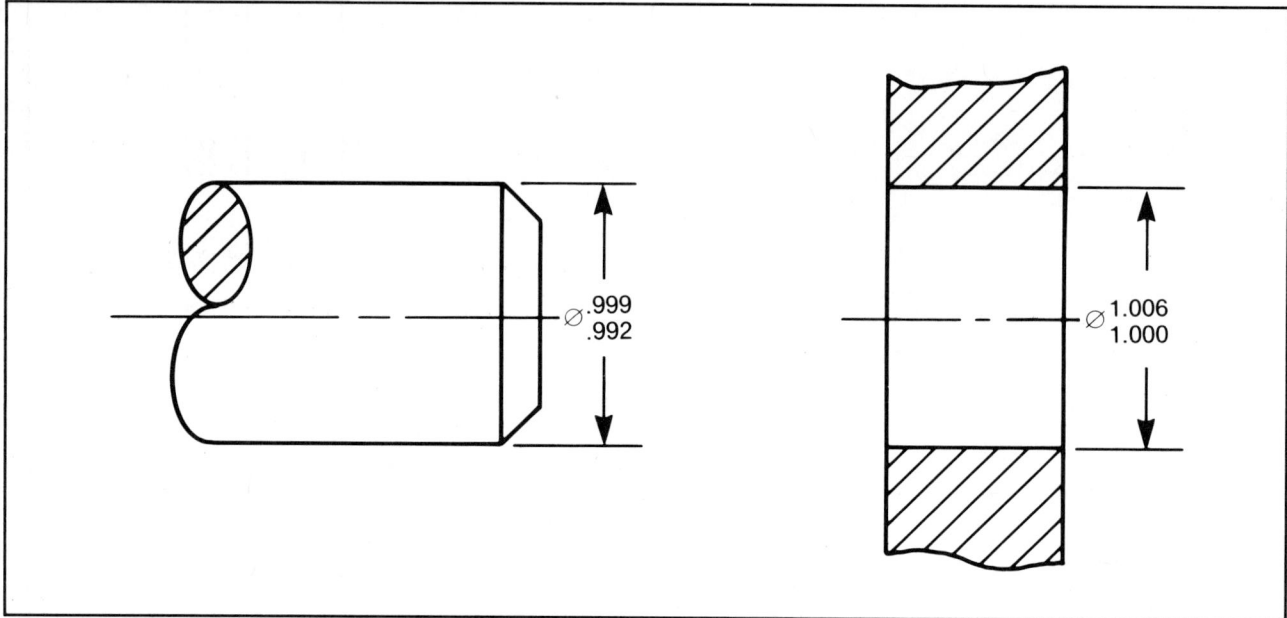

Figure E8-3.

1. What type of tolerance is shown on the shaft? _____

2. What type of tolerance is shown on the hole? _____

3. What are the limits of size of the shaft? _____

4. What are the limits of size of the hole? _____

5. What is the tolerance of the shaft size? _____

6. What is the tolerance of the hole size? _____

7. What is the allowance between the shaft and hole? _____

8. What is the maximum clearance between the shaft and hole? _____

9. What type of fit will result when these two are assembled? _____

10. What type of fit would result if the basic size of the shaft were shown as 1.007? _____

9

DETAIL AND ASSEMBLY PRINTS

OBJECTIVES

After studying this chapter, you will be able to:

- Identify detail and assembly prints.

- Define the characteristics of detail and assembly prints.

- Identify and describe entries made in a materials list.

- Identify and describe entries made in a revisions list.

Prints used in the shop are called *working prints* or *shop prints*. Working prints are copies of multiview drawings. They are drawn with enough detail to allow the object to be machined or assembled as intended. These prints are divided into two general categories: detail prints and assembly prints.

DETAIL PRINTS

A *detail print* is a multiview print of a single part. It shows the general size and shape of the part, as well as the complete dimensions and specifications needed to make the part. Usually only one part is drawn on a detail print. However, sometimes several related parts are shown on one sheet. In both cases, the detail print shows all the information needed to make the part.

The three-view print is the most common type of detail print. Depending on the part, however, two-view or one-view prints may be used instead. The number of views used in a print depends on the complexity of the part. As a general rule, drafters draw only the minimum number of views needed to completely describe the part.

Three-View Detail Prints

Three-view detail prints, Figure 9-1, show three different views of the part. The views selected are usually the front, top, and right-side views. In most cases a three-view print shows enough detail to completely describe the part. Only rarely will more than three views be necessary in a detail print.

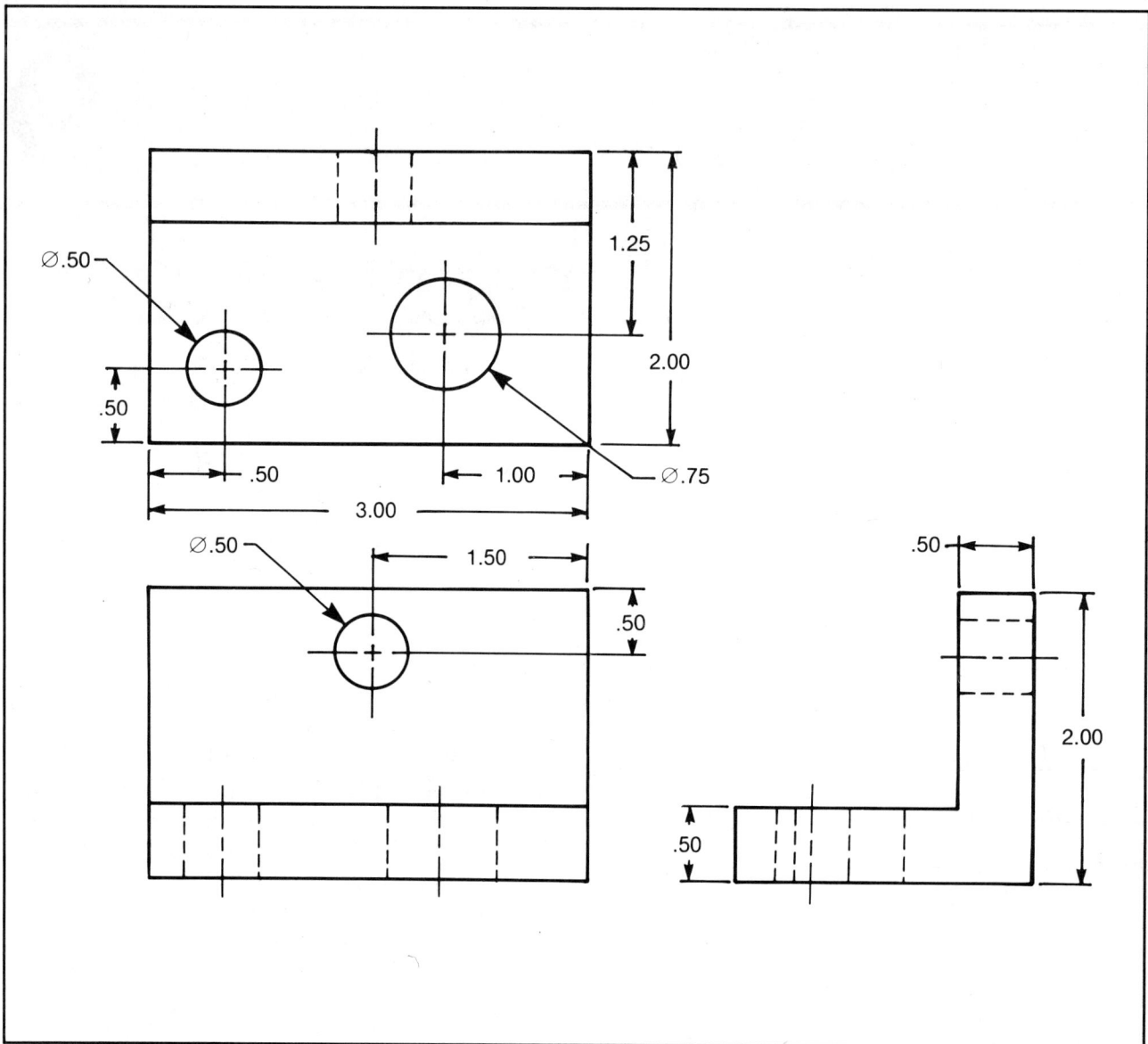

Figure 9-1. Three-view detail print

Two-View Detail Prints

Two-view detail prints, Figure 9-2, are generally used in cases where a third view would simply repeat one of the other views. See Figure 9-3 for another example. For the most part, two-view prints are used for parts with symmetrical features, Figure 9-4. The views generally shown in two-view detail prints are the front and top views, or the front view and one of the side views.

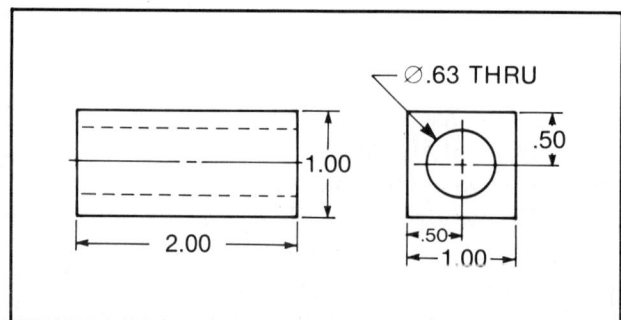

Figure 9-2. Two-view detail print

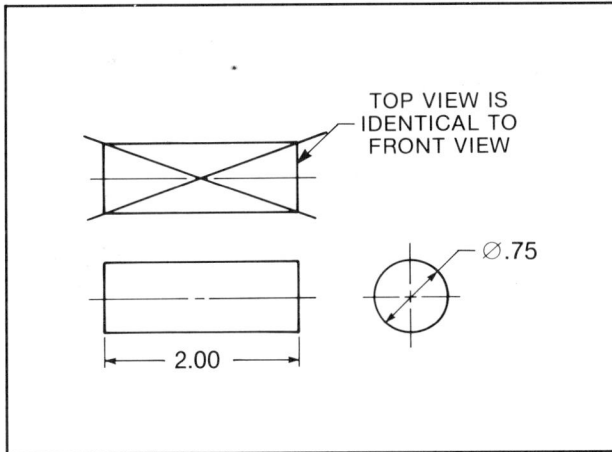

Figure 9-3. Two-view prints can eliminate repetitive views.

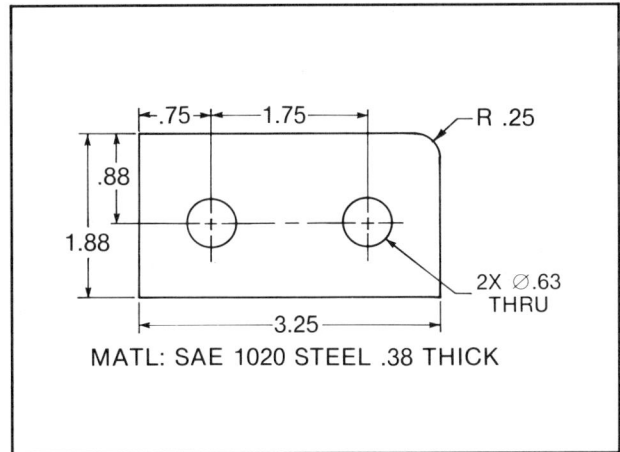

Figure 9-5. One-view detail print

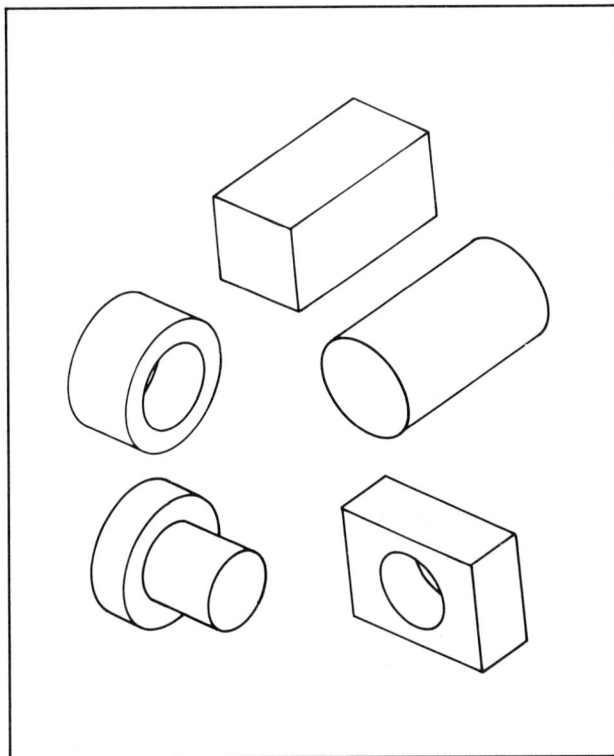

Figure 9-4. Parts with symmetrical features are often shown by two-view detail prints.

One-View Detail Prints

All multiview prints describe a part in at least two views. However, if all but one of these views are replaced with a note, the print becomes a *one-view detail print*. See Figure 9-5. One-view detail prints are used when a second

or third view would not add any important information. As shown in Figure 9-6, the top and right side views could be replaced with a note specifying the part thickness. The only information not included when these views are eliminated is the hidden lines showing the holes.

When the abbreviation *DIA* or the symbol ⌀ is used in a print, Figure 9-7, a diameter is indicated. Showing a diameter this way sometimes eliminates the need for a second view. However, in most cases where diameters must be indicated, the true profile will be shown. The abbreviation *DIA* or the symbol ⌀ will only be used when the true shape of the diameter cannot be determined by the dimensioned view.

ASSEMBLY PRINTS

Assembly prints show the position and relationship of parts in an assembled unit. These prints can be divided into two general types: unit assembly prints and detail assembly prints.

Unit Assembly Prints

Unit assembly prints are used to show the relationship of two or more assembled parts. As shown in Figure 9-8, unit assembly prints show the proper location and position of each part of an assembled unit.

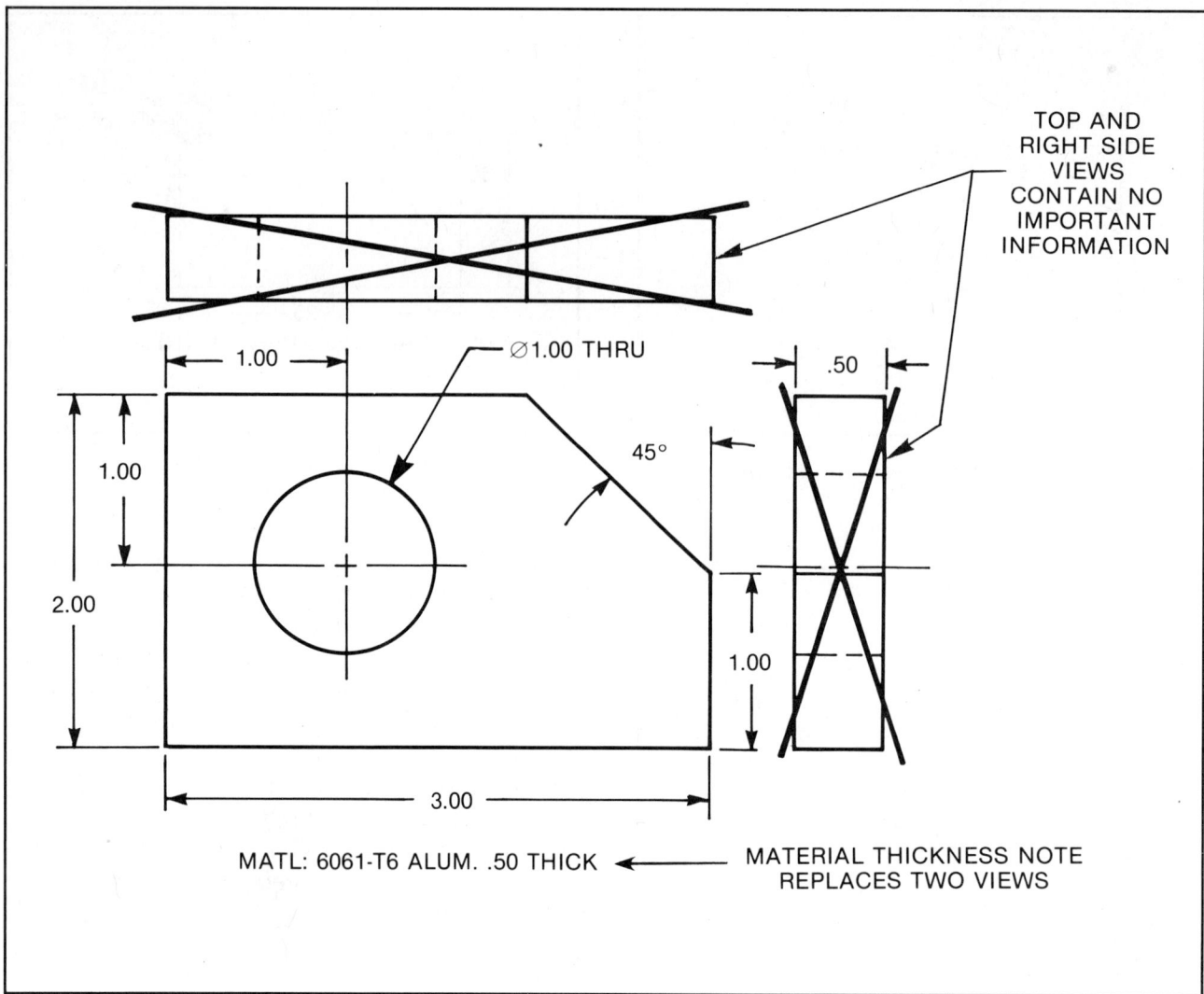

Figure 9-6. This one-view print uses a thickness note to replace two views.

TOP AND RIGHT SIDE VIEWS CONTAIN NO IMPORTANT INFORMATION

1.00

Ø1.00 THRU

45°

1.00

2.00

1.00

.50

3.00

MATL: 6061-T6 ALUM. .50 THICK

MATERIAL THICKNESS NOTE REPLACES TWO VIEWS

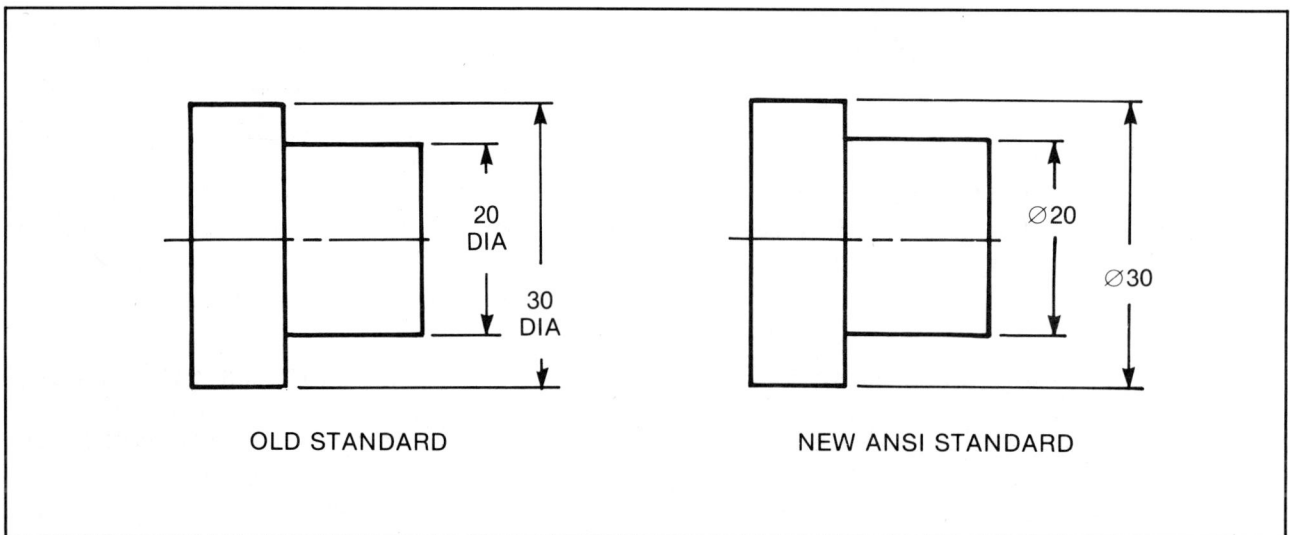

Figure 9-7. Indicating diameters with an abbreviation or a symbol

20 DIA

30 DIA

Ø20

Ø30

OLD STANDARD

NEW ANSI STANDARD

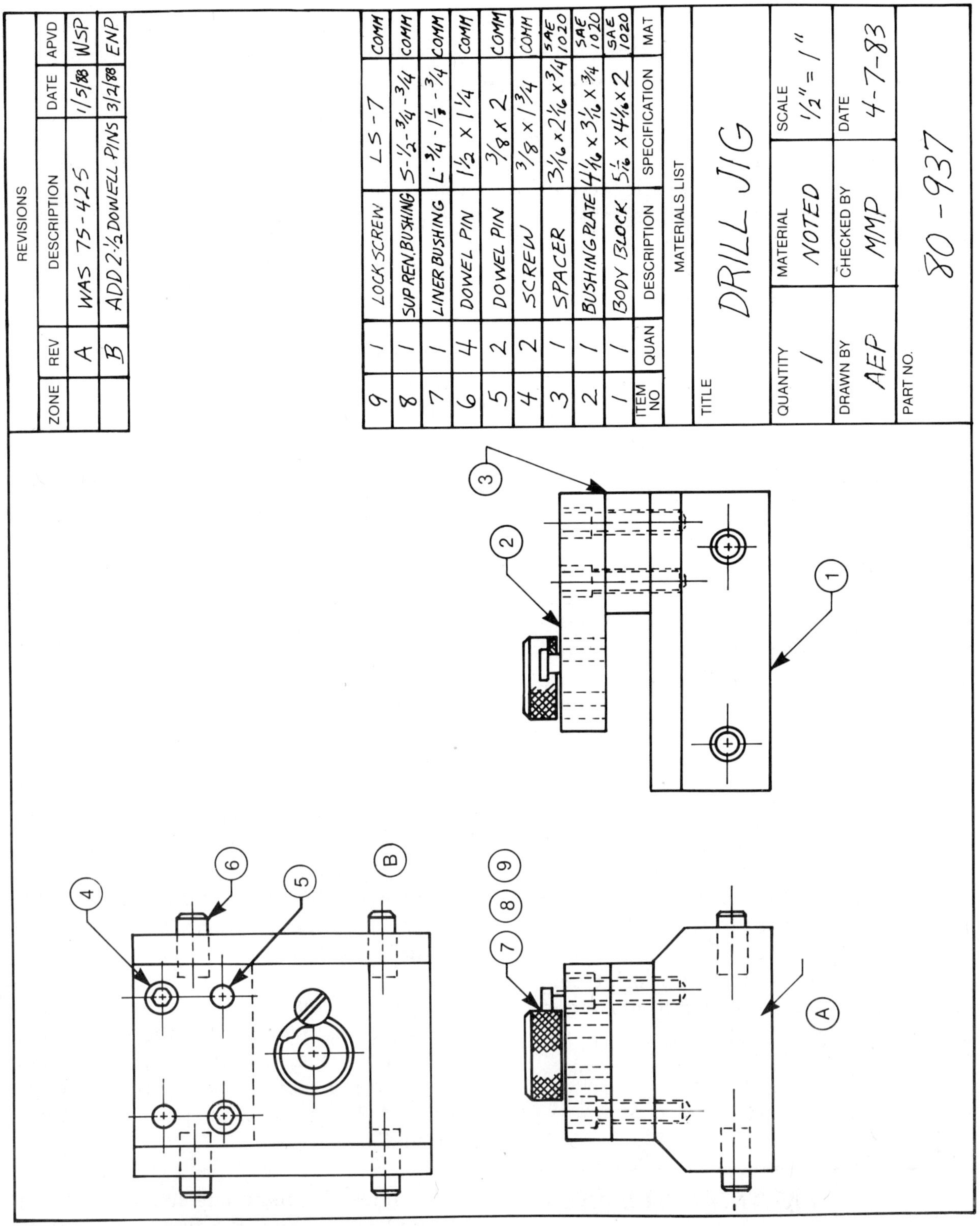
REVISIONS
ZONE | REV | DESCRIPTION | DATE | APVD
A | WAS 75-425 | 1/5/88 | WSP
B | ADD 2½ DOWELL PINS | 3/2/88 | ENP
9 | 1 | LOCK SCREW | LS-7 | COMM
8 | 1 | SUP REN.BUSHING | 5-½-¾-¾ | COMM
7 | 1 | LINER BUSHING | L-¾-1⅛-¾ | COMM
6 | 4 | DOWEL PIN | ½ X 1¼ | COMM
5 | 2 | DOWEL PIN | ⅜ X 2 | COMM
4 | 2 | SCREW | ⅜ X 1¾ | COMM
3 | 1 | SPACER | 3 1/16 X 2 1/16 X ¾ | SAE 1020
2 | 1 | BUSHING PLATE | 4 1/16 X 3 1/16 X ¾ | SAE 1020
1 | 1 | BODY BLOCK | 5 1/16 X 4 1/16 X 2 | SAE 1020
ITEM NO | QUAN | DESCRIPTION | SPECIFICATION | MAT
MATERIALS LIST
TITLE DRILL JIG
QUANTITY 1
MATERIAL NOTED
SCALE ½" = 1"
DRAWN BY AEP
CHECKED BY MMP
DATE 4-7-83
PART NO. 80-937
1
2
3
4
5
6
7
8
9
A
B

Figure 9-8. Unit assembly print

The main purposes of this type of assembly print are to identify the parts of an assembly and to show how each part is related to the entire unit. For these reasons, unit assembly prints, when dimensioned, generally show just the sizes of the complete unit rather than the individual sizes of the parts.

Unit assembly prints can describe either complete assemblies or smaller subassemblies. In both cases, each part in the unit assembly print is identified by a reference number. This reference number matches a number shown in the materials list, and identifies each part by name or number. (The materials list is discussed later in this chapter.)

Detail Assembly Prints

Detail assembly prints are used to show both the positions and sizes of parts in an assembly. For simple assemblies with very few parts, the print often shows the assembled unit with each part dimensioned, as shown in Figure 9-9. More complicated assemblies may require separate detail prints along with the assembly print. These prints are normally combined into sets, with the print of the assembled unit followed by the individual detail prints needed to make each part. Often both the assembled unit and the required detail views are included in a single print. See Figure 9-10. Here the unit is shown assembled, and each part of the assembly is drawn and dimensioned completely.

In any case, a detail assembly print contains all the information needed to make and assemble the complete unit.

MATERIALS LIST

The *materials list* is used primarily with assembly prints. It lists the part names and part numbers and/or the rough stock sizes for each part shown in the print of the assembled unit. When materials lists are used, the material

specifications are not shown in the title block. If the title block has a block for the material specification, the word *NOTED* is normally shown.

The materials list is generally located in the lower right corner of the print, just above the title block. For large assemblies, the materials list may be shown on more than one sheet. In either case, the materials list, Figure 9-11, will usually contain the following information:

1. **Item Number.** This block is used to identify each specific part in the assembly. The letter or number shown in this block matches the reference letter or number assigned to each part.

2. **Quantity.** The quantity block is used to indicate the number of parts needed for each assembled unit.

3. **Description.** This block contains the name of the part. When commercially made parts are specified, the commercial name is used. For rough stock, the name of the part being made is used.

4. **Specification.** The specification block is used to describe the part named in the description block. When a commercial part is specified, a part number will normally be shown in this block. If common hardware is specified, the size is shown. When the part must be specially made, the sizes of the required rough stock are listed in this block.

5. **Material.** This block contains the name of the material specified for the part. When a finished part such as a bolt or screw is to be obtained from a commercial source, the entry will read *COML* to show the part is a commercial item.

Sometimes the materials list will contain more information, but the five items just described are generally the minimum information provided in the materials list. Each company uses a materials list that best suits its needs. Title blocks and materials lists will vary among different companies.

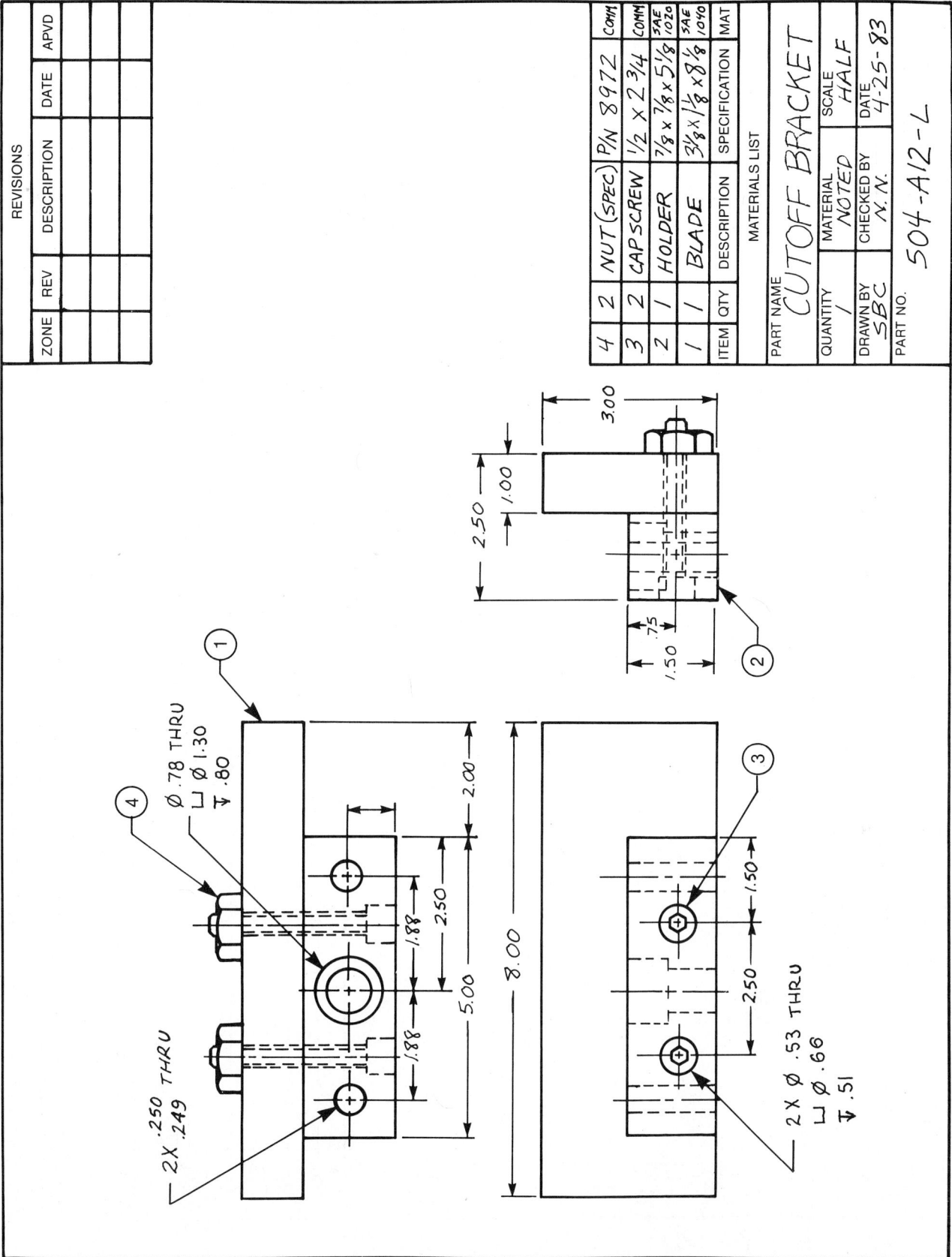

Figure 9-9. Detail assembly prints may be completely dimensioned.

Materials List

ITEM	QTY	DESCRIPTION	SPECIFICATION	MAT
4	2	NUT (SPEC)	P/N 8972	COMM
3	2	CAP SCREW	1/2 × 2 3/4	COMM
2	1	HOLDER	7/8 × 7/8 × 5 1/8	SAE 1020
1	1	BLADE	3/8 × 1 1/8 × 8 1/8	SAE 1040

PART NAME: *CUTOFF BRACKET*

QUANTITY	MATERIAL	SCALE
1	NOTED	HALF

DRAWN BY	CHECKED BY	DATE
SBC	N.N.	4-25-83

PART NO. *504-A12-L*

Ø .78 THRU
⌴ Ø 1.30
⩒ .80

2X .250 THRU
.249

2X Ø .53 THRU
⌴ Ø .66
⩒ .51

3.00
2.50
1.00
.75
1.50

2.00
2.50
1.88
1.88
5.00
8.00
1.50
2.50

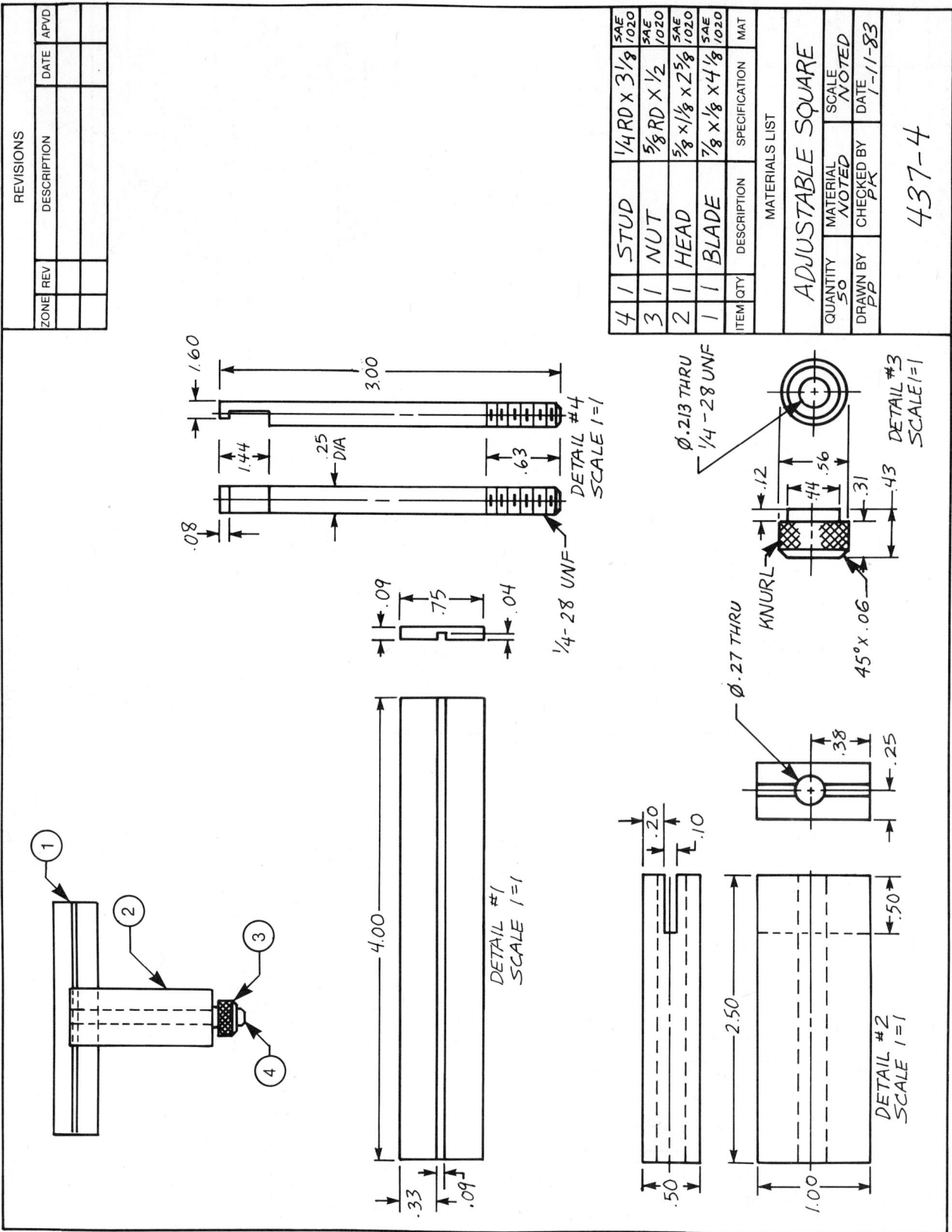

Figure 9-10. Detail assembly print showing both the assembled view and the detail views

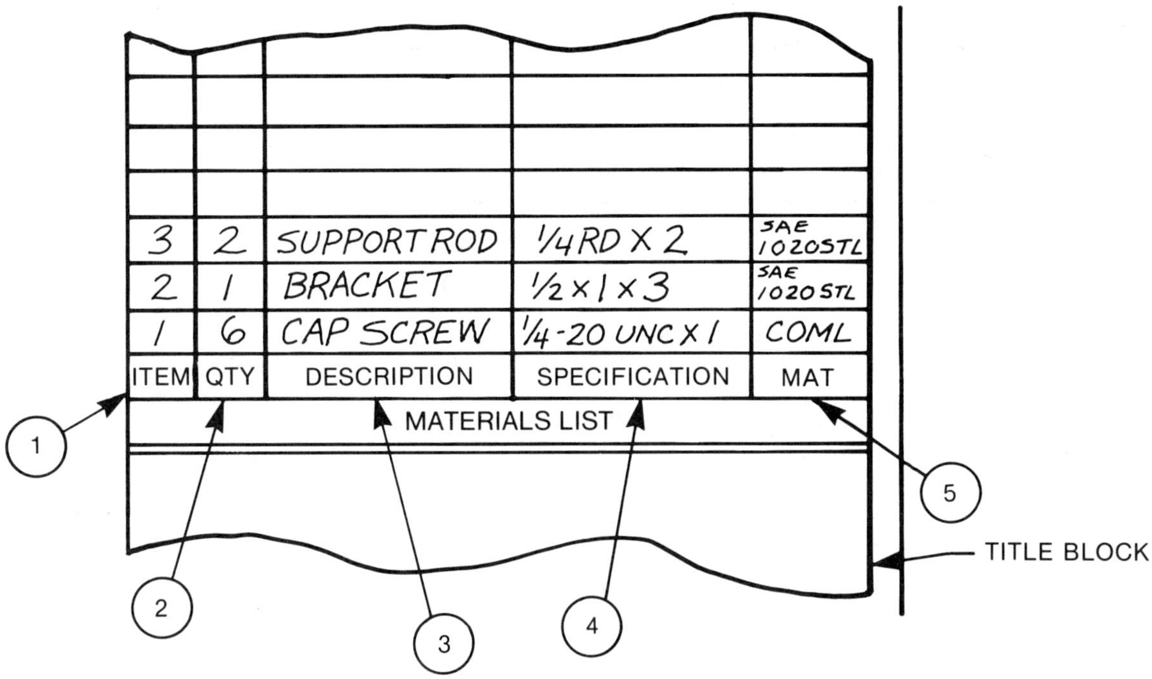

Figure 9-11. Materials list

REVISIONS LIST

The *revisions list*, or *change list*, is used to record all changes made to the engineering drawing since it was originally drawn. The changes are noted in the body of the drawing and listed in the revisions list, as shown in Figure 9-12. The revisions list is usually located either in the upper right corner of the print or near the title block. The revisions list will usually contain the following information:

1. Zone. The zone reference block is used to locate the changes on large drawing sheets. These sheets have letters running horizontally across the top and bottom of the sheet, and numbers running vertically along each side. The zone references use a combination of letters and numbers to accurately locate the specific

area of the print where a change was made. To use zone references, simply match the numbers and letters noted in this block with the corresponding values on the print borders, much the same as with a road map.

2. Revision. This block is used to identify the exact change. The letter used here matches the reference letter used to note the change in the body of the print.

3. Description. The description block contains a brief description of what was changed. In cases where a size is changed, the old size will be noted here. If the change adds something new, then this addition will be shown in this block.

4. Date. This block shows the date the change was made and the date the change became effective.

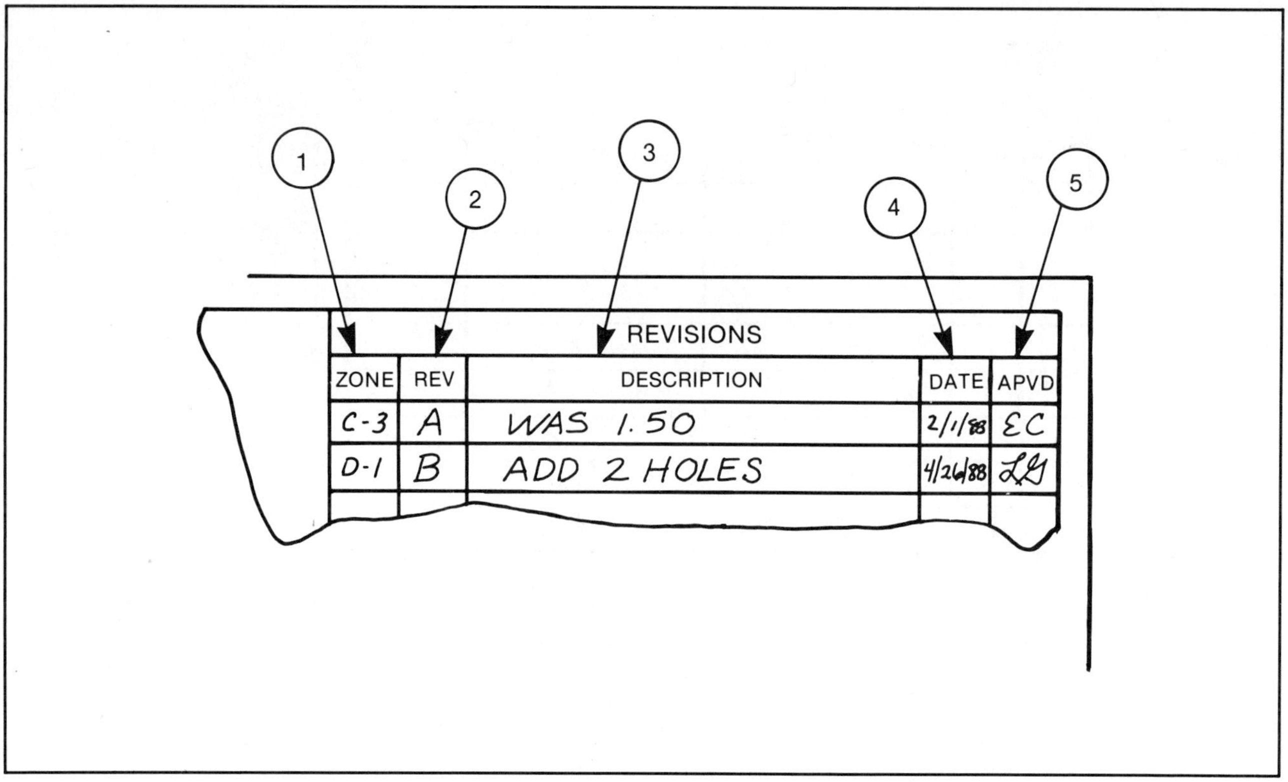

Figure 9-12. Revisions list

5. Approved. This block contains the name or initials of the checker, supervisor, or other authorized person who approved the change to the drawing.

Revisions lists, like title blocks and materials lists, vary from one company to another, but for the most part they will follow the general pattern just described.

Note: Always verify that the print being used is the most current available. Most companies destroy all outdated prints, but sometimes an old print will be overlooked. Therefore, always check to make sure that you have a current print.

KEY TERMS

The following key terms were introduced in this chapter. Be sure you know the meaning of each term before proceeding to the review material.

Assembly Print
Detail Assembly Print
Detail Print
Materials List
Revisions List
Unit Assembly Print

Refer to Figure E9-1 on page 164 to answer the following questions.

1. What type of print is this? _____

    _____

2. What is the part name? _____

3. What is the scale? _____

4. How many units are required? _____

5. Name the following parts.

    a. Part #1: _____

    b. Part #2: _____

    c. Part #3: _____

6. What material is specified for this unit? _____

    _____

7. What does the (A) mean? _____

    _____

    _____

8. What is the main purpose of this type of print? _____

    _____

    _____

9. What does the word *NOTED* in the title block mean? _____

    _____

    _____

10. If this print were completely dimensioned, what type of print would it be? _____

    _____

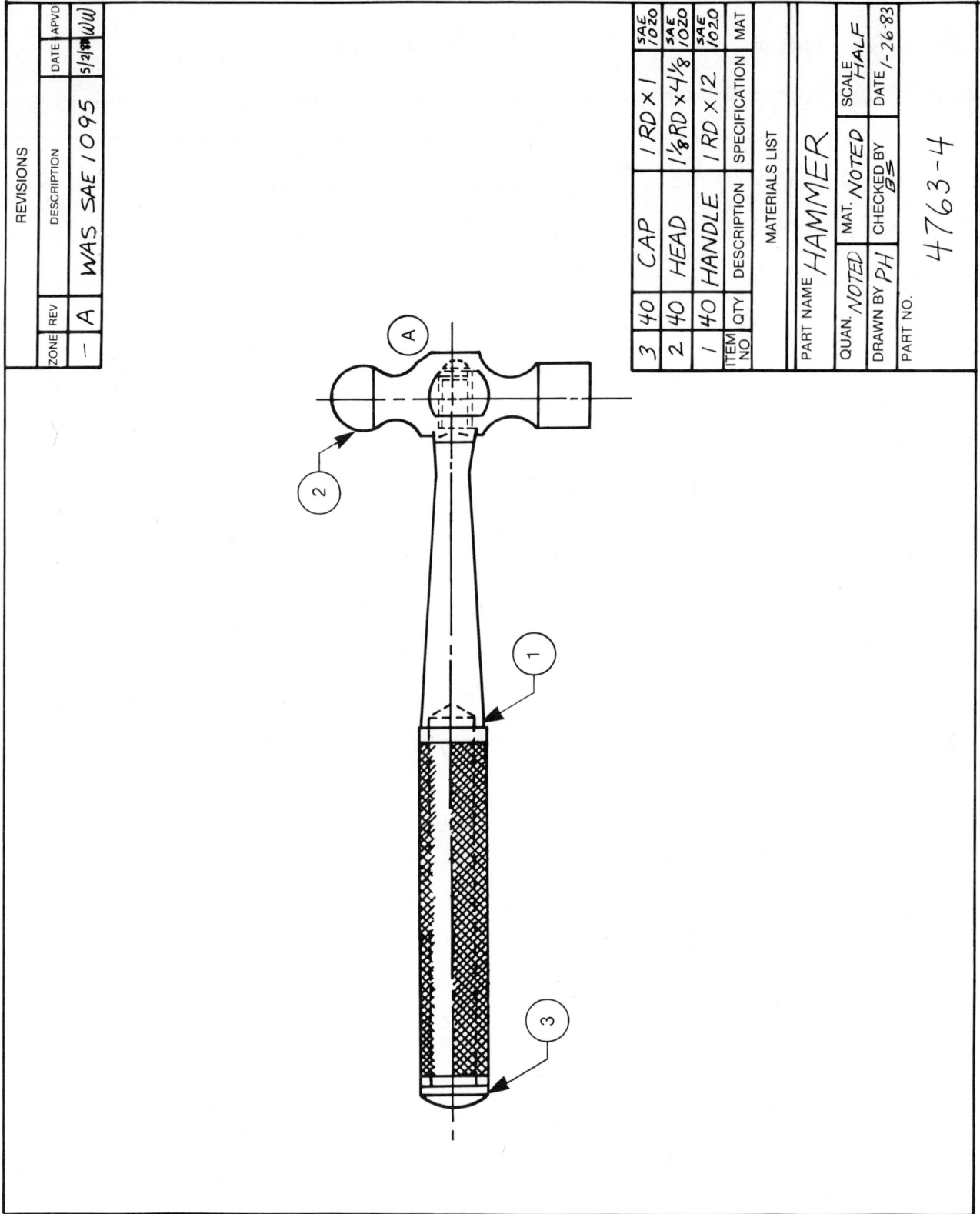

REVISIONS

ZONE	REV	DESCRIPTION	DATE	APVD
—	A	WAS SAE 1095	5/2/58	WW

MATERIALS LIST

ITEM NO	QTY	DESCRIPTION	SPECIFICATION	MAT
3	40	CAP	1 RD × 1	SAE 1020
2	40	HEAD	1⅛ RD × 4⅛	SAE 1020
1	40	HANDLE	1 RD × 12	SAE 1020

PART NAME *HAMMER*

QUAN. *NOTED*	MAT. *NOTED*	SCALE *HALF*
DRAWN BY *PH*	CHECKED BY *BS*	DATE *1-26-83*
PART NO.		

4763-4

Figure E9-1.

Refer to Figure E9-2 on page 166 to answer the following questions.

1. What type of print is this? _____

    _____

2. What is the part name? _____

3. What material is specified for the following items?

    a. Item #1: _____

    b. Item #2: _____

    c. *Item #3:* _____

4. Name the following parts.

    a. Part #1: _____

    b. Part #2: _____

    c. Part #3: _____

5. What change was made to Part #3? _____

    _____

    _____

6. What is the rough stock size for Part #1? _____

    _____

7. What is the purpose of showing view *A*? _____

    _____

8. What type of drawing is used to describe Parts #1, #2, and #3? _____

    _____

9. What is noted by the Ⓐ? _____

    _____

    _____

10. What are the finished sizes of the following parts?

    a. Part #1: _____

    b. Part #2: _____

    c. Part #3: _____

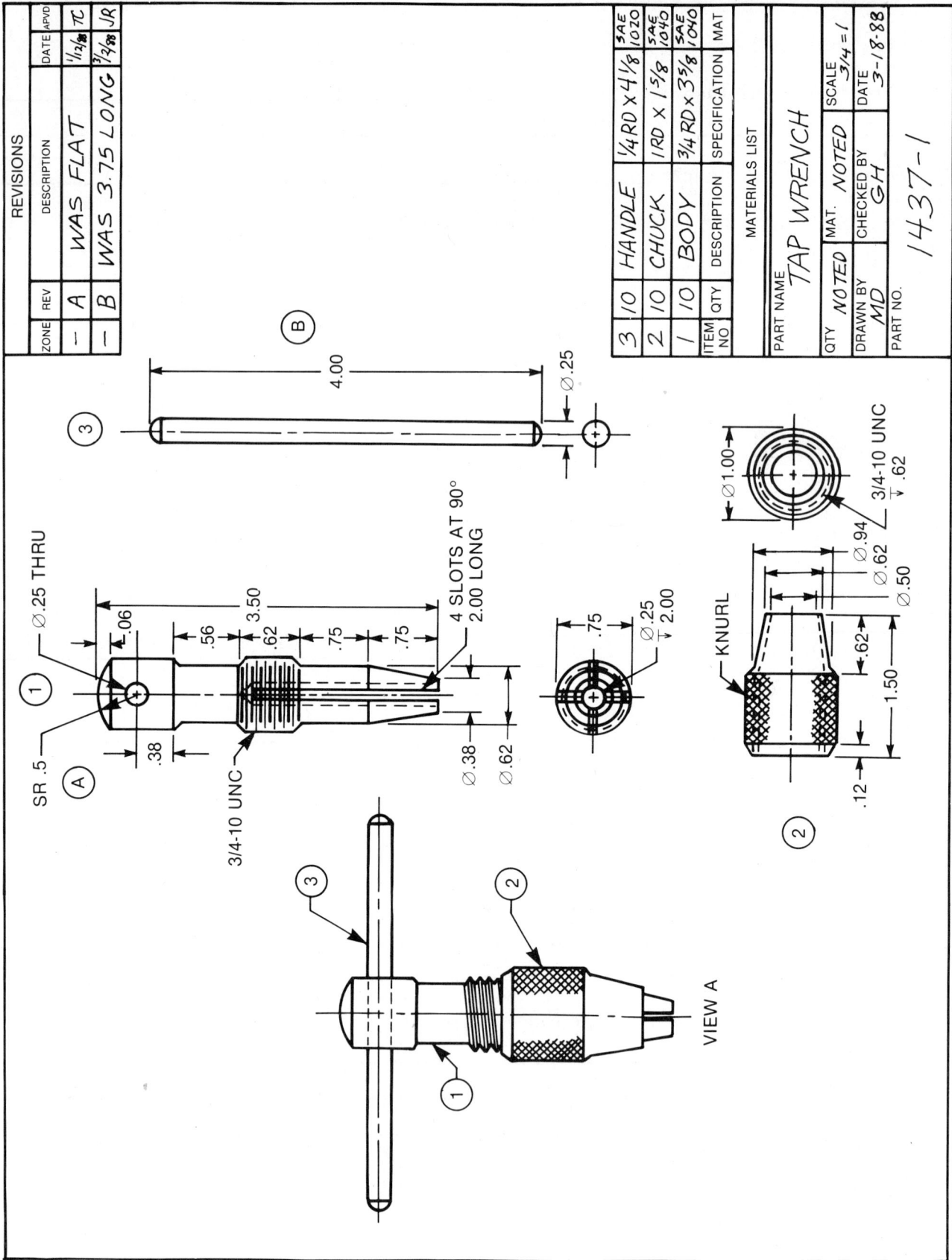

Figure E9-2.

REVISIONS

ZONE	REV	DESCRIPTION	DATE	APVD
–	A	WAS FLAT	11/2/88	TC
–	B	WAS 3.75 LONG	3/2/88	JR

MATERIALS LIST

ITEM NO	QTY	DESCRIPTION	SPECIFICATION	MAT
3	10	HANDLE	1/4 RD x 4 1/8	SAE 1020
2	10	CHUCK	1 RD x 1 5/8	SAE 1040
1	10	BODY	3/4 RD x 3 5/8	SAE 1040

PART NAME TAP WRENCH

QTY NOTED	MAT. NOTED	SCALE 3/4 = 1
DRAWN BY MD	CHECKED BY GH	DATE 3-1-88

PART NO. 1437-1

4.00

∅.25

∅.25 THRU

SR .5

.06

.38

.56

3.50

.62

.75

.75

3/4-10 UNC

4 SLOTS AT 90°
2.00 LONG

∅.38

∅.62

.75

∅.25
2.00

∅1.00

∅.94

∅.62

∅.50

KNURL

3/4-10 UNC
.62

.62

1.50

.12

VIEW A

SECTIONAL AND DETAILED VIEWS

OBJECTIVES

After studying this chapter, you will be able to:

- Identify the methods used to show sectional views.

- Identify and describe the types of sectional views.

- Determine the meaning of the section lines used with sectional views.

- Locate and identify enlarged detailed views.

- Locate and identify sectional detailed views.

- Locate and identify partial views.

The parts made in industry are often too detailed and complicated to describe with standard multiview prints. Small parts and features such as internal details require special views to clearly describe all details. The two types of views normally used to show these features are sectional views and detailed views.

SECTIONAL VIEWS

Sectional views are used to show the internal details of complex parts. Not all parts that have internal details require sectional views. For example, the bushing shown in Figure 10-1 is completely described by visible lines and hidden lines. However, in parts having internal details such as those in Figure 10-2A, hidden lines will not clearly explain the details. A sectional view, Figure 10-2B, should be used in such cases. Depending on the part, sectional views may also be used to replace conventional views in a multiview print.

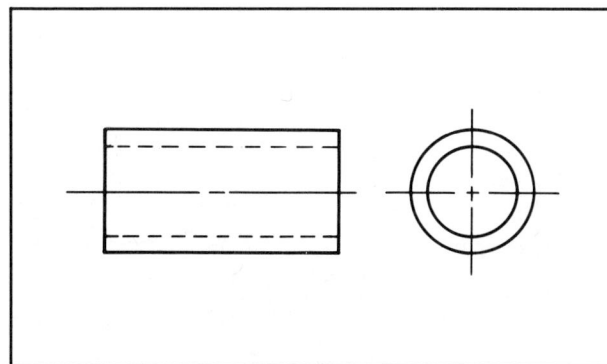

Figure 10-1. Some internal details can be shown completely without a sectional view.

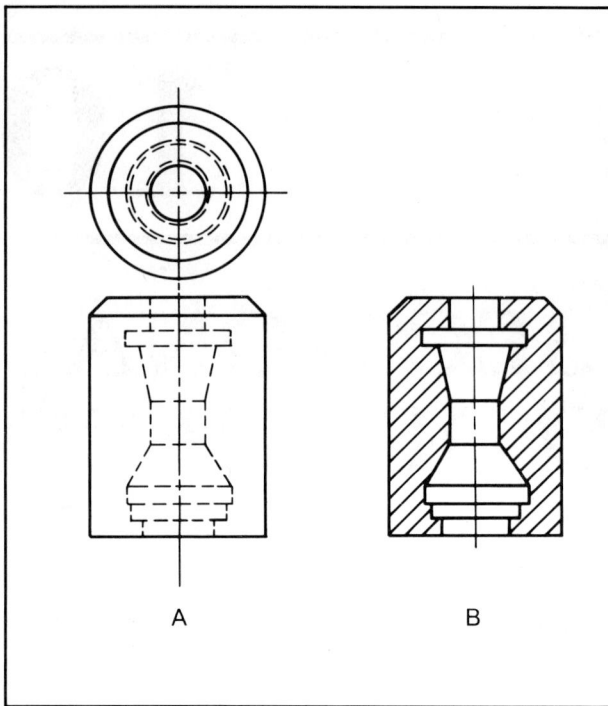

Figure 10-2. If the internal details of a part are not adequately described by hidden lines (A), a sectional view of the part can be shown (B).

Indicating Sectional Views

The location of a sectional view is normally shown on a print with a cutting plane line. *Cutting plane lines* show the exact path and position of an imaginary cut made to form a sectional view. In Figure 10-3A, the section is formed by imagining the cutting plane positioned as shown. The sectional view is then drawn as in Figure 10-3B.

The position of a sectional view is determined by the viewing direction. The arrowheads on the ends of the cutting plane line show the viewing direction. If the cutting plane line were positioned as in Figure 10-4, the section would be shown as a front view. Similarly, the sections in Figures 10-5A and 10-5B are shown as side views. The direction of the arrowheads shows how the sections are viewed. In Figure 10-5A, the arrowheads point outward. This means the section marked A-A is to be viewed as a right-side view. The section marked B-B is to be viewed as a left-side view. The opposite arrangement is shown in Figure 10-5B. Here the

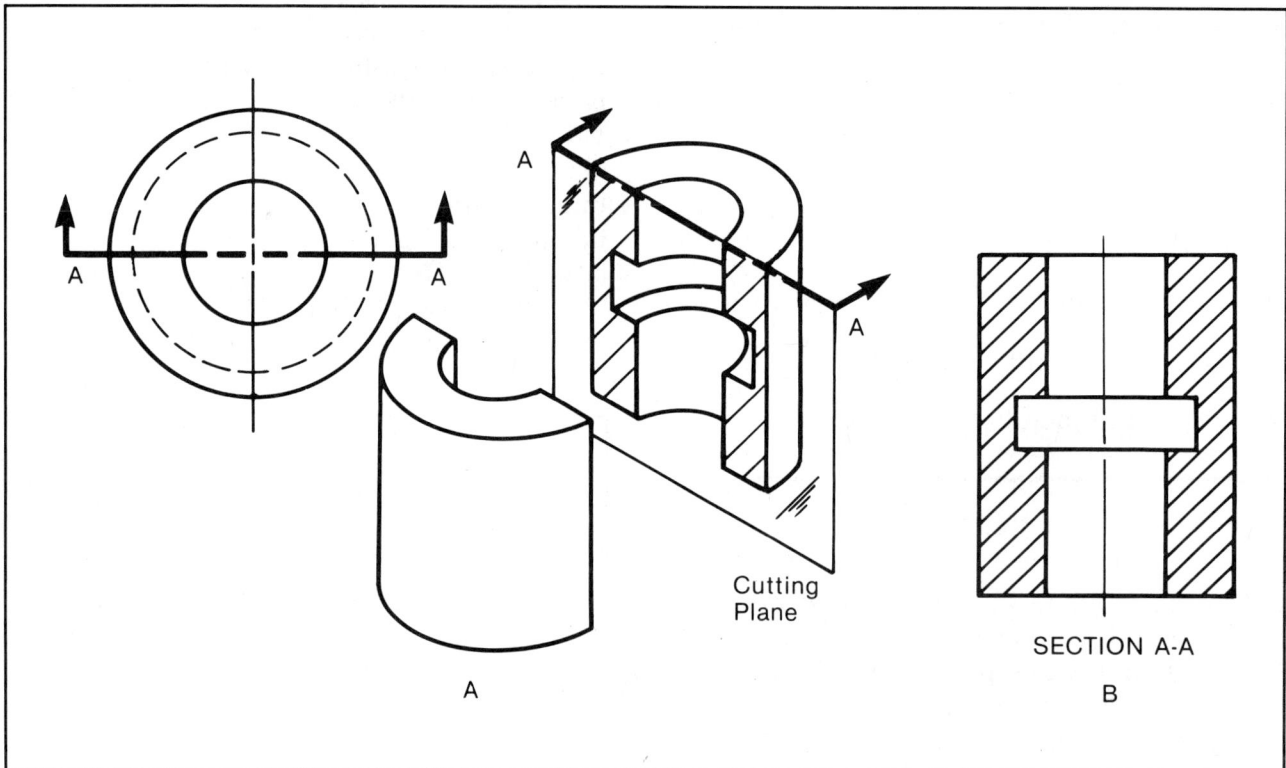

Figure 10-3. Cutting plane lines show the position and path of an imaginary cut (A). A sectional view (B) shows the part after the imaginary cut has been made.

arrowheads are pointing inward, which reverses the positions of the views. The section marked A-A is now viewed as a left-side view, while section B-B is shown as it appears when viewed from the right side.

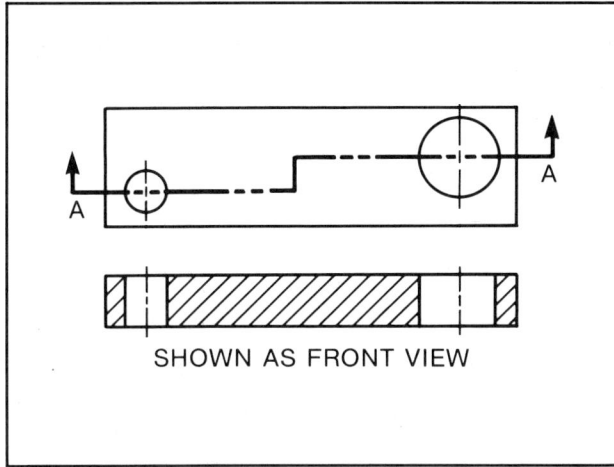

Figure 10-4. The position of a sectional view is determined by the viewing direction.

Whenever possible, sectional views are drawn in line with the surface they represent. The placement of a sectional view is not always dependent on the principal view from which it is taken.

When sectional views are placed on a print, they are usually noted and identified with the same letters that are used to identify the cutting plane line. For example, Figure 10-6 shows a shaft with three cutting plane lines. Each cutting plane line is marked with letters that refer to the specific section it represents.

The letters used to identify cutting plane lines and sectional views follow an alphabetical sequence. The first section is labeled A-A, the second is labeled B-B, and so on. For larger drawings, with more than 26 sections, double letters are used: AA-AA, AB-AB, AC-AC, and so on.

Cutting plane lines are used to indicate all sections, except where the location is obvious, as in Figure 10-7. Generally, details or hidden

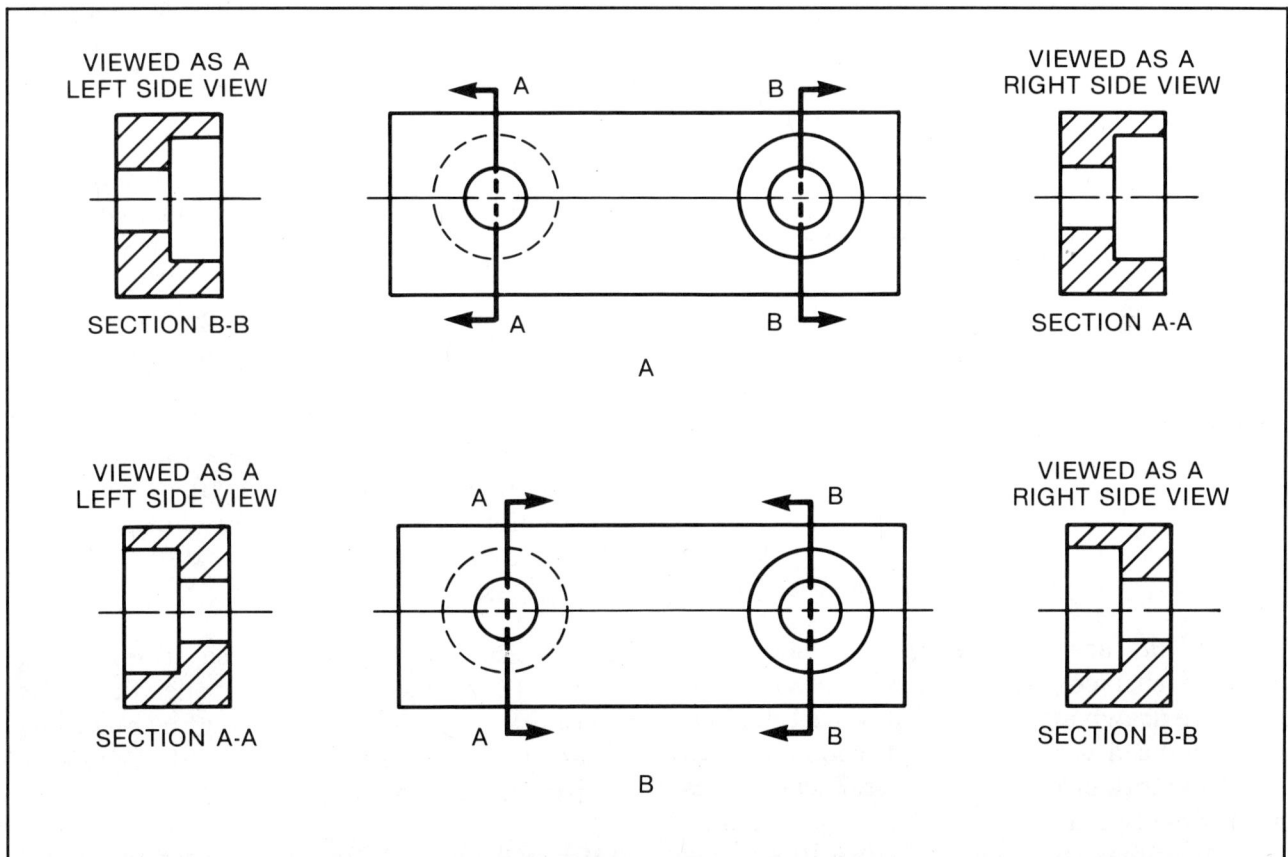

Figure 10-5. The arrowheads on the cutting plane lines show the viewing direction. The arrowheads point outward in (A) and inward in (B).

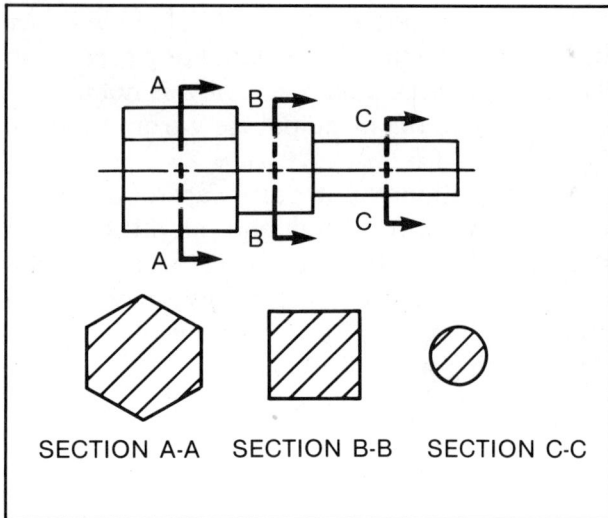

Figure 10-6. Sectional views are identified with letters.

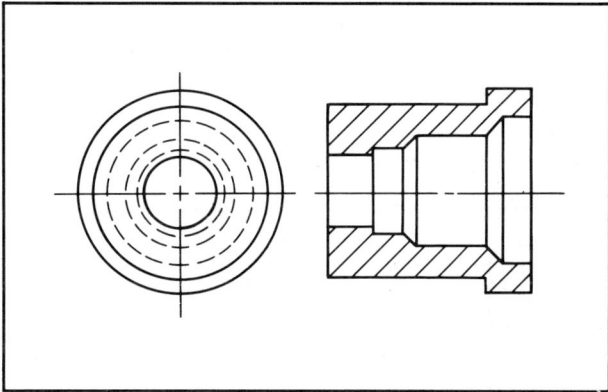

Figure 10-7. Cutting plane lines are eliminated when the location of a section is obvious.

lines that would normally appear behind a sectional view are omitted unless they are needed for clarity or to explain a complex detail.

Section Lines

Section lines are the lines used to show the exposed surfaces of a sectional view. These lines are drawn at an angle. They are intended to show the difference between sectional views and standard, or principal views. Section lines are normally drawn at a 45-degree (45°) angle, but sometimes, as shown in Figure 10-8A, this angle must be altered to suit the part. When two or more parts are sectioned, the angles

of the section lines are alternated to avoid confusion, Figure 10-8B. Other methods of varying the section lines are shown in Figure 10-8C.

Section lines are also used in some prints to identify the materials in the sectional view. The standard forms of section lines used for this purpose are shown in Figure 10-9. For general-purpose section lines, where no material is identified, the 45° section line is used. When it is used to indicate general sectioning, this 45° section line should not be confused with the section lines for cast iron.

Types of Sectional Views

Several different types of sectional views are commonly found on industrial prints. Each has a particular application to which it is best suited. The different types of sectional views are described here.

FULL SECTIONS. These sections are drawn by placing the cutting plane line along the central axis of the part, Figure 10-10. Full sections show the whole detail and are used mainly to show complicated cross sections and assemblies.

HALF SECTIONS. These sectional views are usually used for symmetrical parts, Figure 10-11. Half sections allow both the internal and external details of a part to be shown in a single view. Often half sections are shown with a cutting plane line that has only one arrowhead. This is due to the position of the cutting plane line. In these cases, the one arrowhead shown indicates the viewing direction of the section. In a sectioned view, a center line is used to divide the internal and external features of the part.

BROKEN SECTIONS. These sections, Figure 10-12, are used to point out specific details. They are not indicated with a cutting plane line; instead, a short break line is used to show the portion removed.

OFFSET SECTIONS. These sectional views are drawn with the cutting plane line positioned to show the most detail. Therefore, the cutting

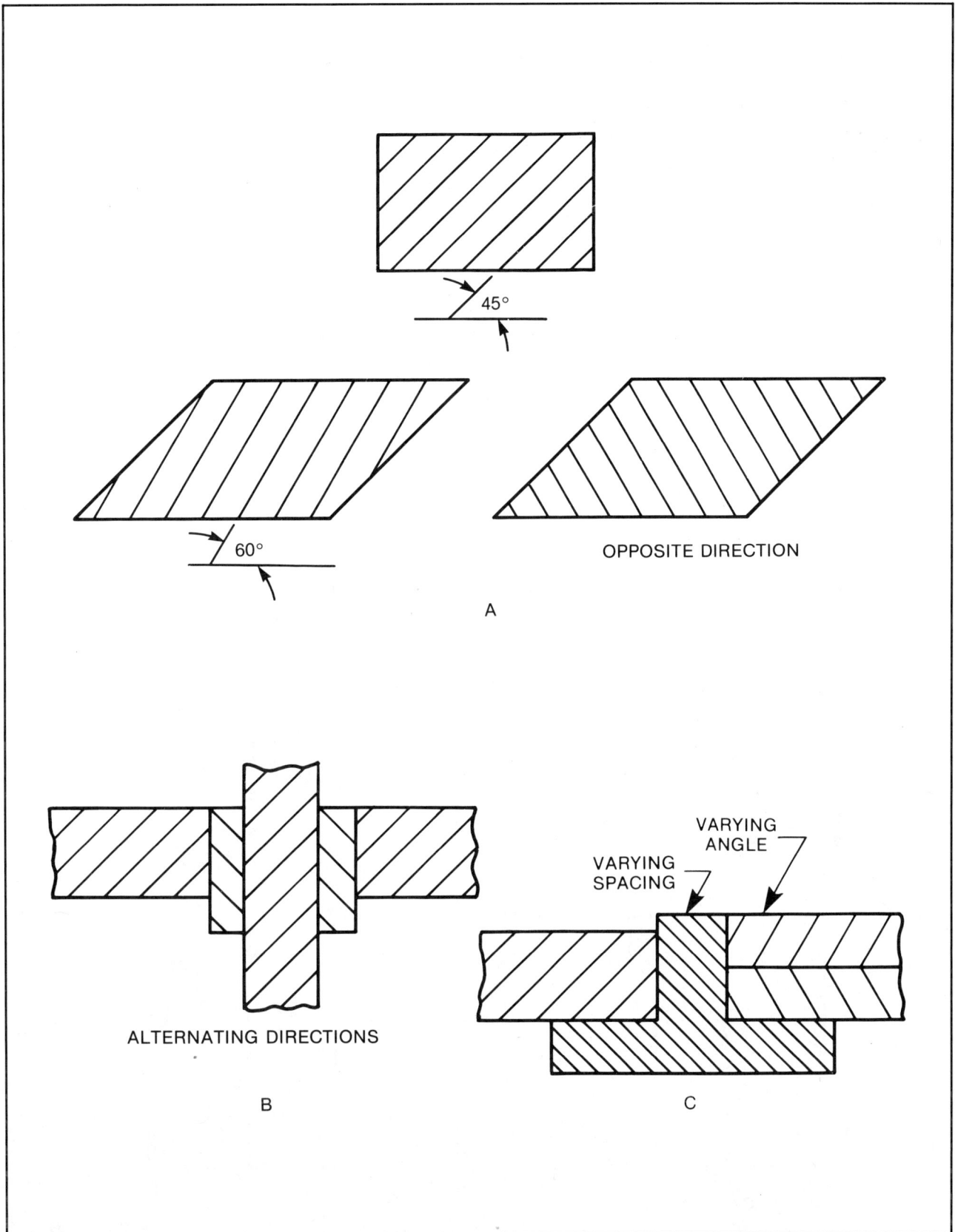

Figure 10-8. The angle, direction, and spacing of section lines may be varied to minimize confusion in section views.

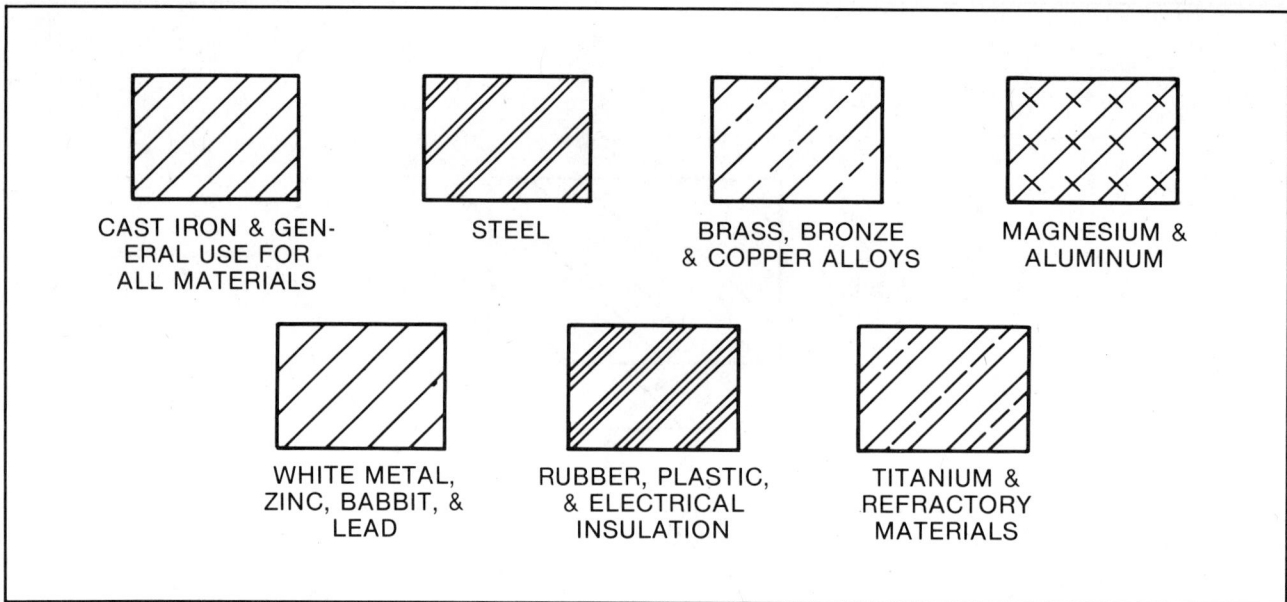

Figure 10-9. Section lines may be used to identify specific materials.

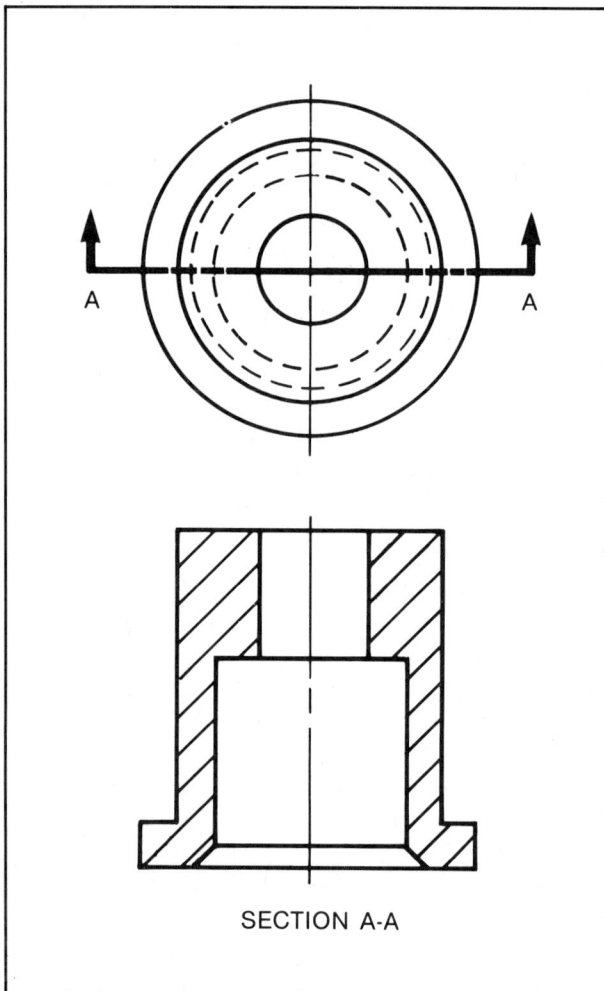

Figure 10-10. Full sectional view

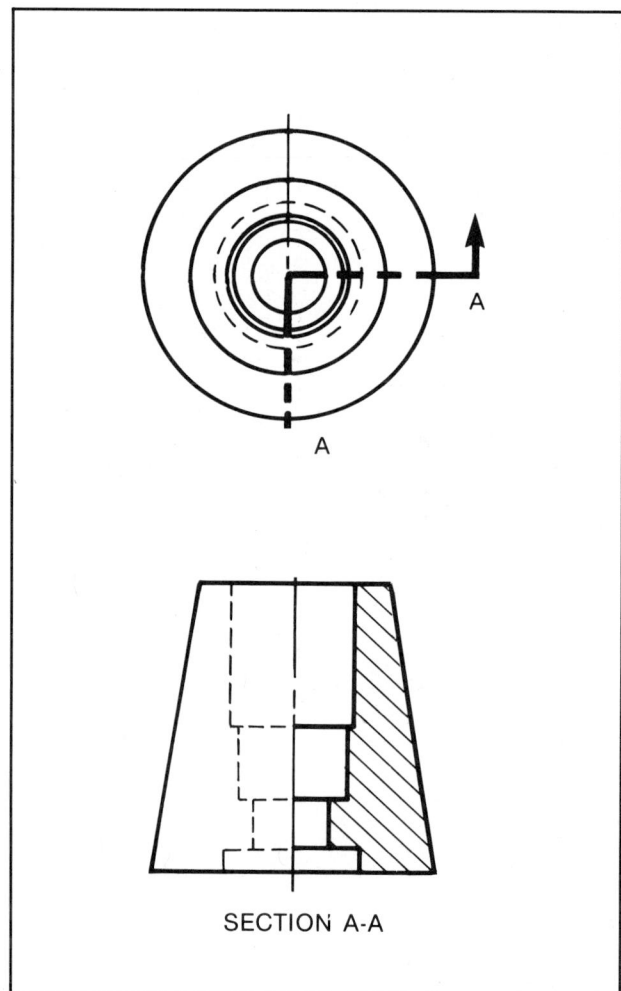

Figure 10-11. Half sectional view

plane line may be placed through only those areas where a sectional view is desired. The part in Figure 10-13 shows how a typical offset section would appear. In this case, the holes are the important features and the cutting plane line is placed to show these details.

REMOVED SECTIONS. These sections are drawn to show sections that are not direct projections of the areas they represent. Removed sections are identified with the letters A-A, B-B, and so on. See Figure 10-14A. When removed sections are projected directly from the part, the letters are omitted and a center line is used to show the area that the section view represents.

Detaching the sectional view as shown in Figure 10-14B permits greater clarity for both the object and the sectional view.

REVOLVED SECTIONS. These sectional views are normally used to show cross-sectional areas of details such as spokes, ribs, and arms. The section is simply turned 90° and is either superimposed on the object or drawn in a broken-out area, Figure 10-15.

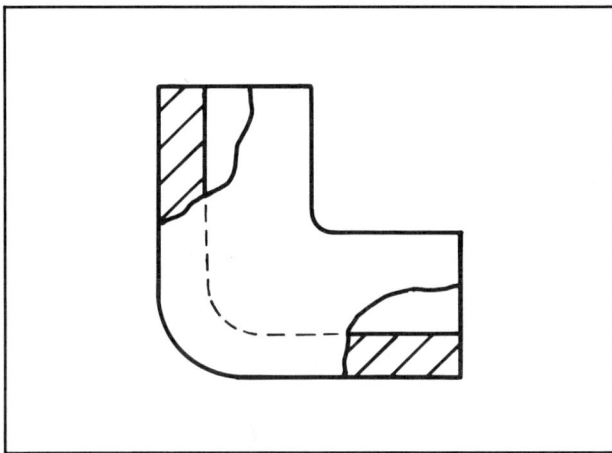

Figure 10-12. Broken sectional views

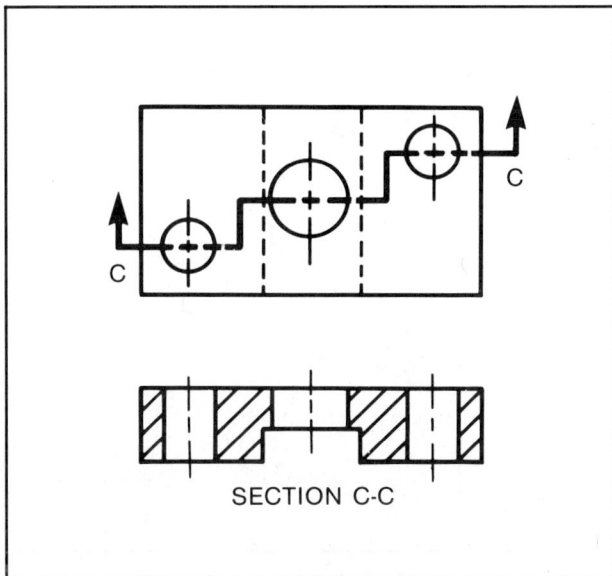

SECTION C-C

Figure 10-13. Offset sectional view

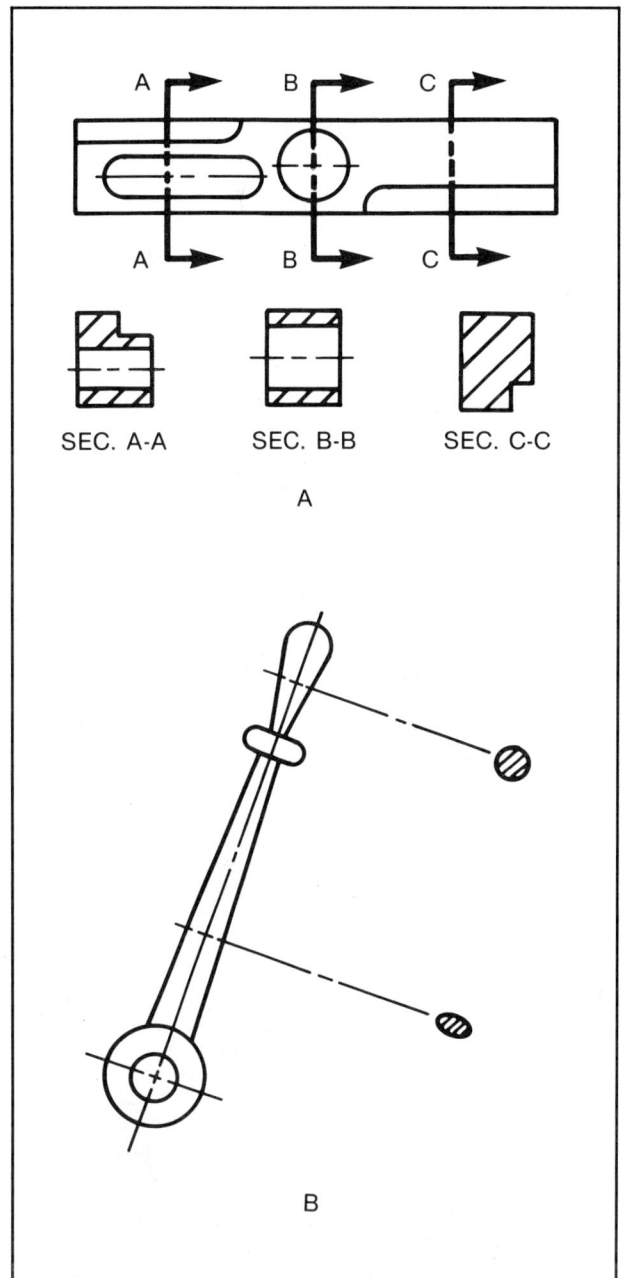

SEC. A-A   SEC. B-B   SEC. C-C

A

B

Figure 10-14. Removed sections may be (A) identified by letters or (B) projected directly from the part.

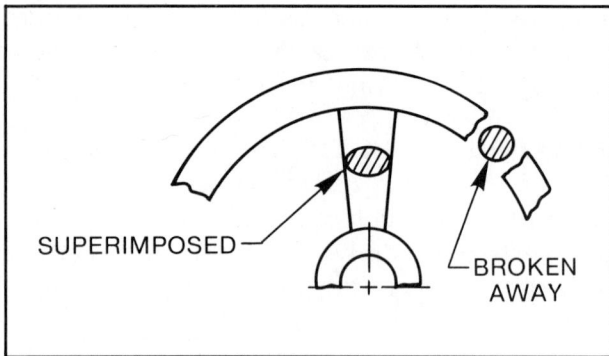

Figure 10-15. Revolved sectional views may be superimposed or broken away.

When sectional views are drawn, section lines are normally used on all areas cut by the cutting plane line. The exceptions to this rule are features such as ribs, spokes, and similar details, Figure 10-16. The only time these features are sectioned is when they are the object of the sectional view. Another point to remember about sectional views is their rotation. When features such as the wheel in Figure 10-17 are drawn, the spoke is revolved to a true vertical position. These features are not drawn with true projection if their size is distorted. This principle is applied to all parts that have similar features.

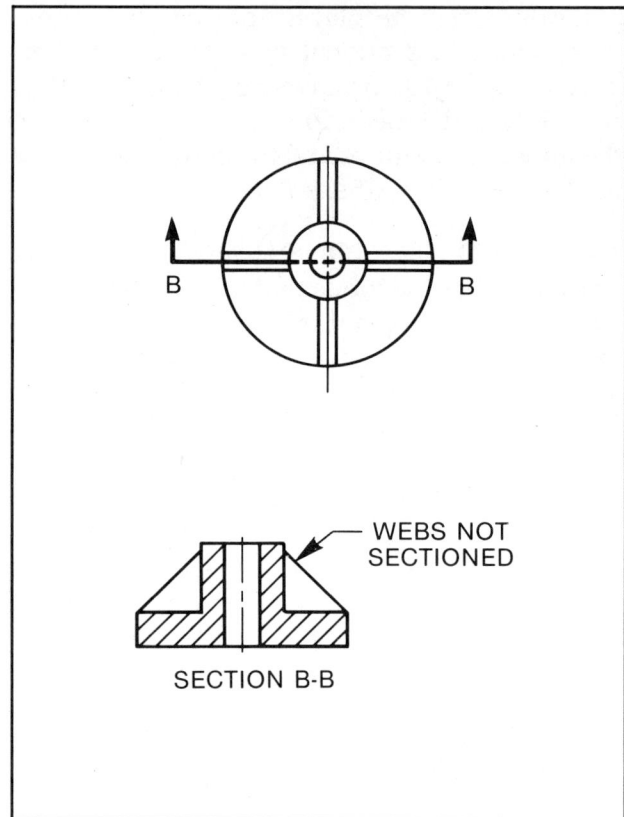

Figure 10-16. Ribs, gussets, and spokes are sectioned only when they are the object of the sectional view.

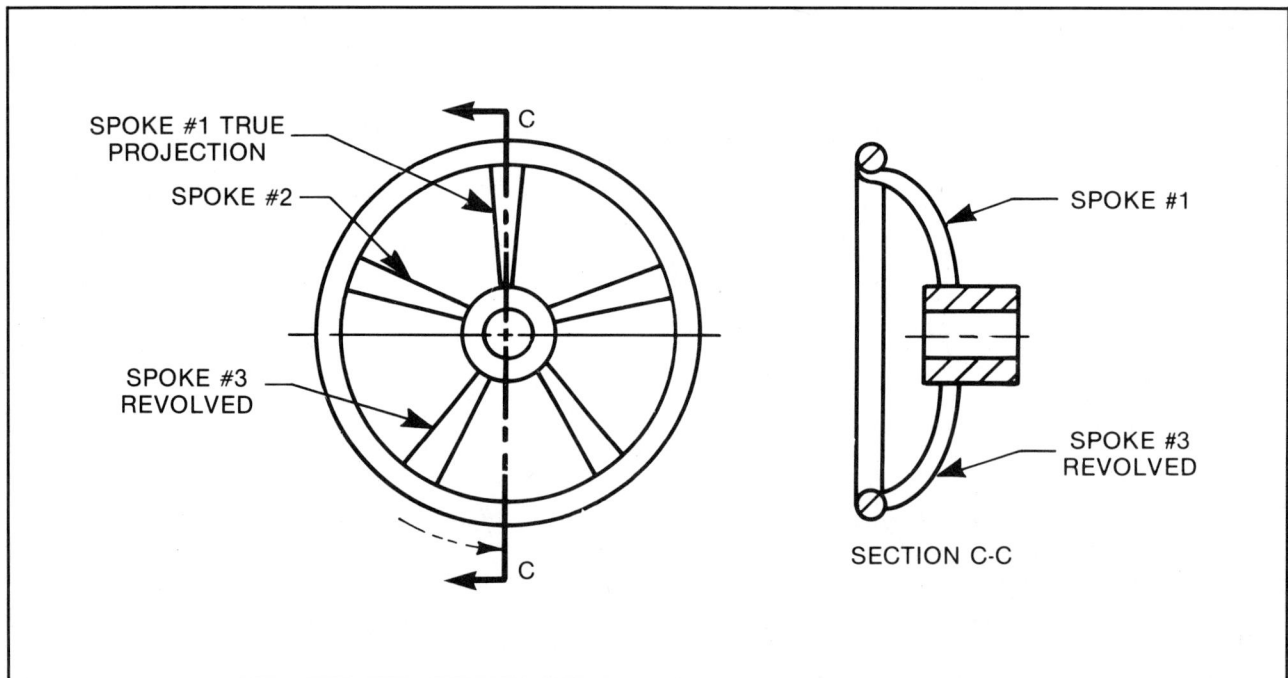

Figure 10-17. Spokes and similar details are usually rotated to make the sectional view appear visually correct.

DETAILED VIEWS

Part features and details that require special attention are often shown as detailed views. *Detailed views* allow the drafter to draw only those parts of the object needed for clarity or special instructions. The principal forms of detailed views are enlarged, sectional, and partial views.

Enlarged Detailed Views

Enlarged detailed views are used to identify and isolate special features of a part. For example, the part in Figure 10-18 has grooves that must be cut to a special shape. Since this cannot be seen in the principal view, an enlarged detailed view is used to relay the specific information.

Enlarged detailed views are noted on the print in several ways. Figure 10-19 shows the most commonly accepted methods. Either a dashed line or a heavy phantom line may be used to circle the area shown in the enlarged view. Occasionally the word *view* will be replaced with the word *detail* in these notes. In most cases, the scale of the enlarged view will also be shown close to the view.

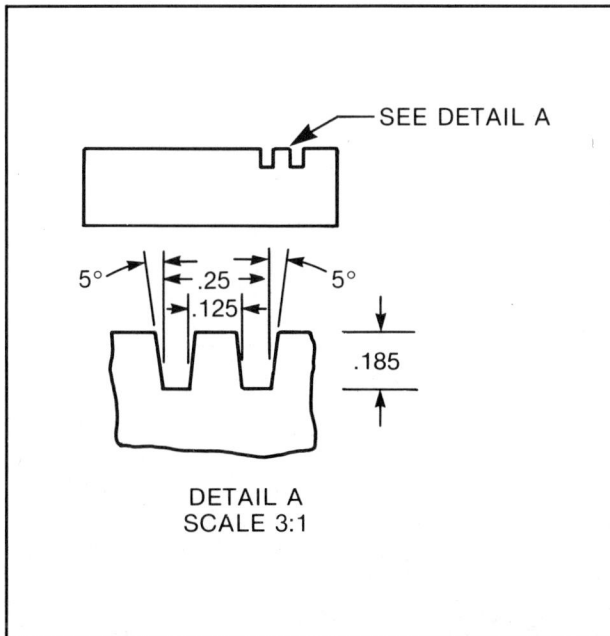

Figure 10-19. Methods used to indicate detailed views

Sectional Detailed Views

Sectional detailed views are used to call attention to special internal features of a part. The part in Figure 10-20 has a hole with a special shape that is not readily seen in the principal view. A sectional detailed view is used to enlarge the area of the hole so that all details can be clearly seen and dimensioned.

Sectional detailed views are identified on a drawing in the same way as enlarged detailed views. The scale of the enlargement depends on the part and the details that must be shown. The view is normally enlarged to the point where all details are clear and understandable. The scale size of the enlarged view is also shown below the view.

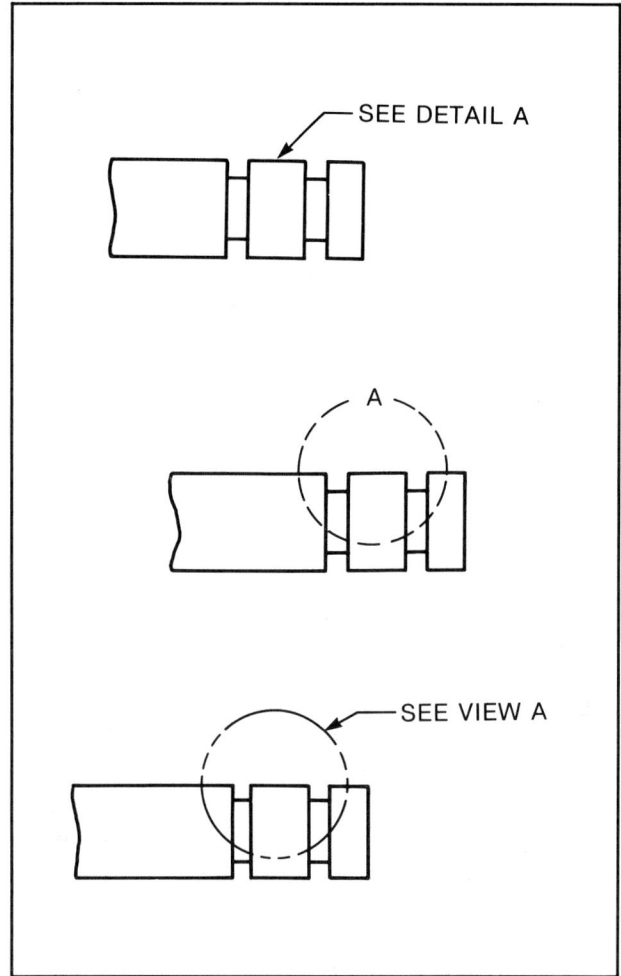

Figure 10-18. Enlarged detailed view

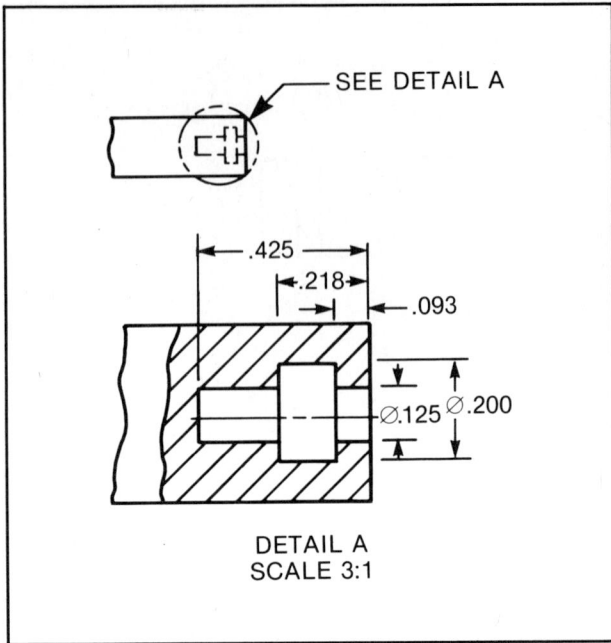

Figure 10-20. Sectional detailed view

Partial Views

Partial views are used in place of full views to save space and time, and to simplify the drawing. Partial views show only the important features that are not shown clearly in a principal view. Figure 10-21 shows several examples of partial views.

KEY TERMS

The following key terms were introduced in this chapter. Be sure you know the meaning of each term before proceeding to the review material.

Broken Section
Detailed View
Enlarged Detailed View
Full Section
Half Section
Offset Section
Partial View
Removed Section
Revolved Section
Sectional Detailed View
Sectional Views

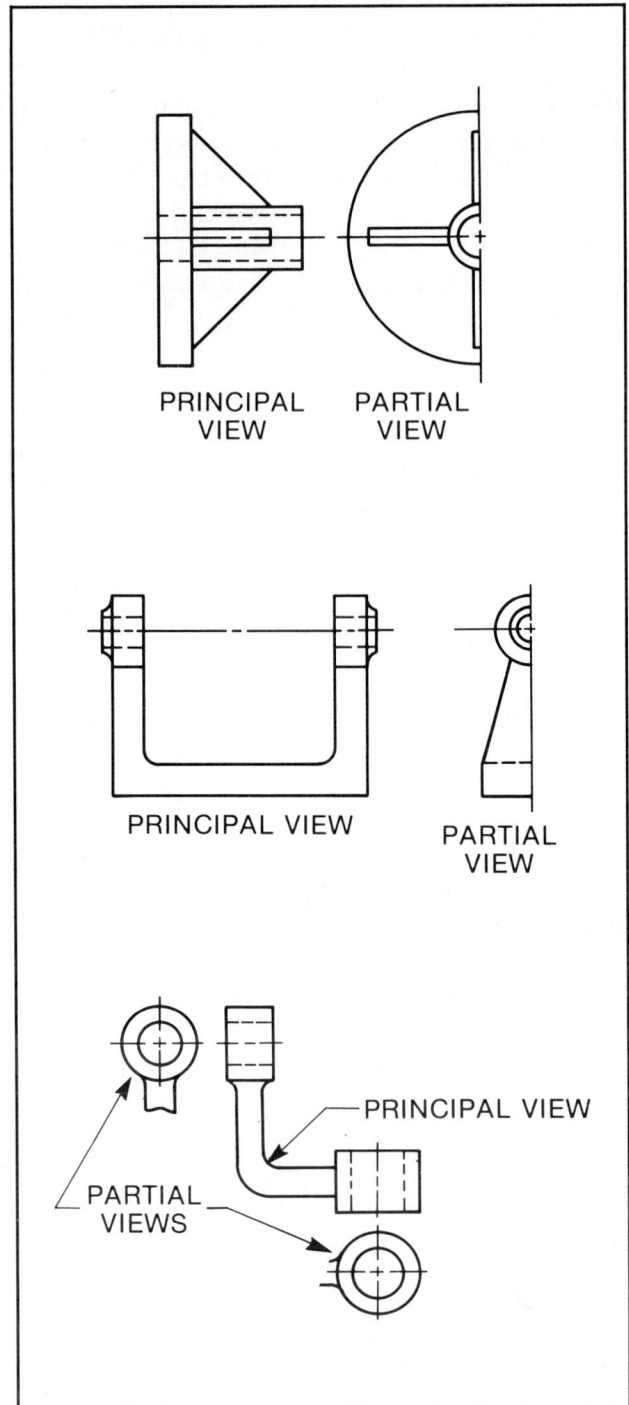

Figure 10-21. Partial views

REVIEW

Test your knowledge with this reinforcement study material. Write your answers to the questions in the spaces provided.

1. How is the location of a sectional view shown on a print? _____
_____
_____

2. What is the purpose of a sectional view? _____
_____
_____

3. What type of line is used to show the position and path of the imaginary cut made to form a section? _____

4. How is the viewing direction of a sectional view noted? _____
_____
_____

5. What is the purpose of using letters with sectional views and cutting plane lines?
_____
_____

6. What determines the position of a sectional view? _____
_____
_____

7. What purpose do section lines serve? _____
_____
_____

8. How do general-purpose section lines appear on a print? _____
_____

9. How are the section lines placed to avoid confusion among two or more parts?
_____
_____

10. What type of sectional view shows the complete internal details of a part? _____
_____

11. Which sectional view is superimposed on the part? _____
_____

12. What type of sectional view sometimes uses a single arrowhead? _____

_____

13. Which type of section uses a short break line rather than a cutting plane line to show the location of a section? _____

14. How can the part material be identified in a sectional view? _____

_____

_____

15. When a removed section is projected directly from a part, how is the area that the section represents noted? _____

_____

16. What type of view is used to identify and isolate external areas of a part not clearly shown in a principal view? _____

17. Which type of view is intended to show enlarged internal areas of a part? _____

_____

18. What determines the scale of an enlarged detailed view? _____

_____

19. Where is the scale of a detailed view noted? _____

_____

20. What type of view is used to save space and to simplify a drawing? _____

_____

Refer to Figure E10-1 on page 180 to answer the following questions.

1. What is the part name? _____

2. What type of sectional view is A-A? _____
   _____

3. What type of sectional view is B-B? _____
   _____

4. What type of sectional view is shown at *K*? _____
   _____

5. What type of line is shown at *E*? _____
   _____

6. List the tolerances of the following dimensions.
   a. *B*: _____
   b. *F*: _____
   c. *G*: _____
   d. *L*: _____

7. List the material indicated in the print at the following.
   a. Material at *H*: _____
   b. Material at *I*: _____
   c. Material at *J*: _____

8. What material is specified in section *K*? _____
   _____

9. What is the value of dimension *C*? _____
   _____

10. What are the rough stock sizes for items #2 and #3? _____
    _____

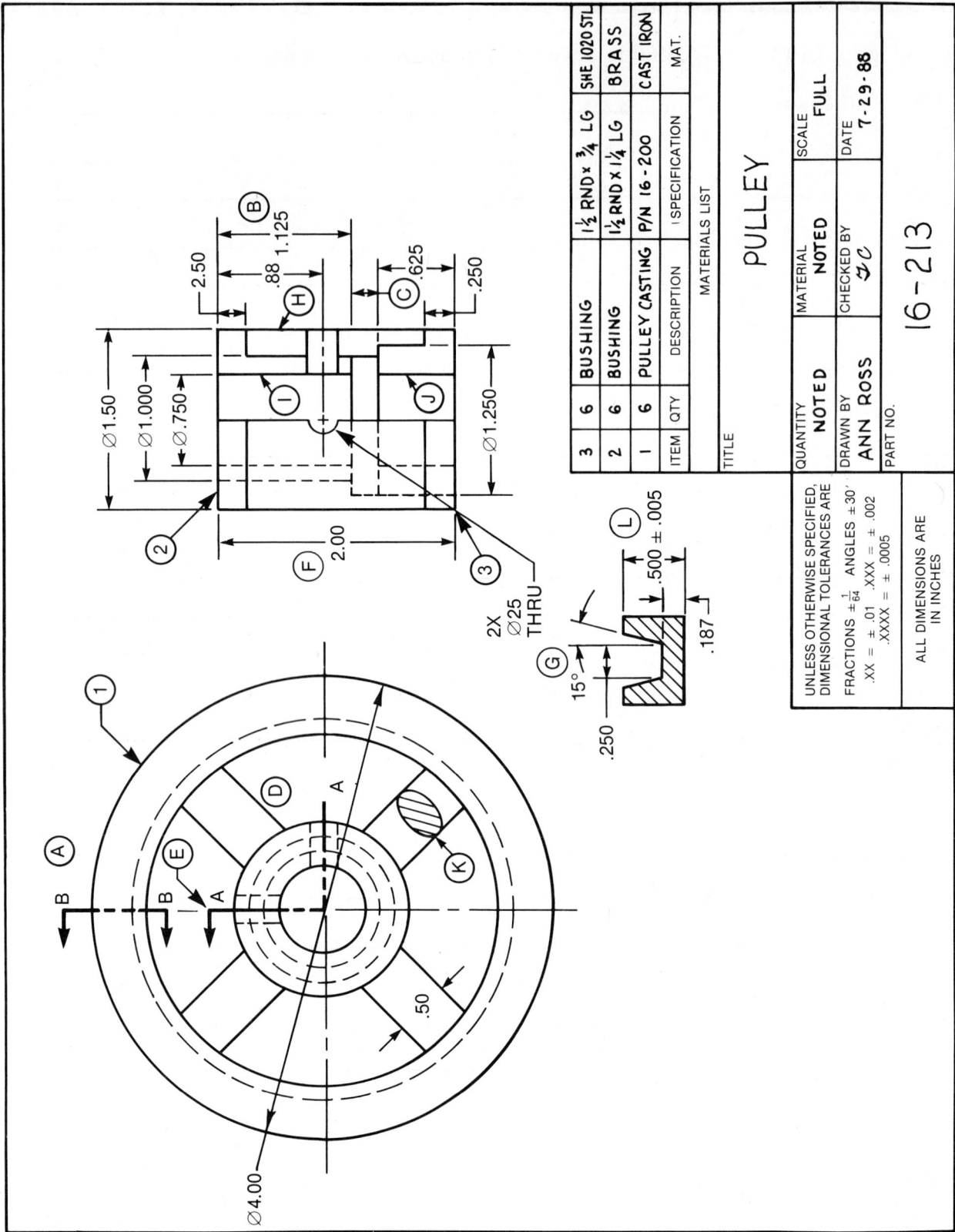

Figure E10-1.

Refer to Figure E10-2 on page 182 to answer the following questions.

1. What is the part name? _____

2. What type of print is this? _____
_____

3. What materials are specified? _____
_____
_____

4. What type of view is shown at *C*? _____
_____

5. What type of view is shown at *B*? _____
_____

6. What is shown at *H*? _____
_____

7. Find the following dimensions.
   a. *E* = _____
   b. *F* = _____
   c. *G* = _____

8. What type of view is shown at *A*? _____
_____

9. What type of view is shown at *D*? _____
_____

10. What is the overall length and width of the slot at the top of the part? _____
_____

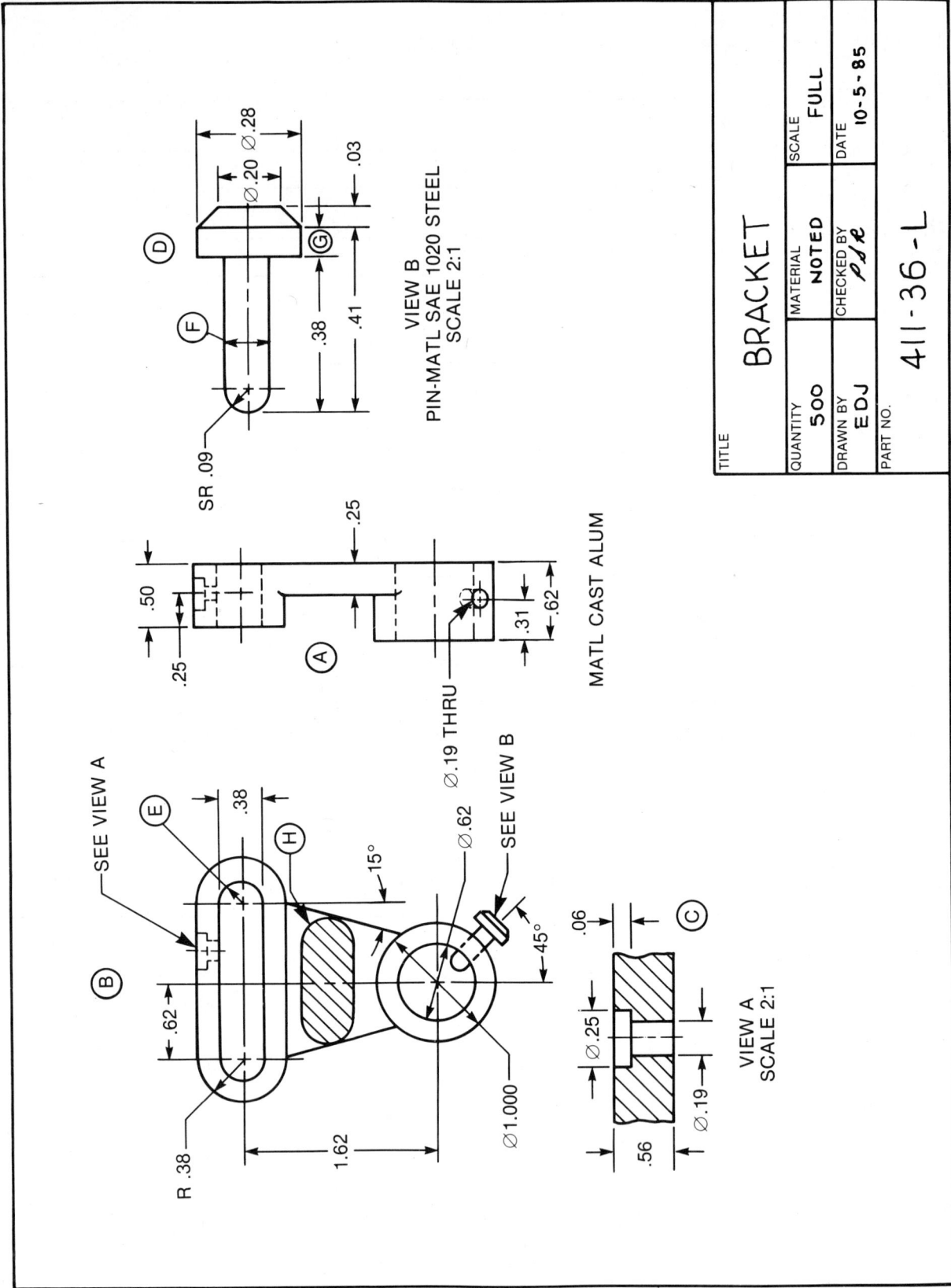

VIEW B
PIN-MATL SAE 1020 STEEL
SCALE 2:1

Ø.28
Ø.20
.03
.38
.41
SR .09

MATL CAST ALUM

.25
.50
.25
.31
.62
Ø.19 THRU
Ø.62
SEE VIEW B

SEE VIEW A
.38
15°
45°
Ø.62
Ø1.000
1.62
.62
R .38

VIEW A
SCALE 2:1
.06
.25
.19
.56

BRACKET

TITLE			
QUANTITY 500	MATERIAL NOTED	SCALE FULL	
DRAWN BY EDJ	CHECKED BY Plk	DATE 10-5-85	
PART NO. 411-36-L			

Figure E10-2.

SPECIFYING MACHINED DETAILS AND ELEMENTS

OBJECTIVES

After studying this chapter, you will be able to:

- Identify standard variations of tapers and calculate taper sizes.
- Identify standard forms of keys, keyways, and keyseats.
- Determine key sizes and types from print specifications.
- Identify standard thread forms.
- Identify and define standard thread designations and representations.
- Identify and define common hardware items and specifications.
- Identify the types and parts of standard gears.
- Determine the required values needed to make gears from print specifications.
- Identify and interpret surface texture symbols.
- Determine the production methods needed to produce a desired finish.

Machined details and elements are frequently found on manufactured products. These details include tapers, keyed assemblies, threads, and a variety of gears. To properly interpret a print, it is important to be familiar with the methods drafters use to specify these details. You should also know how to read and interpret the surface finish designations frequently used to specify surface textures on machined parts.

TAPERS

A *taper* is a uniform change in width or diameter over a specific length. The two primary forms of tapers are conical and flat, Figure 11-1.

Conical tapers are machined on a shaft or in a hole. This type of taper is basically a self-centering cone that is either self-holding or self-releasing.

Self-holding tapers have a 2° to 3° angle of taper, Figure 11-2. They are used to centrally locate and hold tools such as drills, reamers, and lathe centers. *Self-releasing tapers* are used to centrally locate tools such as collets, lathe chucks, and milling machine arbors. Self-releasing tapers have a taper angle of more than 8° and must be held in place with a draw nut or drawbar. See Figure 11-2.

Flat tapers are normally used as mechanical locks or wedges. Like conical tapers, flat tapers with angles between 2° and 3° tend to be self-holding. Flat tapers greater than 8° tend to be self-releasing.

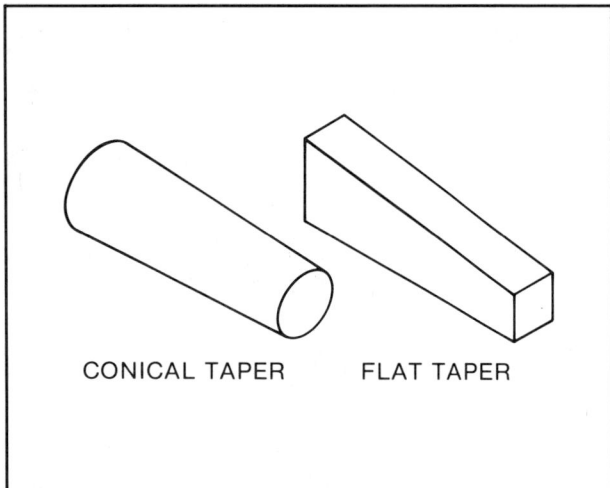

Figure 11-1. Primary forms of tapers

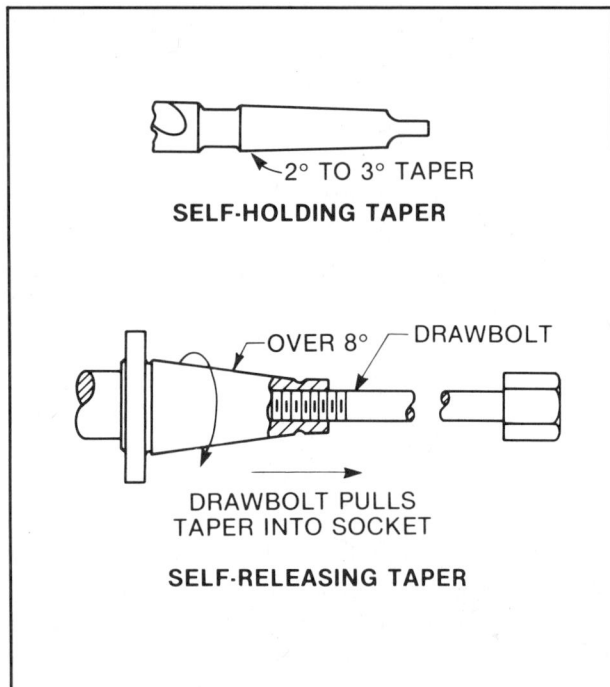

Figure 11-2. Self-holding and self-releasing tapers

Taper Specifications

Tapers are usually shown on a print in units of taper per inch or taper per foot.

Taper per inch (TPI) is the amount of difference in the end diameters of the large and small ends of the part, per inch of taper length, Figure 11-3A.

Taper per foot (TPF) is the amount of difference in the end diameters of the large and small ends of the part, per foot of taper length, Figure 11-3B. A part does not have to be twelve inches long to use TPF for dimensioning. When TPF is specified, it simply means the part would have that much taper if it were a foot long.

Taper per inch and taper per foot are directly proportional to each other. Therefore,

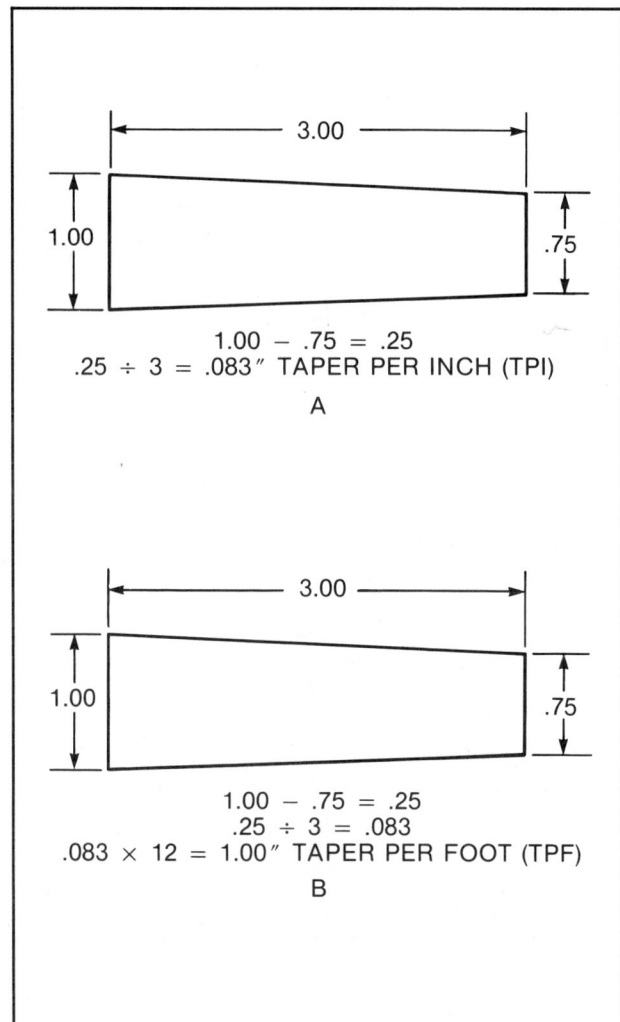

Figure 11-3. Tapers are specified on a print as (A) taper per inch or (B) taper per foot.

a taper of .10″ TPI is identical to a taper of 1.20″ TPF (.10″ × 12 = 1.20″). A taper of 3.00″ TPF is identical to a taper of .25″ TPI (3.00″ ÷ 12 = .25″).

In addition to taper per inch and taper per foot, tapers can be dimensioned in several other ways. See Figure 11-4. The symbol used to denote conical tapers (▷) is also used on newer prints. When used, the values shown specify the amount of taper per unit. For example, in .05:1, the .05 shows the amount of taper (50 thousandths), and the 1 specifies the unit (1 inch). When a flat taper is shown, the slope symbol (◺) is used. Here again, the dimension shown indicates the amount of slope per unit.

In cases where required dimensions are not given, they must be calculated. Figure 11-5 lists the common formulas used for taper conversions and calculations. When tapers are specified by a standard taper number, Figure 11-6, check a handbook to find the proper dimensions for each standard size.

KEYWAYS

The term *keyway* is sometimes used incorrectly to refer to a complete keyed assembly. Actually, there are three parts to any keyed assembly: the keyway, the keyseat, and the key. See Figure 11-7.

The *keyway* is the slot, or groove, that is cut into the external part of the assembly. The *keyseat* is the slot, or groove, that is cut into the shaft, or internal part of the assembly. The *key* is the solid coupler that fits between the keyway and keyseat.

The most common types of keys are the *square, Woodruff, Pratt & Whitney*, and *flat*. See Figure 11-8. These keyed assemblies are usually dimensioned as shown in Figure 11-9. The depths of the keyway and keyseat, when shown, are dimensioned from the opposite side of the hole or shaft. When keyseats and keyways are dimensioned with simple notes, an appropriate handbook should be used to find the exact sizes and tolerances for each dimension.

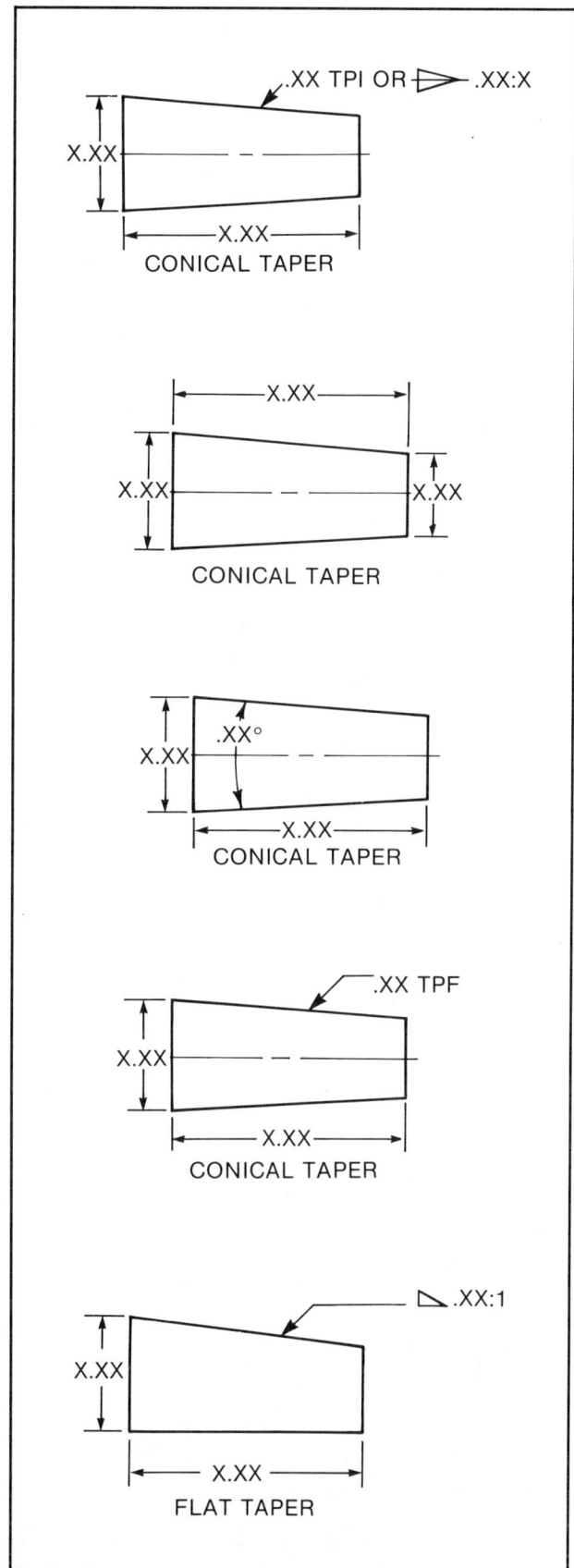

Figure 11-4. Methods of dimensioning tapers

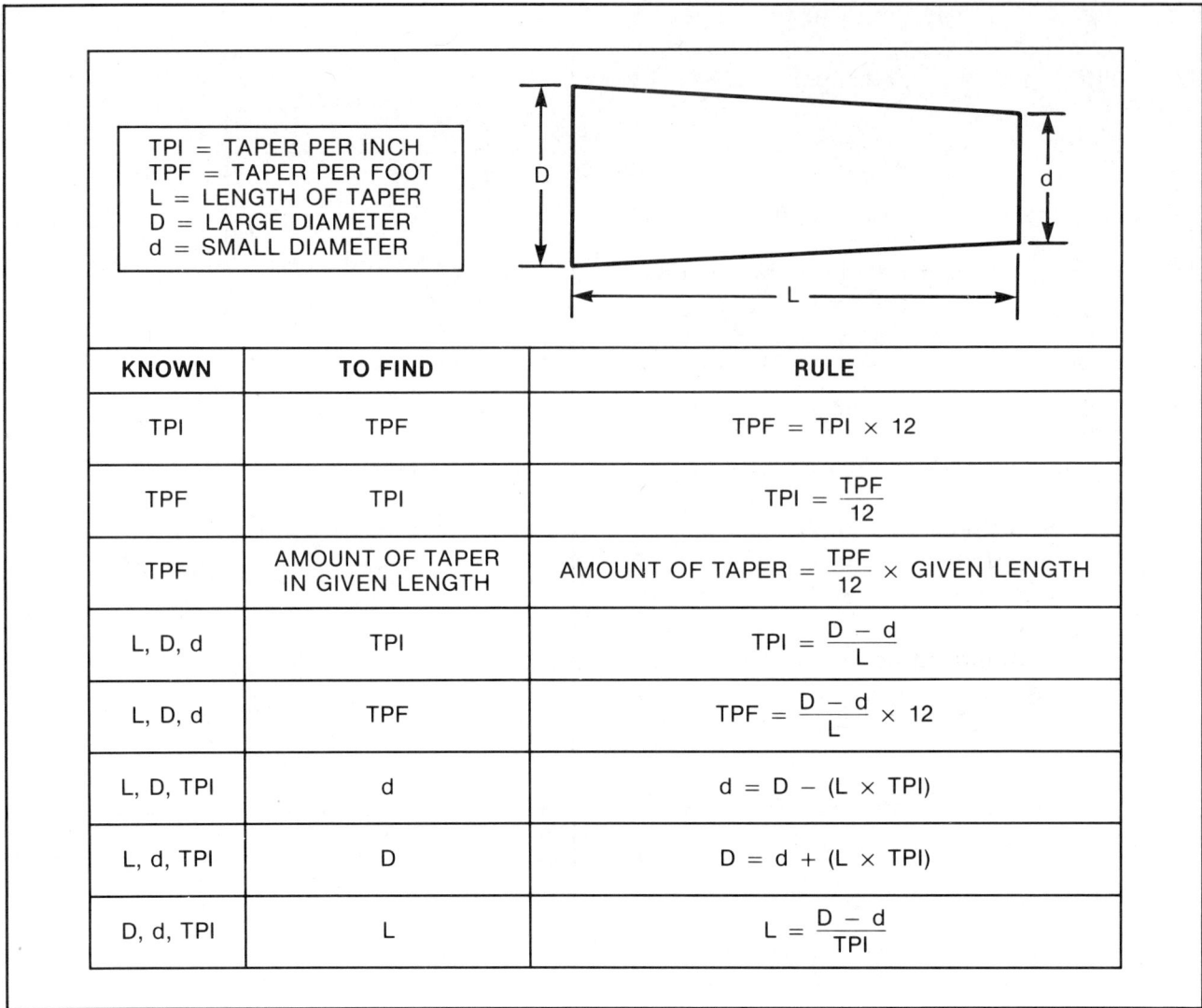

KNOWN	TO FIND	RULE
TPI	TPF	$\text{TPF} = \text{TPI} \times 12$
TPF	TPI	$\text{TPI} = \dfrac{\text{TPF}}{12}$
TPF	AMOUNT OF TAPER IN GIVEN LENGTH	$\text{AMOUNT OF TAPER} = \dfrac{\text{TPF}}{12} \times \text{GIVEN LENGTH}$
L, D, d	TPI	$\text{TPI} = \dfrac{D - d}{L}$
L, D, d	TPF	$\text{TPF} = \dfrac{D - d}{L} \times 12$
L, D, TPI	d	$d = D - (L \times \text{TPI})$
L, d, TPI	D	$D = d + (L \times \text{TPI})$
D, d, TPI	L	$L = \dfrac{D - d}{\text{TPI}}$

TPI = TAPER PER INCH
TPF = TAPER PER FOOT
L = LENGTH OF TAPER
D = LARGE DIAMETER
d = SMALL DIAMETER

Figure 11-5. Formulas for calculating taper values

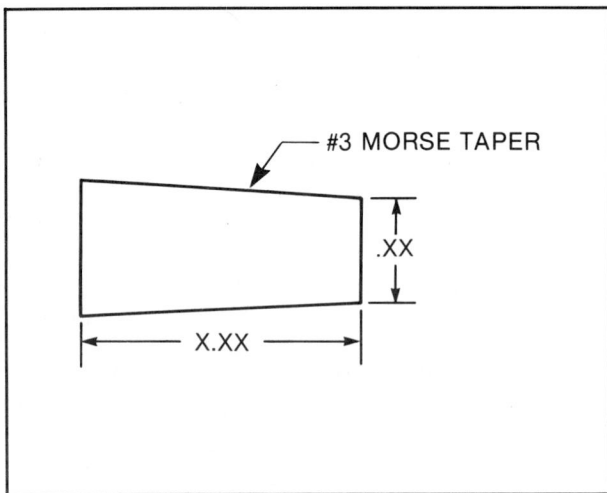

Figure 11-6. Tapers may be specified by a standard taper number.

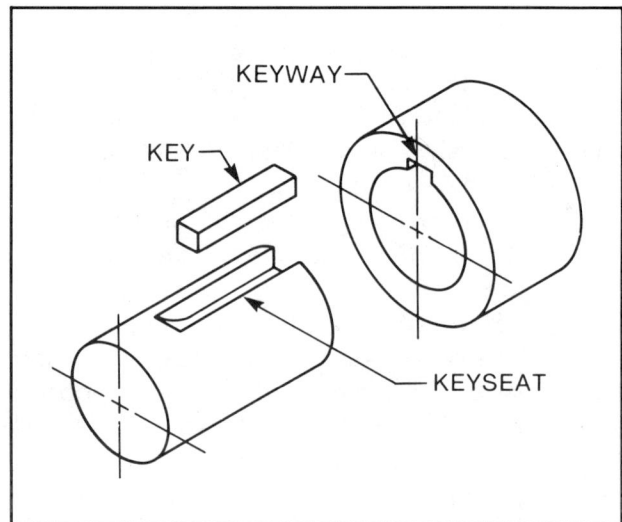

Figure 11-7. Keyed assembly

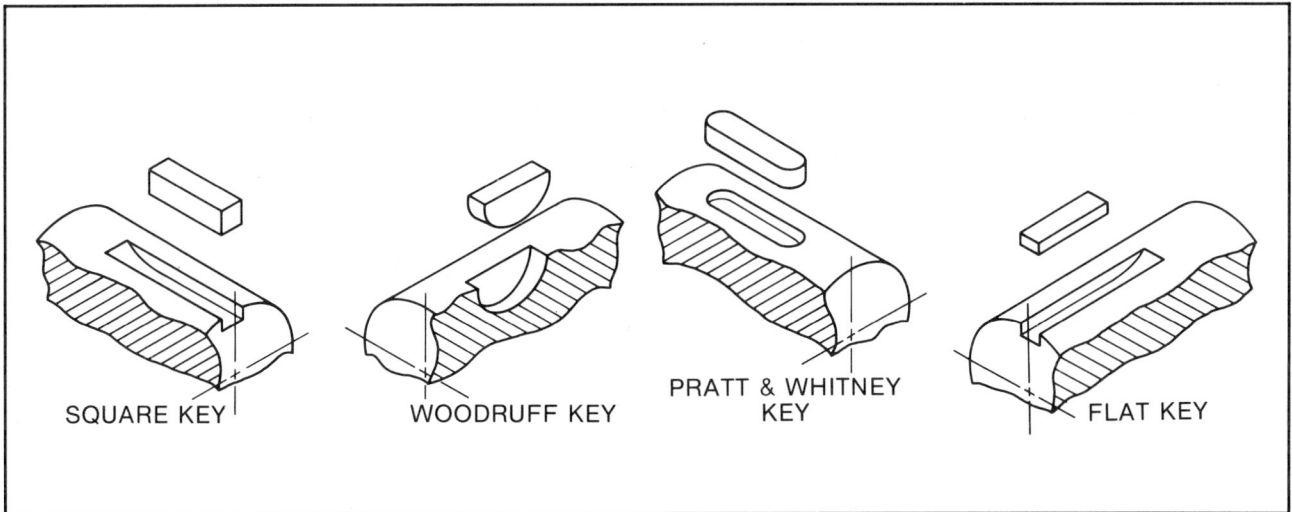

Figure 11-8. Types of keys

SQUARE KEY

WOODRUFF KEY

PRATT & WHITNEY KEY

FLAT KEY

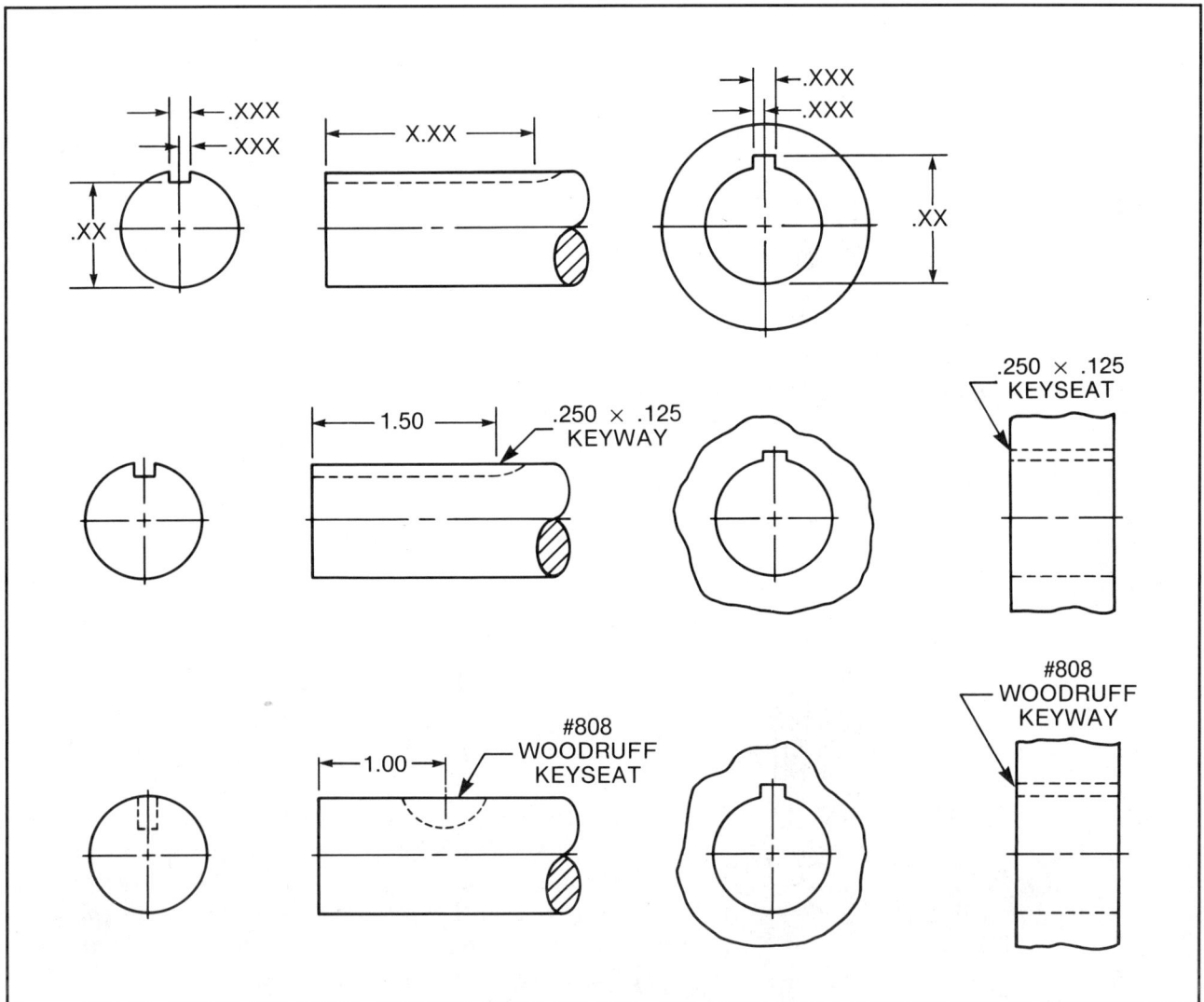

.XXX
.XXX

X.XX

.XXX
.XXX

.XX

.XX

1.50

.250 × .125
KEYWAY

.250 × .125
KEYSEAT

#808
WOODRUFF
KEYSEAT

1.00

#808
WOODRUFF
KEYWAY

Figure 11-9. Dimensioning keyways and keyseats

SCREW THREADS

Screw threads and threaded devices are used for a wide variety of applications. To meet the wide range of uses, several styles and types of thread forms have been developed.

Thread Forms

The most common thread form used today is the *Unified thread*, Figure 11-10. This thread has a 60° thread angle and a rounded root. Another thread, which closely resembles the Unified thread, is the *ISO metric thread* form. The only major differences between these two threads are the sizes and specifications. Other variations of the 60° thread include the *American National thread* and the *sharp V thread*, Figure 11-11. The 60° thread form is primarily used for threaded fasteners such as bolts, studs, and screws.

When a thread is needed to transmit motion or power, the thread forms shown in Figure 11-12 are generally used. The *square thread* has a 90° thread angle and is used to transmit heavy loads, torque, and force. The *buttress thread* has a 45° to 52° thread angle and is generally used to transmit power in only one direction, such as in vise screws. The *Acme thread* combines the advantages of the square and buttress threads into a single thread form. The Acme is easier to make than either of the others, and for

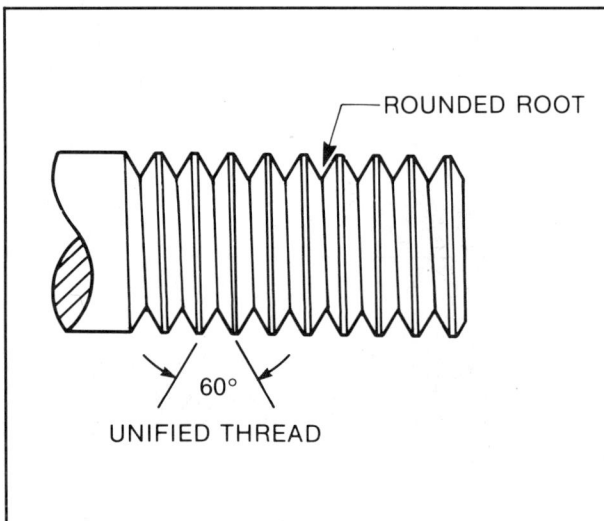

Figure 11-11. American National and sharp V thread forms

all practical purposes it is just as strong. The 29° Acme thread form can also be adjusted to compensate for wear. This feature makes the Acme thread ideal for feed and adjusting screws for machine tools.

Another thread form commonly found in mechanical assemblies is the *Pipe thread*. See Figure 11-13. The pipe thread has a 60° thread angle and is either tapered or straight. *Tapered pipe threads* have a taper of $\frac{3}{4}''$ TPF. They are used for self-sealing, leakproof joints in water pipes and air lines. *Straight pipe threads* are used where a sealed joint is not required, such as conduit for routing electrical wires. The only difference between these two pipe threads is the taper; otherwise they are identical.

Figure 11-10. Unified thread form

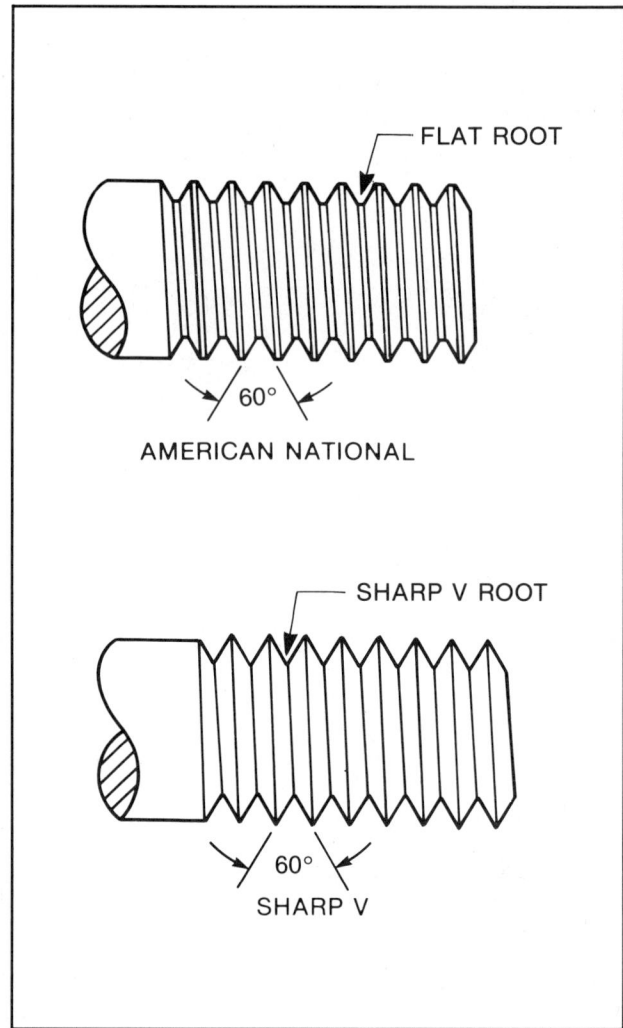

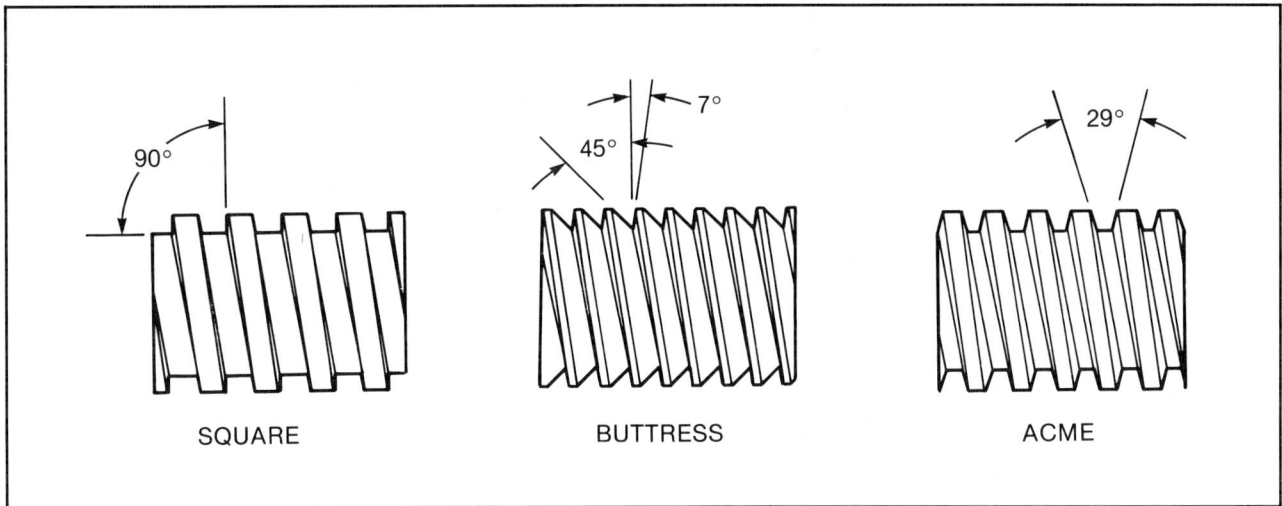

Figure 11-12. Square, buttress, and Acme thread forms

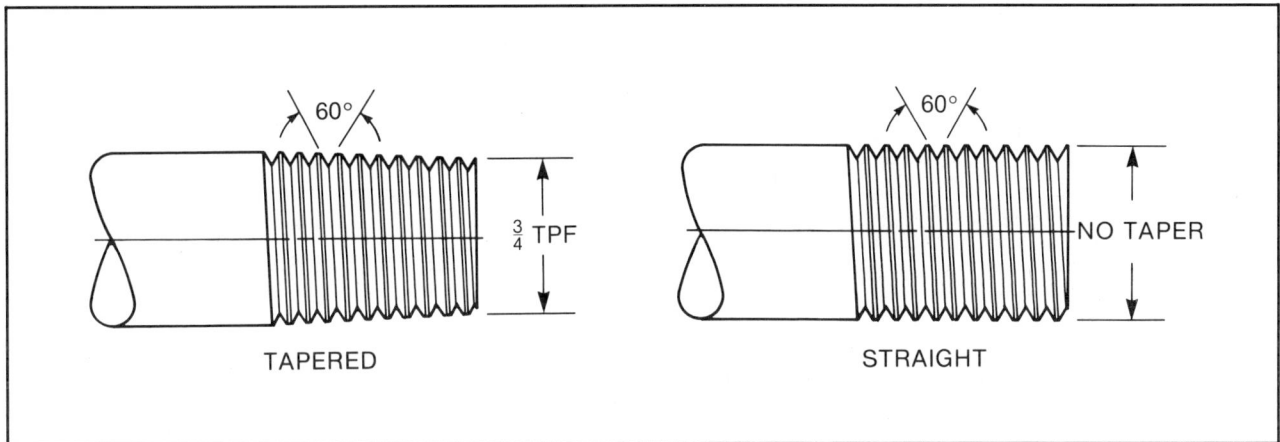

Figure 11-13. Pipe thread forms

Thread Designations

Screw threads, like everything a machinist makes, must conform to specific dimensions in order to work properly. Knowledge of how threads are specified on a print is necessary to correctly interpret thread dimensions. Since the Unified, Acme, pipe, and ISO metric are the most common thread forms, the discussion here is limited to these four forms. Information regarding the other forms can be found in handbooks.

UNIFIED. The size of a Unified thread is shown with a series of numbers and letters. This series follows the pattern shown in Figure 11-14 and must be read in the order shown. The first number set in the series ($\frac{1}{2}$) shows the nominal diameter of the thread. The number shown is either the fractional diameter, decimal diameter, or the screw size. The second number set (*13*) indicates the number of threads per inch. The next group of letters (*UNC*) identifies the thread series. In this example, Unified National Coarse is indicated.

The next number set (*2*) indicates the thread class, or *class of fit*. The standard classes are *1*, *2*, and *3*. Number 1 is the loosest fit; number 3 is the tightest fit. The letter used at the end indicates whether the thread is external or internal. The *A* means the thread is external. When the letter *B* is used, it means the thread is internal.

In some cases other information is added to the standard designation. See Figure 11-15. Here the *LH* means left hand. All threads are considered to be right hand unless the LH designation is used. The last entry in this thread designation indicates the length of thread. Length of thread may be noted in the thread designation or as a separate dimension on the print.

Figure 11-16 shows the standard sizes of threads used in the Unified series. The three standard thread series in the Unified system are Unified Coarse (UNC), Unified Fine (UNF), and Unified Extra Fine (UNEF). The numbers designate the standard number of threads per inch for each series and nominal diameter.

ACME. Acme threads are shown on prints in basically the same way as Unified threads. The principal differences are the use of the word *ACME* and the different thread classes, Figure 11-17. The thread classes commonly used for Acme threads are *2, 3, 4,* and *5,* with class 2 being the loosest and class 5 the tightest. The 6 indicates a general-purpose thread. If a *C* is used, it indicates a centralizing Acme thread. An *LH*, when used, means the thread is left hand.

PIPE. Pipe threads are specified on a print by the nominal size and threads per inch. See Figure 11-18. The letters *NPT* are used to specify taper pipe. The letters *NPS* are used to specify straight pipe.

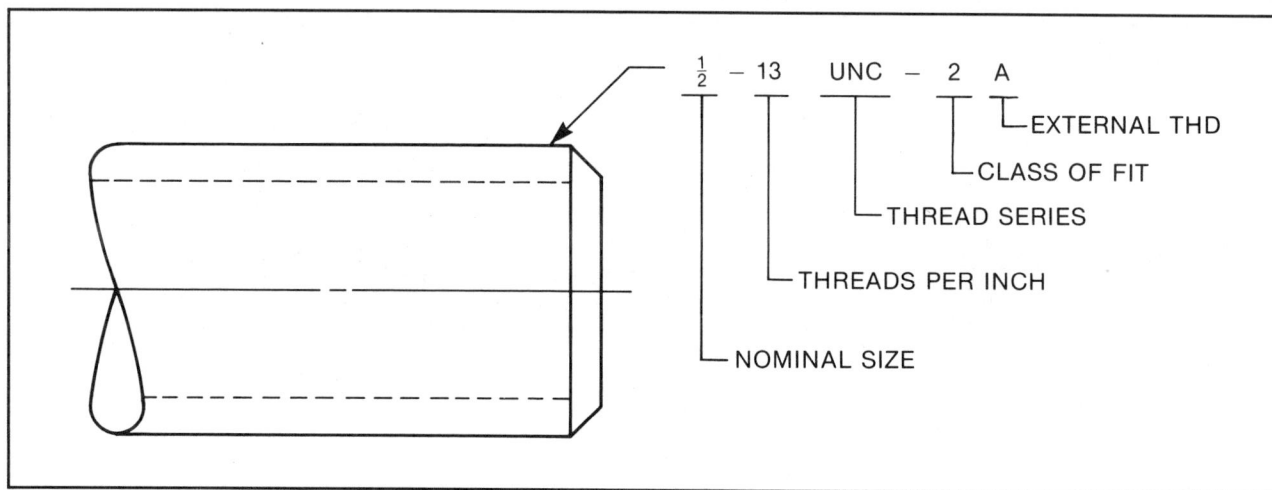

Figure 11-14. Standard thread designation

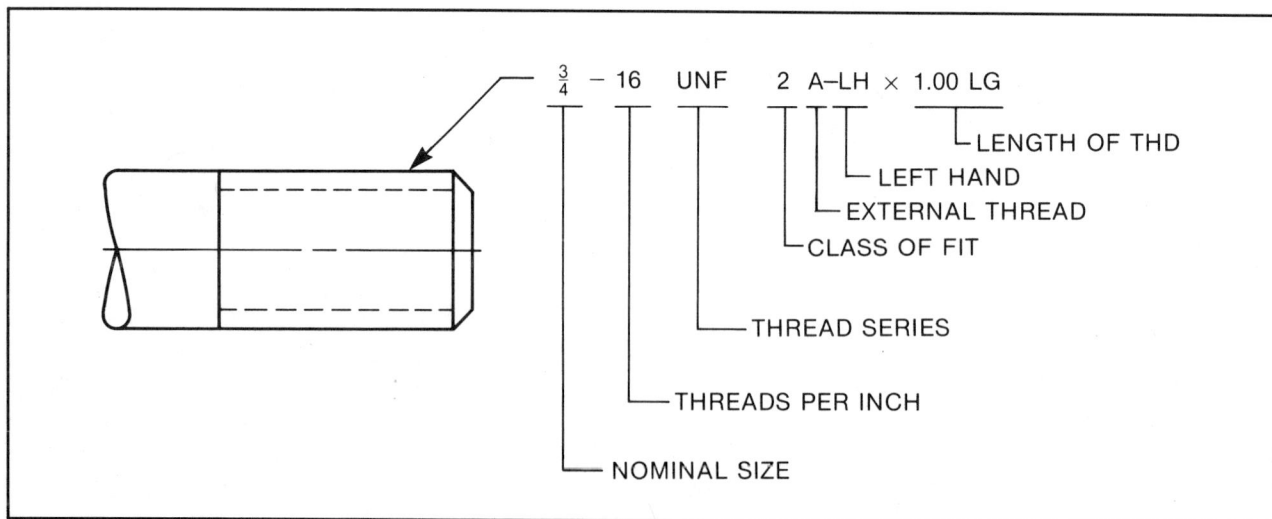

Figure 11-15. Thread designation with additional information

STANDARD THREADS FOR EACH UNIFIED THREAD SERIES			
NOMINAL SIZE	**UNC**	**UNF**	**UNEF**
0		80	—
1	64	72	—
2	56	64	—
3	48	56	—
4	40	48	—
5	40	44	—
6	32	40	—
8	32	36	—
10	24	32	—
12	24	28	32
$\frac{1}{4}$	20	28	32
$\frac{5}{16}$	18	24	32
$\frac{3}{8}$	16	24	32
$\frac{7}{16}$	14	20	28
$\frac{1}{2}$	13	20	28
$\frac{9}{16}$	12	18	24
$\frac{5}{8}$	11	18	24
$\frac{3}{4}$	10	16	20
$\frac{7}{8}$	9	14	20
1″	8	12	20
UNC—UNIFIED COARSE UNF—UNIFIED FINE UNEF—UNIFIED EXTRA FINE			

Figure 11-16. Standard sizes for Unified thread series

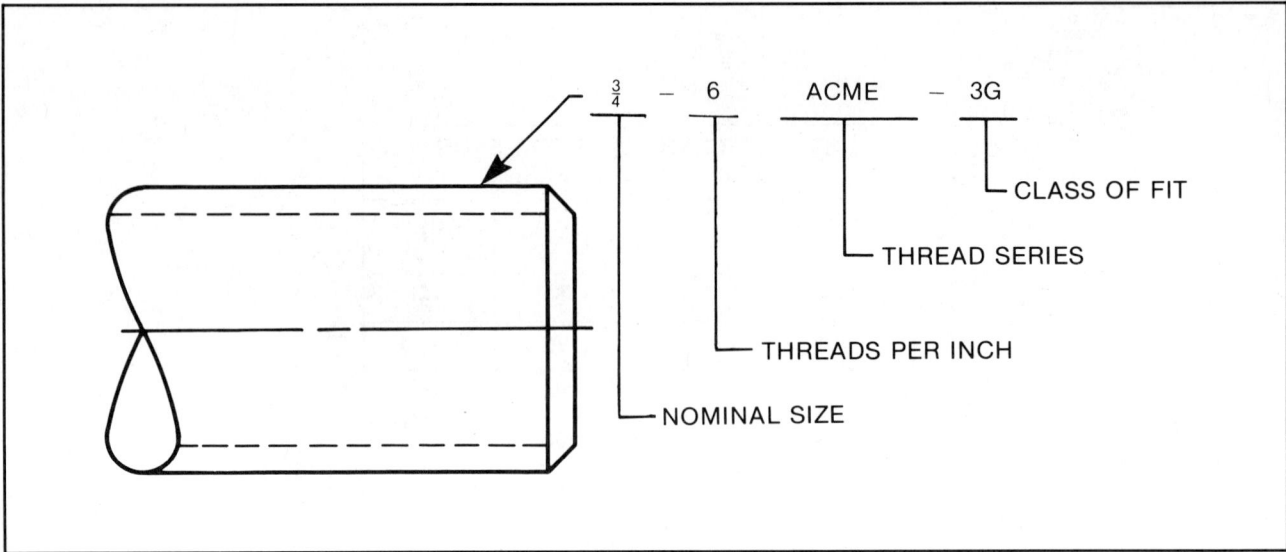

Figure 11-17. Acme thread designation

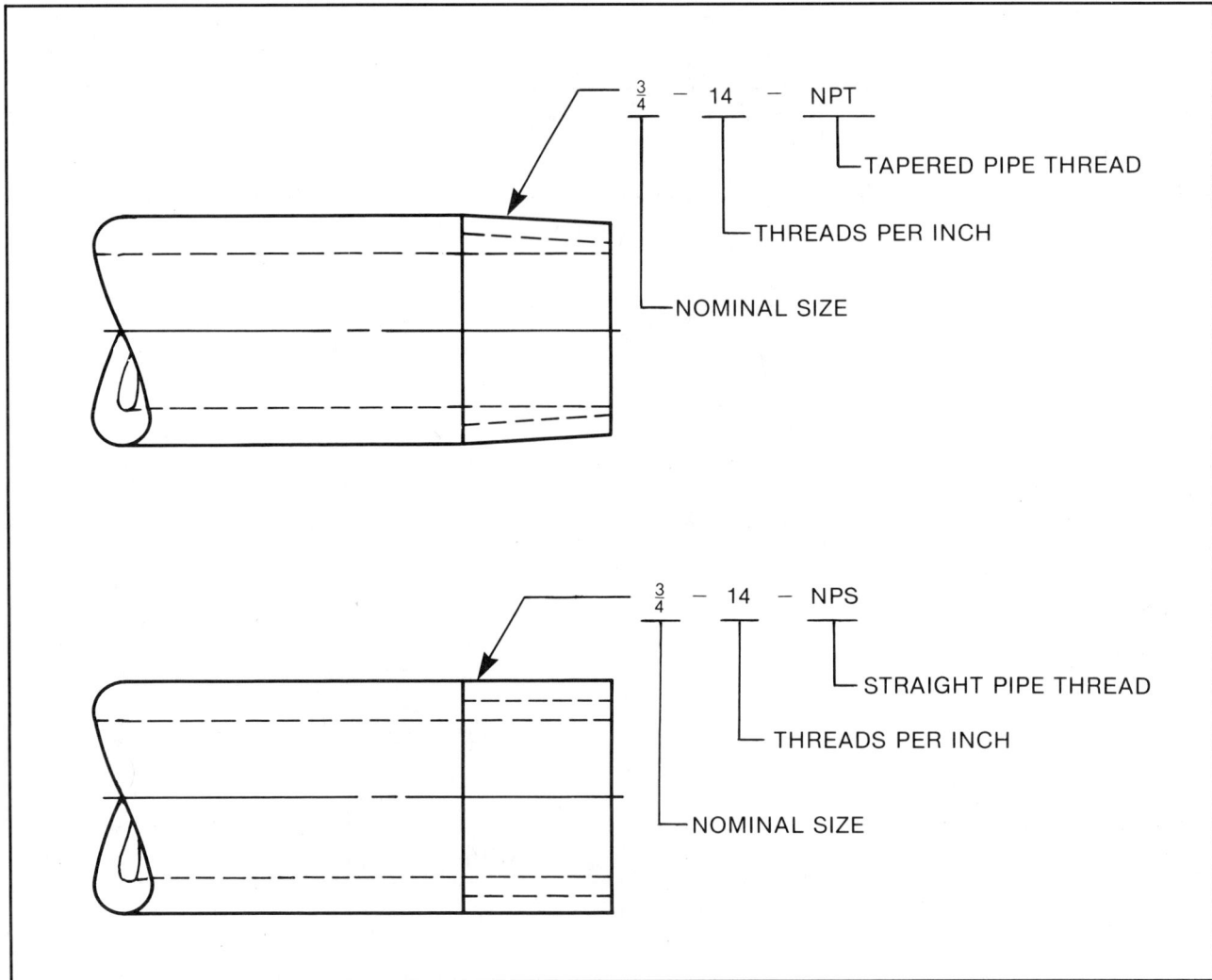

Figure 11-18. Pipe thread designations

ISO METRIC. ISO metric threads are shown on a print in much the same way as Unified threads. As shown in Figure 11-19, the letter *M* is used to identify all ISO (International Standards Organization) metric threads. The first number set (*8*) indicates the nominal diameter of the thread. The second number set (*1.25*) shows the pitch, or distance between adjacent threads, in millimeters. The last number and letter set (*4h*) indicate the class of fit. The designations *8g*, *6g*, and *4h* are used for external threads, and *7H*, *6H*, and *5H* show the class of fit for internal threads. The designations 4h and 5H are used for close fits, and 6g and 6H for medium fits. The designations 8g and 7H indicate free, or loose, fits.

Several other thread designations are in common use today, besides those discussed here. To find specific information about these thread types, consult an appropriate ISO handbook.

Representing Screw Threads

Screw threads are shown on prints by using one of three methods, Figure 11-20A. *Pictorial representation* shows threads as they actually appear. *Schematic representation* shows threads by a simpler variation of the pictorial form.

Simplified representation shows threads with a series of dashes. The schematic and simplified representations are the quickest and easiest methods of drawing screw threads. Therefore, these are the methods most frequently used.

Internal threads are designated by the same system. The only exception to this is the use of the letter *B* to indicate an internal thread. On a print, internal threads are normally shown as threaded through, threaded partway, or threaded to the bottom of the hole, Figure 11-20B. Each of these variations usually includes a note or dimension to indicate the depth of thread. When internal threads are specified, the hole that is drilled is called the *tap drill*. The standard tap drill sizes are shown in Appendix A of this book.

COMMON HARDWARE

Common hardware items are used throughout every mechanical assembly. These items include bolts, screws, nuts, washers, and pins. To properly interpret common hardware designations, it is necessary to be familiar with the hardware items commonly used for assemblies.

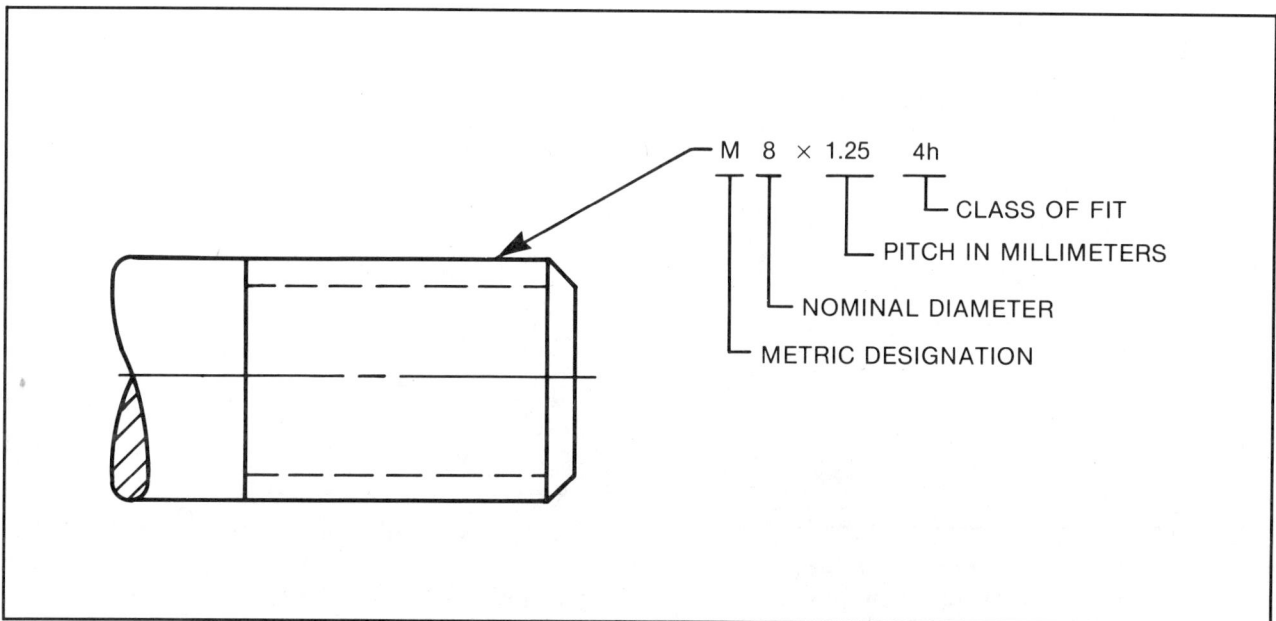

Figure 11-19. ISO metric thread designation

M 8 × 1.25 4h
— CLASS OF FIT
— PITCH IN MILLIMETERS
— NOMINAL DIAMETER
— METRIC DESIGNATION

Bolts and Screws

Bolts and *screws* are generally specified on a print by nominal size, threads per inch, thread series, head style, and length, Figure 11-21A. In special situations the thread class may also be specified. When no specification is given for the thread class, a class 2 is normally assumed, except for socket head cap and set screws which are usually 3A.

The lengths of bolts and screws are dimensioned from the bearing surface to the end of the threaded area, Figure 11-21B. The head styles most commonly used for mechanical assemblies are shown in Figure 11-22.

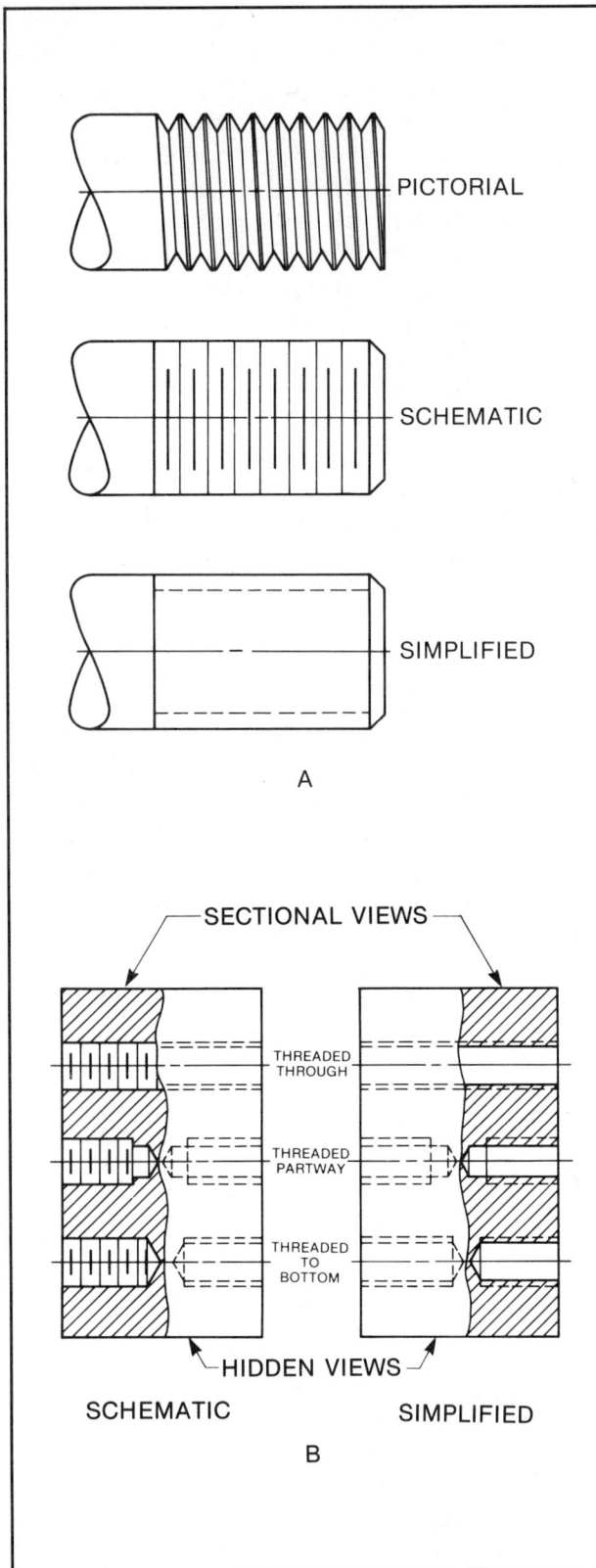

Figure 11-20. Threads are shown on prints using a pictorial, schematic, or simplified representation (A). Internal threads also show the depth of the thread (B).

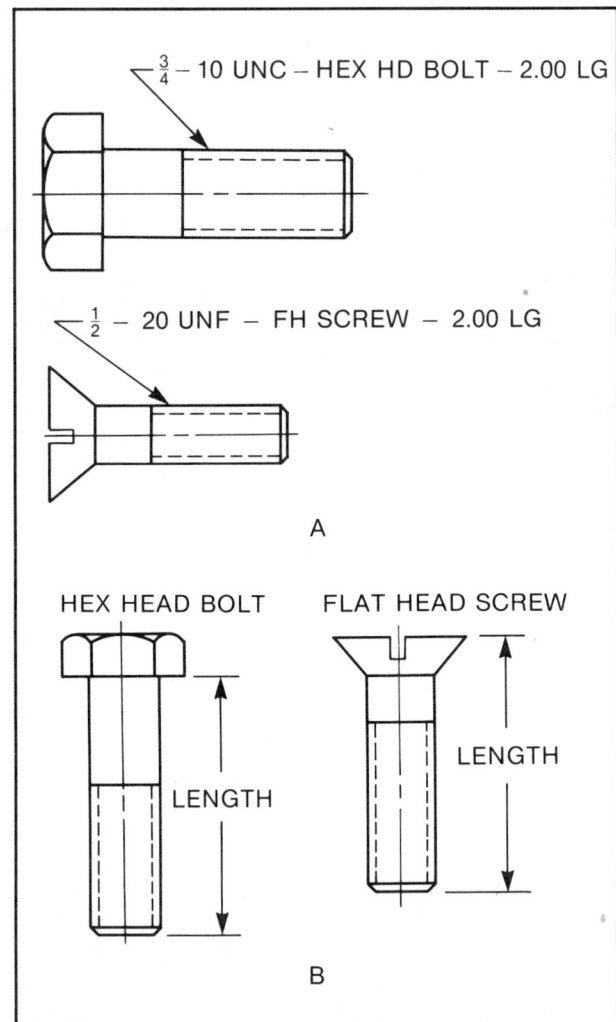

Figure 11-21. Bolts and screws are specified by nominal size, threads per inch, thread series, head style, and length (A). The lengths of bolts and screws are measured from the bearing surface to the end of the thread (B).

Nuts and Washers

Nuts and *washers* are normally specified on a print with either a simple note or drawn detail. When special nuts or washers are required, they are usually specified with a part number or other form of exact identification. Figure 11-23 shows the most common forms of nuts and washers used in mechanical assemblies.

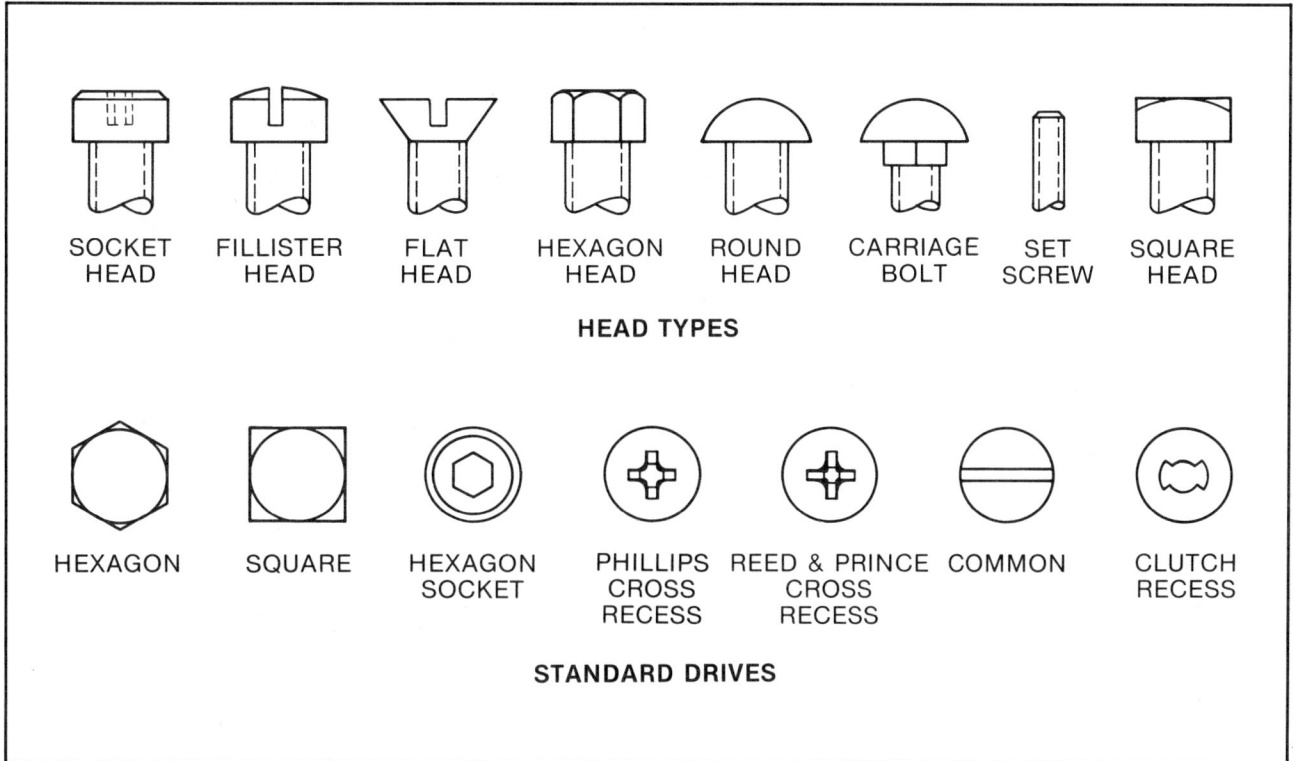

Figure 11-22. Common head styles for bolts and screws

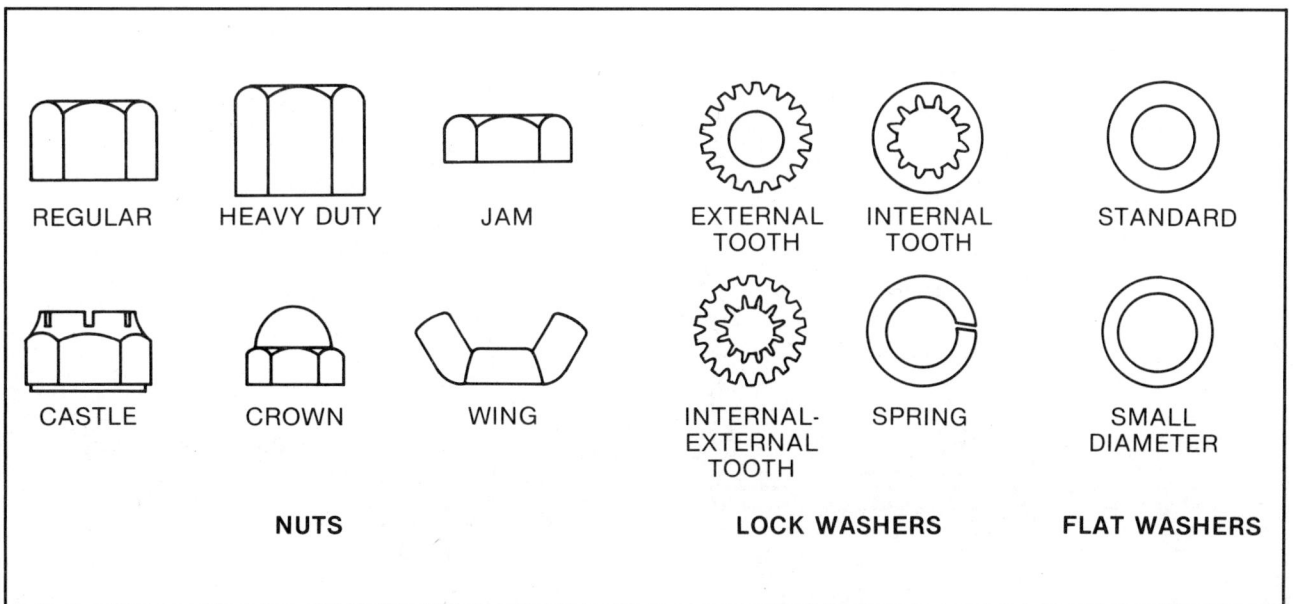

Figure 11-23. Common forms of nuts and washers

Pins

Pins are generally shown on a print by diameter, length, and style, Figure 11-24. The most common forms of pins used in the shop are shown in Figure 11-25.

Gears are a very important part of many mechanical assemblies. They transmit power and torque, increase or decrease speed, and provide

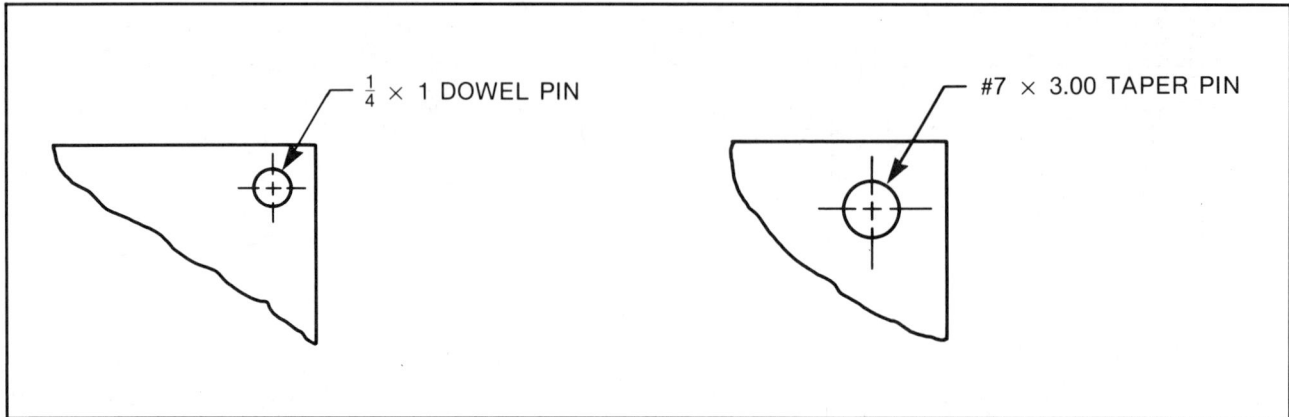

Figure 11-24. Specifying pins on a print

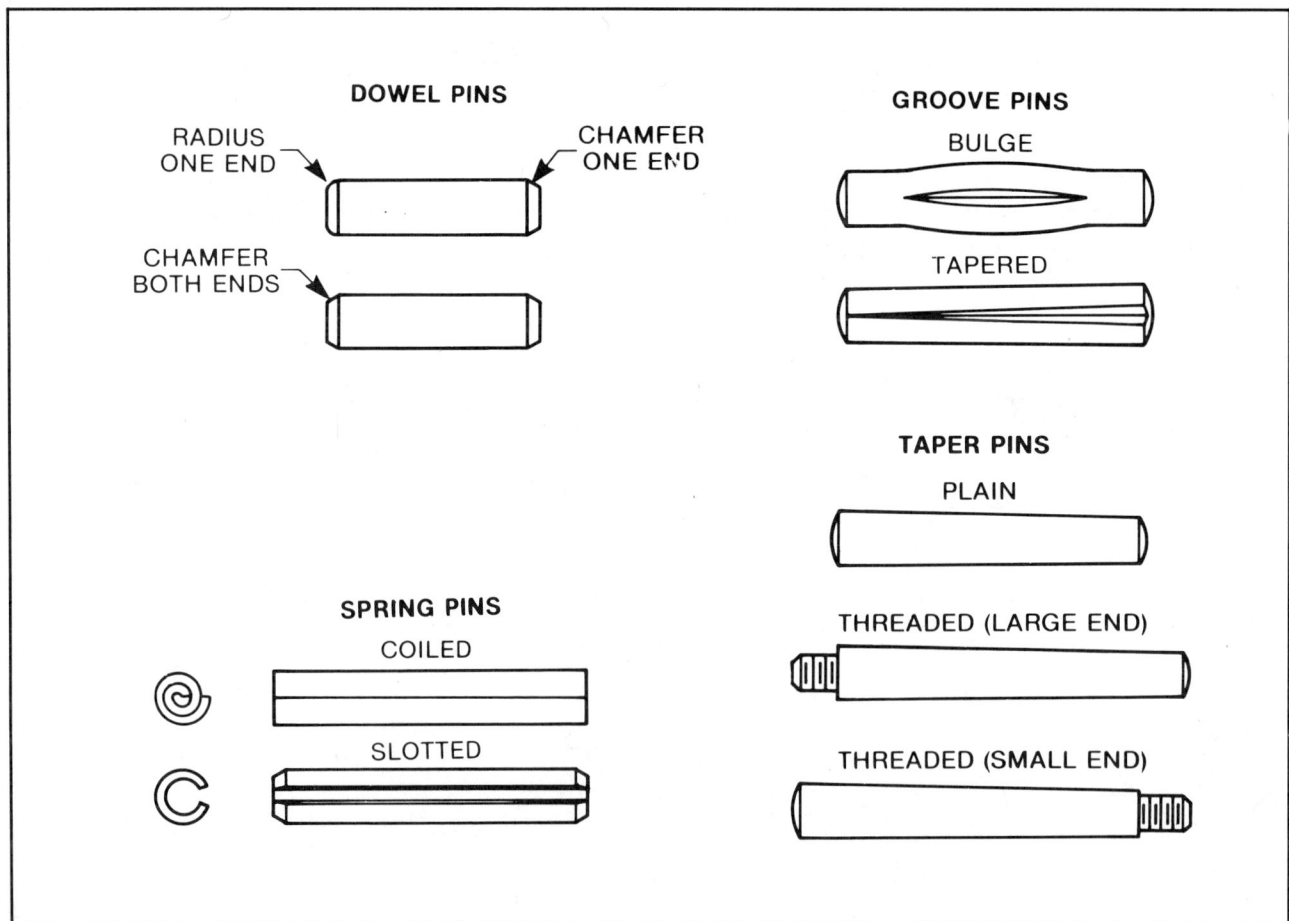

Figure 11-25. Common types of pins

a constant slip-free rate of speed. The most common types of gears are spur gears, helical gears, bevel gears, and worm gears. Machinist must be familiar with the basic types of gears, as well as the terms used to describe their values, to properly interpret gearing prints.

Spur Gears

Spur gears, Figure 11-26, are the simplest and most common type of gears used in industry today. These gears have straight teeth cut parallel with the axis of rotation of the gear body. Terms used to identify and describe spur gear values are defined in this section.

OUTSIDE DIAMETER. The distance across the center of the gear from one edge to the other. This is also the diameter of the addendum circle (Figure 11-27).

PITCH POINT. The point where meshing (running together) gears contact each other (Figure 11-27).

Figure 11-26. Spur gears

PITCH DIAMETER. The distance across the center of the gear from the pitch point on one side to the pitch point on the opposite side. This is the diameter from which most of the other gear values are calculated (Figure 11-27).

PITCH CIRCLE. An imaginary circle formed by the pitch diameter (Figure 11-27).

ROOT DIAMETER. The distance across the center of a gear from the bottom of the tooth on one side to the bottom of a tooth on the opposite side (Figure 11-27). This is also the diameter of the dedendum circle.

ADDENDUM. The radial distance between the pitch circle and the top of each tooth (Figure 11-28A).

DEDENDUM. The radial distance between the pitch circle and the bottom of each tooth (Figure 11-28A).

CIRCULAR THICKNESS. The thickness of a tooth measured along the pitch circle (Figure 11-28A).

CHORDAL THICKNESS. The thickness of a tooth measured along a straight line, or *chord*, that connects the two points where the pitch circle contacts the finished contour of the tooth on both sides (Figure 11-28B). This value is normally found in handbook charts.

CHORDAL ADDENDUM. The length of a straight line between the top of each tooth and the chordal thickness line (Figure 11-28B). This value is usually found in handbook charts.

CLEARANCE. The amount by which the dedendum of a gear is larger than the addendum of its mating gear. Also, the space between the top of one tooth and the bottom of the tooth space in meshing gears (Figure 11-28C).

WHOLE DEPTH. The depth at which the gear is cut, or the total value of the addendum plus the dedendum (Figure 11-28C).

WORKING DEPTH. The depth of engagement between meshing gears. It is equal to the addendum plus the dedendum minus the clearance (Figure 11-28C).

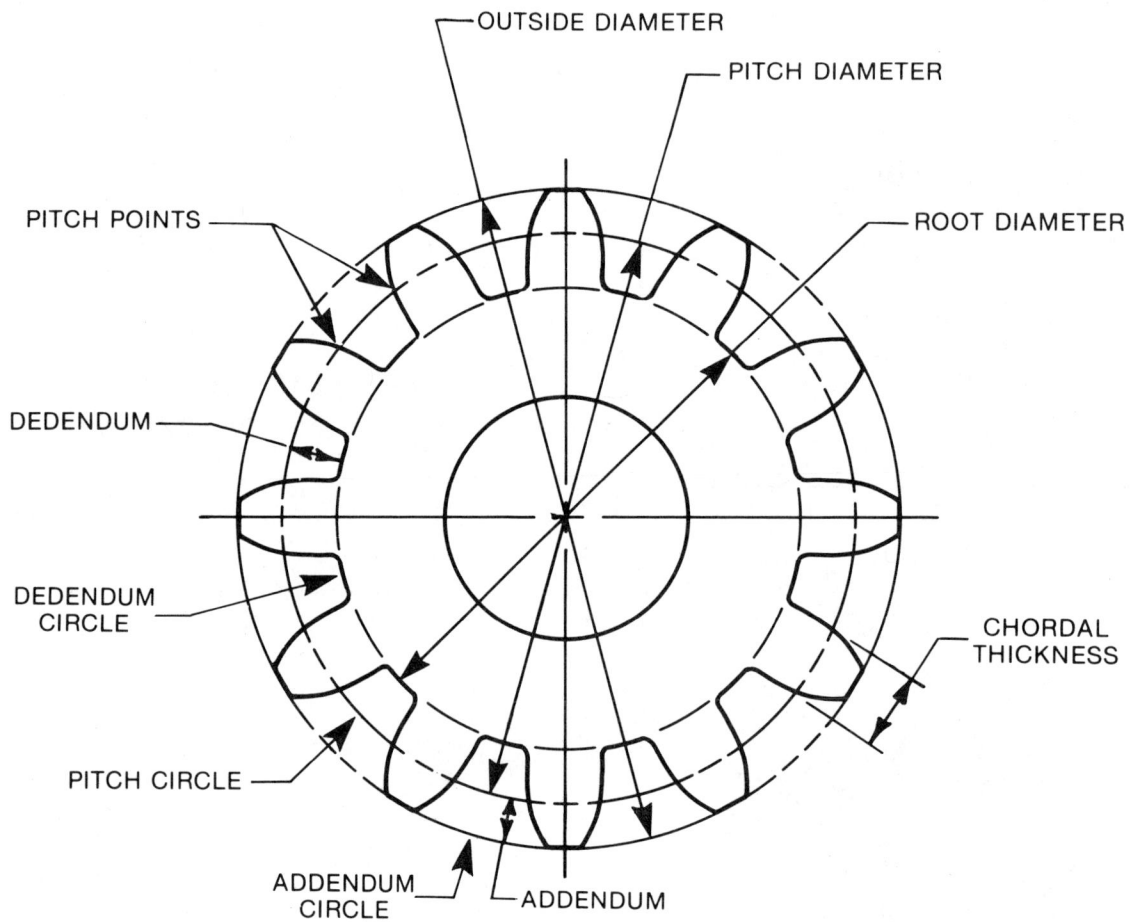

Figure 11-27. Spur gear terms

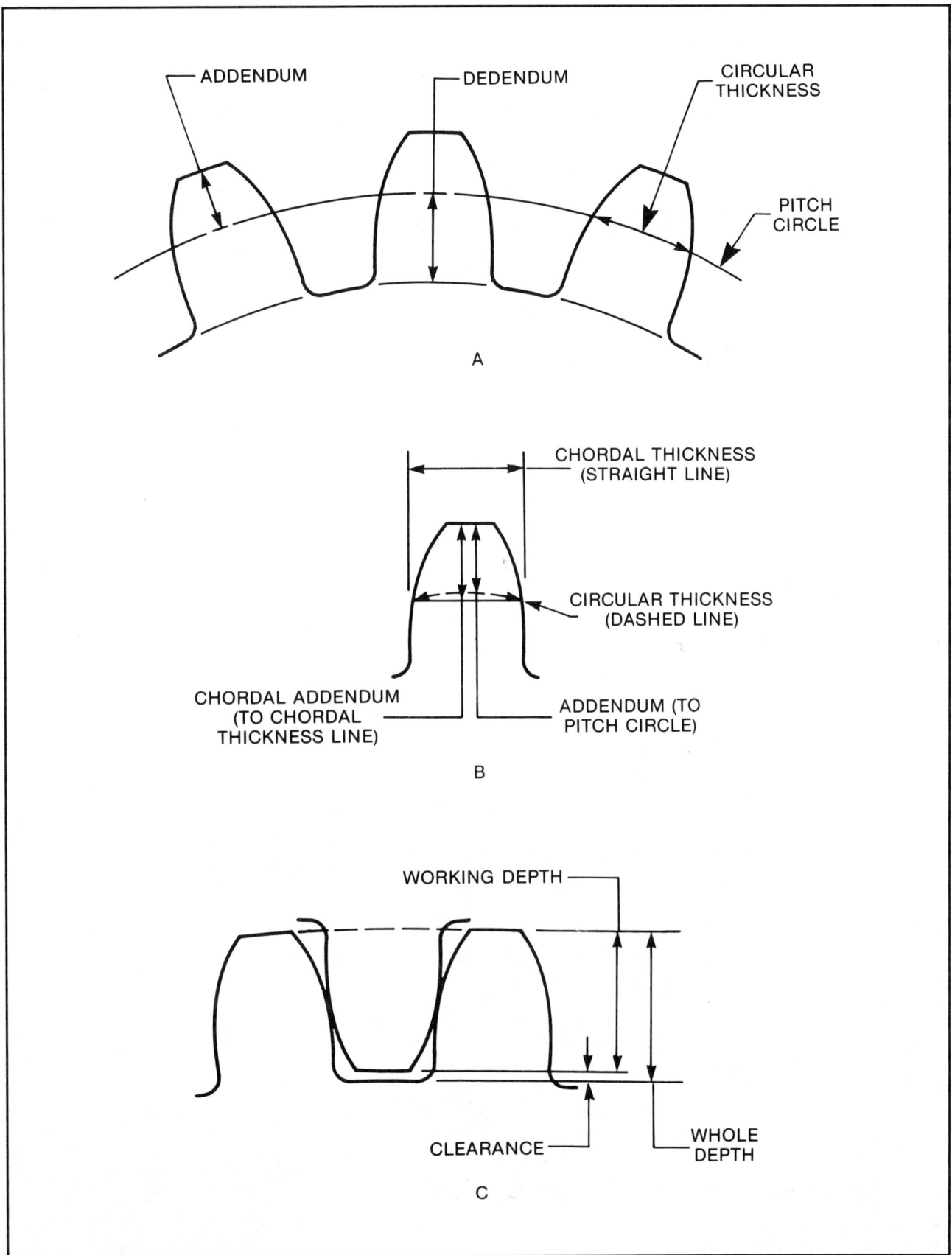

Figure 11-28. Gear tooth terms

CENTER DISTANCE. The distance between the center axes of mating gears (Figure 11-29).

CIRCULAR PITCH. The distance from a point on one tooth to a corresponding point on the next adjacent tooth (Figure 11-30).

DIAMETRAL PITCH. The ratio of the number of teeth to the pitch diameter. If a gear has 12 teeth and a pitch diameter of 3.00″, the diametral pitch is $\frac{12}{3} = 4$ (Figure 11-30). This is also the factor that determines the size of the teeth on a gear. The larger the diametral pitch number, the smaller the size of the teeth.

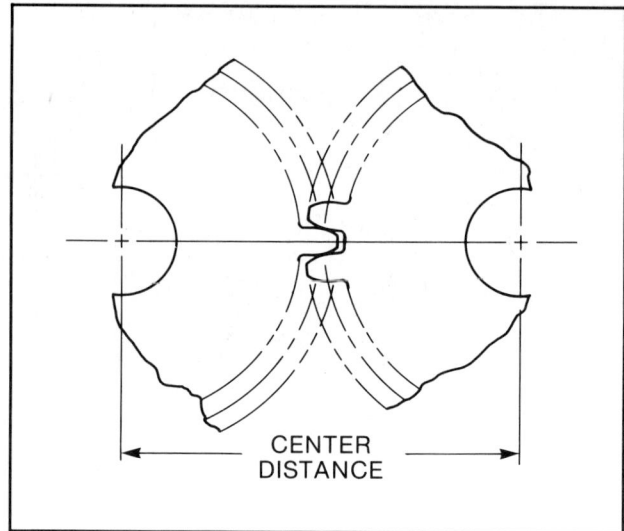

Figure 11-29. Center distance between gears

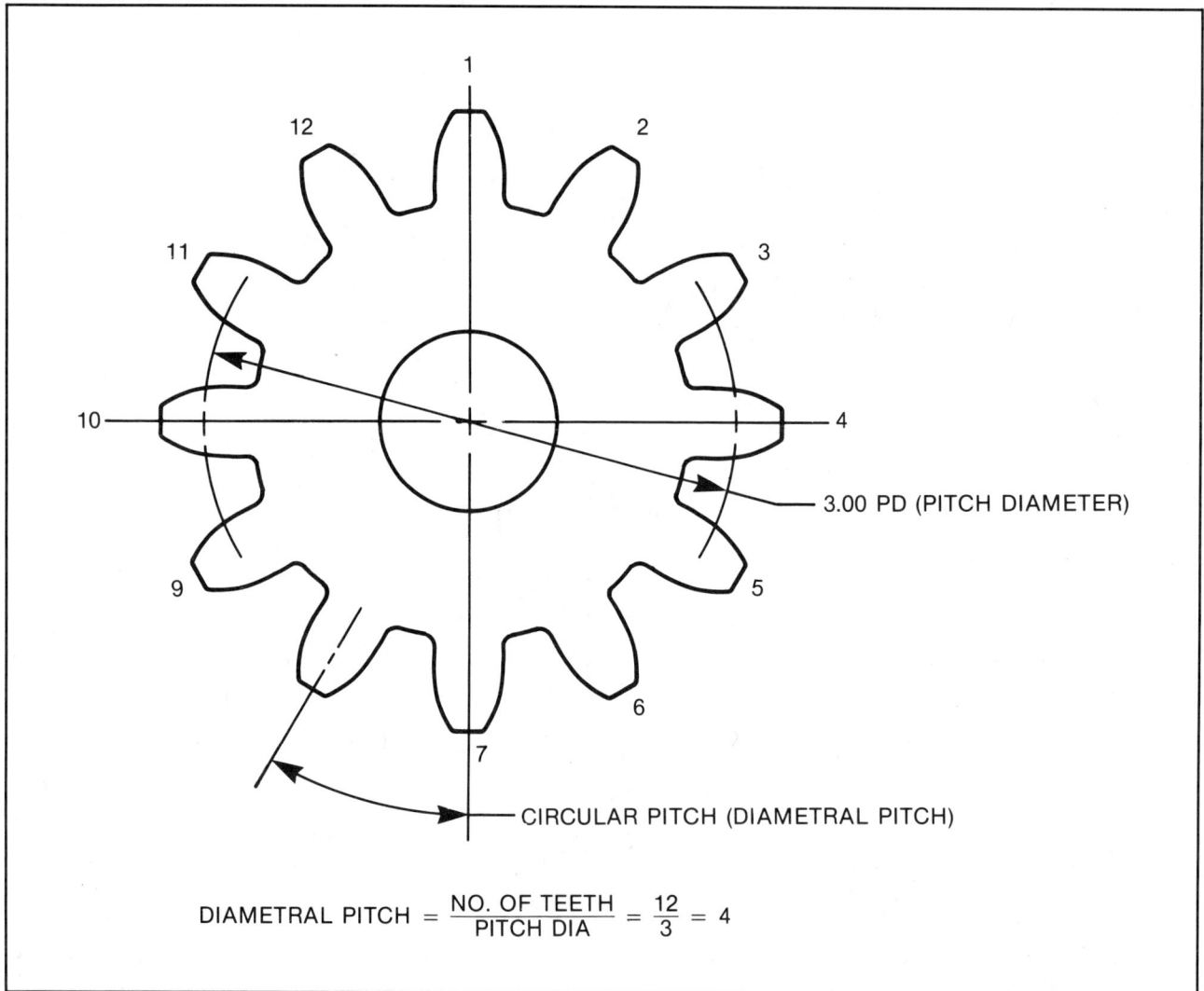

$$\text{DIAMETRAL PITCH} = \frac{\text{NO. OF TEETH}}{\text{PITCH DIA}} = \frac{12}{3} = 4$$

Figure 11-30. Circular pitch and diametral pitch

Spur Gear Specifications

Spur gear dimensions are generally divided into two groups: gear blank and gear cutting dimensions. *Gear blank dimensions* deal directly with the gear blank and are usually shown on the drawn detail. The *gear cutting dimensions* refer to the actual cutting of the gear, and are generally shown in a data block on the print, Figure 11-31. Often the gear cutting dimensions will include the note *REF* in parentheses. In these cases, the note is used to indicate a calculated dimension that cannot be measured directly.

Spur gears are usually shown in prints by a front view and a full section side view. Complete gear profiles are rarely shown on prints. Generally the addendum circle and the dedendum circle are shown with phantom lines, and the pitch circle is shown with a center line. Sometimes a few gear teeth are drawn for clarity and to avoid confusion.

In addition to the external spur gears already discussed, there are two other spur gear forms: the internal gear and the rack. An *internal spur gear*, Figure 11-32A, is basically the same as an external gear with the exception of the location and the shape of the teeth. Internal spur gear teeth have the same shape as the spaces between the teeth of the mating external spur gear. A *rack gear* is simply a flat gear. The rack and gear is often used to transfer the rotary motion of an external spur gear into linear motion, Figure 11-32B.

Helical Gears

A *helical gear*, Figure 11-33, is a variation of the spur gear. Helical gears have teeth cut at an angle to the axis of rotation. These gears are cut by advancing a cutter across the gear blank while the blank slowly revolves. This method produces a tooth form that runs more smoothly and quietly than normal spur gears. While spur gears engage across the entire face of the tooth, helical gears engage on a small area that moves across the face of the tooth.

Helical gears may be used for parallel gear shafts, or for shafts that are at an angle to each other. Helical gears are either right hand or left hand. The hand of a gear may be determined by looking at the tooth. If the tooth runs down on the right side, it is a right-hand helix. If the tooth runs down on the left side, it is a left-hand helical gear. See Figure 11-34.

The values and terms associated with helical gears are similar to those used for spur gears. The following list defines a few different terms.

NORMAL CIRCULAR PITCH. The distance from the center of one tooth to the center of the next, measured at the pitch line at a right angle to the tooth (Figure 11-35).

NORMAL DIAMETRAL PITCH. The diametral pitch of the cutter needed to cut the teeth to the proper size and form.

HELIX ANGLE. The angle of the teeth to the axis of the gear bore (Figure 11-35).

LEAD OF HELIX. The linear distance the tooth would travel in one 360° revolution if it were free to move axially.

Helical Gear Specifications

Helical gears are usually specified in much the same way as spur gears. The only major difference is in the additions to the gear cutting data block. In most cases the drawn view of the helical gear is the same as that used for a spur gear.

Bevel Gears

Bevel gears, Figure 11-36 (page 204), are characterized by a variable pitch diameter, tooth depth, and tooth thickness. Like helical gears, bevel gears are used for applications where the gears must be positioned at an angle to each other. The most common type of bevel gears run at a 90° angle. These gears are called *miter gears*.

Bevel gears can best be compared to cones that have teeth. Unlike spur gears and helical gears, which engage on their outer surfaces, bevel gears engage, or mesh, along the angular side of their conical form.

CUTTING DATA

NO. OF TEETH	—	16
DIAMETRAL PITCH	—	5
PITCH DIA	—	Ø3.200
CIRC. THICKNESS	—	(.315)
ADDENDUM	—	(.200)
WHOLE DEPTH	—	.431
CHORDAL THICKNESS	—	.314
CHORDAL ADDENDUM	—	.208

Ø 3.432 / 3.431

Ø 1.002 / 1.000

4X R .12

2X 45° × .06

Ø 2.50 / 2.49

Ø 1.88 / 1.87

.25

.50 / .49

.750 / .748

.251 / .249

1.114 / 1.112

R .02 MAX

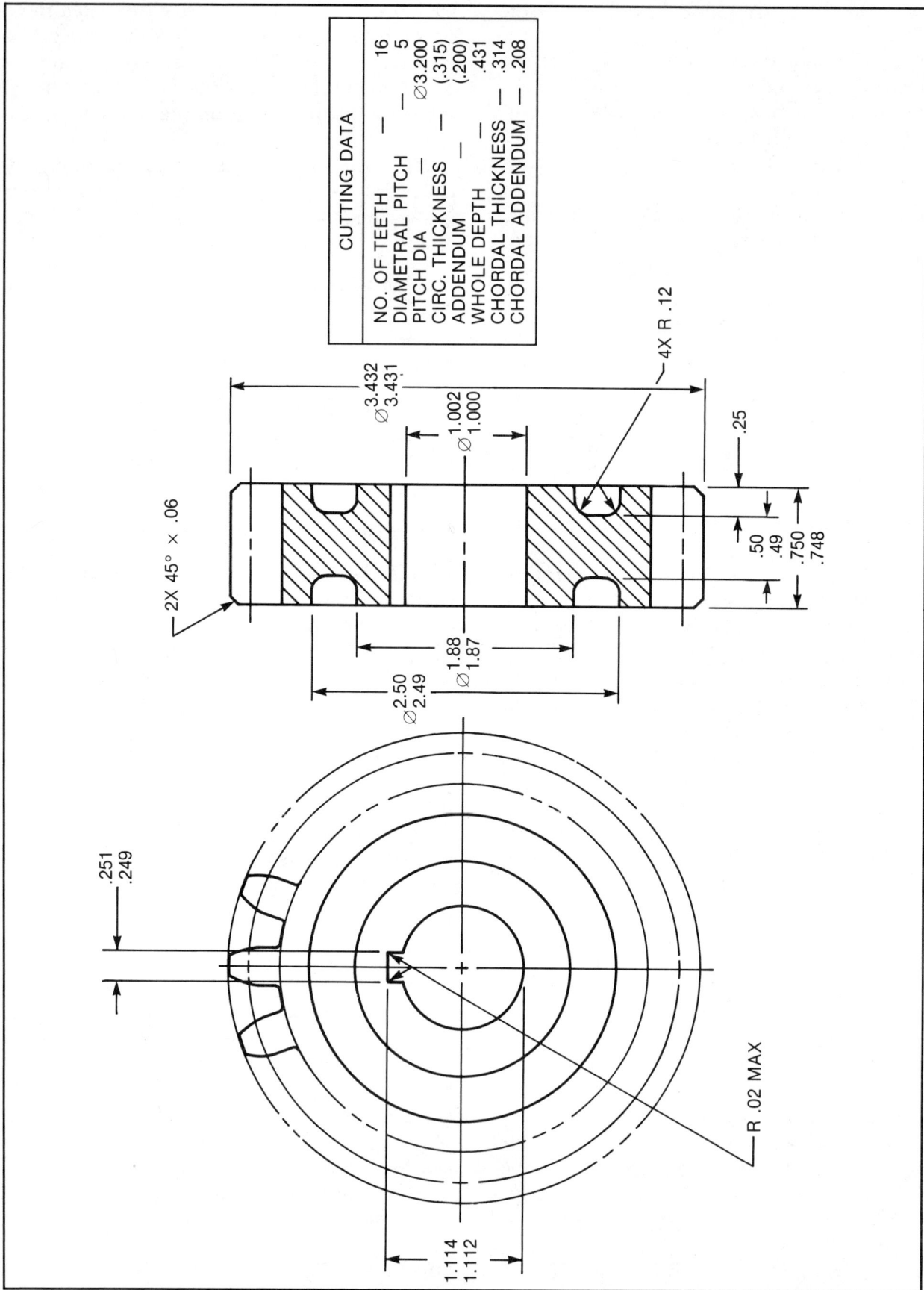

Figure 11-31. Spur gear print

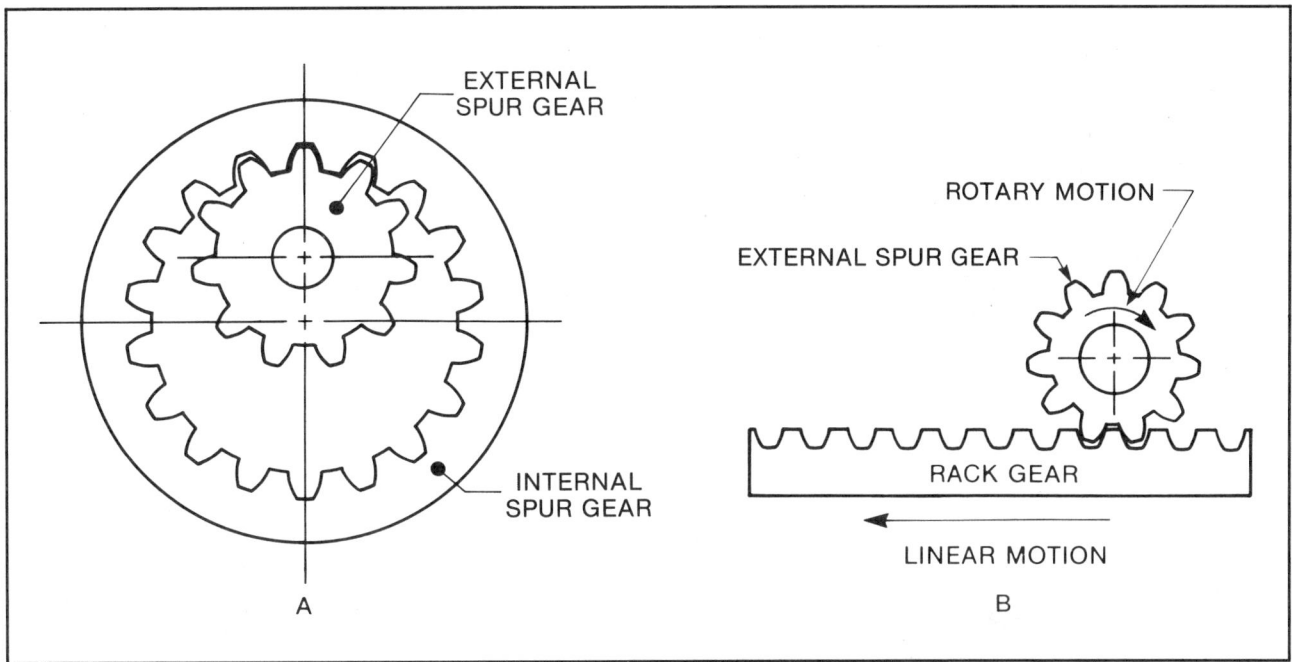

Figure 11-32. In addition to external spur gears, two other spur gear forms are (A) the internal spur gear and (B) the rack spur gear.

Figure 11-33. Helical gears

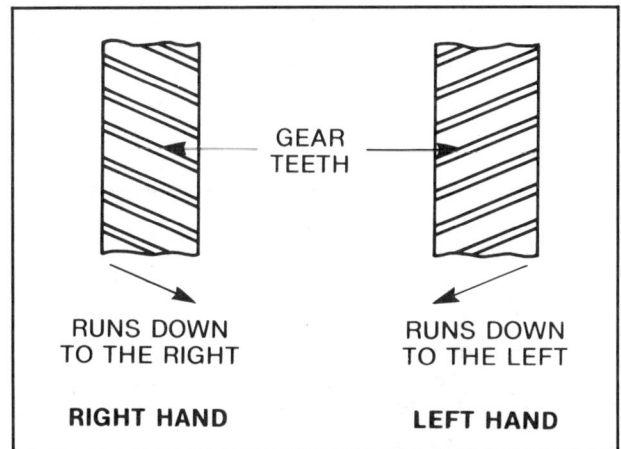

Figure 11-34. Determining the hand of a helical gear

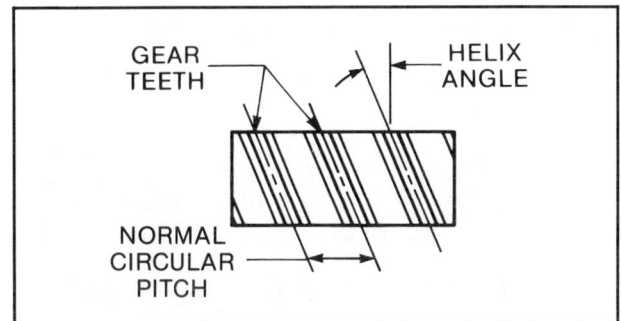

Figure 11-35. Helical gear terms

Figure 11-36. Bevel gears

While bevel gears are similar to other forms of gears, there are a few terms that apply only to bevel gears. These terms are defined here, and are illustrated in Figure 11-37.

PITCH CONE DISTANCE. The radius of the imaginary cone formed by the gear face.

ADDENDUM ANGLE. The angular difference in the addendum from the small end to the large end of the tooth.

DEDENDUM ANGLE. The angular difference in the dedendum from the small end to the large end of the tooth.

FACE ANGLE. The angle of the face of the teeth measured from the center axis of the gear bore.

CUTTING ANGLE. The angle on which the teeth are actually cut, measured from the center axis of the gear bore.

ANGULAR ADDENDUM. The addendum measured parallel to the axis of the gear bore.

PITCH CONE ANGLE. The angle of the gear in relation to the mating gear.

FACE WIDTH. The width of the teeth measured along the pitch cone radius.

Bevel Gear Specifications

Bevel gears are normally shown on a print as full sections. When two mating bevel gears are required, they may both be shown in their

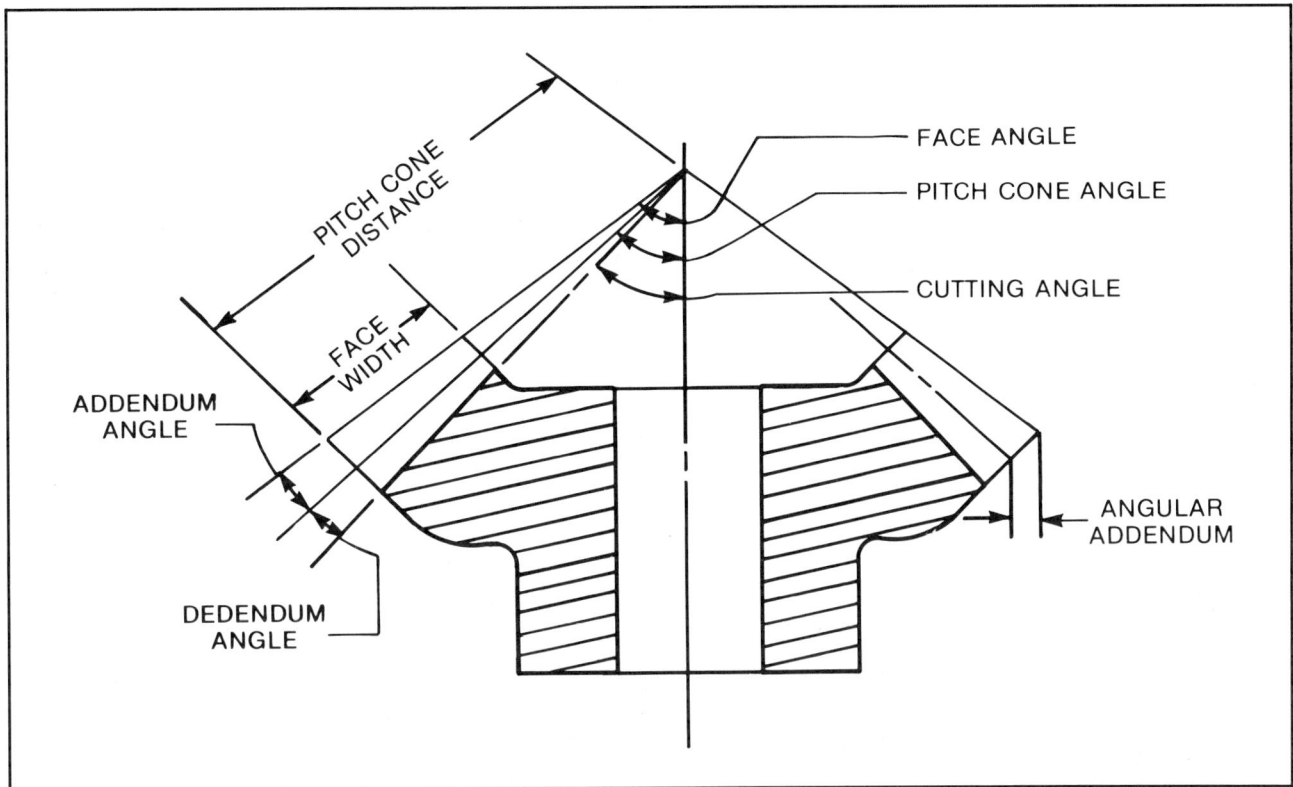

Figure 11-37. Bevel gear terms

operational position. In this case, the gear cutting data block is divided into two sections, one for the driver gear and one for the driven gear. Depending on their construction, they may also be called a pinion (driver) gear and a ring (driven) gear.

Worm Gears

Worm gearing is primarily used to obtain great reduction in speed and an increase in power. The two basic parts that are included in a worm drive are the *worm* and the *worm gear*, Figure 11-38. Due to their basic design, the worm is always the driver and the worm gear is always the driven gear in this system. Worm gearing can be compared to an endless screw. As the worm rotates, the worm gear revolves, much the same as the movement of a nut along a bolt. The helix angle of the thread is what causes the worm to drive the worm gear. Therefore, the greater the helix angle, the faster the movement.

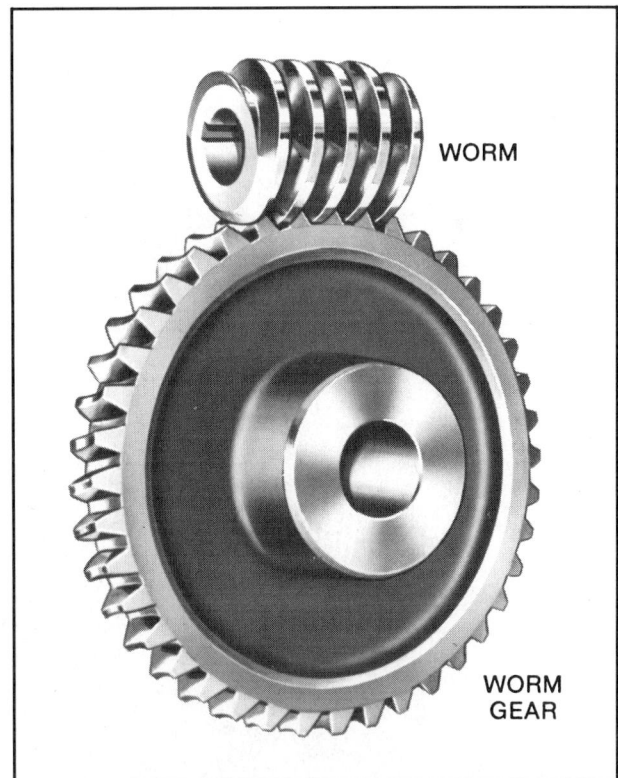

Figure 11-38. Worm gearing

Worms and worm gears, like the other gear forms, have their own special terms and values. The following list defines the terms associated with worms and worm gears. The terms are illustrated in Figure 11-39.

THROAT DIAMETER. The diameter of an imaginary circle connecting the tops of the worm gear teeth at their center points.

RADIUS OF CURVATURE. The size of the radius formed by the curvature of the worm gear tooth.

LINEAR PITCH. The distance between corresponding points on adjacent threads on the worm. This value is equal to the circular pitch of the worm gear.

LEAD. The distance a thread advances in one revolution. With single lead worms, the lead is equal to the linear pitch. With multiple lead worms, the lead is a multiple of the number of the lead. For example, a double lead thread has a lead equal to twice the linear pitch. A triple lead worm has a lead that is three times the size of the linear pitch.

LEAD ANGLE. The angle of the thread measured perpendicular to the thread. This value is also called the *helix angle*.

Worm Gear Specifications

Worms and worm gears are sometimes shown as individual details in a print. In some cases, only the worm gear is shown, and the cutting data of the worm is shown in the data block along with the data necessary to make the worm gear. As is the case with other forms of gears, the drawn views are normally intended to show the dimensions needed to make the gear blank. All relevant gear cutting data is contained in a special data block on the print.

Metric Gears

Metric gears use basically the same values as the inch type of gears. The only major difference is the use of the term *module* in place of the terms *diametral pitch* and *circular pitch*. While diametral pitch is defined as the ratio of the number of teeth per inch of pitch diameter, the module is equal to the length of pitch diameter per tooth, expressed in millimeters. Therefore, rather than being a ratio, the module is a dimension.

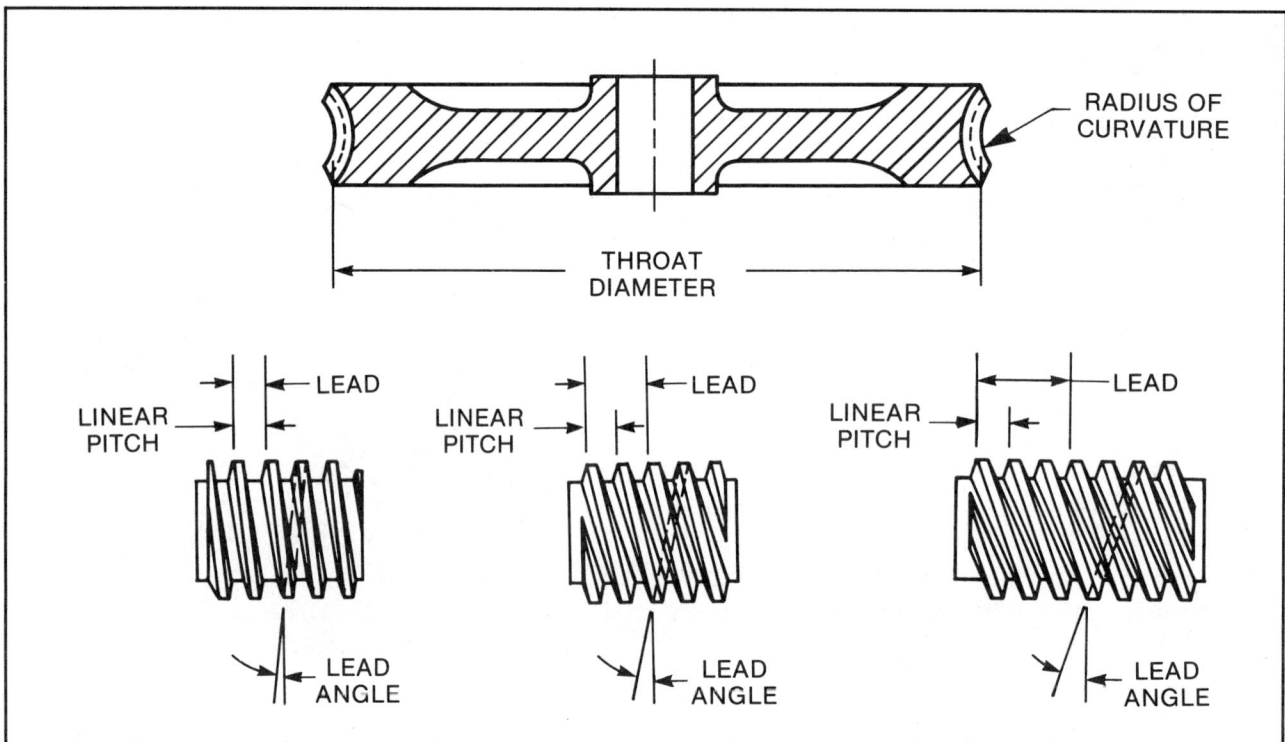

Figure 11-39. Worm gearing terms

The formulas below can be used to convert diametral and circular pitch values into modules.

$$\text{Module} = \frac{\text{Pitch Diameter in Millimeters}}{\text{Number of Teeth}}$$

OR

$$\text{Module} = \frac{25.4 \text{ (Conversion Value)}}{\text{Diametral Pitch}}$$

SURFACE FINISH DESIGNATIONS

Surface finish designations are symbols used to convey specific instructions from the designer to the machinist concerning the surface finish of a part. Often surfaces must be machined to certain specifications in order to perform properly. These symbols relate all necessary data concerning surface finish and texture.

Surface Texture Symbols

The basic surface texture symbol, Figure 11-40, is a V-shaped checkmark. This symbol only indicates that the referenced surface is to be finished. It does not indicate *how* it is to be finished. Two variations of the basic symbol are shown in Figure 11-41. The symbol shown in Figure 11-41A indicates that the surface is to be finished and that material has been provided for finishing. The number to the left of the symbol shows the amount of material to be removed when finishing. The symbol shown in Figure 11-41B indicates a finished surface, one in which no material may be removed from the referenced surface. This symbol is typically used when a surface has been finished by die casting, forging, or a similar process.

When a number is placed on the surface texture symbol, Figure 11-42A, it indicates the maximum roughness height value. When two numbers are used, Figure 11-42B, they represent the maximum and minimum roughness height values.

These symbols are the symbols most commonly used to indicate surface finishes. However, when more control is needed, other values are added to the basic symbol, Figure 11-43. To

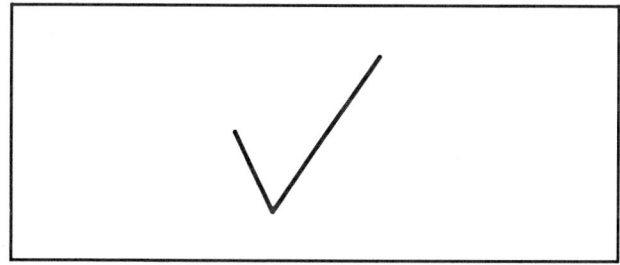

Figure 11-40. Basic surface texture symbol

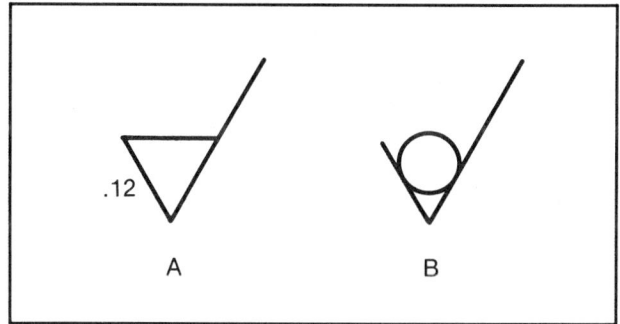

Figure 11-41. Variations of the surface texture symbol

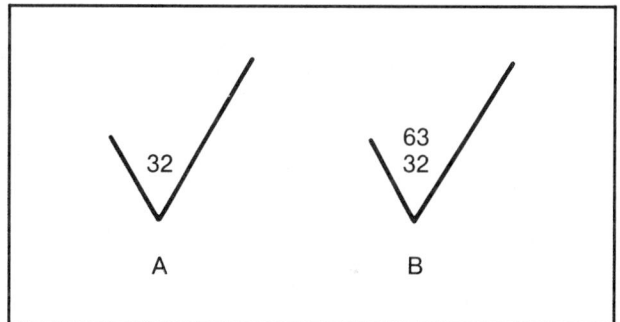

Figure 11-42. Specifying (A) the maximum roughness height value and (B) the maximum and minimum roughness height values on the surface texture symbol

thoroughly understand surface finish specifications, it is necessary to be familiar with the meanings of these additional values.

Surface Texture Terms

The following list defines the terms used to describe the different values shown in a surface texture symbol.

ROUGHNESS. The fine irregularities in the surface finish caused by the machining processes used to produce the finish.

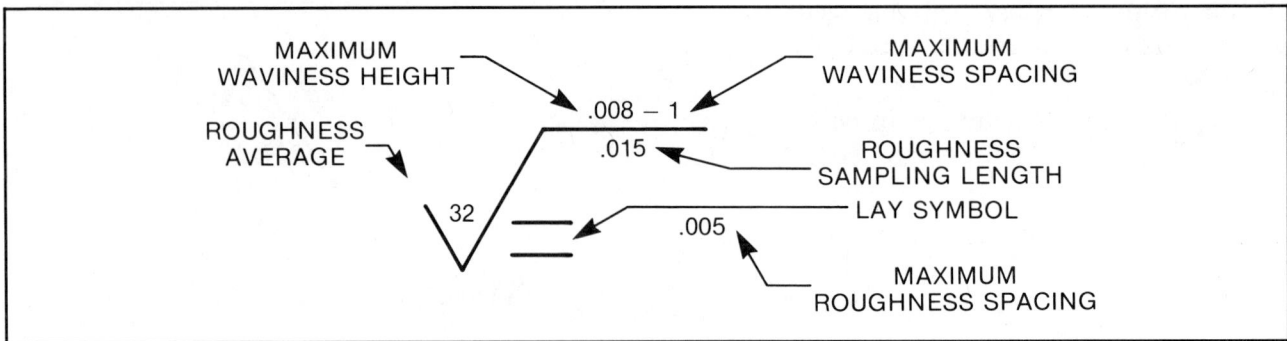

Figure 11-43. Complete surface texture symbol and entries

WAVINESS. The larger, more widely spaced irregularities of surface texture. Roughness is often thought of as being superimposed on the wavy surface.

LAY. The direction of the predominant marks of the surface pattern.

WAVINESS HEIGHT. The peak-to-valley height of the waviness within the waviness spacing (Figure 11-44A).

WAVINESS SPACING. The average spacing between the adjacent peaks of waviness (Figure 11-44A).

ROUGHNESS AVERAGE. The arithmetical average of the roughness of a surface within a certain roughness sampling length. This value is measured from a center line and is expressed in microinches. One microinch equals one-millionth (.000001) of an inch. See Figure 11-44B.

ROUGHNESS SAMPLING LENGTH. The length of the sample used to determine the roughness average (Figure 11-44B).

ROUGHNESS SPACING. The average spacing between adjacent peaks of the roughness profile (Figure 11-44B).

LAY SYMBOLS. The symbols used to identify the desired pattern of surface roughness (Figure 11-45).

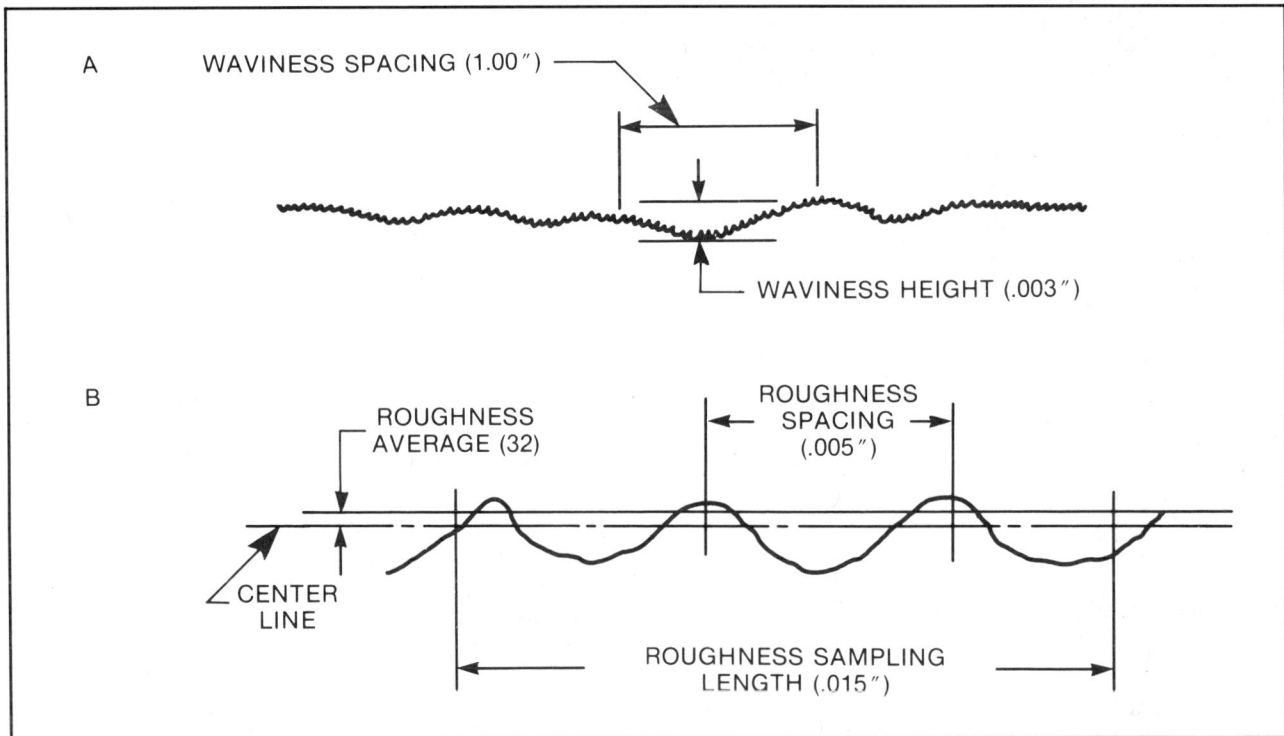

Figure 11-44. Surface texture terms

LAY SYMBOL	MEANING	EXAMPLE SHOWING DIRECTION OF TOOL MARKS
=	Lay approximately parallel to the line representing the surface to which the symbol is applied	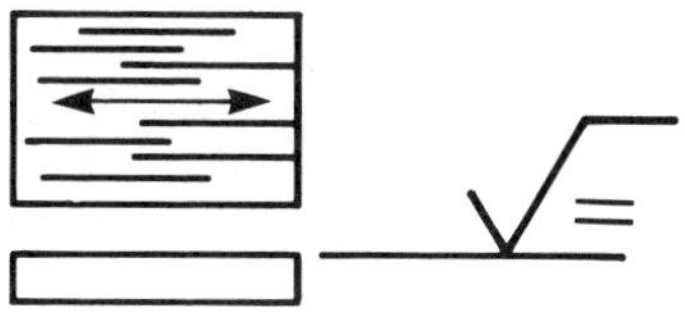
⊥	Lay approximately perpendicular to the line representing the surface to which the symbol is applied	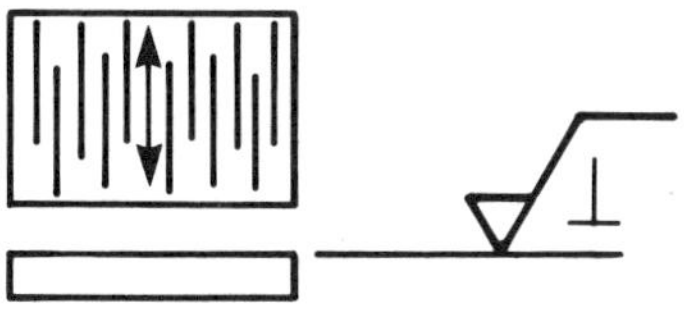
X	Lay angular in both directions to the line representing the surface to which the symbol is applied	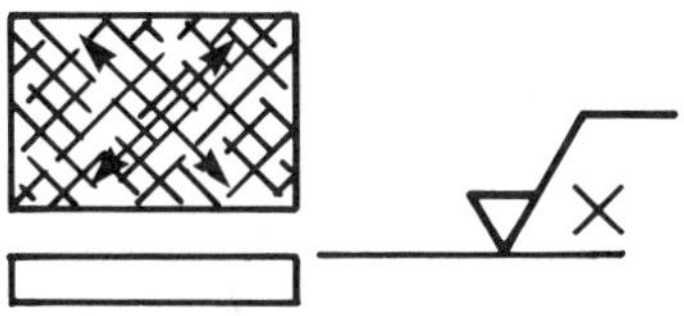
M	Lay multidirectional	
C	Lay approximately circular relative to the center of the surface to which the symbol is applied	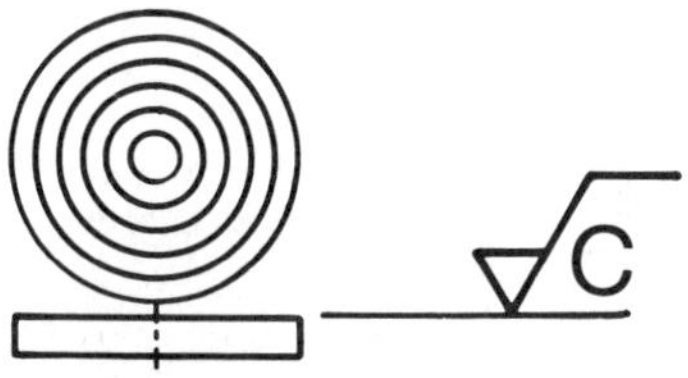
R	Lay approximately radial relative to the center of the surface to which the symbol is applied	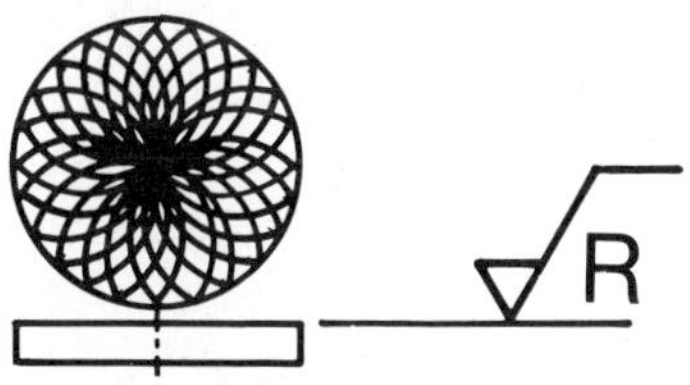
P	Lay particulate, nondirectional, or protuberant	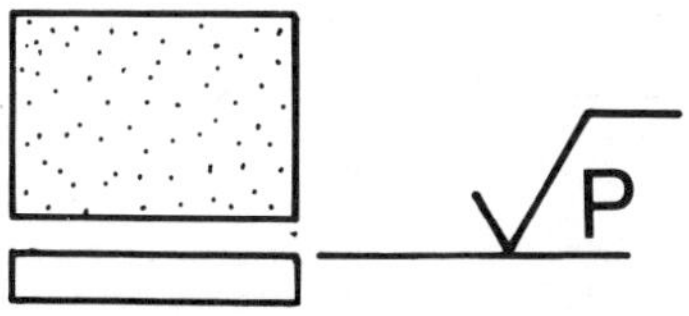

Figure 11-45. Lay symbols

Using Surface Texture Symbols

Surface texture symbols are always placed on a line that represents the surface to be finished. The line may be the surface itself, an extension line from the surface, a leader line to the surface, or included as part of a diameter dimension, Figure 11-46.

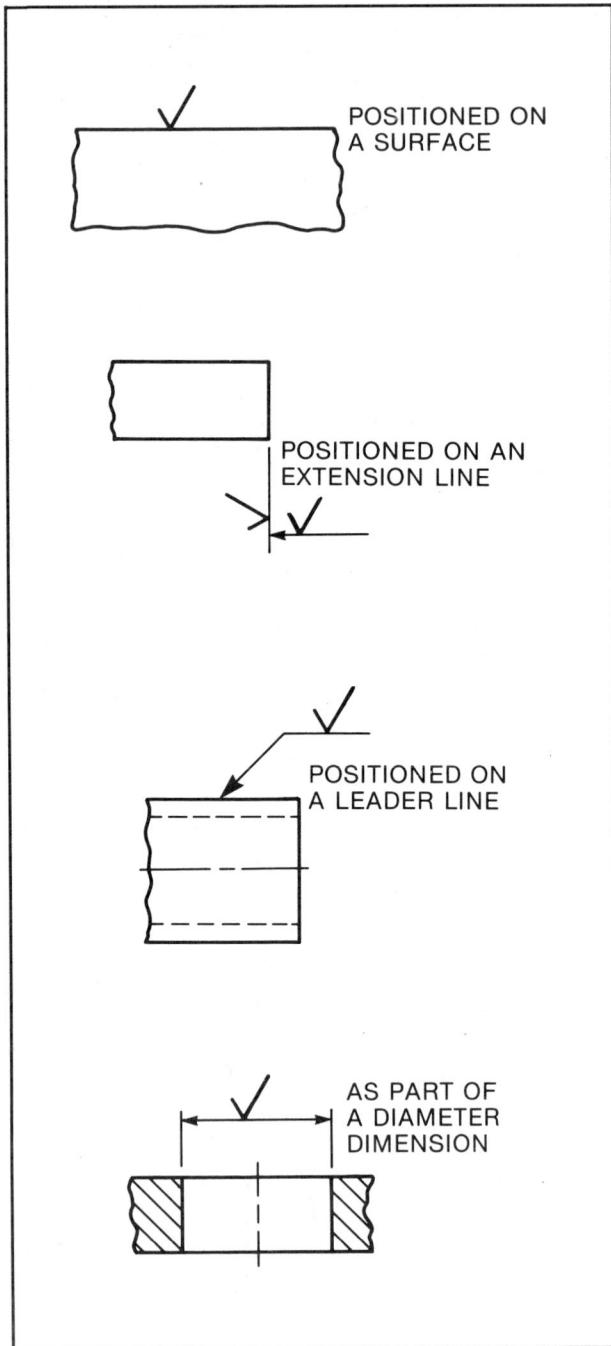

POSITIONED ON A SURFACE

POSITIONED ON AN EXTENSION LINE

POSITIONED ON A LEADER LINE

AS PART OF A DIAMETER DIMENSION

Figure 11-46. Positioning surface texture symbols

When several surfaces are to be finished, a general note may be used, Figure 11-47. In these cases, the surface texture symbol is used without the roughness average number. The roughness average is given in a general note. If any surface on the part is to have a roughness value other than that shown in the general note, the surface will have a roughness average number included in its symbol.

Figure 11-48 shows the roughness averages for several common manufacturing processes. When very smooth surfaces are specified, several processes are used to obtain the desired finish. When specific roughness averages are called for on a surface, the value shown represents the highest allowable value. Any surface finish less than that specified is acceptable.

When machining any surface, the surface must always conform to the specification. Do not try to make it smoother than is specified. Surface texture symbols, like tolerances, must be maintained. However, trying to make the surface smoother than is specified adds cost, not value to the part.

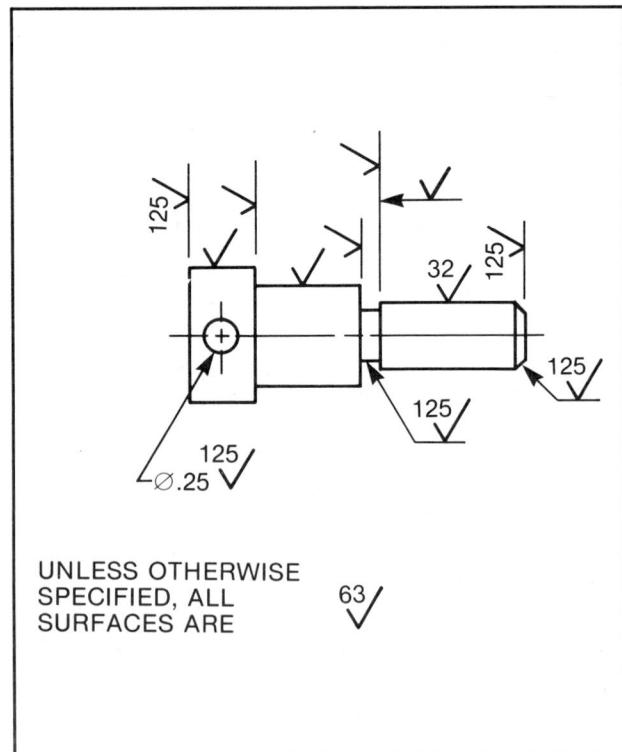

UNLESS OTHERWISE SPECIFIED, ALL SURFACES ARE 63

Figure 11-47. Specifying the roughness value with a general note

ROUGHNESS AVERAGE, R_a — MICROINCHES μ in.

Surface roughness produced by common production methods. Key: ■ Average application; ▨ Less frequent application.

PROCESS	2,000	1,000	500	250	125	63	32	16	8	4	2	1	.5
FLAME CUTTING	L	A	L										
SNAGGING	L	A	A	L									
SAWING	L	A	A	A	A	L							
PLANING, SHAPING		L	A	A	A	A	L						
DRILLING				L	A	A	L						
CHEMICAL MILLING				L	A	A	L						
ELECT. DISCHARGE MACH.				L	A	A	L						
MILLING		L	A	A	A	A	A	L					
BROACHING					L	A	A	L					
REAMING					L	A	A	L					
ELECTRON BEAM					A	A	A	A	L				
LASER					A	A	A	A	L				
ELECTRO-CHEMICAL				L	A	A	A	A	L	L			
BORING, TURNING		L	A	A	A	A	A	A	A	L	L		
BARREL FINISHING					L	A	A	A	A	L	L		
ELECTROLYTIC GRINDING							L	A	L				
ROLLER BURNISHING								A	L				
GRINDING					L	A	A	A	A	A	L	L	
HONING						L	A	A	A	A	L	L	
ELECTRO-POLISH								L	A	A	A	L	L
POLISHING								L	A	A	A	L	
LAPPING									L	A	A	A	L
SUPERFINISHING									L	A	A	A	L
SAND CASTING	L	A	L										
HOT ROLLING	L	A	L										
FORGING		L	A	A	L								
PERM MOLD CASTING				L	A	A	L						
INVESTMENT CASTING					L	A	L						
EXTRUDING				L	A	A	L						
COLD ROLLING, DRAWING					L	A	A	L					
DIE CASTING					L	A	A	L					

The ranges shown above are typical of the process listed. Higher or lower values may be obtained under special conditions.

KEY
■ Average application
▨ Less frequent application

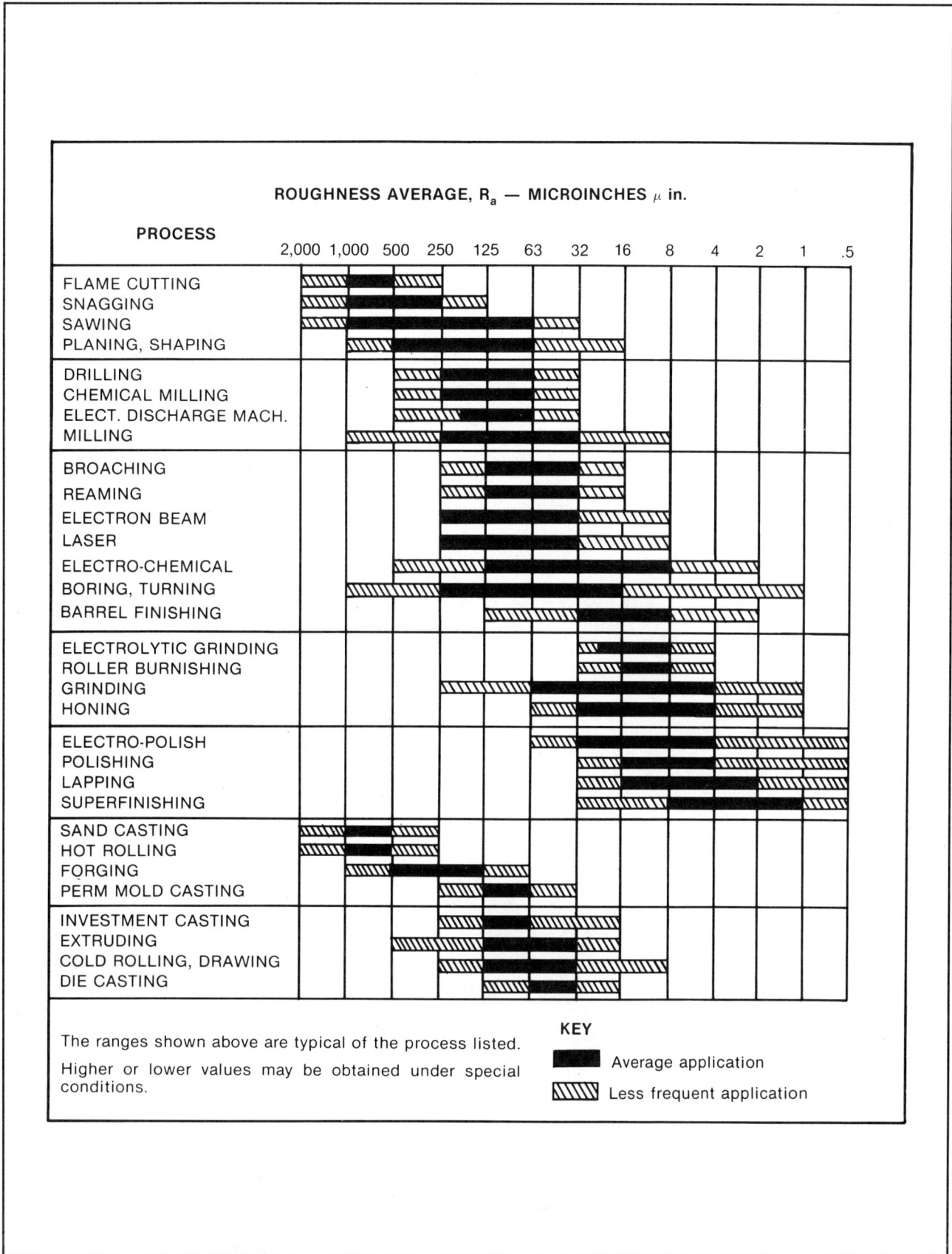

Figure 11-48. Surface roughness produced by common production methods

KEY TERMS

The following key terms were introduced in this chapter. Be sure you know the meaning of each term before proceeding to the review material.

Addendum
Addendum Angle
Angular Addendum
Center Distance
Chordal Addendum
Chordal Thickness
Circular Pitch
Circular Thickness
Clearance
Cutting Angle
Dedendum
Dedendum Angle
Diametral Pitch
Face Angle
Face Width
Helix Angle
Key
Keyseat
Keyway
Lay Symbols
Lead
Lead Angle
Lead of Helix

Linear Pitch
Microinch
Normal Circular Pitch
Normal Diametral Pitch
Outside Diameter
Pitch Circle
Pitch Cone Angle
Pitch Cone Distance
Pitch Diameter
Pitch Point
Radius of Curvature
Root Diameter
Roughness
Roughness Average
Roughness Sampling Length
Roughness Spacing
Taper
Taper Per Foot
Taper Per Inch
Thread Class
Thread Designation
Thread Form
Thread Series
Throat Diameter
Waviness
Waviness Height
Waviness Spacing Lay
Whole Depth
Working Depth

Test your knowledge with this reinforcement study material. Write your answers to the questions in the spaces provided.

1. What are the two forms of conical tapers? _____

_____

2. What units are commonly used to show tapers on a print? _____

_____

3. Name the three parts of a keyed assembly. _____

_____

4. What are the four most common types of keys?

   a. _____

   b. _____

   c. _____

   d. _____

5. What is the thread angle of a Unified thread? _____

6. List the thread forms that resemble the basic form of the Unified thread.

   a. _____

   b. _____

   c. _____

7. What type of thread has a 90° thread angle? _____

_____

8. What is the thread angle of an Acme thread? _____

9. What is the principal difference between the two forms of pipe threads? _____

_____

_____

10. What does the first number set in a Unified thread designation indicate? _____

_____

_____

11. What does the second number set in a Unified thread designation indicate? _____

_____

_____

12. What do the letters *UNC, UNF,* and *UNEF* mean?

    a. *UNC:* _____

    b. *UNF:* _____

    c. *UNEF:* _____

13. What do the letters *A* and *B* indicate at the end of a thread designation? _____

_____

14. What do the numbers *1, 2,* and *3* indicate in a thread designation? _____

_____

_____

15. What do the letters *LH* mean? _____

_____

16. What letter is used at the beginning of every metric thread designation? _____

17. Name the three methods used to show threads on a print.

    a. _____

    b. _____

    c. _____

18. How is the length of bolts and screws shown? _____

_____

19. Which type of gear has teeth cut parallel to its axis of rotation? _____

20. What term is used to describe the ratio of the number of teeth to the pitch diameter of a gear?

_____

21. What term is used to describe the distance between corresponding points on two adjacent gear teeth? _____

22. Which type of gear resembles a spur gear but has teeth cut at an angle across the face of the gear? _____

23. Which type of gear has a variable tooth thickness and depth? _____

_____

24. What do the numbers above the short leg of the surface texture symbol indicate?

_____

_____

25. What are the patterned marks of the surface pattern called? _____

_____

Refer to Figure E11-1 on page 217 to answer the following questions.

1. What material is specified? _____

2. Find the following dimensions.

   a. $A =$ _____

   b. $D =$ _____

3. What term, or word, describes the dimensions at B and C? _____
   _____

4. What is the length of taper for the taper on the left? _____

5. What is the taper per inch of the taper on the left? _____

6. What is the tolerance for angles? _____

7. What is the tolerance of the 2.50 dimension? _____

8. What is the taper per foot of the taper on the right? _____

9. What is the diameter of the part at the bottom of the groove on the right? _____

10. What type of taper is shown on the left? _____
    _____

11. What type of taper is shown on the right? _____
    _____

12. What are the tolerances of the dimensions shown at B and C? _____
    _____

13. What is the taper per inch of the taper shown on the right? _____

14. What is the depth of the taper angle for the taper shown on the right? _____

15. How many of these parts are required? _____

16. What views are shown in this print? _____
    _____

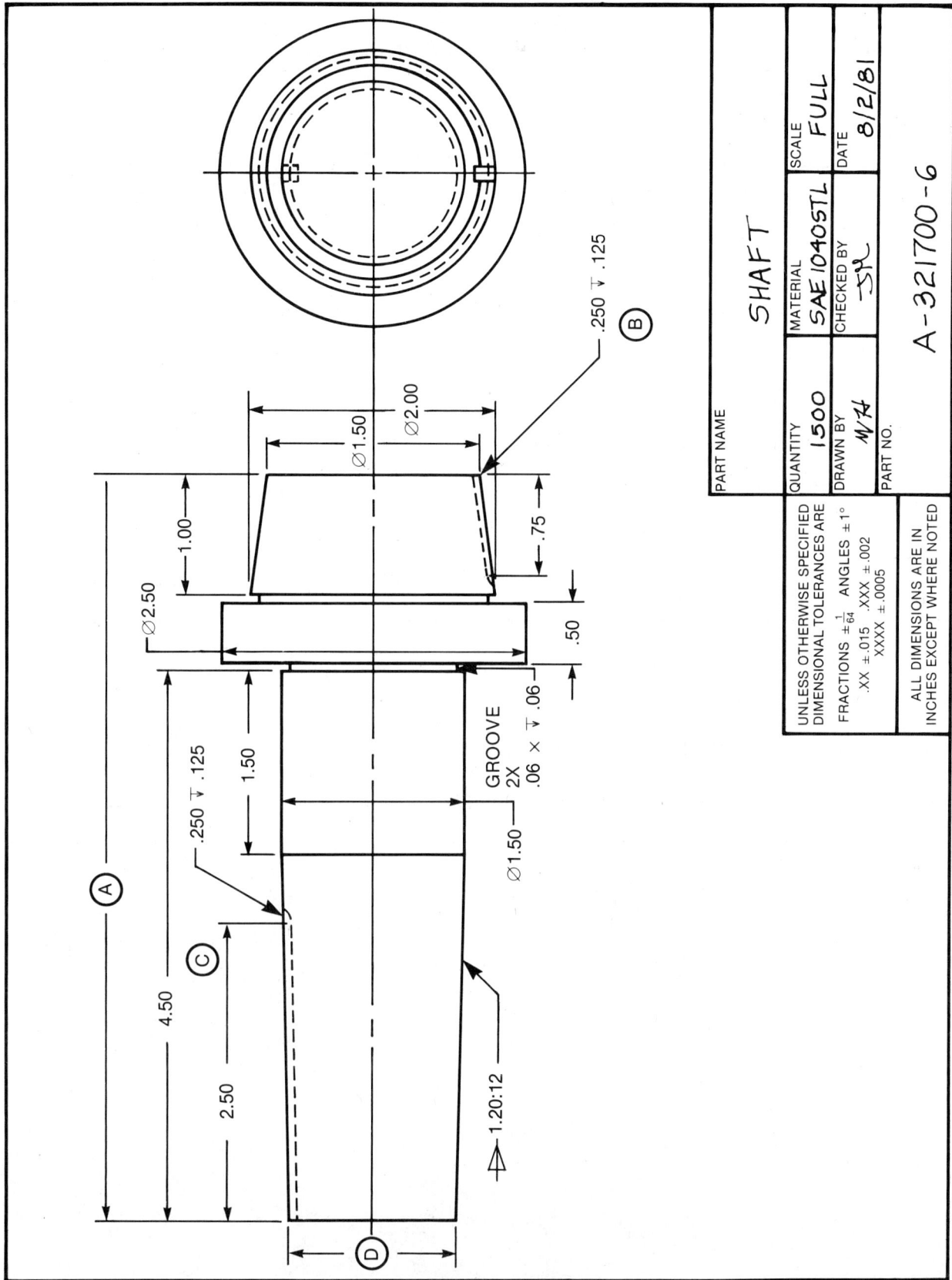

Figure E11-1.

PART NAME: SHAFT

MATERIAL	SAE 1040STL
SCALE	FULL
QUANTITY	1500
CHECKED BY	
DRAWN BY	MH
DATE	8/2/81
PART NO.	A-321700-6

UNLESS OTHERWISE SPECIFIED DIMENSIONAL TOLERANCES ARE

FRACTIONS ±$\frac{1}{64}$ ANGLES ±1°
.XX ±.015 .XXX ±.002
.XX ±.015 XXXX ±.0005

ALL DIMENSIONS ARE IN INCHES EXCEPT WHERE NOTED

Ø1.50
Ø2.00
.250 ⊽ .125
1.00
.75
Ø2.50
.50
.250 ⊽ .125
1.50
GROOVE
2X
.06 × ⊽ .06
Ø1.50
A
C
4.50
2.50
1.20:12
D

Refer to Figure E11-2 on page 219 to answer the following questions.

1. What is the overall length of this part (Dimension *B*)? _____

2. Find the following dimensions.

   a. *A* = _____

   b. *C* = _____

3. Which thread on this part has the tightest fit? _____

   _____

4. What is the meaning of the letter *A* after the thread class number? _____

   _____

5. What is the meaning of the letter *B* after the thread class number? _____

   _____

6. How many threaded areas are there on this part? _____

7. What type of representation is used to show the threads on this part? _____

   _____

8. What are the tolerances of the unthreaded diameters of the part? _____

   _____

9. What are the tolerances of the length dimensions? _____

10. How many undercut grooves are there in this part? _____

   _____

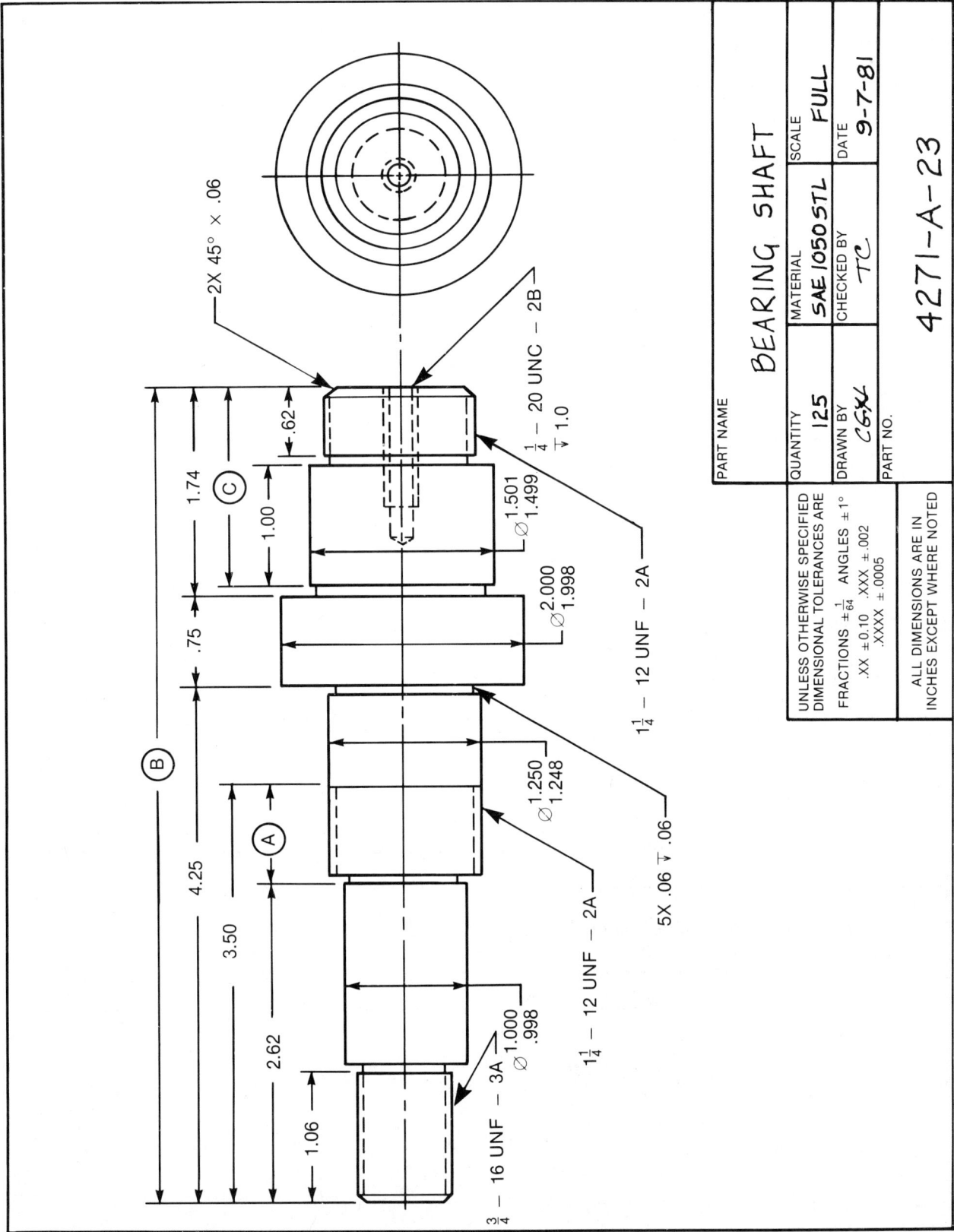

Figure E11-2.

BEARING SHAFT

PART NAME		
QUANTITY	MATERIAL	SCALE
125	SAE 1050 STL	FULL
DRAWN BY	CHECKED BY	DATE
CGM	TC	9-7-81
PART NO.		

4271-A-23

UNLESS OTHERWISE SPECIFIED
DIMENSIONAL TOLERANCES ARE

FRACTIONS $\pm\frac{1}{64}$ ANGLES $\pm 1°$
.XX ± 0.10 .XXX $\pm .002$
.XXXX $\pm .0005$

ALL DIMENSIONS ARE IN
INCHES EXCEPT WHERE NOTED

2X 45° × .06

$\frac{1}{4}$ – 20 UNC – 2B
$\frac{1}{4}$ ▼ 1.0

.62

1.74

Ⓒ

1.00

∅ 1.501
1.499

∅ 2.000
1.998

$1\frac{1}{4}$ – 12 UNF – 2A

.75

Ⓑ

∅ 1.250
1.248

5X .06 ▼ .06

4.25

Ⓐ

3.50

2.62

$1\frac{1}{4}$ – 12 UNF – 2A

1.06

∅ 1.000
.998

$\frac{3}{4}$ – 16 UNF – 3A

Refer to Figure E11-3 on page 221 to answer the following questions.

1. What is the name of this part? _____

_____

2. What is the part number? _____

3. What type of print is this? _____

_____

4. Which type of thread representation is used on detail #5? _____

_____

5. What is detail #5? _____

_____

6. What are the names of the following hardware items?

   a. Item #6: _____

   b. Item #7: _____

   c. Item #8: _____

   d. Item #9: _____

7. What type of drive is used for details #6 and #7? _____

_____

8. What type of drive is used for detail #9? _____

_____

9. What type of drive is used for detail #5? _____

_____

10. What is the meaning of the abbreviation *COMM* in the materials list? _____

_____

_____

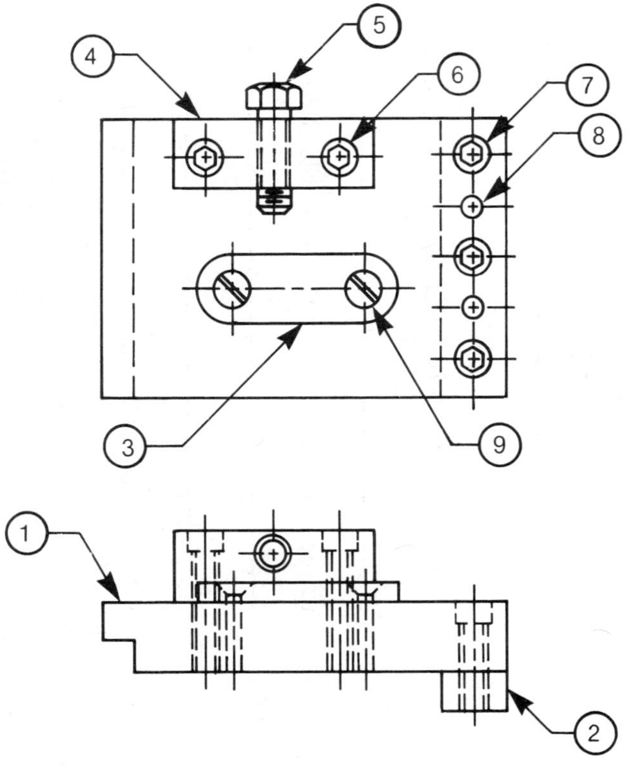

ITEM NO	QTY	DESCRIPTION	SPECIFICATION	MAT.
9	4	SCREW	1/4-20 UNC-1.25LG	COMM
8	4	DOWEL	1/4 X 1 1/2 HDN.	COMM
7	6	CAP SCREW	1/4-20 UNC-1.25LG	COMM
6	4	CAP SCREW	1/4-20 UNC-2.00LG	COMM
5	2	BOLT	1/2-13 UNC-1.50 LG	COMM
4	2	BLOCK	1 X 1 X 3 LG	SAE 1020 STL
3	2	LOCATOR PAD	1 X 1/2 X 3 LG	SAE 1020 STL
2	2	KLEET	1 X 1/2 X 4 LG	SAE 1020 STL
1	2	BASE PLATE	4 X 6 X 1 PLATE	SAE 1020 STL

MATERIALS LIST

PART NAME

ASSEMBLY FIXTURE

QUANTITY	MATERIAL	SCALE
2	NOTED	1/2 = 1

DRAWN BY	CHECKED BY	DATE
TP	BG	10-12-81

PART NO.

T-472-66-A

Figure E11-3.

Refer to Figure E11-4 on page 223 to answer the following questions.

1. What is the outside diameter of this gear? _____

2. What type of gear is this? _____

3. What function is served by the dimensions shown on the gear blank? _____

_____

_____

4. Where are the gear cutting dimensions shown? _____

_____

5. What views are shown in this print? _____

_____

6. What is the purpose of the phantom lines in the right view? _____

_____

_____

7. How many teeth are there on this gear? _____

8. What is the pitch diameter of the gear? _____

9. How deep should the teeth be cut? _____

10. What does the note *(REF)* indicate? _____

_____

_____

11. What should be the measured thickness of each gear tooth? _____

12. What is the hole diameter in the gear? _____

13. What tolerance should be applied to the outside diameter of the gear blank? _____

_____

14. What is the diametral pitch of this gear? _____

15. What size of keyway is specified? _____

R 0.30 ± .005

.250

.250 × .106
KEYWAY

1.606

Ø1.500

1.50

.12 × 45° CHAM

Ø3.667

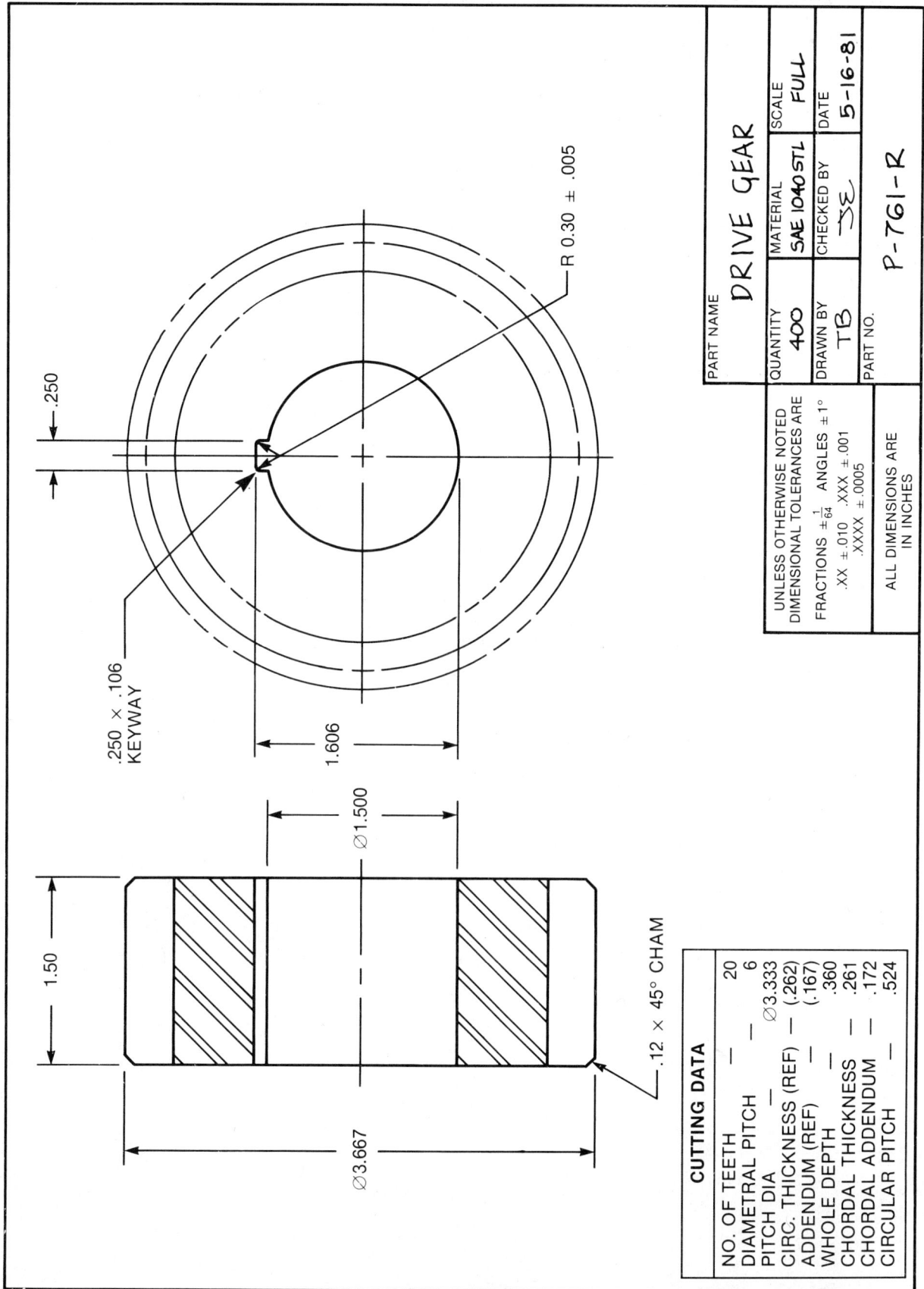

PART NAME		
DRIVE GEAR		
QUANTITY 400	MATERIAL SAE 1040 STL	SCALE FULL
DRAWN BY TB	CHECKED BY SE	DATE 5-16-81
PART NO.		
P-761-R		

UNLESS OTHERWISE NOTED
DIMENSIONAL TOLERANCES ARE

FRACTIONS ±$\frac{1}{64}$ ANGLES ±1°
.XX ±.010 .XXX ±.001
.XXXX ±.0005

ALL DIMENSIONS ARE
IN INCHES

Figure E11-4.

Refer to Figure E11-5 on page 225 to answer the following questions.

1. What type of gear is shown in this print? _____
   _____

2. What is the outside diameter of the gear? _____

3. How many teeth are on this gear? _____

4. What is the helix angle? _____

5. What is the hand of the helix? _____

6. What is the pitch diameter of this gear? _____

7. What is the normal diametral pitch? _____

8. Explain the term *normal diametral pitch.* _____
   _____
   _____

9. How deep should the teeth in this gear be cut? _____

10. Which values are used when measuring the tooth size? _____
    _____

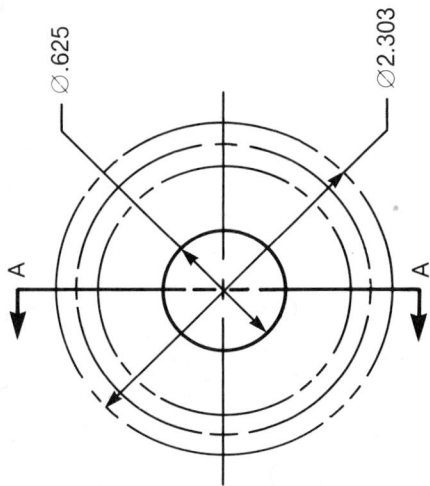

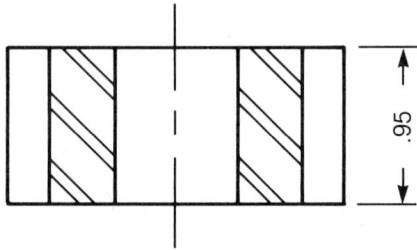

SECTION A-A

.95

Ø.625

Ø2.303

A

A

CUTTING DATA		
NUMBER OF TEETH	—	20
DIAMETRAL PITCH (NORMAL)	—	10
HAND OF HELIX	—	RIGHT
PITCH DIAMETER	—	Ø2.103
ADDENDUM (REF.)	—	(.100)
CHORDAL ADDENDUM	—	.103
THICKNESS	—	(.157)
CHORDAL THICKNESS	—	.156
WHOLE DEPTH	—	.216
HELIX ANGLE	—	18°

PART NAME HELICAL GEAR		SCALE FULL	
QUANTITY 50	MATERIAL SAE 1050		
	CHECKED BY		
DRAWN BY		DATE 5-4-83	
PART NO. 55139~T72			

Figure E11-5.

Refer to Figure E11-6 on page 227 to answer the following questions.

1. What types of gears are shown in this print? _____
_____

2. Define the term *miter gears*. _____
_____
_____

3. What is the face width of these gears? _____
_____

4. What is the cutting angle of these gears? _____
_____

5. What is the tooth form? _____
_____

6. What is the pitch cone angle? _____
_____

7. What is the face angle of these gears? _____
_____

8. How many teeth are there on the pinion? _____
_____

9. What is the pitch diameter of these gears? _____
_____

10. What is the outside diameter of the gears? _____
_____

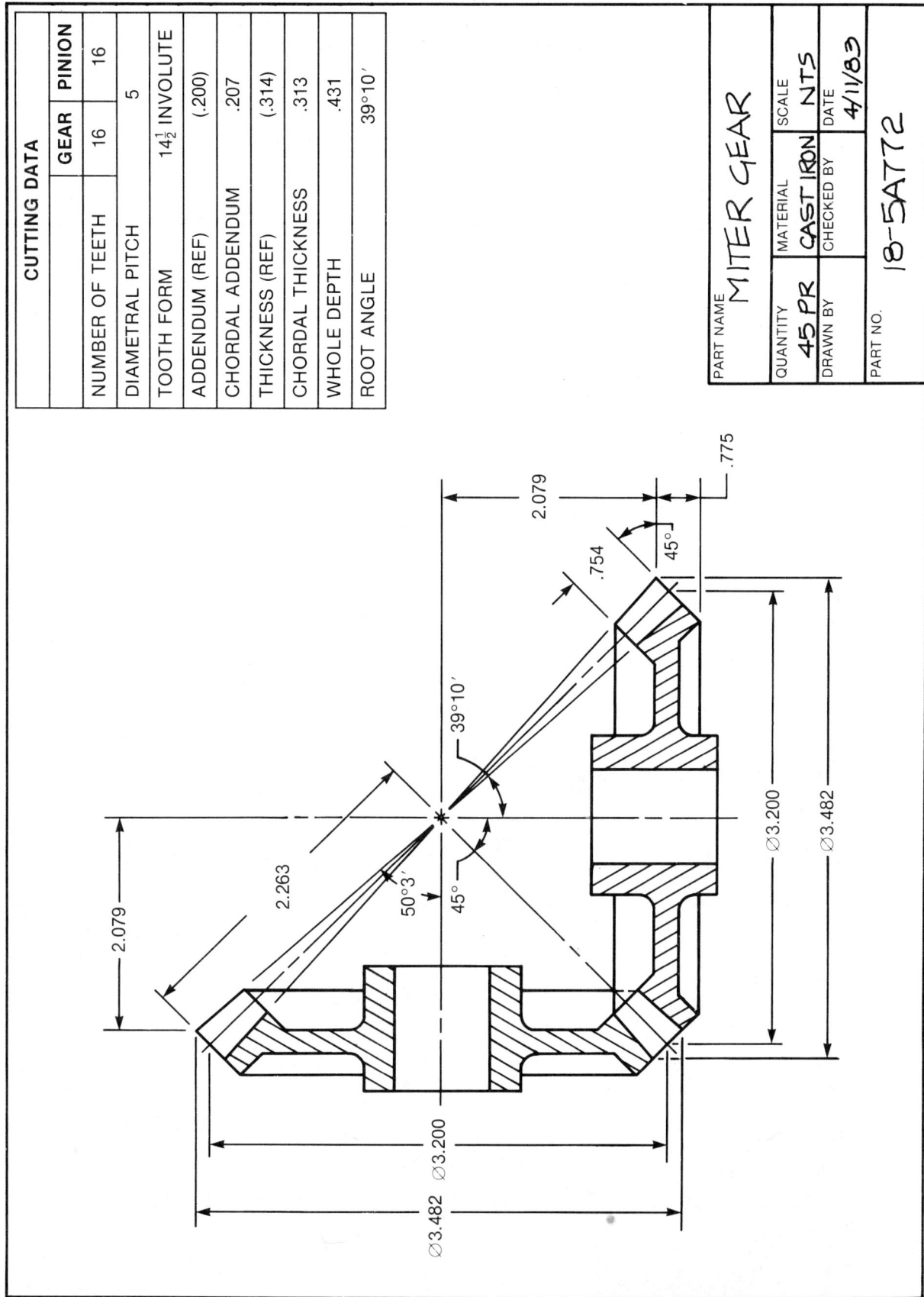

CUTTING DATA		
	GEAR	PINION
NUMBER OF TEETH	16	16
DIAMETRAL PITCH	5	
TOOTH FORM	14½ INVOLUTE	
ADDENDUM (REF)	(.200)	
CHORDAL ADDENDUM	.207	
THICKNESS (REF)	(.314)	
CHORDAL THICKNESS	.313	
WHOLE DEPTH	.431	
ROOT ANGLE	39°10'	

PART NAME	MITER GEAR	
QUANTITY 45 PR	MATERIAL CAST IRON	SCALE NTS
DRAWN BY	CHECKED BY	DATE 4/11/83
PART NO. 18-5A772		

Figure E11-6.

Refer to Figure E11-7 on page 229 to answer the following questions.

1. What types of gears are shown in this print? _____

   _____

2. What is the outside diameter of the worm wheel? _____

   _____

3. What is the radius of curvature? _____

   _____

4. What is the linear pitch of the worm? _____

   _____

5. What is the throat diameter of the worm wheel? _____

   _____

6. What is the outside diameter of the worm? _____

   _____

7. How many teeth are there on the worm wheel? _____

   _____

8. What is the pitch diameter of the worm? _____

   _____

9. What is the center distance between the worm and the worm wheel? _____

   _____

10. What is the helix angle of the worm? _____

   _____

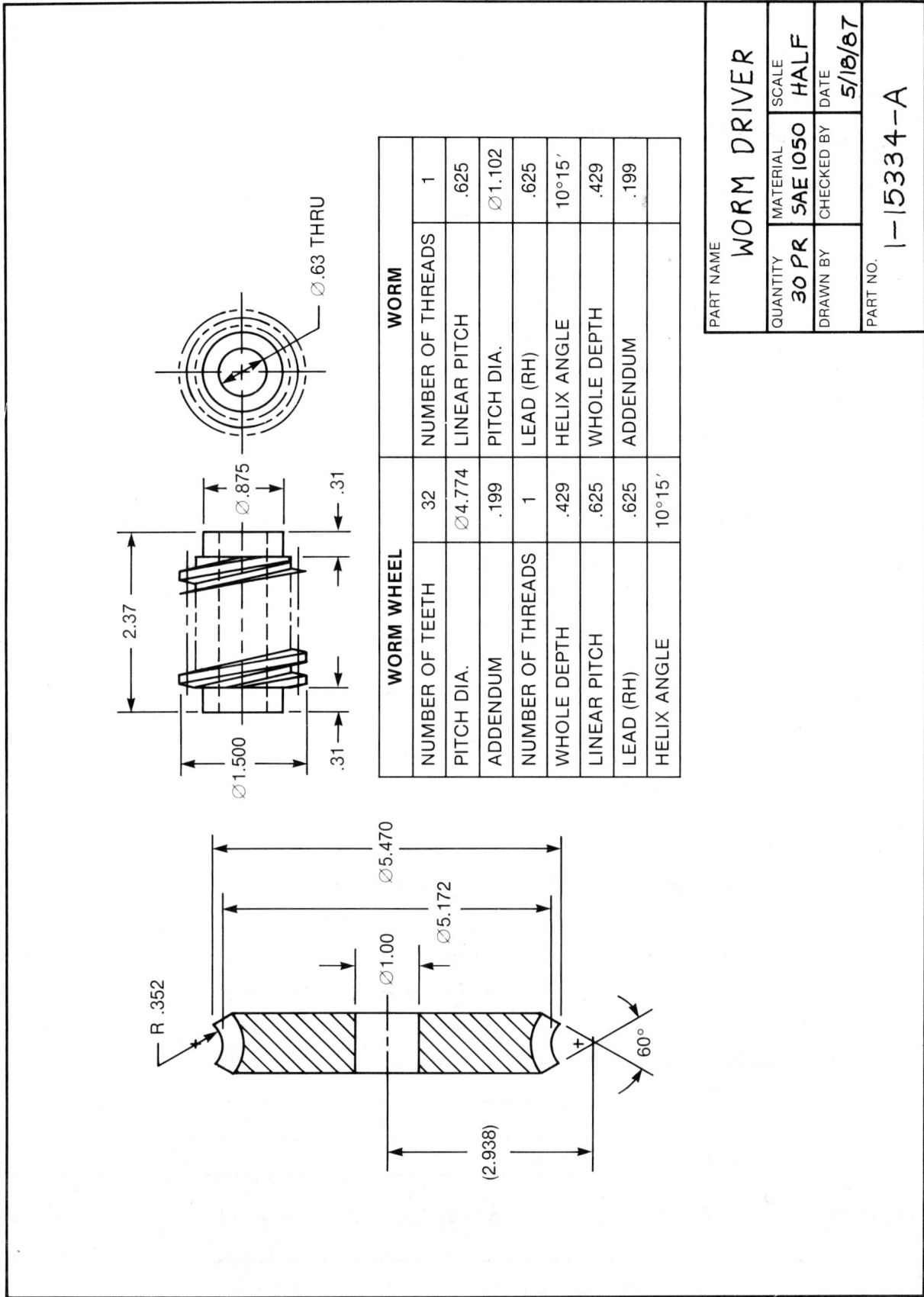

WORM WHEEL	
NUMBER OF TEETH	32
PITCH DIA.	⌀4.774
ADDENDUM	.199
NUMBER OF THREADS	1
WHOLE DEPTH	.429
LINEAR PITCH	.625
LEAD (RH)	.625
HELIX ANGLE	10°15'

WORM	
NUMBER OF THREADS	1
LINEAR PITCH	.625
PITCH DIA.	⌀1.102
LEAD (RH)	.625
HELIX ANGLE	10°15'
WHOLE DEPTH	.429
ADDENDUM	.199

PART NAME
WORM DRIVER

QUANTITY 30 PR	MATERIAL SAE 1050	SCALE HALF
DRAWN BY	CHECKED BY	DATE 5/18/87

PART NO. 1–15334–A

Figure E11-7.

Refer to Figure E11-8 on page 231 to answer the following questions.

1. What material is specified for this part? _____

2. Identify the maximum roughness height values for the following surfaces.

   a. Surface *A*: _____   **d.** Surface *D*: _____

   b. Surface *B*: _____   **e.** Surface *E*: _____

   c. Surface *C*: _____

3. What does the symbol shown on surface *F* mean? _____

   _____

4. What does the symbol shown on surface *B* indicate? _____

   _____

5. What does the number to the left of the symbol shown on surface *B* mean? _____

   _____

6. Find the following dimensions.

   a. *D* = _____   **c.** *I* = _____

   b. *H* = _____   **d.** *J* = _____

7. What is the tolerance of the hole size? _____

8. What size are the radii shown in the front view? _____

9. What tolerance is applied to the radii? _____

10. How many surfaces on this part are to be finished? _____

11. Identify the process (*grinding, milling, drilling, honing, lapping,* or *none*) which should be used to provide the specified finish value for each of the following surfaces.

    a. Surface *A*: _____

    b. Surface *B*: _____

    c. Surface *C*: _____

    d. Surface *D*: _____

    e. Surface *E*: _____

    f. Surface *F*: _____

    g. Surface *G*: _____

12. What is the general finish requirement for all symbols that are not noted with a maximum roughness height? _____

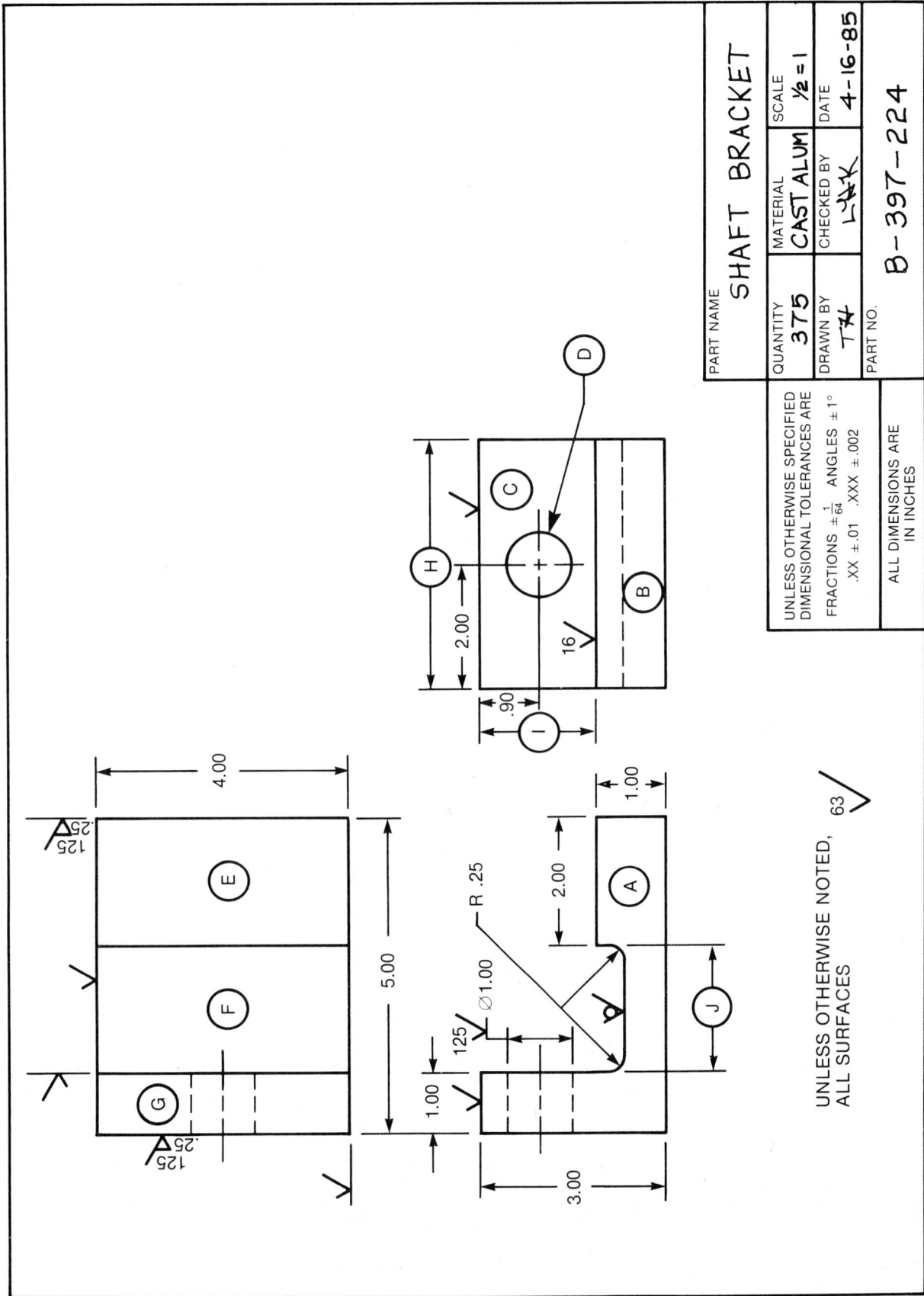

Figure E11-8.

12

GEOMETRIC DIMENSIONING AND TOLERANCING

OBJECTIVES

After studying this chapter, you will be able to:

■ Identify and define feature control symbols.

■ Identify and interpret the meaning of geometric characteristic symbols.

■ Identify and define datums, supplementary symbols, and modifiers.

■ Interpret print specifications involving geometric dimensions and tolerances.

Geometric dimensioning and tolerancing is a method used to control the form, location, and relationship of part features. It specifies the maximum amount a part feature can vary from the desired size and still function as intended. Geometric dimensioning and tolerancing also simplifies complicated instructions by presenting data in easy-to-understand symbols rather than lengthy notes. To understand and interpret geometric dimensioning and tolerancing, it is necessary to know how to read and interpret the feature control frame.

FEATURE CONTROL SYMBOLS

Geometric dimensioning and tolerancing is a form of graphic shorthand. Standard symbols are used to specify complicated details. In this way almost any geometric shape can be controlled with little chance of misinterpretation. The main symbol used in this dimensioning system is the feature control frame.

The *feature control frame*, Figure 12-1, is actually a combination of symbols. It consists of a frame and several elements. A feature control frame normally contains a geometric characteristic symbol, a tolerance value, and a datum reference. However, these components may vary, depending on the application.

Geometric Characteristic Symbols

Geometric characteristic symbols, Figure 12-2, are used to show the characteristic to be controlled. The first box of the feature control frame contains a geometric characteristic symbol. There are three main groups of geometric characteristic symbols:

1. Those that control individual features;
2. Those that control either individual features or related features; and

3. Those that control only related features.

These symbols are also divided into five other groups that control form tolerances, profile tolerances, orientation tolerances, locational tolerances, and runout tolerances.

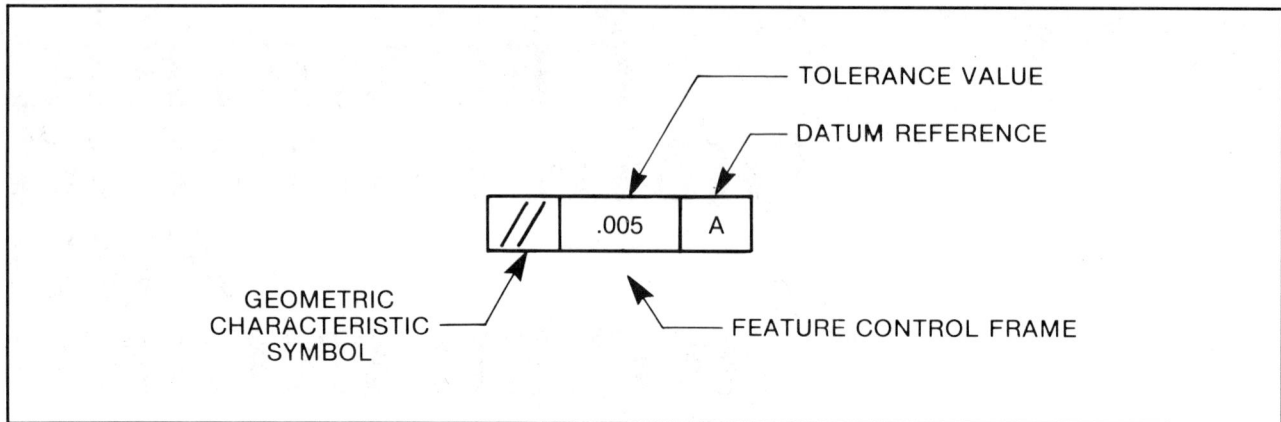

Figure 12-1. The whole symbol is called the feature control frame.

APPLICATION	TYPE OF TOLERANCE	CHARACTERISTIC	SYMBOL
FOR INDIVIDUAL FEATURES	FORM	STRAIGHTNESS	—
		FLATNESS	
		CIRCULARITY (ROUNDNESS)	O
		CYLINDRICITY	
FOR INDIVIDUAL OR RELATED FEATURES	PROFILE	PROFILE OF A LINE	
		PROFILE OF A SURFACE	
FOR RELATED FEATURES	ORIENTATION	ANGULARITY	
		PERPENDICULARITY	⊥
		PARALLELISM	//
	LOCATION	POSITION	⊕
		CONCENTRICITY	◎
	RUNOUT	CIRCULAR RUNOUT	
		TOTAL RUNOUT	

Figure 12-2. Geometric characteristic symbols

234 *Geometric Dimensioning and Tolerancing*

Tolerance Values

The *tolerance value* shown in the feature control frame represents the total amount a part feature may vary from the stated dimension. The tolerance value follows the geometric characteristic symbol in the feature control frame.

Datum References

A *datum* is a specific surface, line, plane or feature that is assumed to be exact. The datum is used as a reference point for dimensions. Datums are identified by letters, and are shown on a print with *datum identifying symbols*. Datum identifying symbols are related to the *datum reference* in the feature control frame. See Figure 12-3. In this example, the letter *A* in the feature control frame refers to datum *A* on the print.

The datum identifying symbol is shown as a framed letter with a dash on both sides. Any letters in the alphabet may be used to identify datums except the letters *I*, *O*, and *Q*. These letters are not used because they could easily be confused with numbers. When more than 23 datums are required, double letters (*AA*, *AB*, *AC*, etc.) are used.

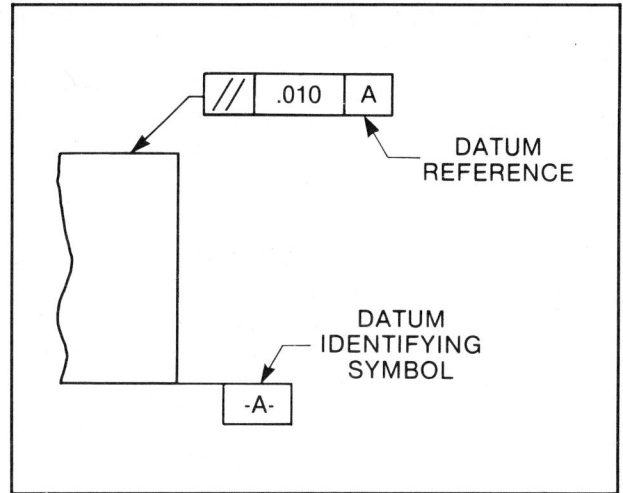

Figure 12-3. Datum reference and datum identifying symbol

SUPPLEMENTARY SYMBOLS

Supplementary symbols are generally used in addition to the basic elements in the feature control frame. These symbols are used to further define and clarify the meaning of the basic feature control symbols. The seven symbols commonly used with geometric dimensioning and tolerancing are shown in Figure 12-4, and are explained in this section.

TERM	ABBREVIATION	SYMBOL
MAXIMUM MATERIAL CONDITION	MMC	Ⓜ
LEAST MATERIAL CONDITION	LMC	Ⓛ
REGARDLESS OF FEATURE SIZE	RFS	Ⓢ
DIAMETER	DIA	⌀
PROJECTED TOLERANCE ZONE	TOL ZONE PROJ	Ⓟ
REFERENCE	REF	(1.250)
BASIC DIMENSION	BSC	3.875

Figure 12-4. Supplementary symbols

Maximum Material Condition

The *maximum material condition*, or *MMC*, is the condition where the part feature has its maximum amount of material within the stated limits of size. This value shows the largest size of an external feature and the smallest size of an internal feature. In other words, the MMC for a hole is the smallest allowable hole, leaving the *maximum* amount of material. The MMC for a shaft is the largest allowable diameter for the shaft, leaving the maximum amount of material. See Figure 12-5.

The MMC modifier is shown by the symbol Ⓜ. This modifier is only used on features that can vary in size. The MMC modifier may be used with either the tolerance value or the datum reference, or with both, Figure 12-6.

Least Material Condition

The *least material condition*, or *LMC*, is the opposite of the maximum material condition. That is, the least material condition shows the least material with the stated limits. The LMC value is the smallest size of an external feature or the largest size of an internal feature. In other words, the LMC for a hole is the largest allowable hole, leaving the *minimum* amount of material. The LMC for a shaft is the smallest allowable diameter for the shaft, leaving the minimum amount of material. See Figure 12-5.

Like the MMC modifier, the LMC modifier is only used for features that can vary in size. The least material condition modifier is shown by the symbol Ⓛ. The LMC modifier may be used with either the tolerance value or the datum reference, or with both, Figure 12-6.

Regardless of Feature Size

The *regardless of feature size*, or *RFS*, indicates that a geometric tolerance or datum reference will apply to any increment of size of the item's feature within its size tolerance. The RFS modifier, shown by the symbol Ⓢ, can also be applied to either the tolerance value or the datum reference, or to both, Figure 12-6. When this modifier is used, it means that the stated tolerance value applies, regardless of the actual size of the feature.

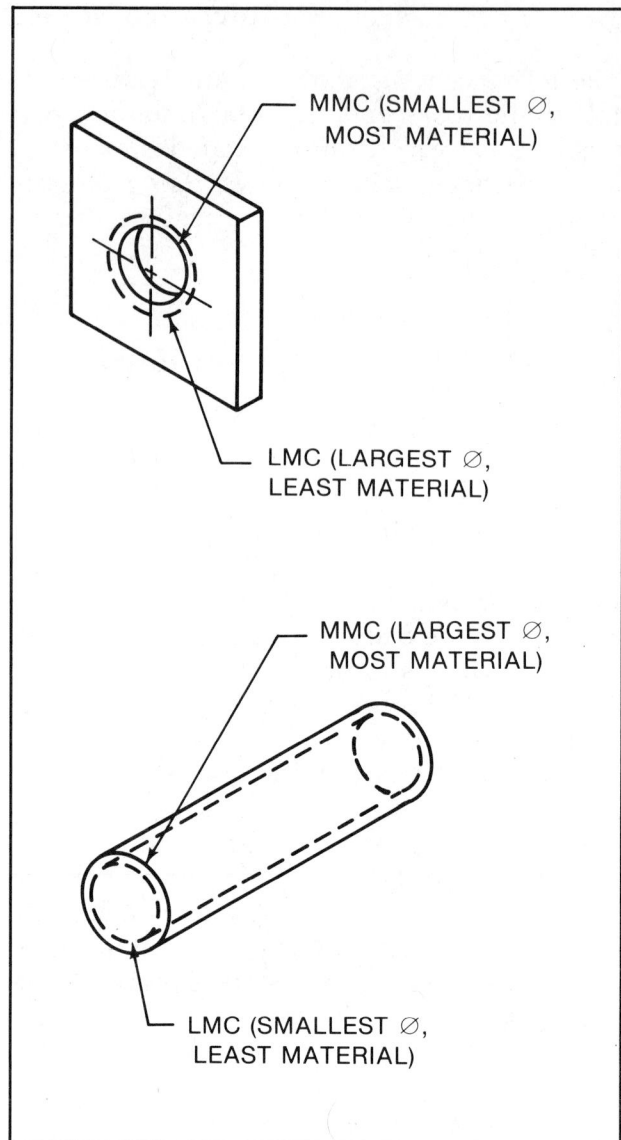

Figure 12-5. Maximum material condition and least material condition

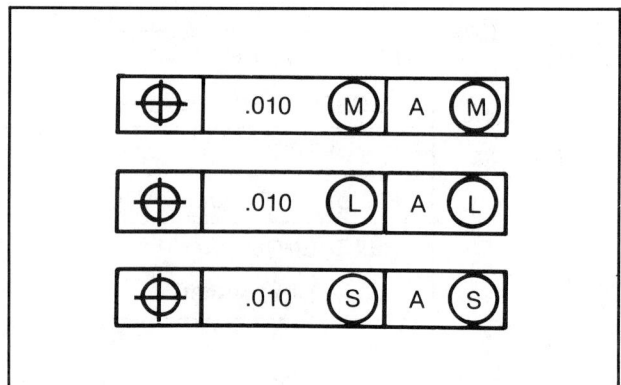

Figure 12-6. Using the MMC, LMC, and RFS modifiers in the feature control frame

Projected Tolerance Zone

The *projected tolerance zone*, shown by the symbol Ⓟ, ensures the alignment between two mating holes. This modifier is positioned below the feature control frame. See Figure 12-7.

Diameter

A *diameter* is indicated by a circle with a slash (∅). In a feature control frame, it is placed before the tolerance value. This symbol is used in place of the old standard symbol *DIA*, which may be seen on old prints. It is placed before the dimensional value, Figure 12-8.

Reference and Basic Dimensions

Reference and *basic dimensions* are treated the same way in geometric dimensioning and tolerancing as they are in standard dimensioning and tolerancing.

A reference dimension, shown by a number enclosed in parentheses, is used only for information and convenience. It has no binding effect on the print. A basic dimension, shown by a number enclosed in a frame, is used as a base dimension. This dimension represents the perfect feature size. All tolerance values are applied to the feature at this size, Figure 12-9.

INTERPRETING FORM TOLERANCES

A *form tolerance* controls the specific geometric shape of a part feature. Interpreting a form tolerance is simply a matter of identifying the geometric characteristic symbol and deciding how it relates to the specified datum within the tolerance value shown. The geometric characteristic symbols that represent form are straightness, flatness, circularity (roundness), and cylindricity.

Straightness

Straightness is a control applied to an individual feature. No datum reference is used. The symbol for straightness is — . When straightness

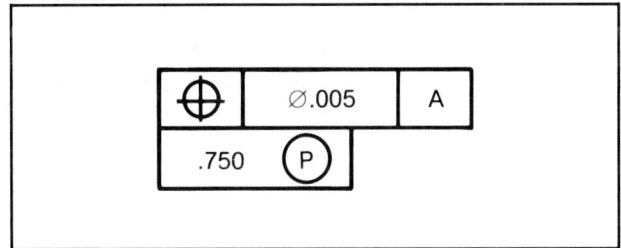

Figure 12-7. Using the projected tolerance zone modifier with the feature control frame

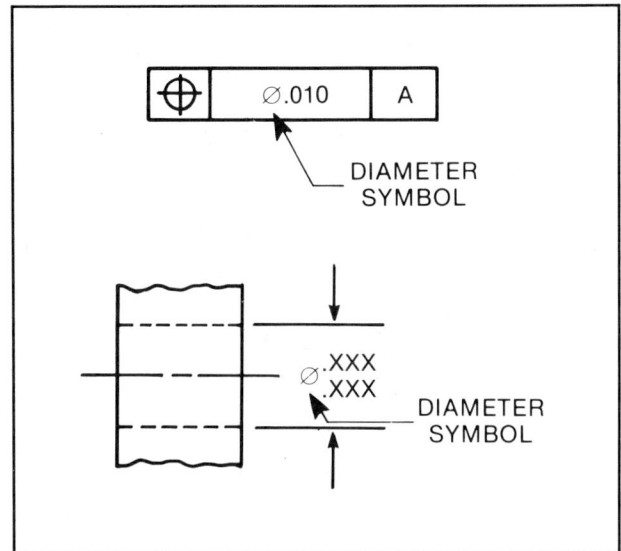

Figure 12-8. Using the diameter symbol

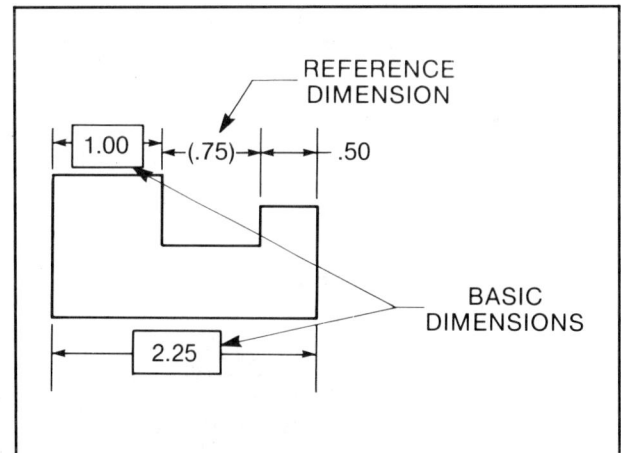

Figure 12-9. Basic dimensions and reference dimensions

is specified, it means that the feature must not only be within the specified size tolerance, but also within the straightness tolerance. In Figure 12-10A, the combined variation in size and

SYMBOL:

.500
.495

| — | .002 |

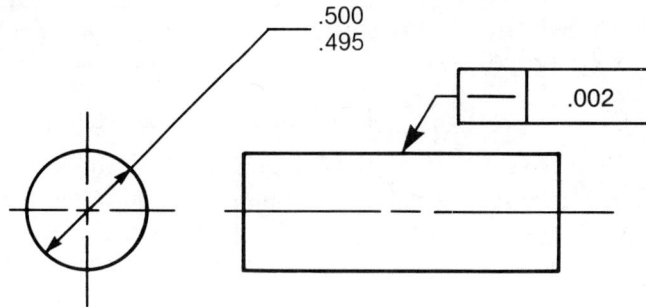

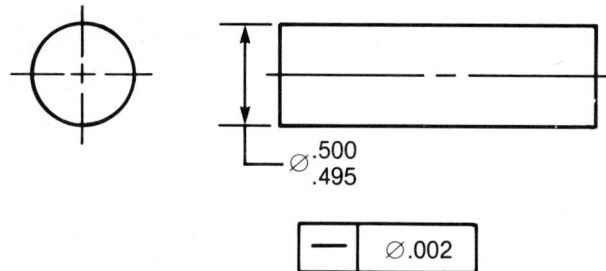

INTERPRETATION:

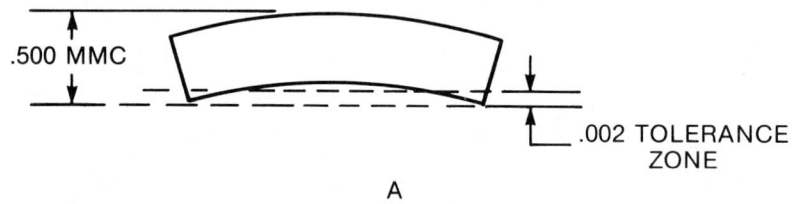

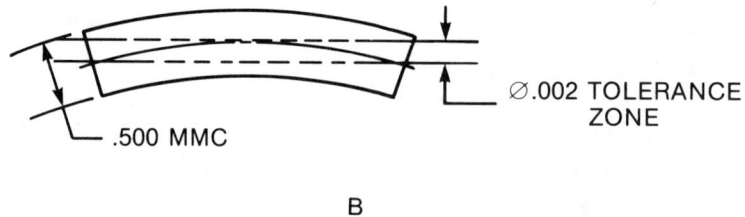

.500 MMC

.002 TOLERANCE
ZONE

A

SYMBOL:

⌀ .500
.495

| — | ⌀.002 |

INTERPRETATION:

⌀.002 TOLERANCE
ZONE

.500 MMC

B

Figure 12-10. Interpreting the straightness symbol (A) without a diameter symbol and (B) with a diameter symbol

straightness cannot be more than the specified MMC size of the feature. One way to visualize any straightness tolerance is to imagine two parallel lines spaced at a distance equal to the tolerance. Each element of the feature must be contained between these imaginary lines, or tolerance zone, to allow the part to pass inspection.

When the diameter symbol is used with the straightness control, Figure 12-10B, the tolerance zone is a cylinder. The diameter of this cylindrical tolerance zone is determined by the tolerance value. Every element of the center line of the part must be contained within this imaginary cylinder.

Flatness

Flatness is another control applied to features. The symbol for flatness is ▱. When it is specified, no datum reference is used. When flatness is indicated, it simply means that every element of a surface must be contained within the imaginary parallel planes representing the tolerance zone, Figure 12-11.

Circularity (Roundness)

Circularity is also specified without a datum reference. The symbol for circularity is ○.

Circularity indicates that every element of the part feature must be contained within two imaginary concentric circles that represent the tolerance zone. The distance between these lines is equal to the tolerance value shown. In Figure 12-12, circularity is based on the radius rather than the diameter of the feature.

Cylindricity

Cylindricity is a control of both roundness and straightness. The symbol for cylindricity is ⌭. Cylindricity is shown without a datum reference and means that every element of the feature msut be held within two imaginary concentric cylinders. See Figure 12-13.

INTERPRETING PROFILE TOLERANCES

A *profile tolerance* specifies the permissible deviation from a desired profile. This control is normally used for irregular surfaces or lines that cannot be properly controlled with other geometric controls. There are two types of profile controls: profile of a line and profile of a surface.

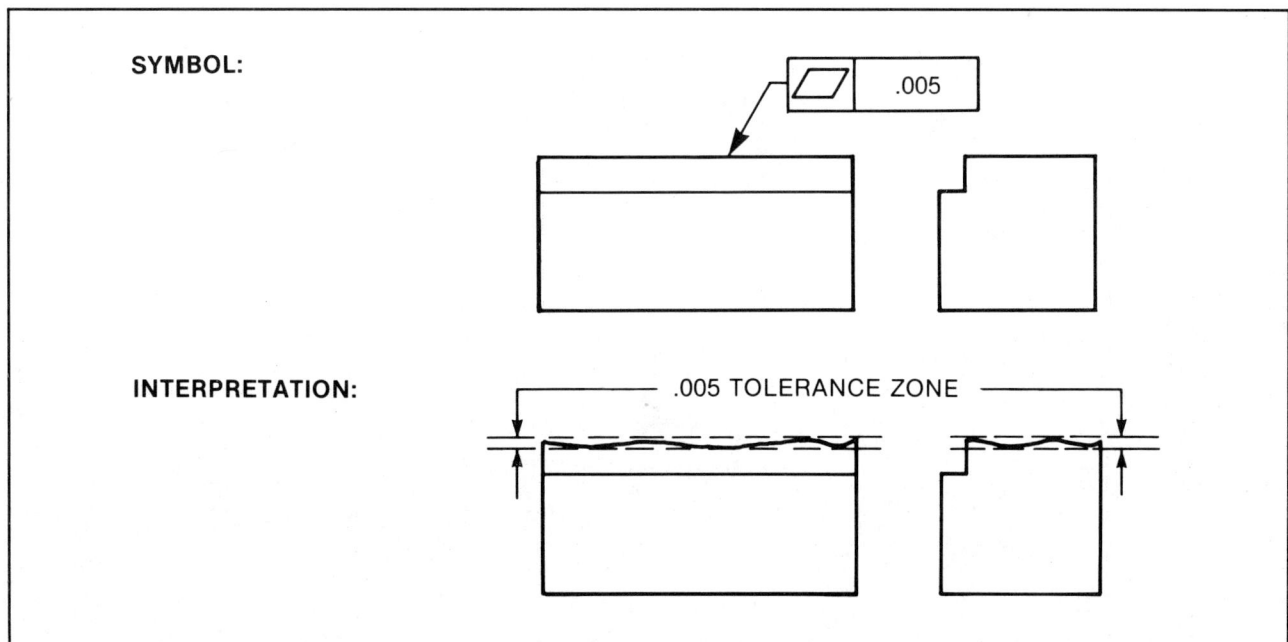

Figure 12-11. Interpreting the flatness symbol

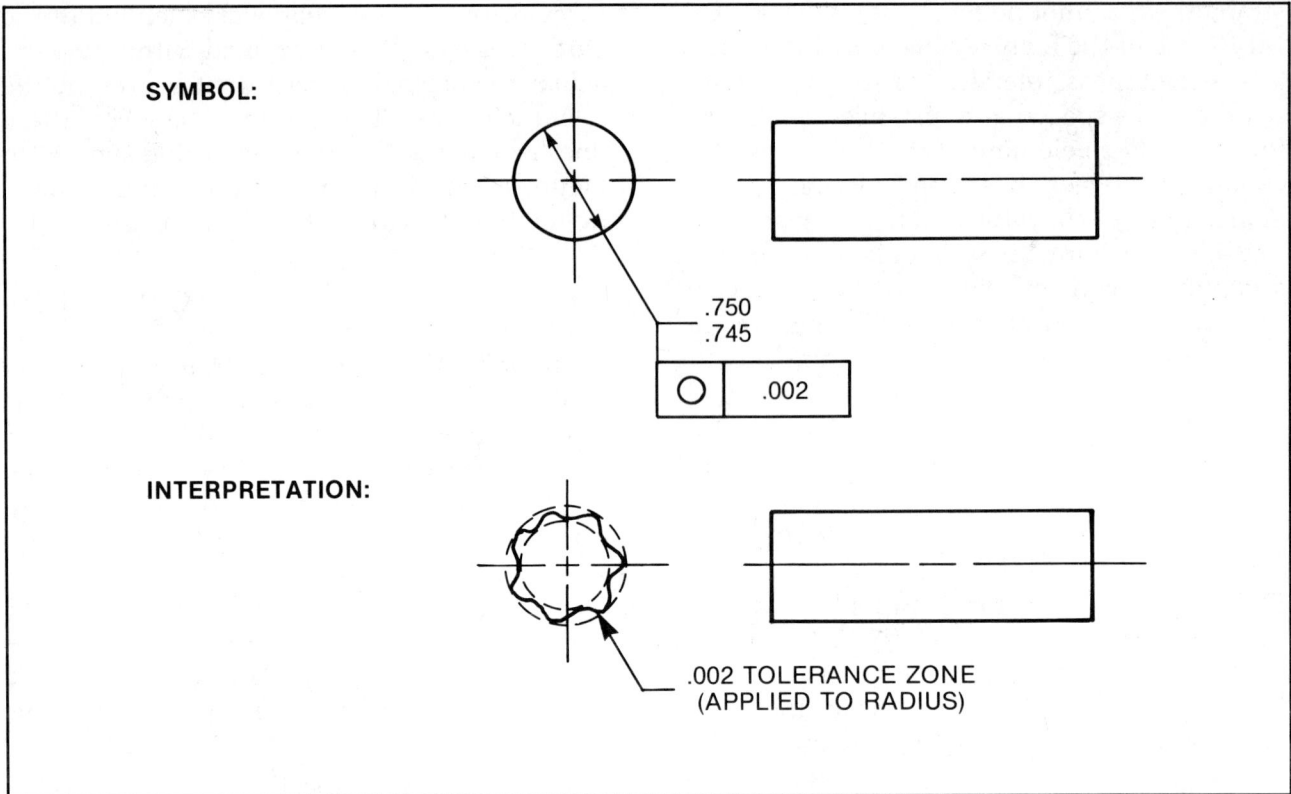

Figure 12-12. Interpreting the circularity (roundness) symbol

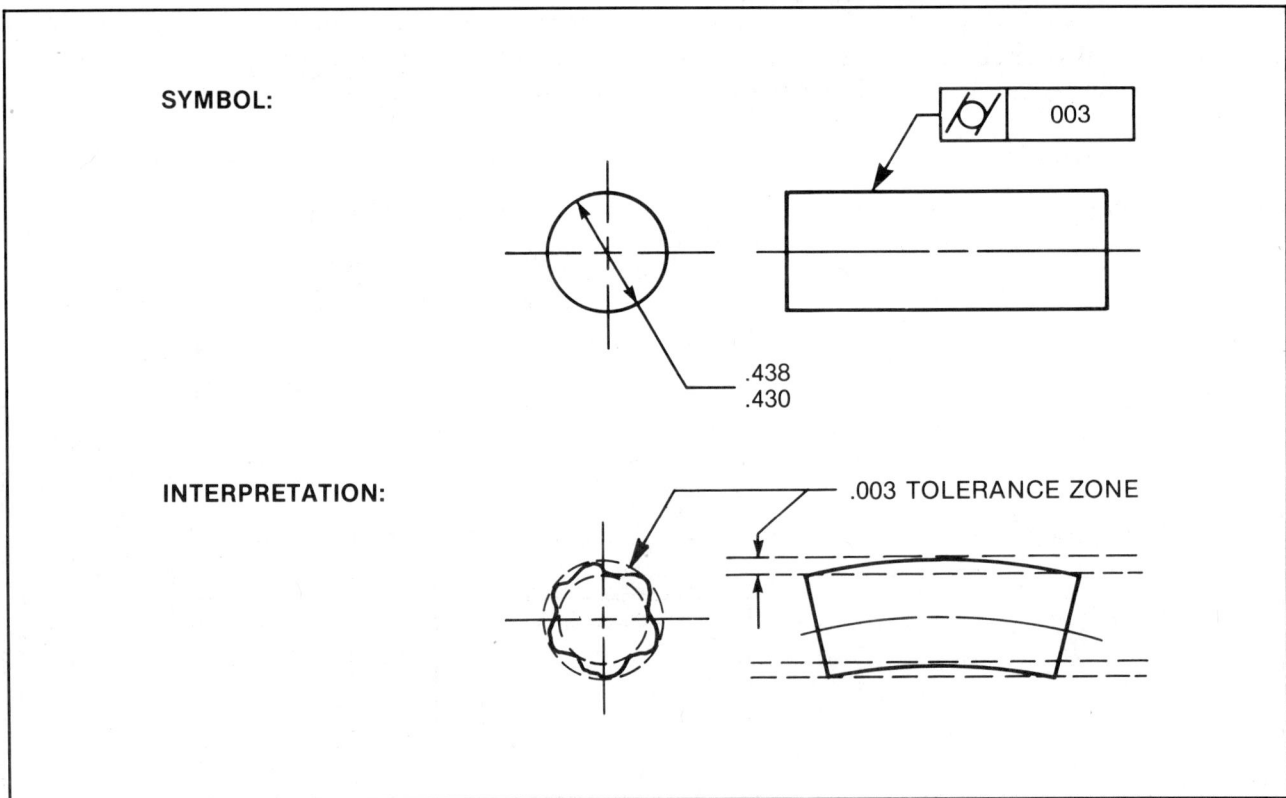

Figure 12-13. Interpreting the cylindricity symbol

240   *Geometric Dimensioning and Tolerancing*

Profile

Profile of a line (symbol ⌒) controls the shape of a part feature along a single line or edge, Figure 12-14. The *profile of a surface* symbol (⌓) is used to control a complete surface, Figure 12-15.

Profile controls can be used for either individual or related features. Where profile of a line or surface is specified, and not related to another feature, no datum reference is used. However, a datum reference can be used when the profile of a line or surface is relative to another feature.

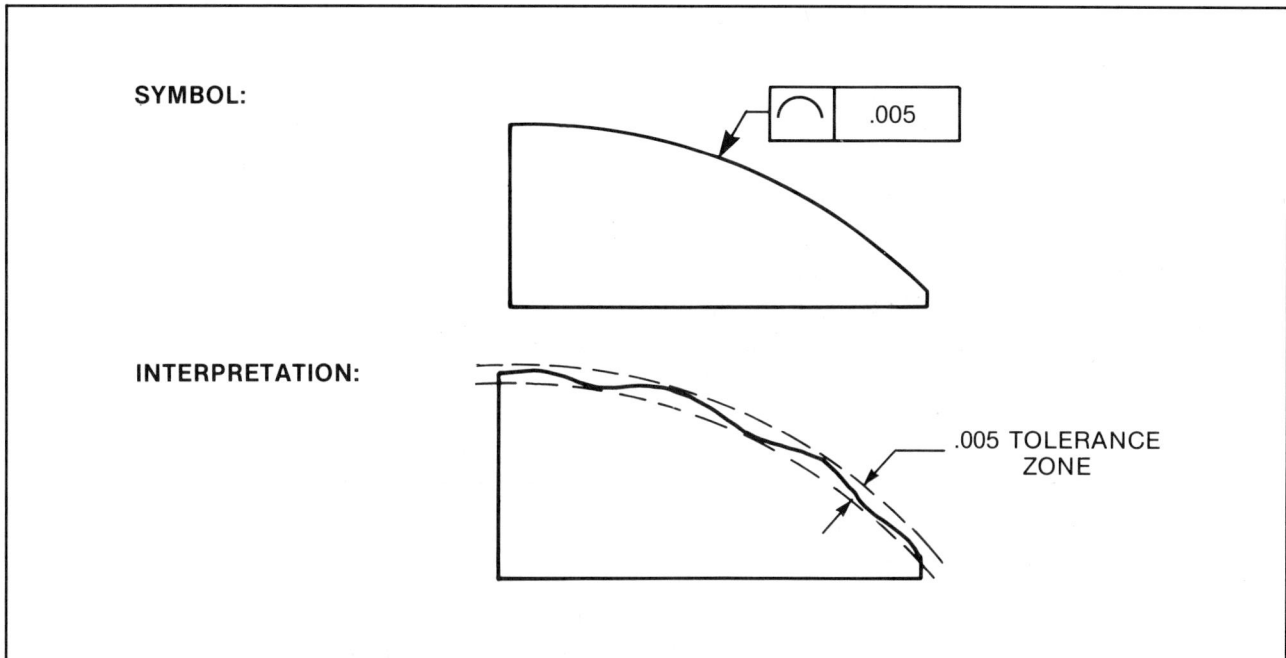

Figure 12-14. Interpreting the profile of a line symbol

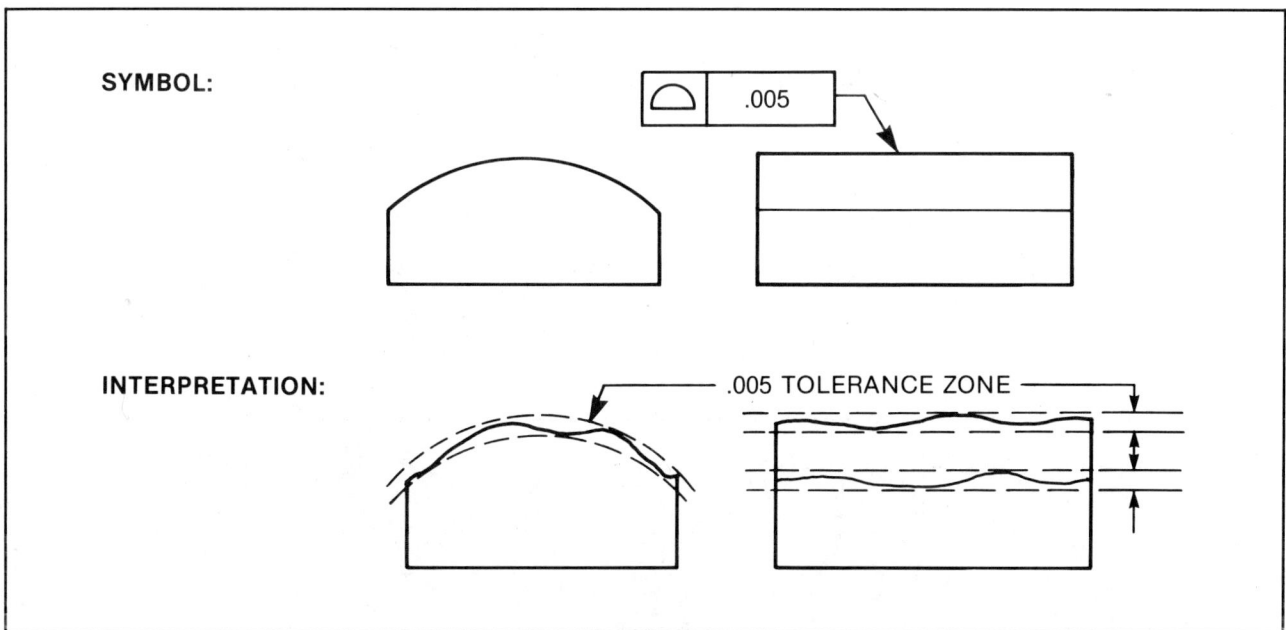

Figure 12-15. Interpreting the profile of a surface symbol

INTERPRETING ORIENTATION TOLERANCES

An *orientation tolerance* is a control that specifies a relationship between features. Orientation tolerance controls include angularity, perpendicularity, and parallelism.

Angularity

Angularity (∠) is a control applied to angular features other than 90°. Angularity indicates that each element of the feature is to be contained within two imaginary parallel lines that represent the tolerance zone, Figure 12-16. When angularity is specified, both a datum reference and a basic dimension for the angular value must be specified.

Perpendicularity

Perpendicularity is a control used to ensure that a controlled feature is perpendicular to the datum within the tolerance value. The symbol for perpendicularity is ⊥. In Figure 12-17, every element of the feature must lie within the imaginary zone formed by the tolerance value. When perpendicularity is specified, a datum reference must be used.

Parallelism

Parallelism (//) is a control used to ensure that the controlled feature is parallel to the datum within the tolerance value. In Figure 12-18, every element of the controlled feature must be within the imaginary tolerance zone. Since parallelism is a related feature, a datum reference is required.

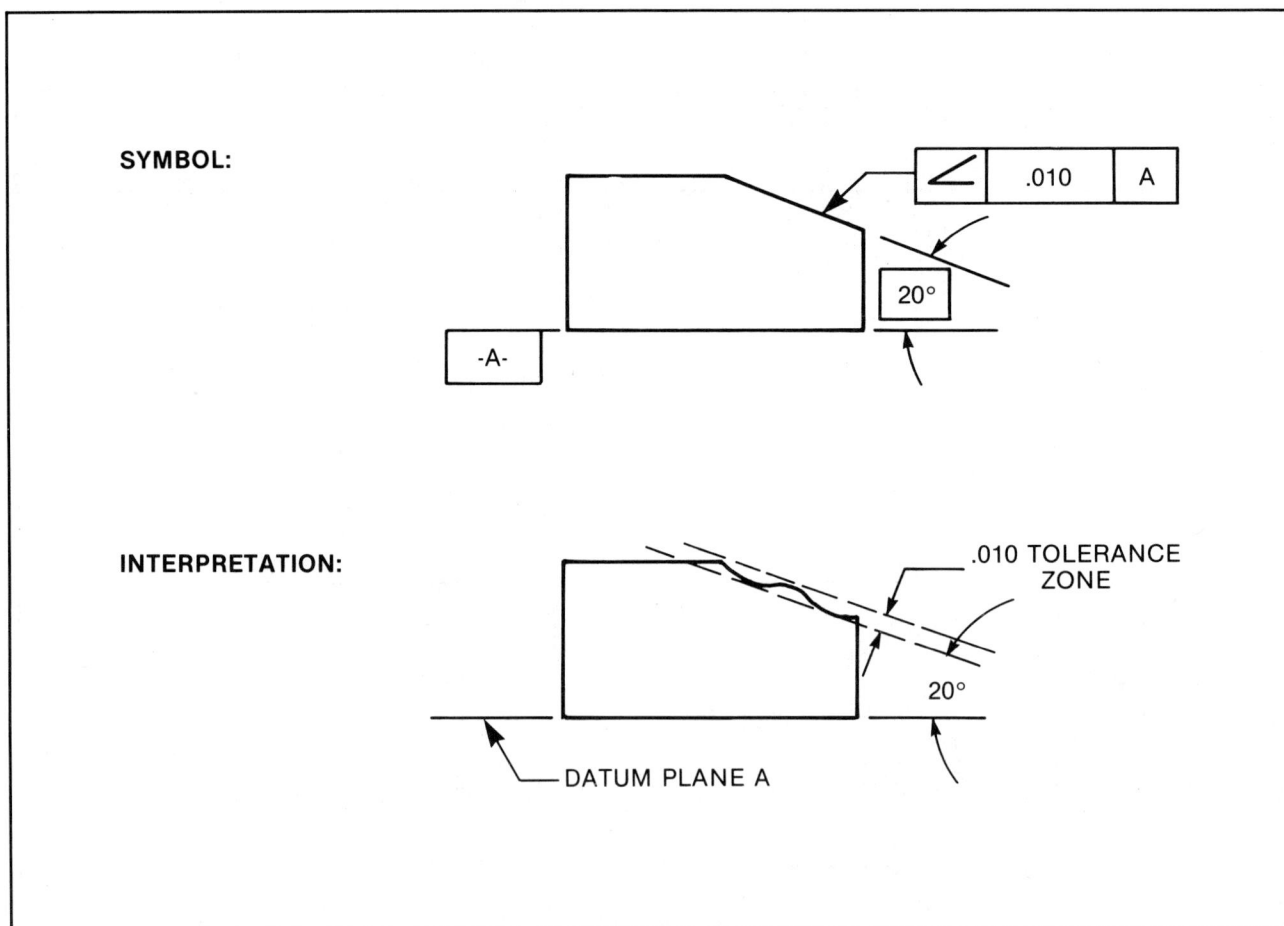

SYMBOL:

∠ | .010 | A

20°

-A-

INTERPRETATION:

.010 TOLERANCE ZONE

20°

DATUM PLANE A

Figure 12-16. Interpreting the angularity symbol

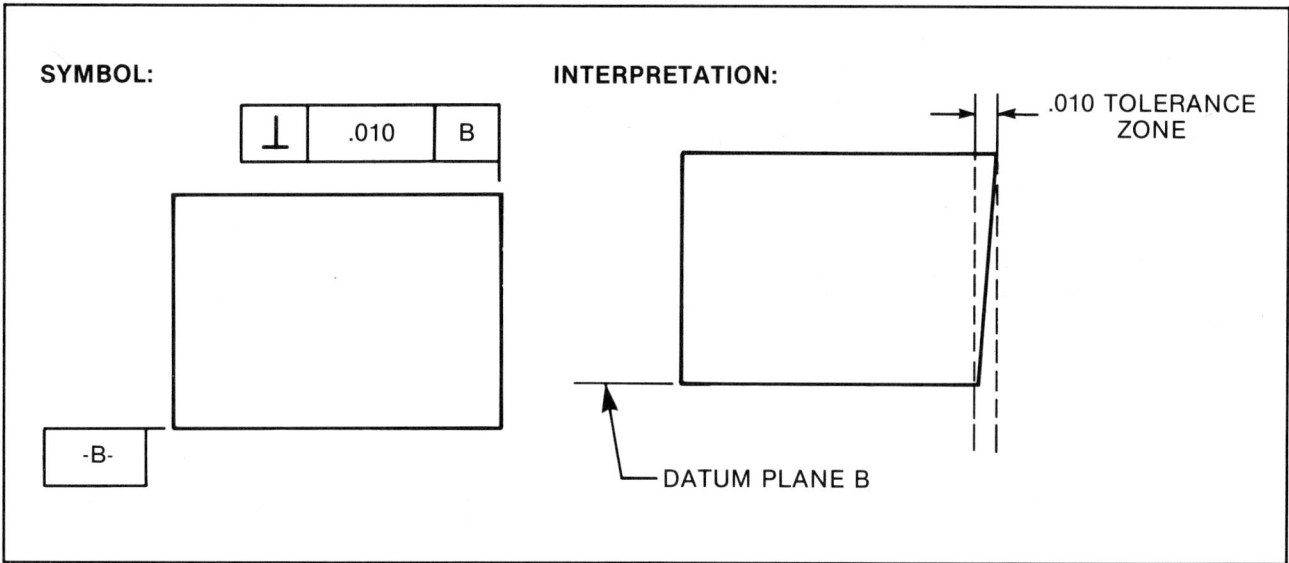

Figure 12-17. Interpreting the perpendicularity symbol

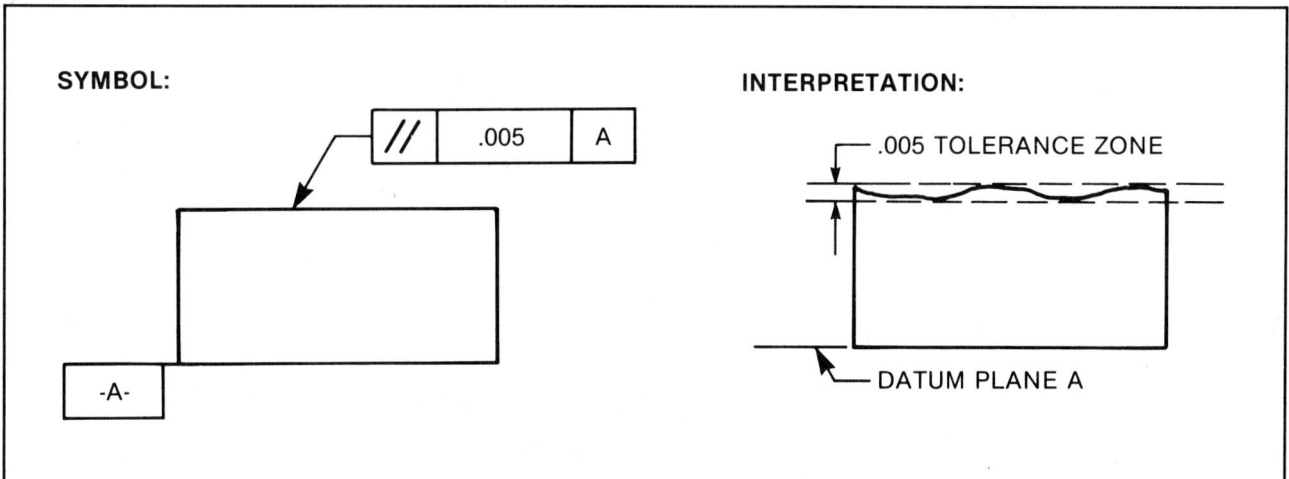

Figure 12-18. Interpreting the parallelism symbol

INTERPRETING LOCATIONAL TOLERANCES

Locational tolerances control the locational relationships of part features. These tolerances include position and concentricity.

Position

Position tolerance (⊕) is the most widely used form of locational tolerance. Position tolerance controls the location of holes or the position of part features, Figure 12-19. Position is also used to specify symmetry. *Symmetry* is the condition in which a feature or features are equally disposed about a center plane or center line.

Concentricity

Concentricity, symbolized by ◎, controls the functional relationship of two or more cylindrical features. When this control is specified, it indicates that the controlled feature must be concentric to the datum feature within the tolerance value, Figure 12-20.

SYMBOL:

2X .502/.498

⊕ | ⌀.010 | A | B

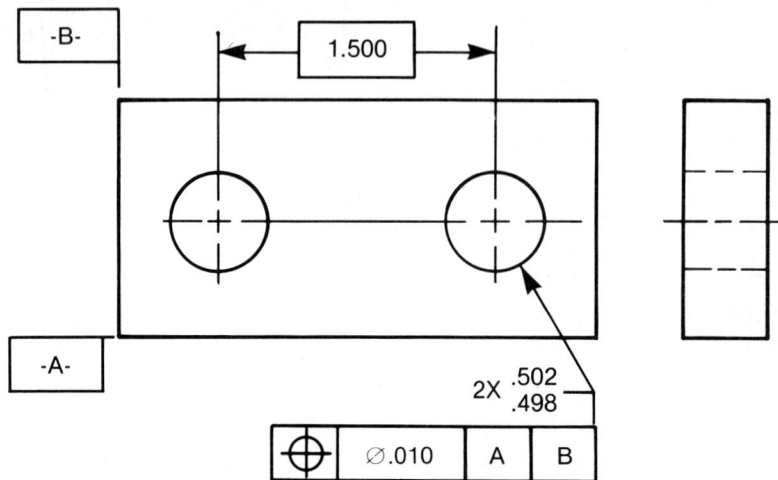

INTERPRETATION:

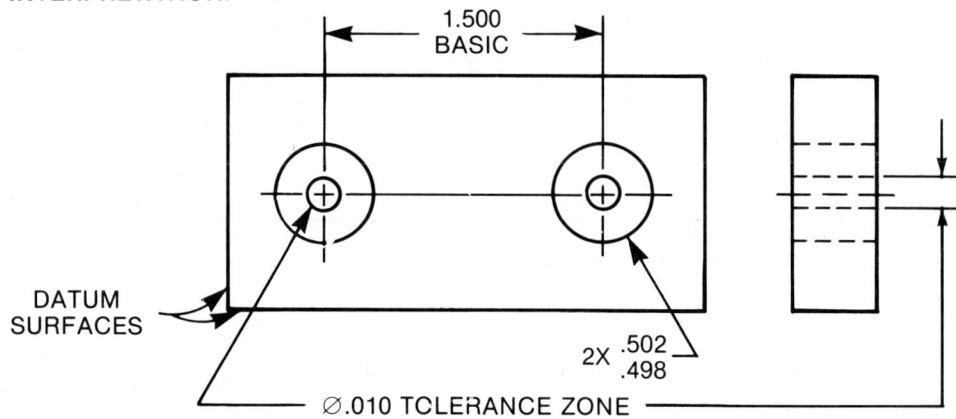

1.500
BASIC

DATUM
SURFACES

2X .502/.498

⌀.010 TOLERANCE ZONE

(HOLE CENTERS CAN BE ANYWHERE
WITHIN THESE TOLERANCE ZONES)

Figure 12-19. Interpreting the position symbol

SYMBOL:

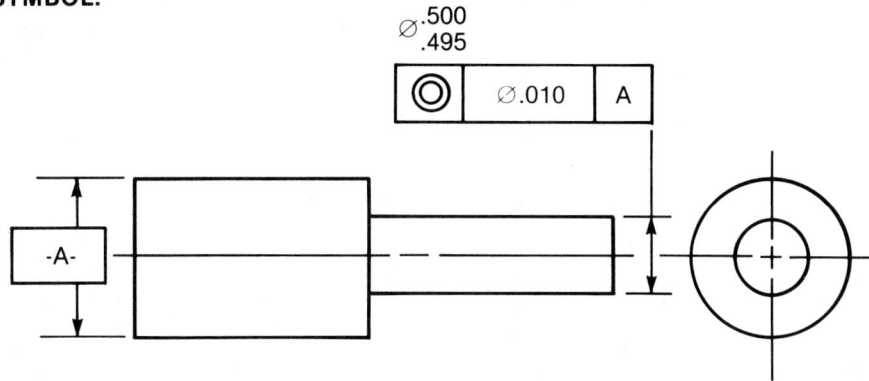

INTERPRETATION:

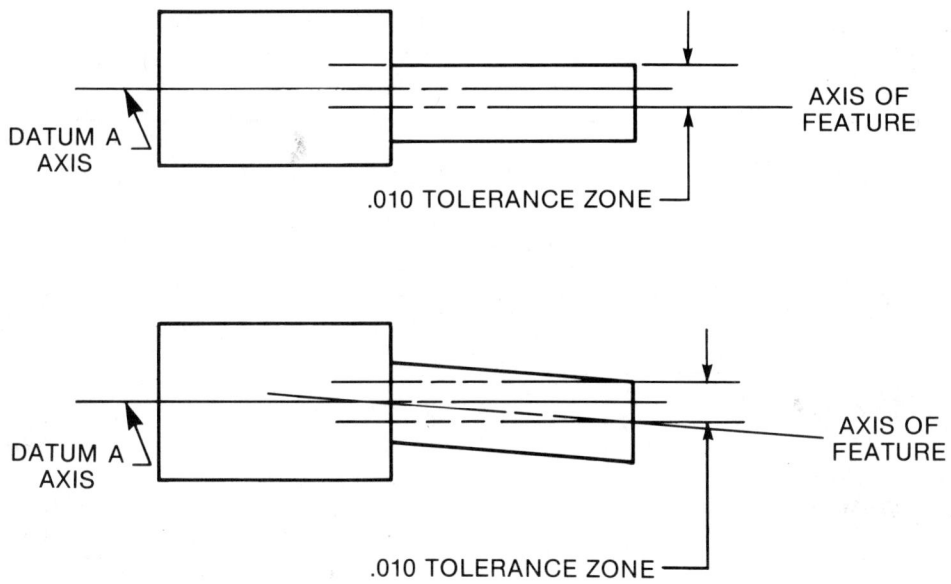

Figure 12-20. Interpreting the concentricity symbol

INTERPRETING RUNOUT TOLERANCES

Runout tolerances are intended to control the functional relationship of one or more features to the datum axis of the part. The two forms of runout tolerances are circular and total.

Circular Runout

Circular runout controls the total effect of form and locational variations on circular elements of the part, Figure 12-21. Circular runout controls only single elements or single indicator positions. The symbol of circular runout is ↗. The abbreviation *FIM* in Figure 12-21 means *full indicator movement*, and indicates the total amount of indicator travel permitted. The abbreviation *TIR* may also be used. *TIR* means *total indicator reading*, and has essentially the same meaning as FIM.

Total Runout

Total runout (↗↗) controls the total effect of form and locational variations over a complete surface, Figure 12-22.

INTERPRETING MODIFIERS

The modifiers used in geometric dimensioning and tolerancing include the maximum and least material conditions, regardless of feature size, and the projected tolerance zone. Each of these modifiers has a specific meaning, and each affects the feature control frame and part tolerance differently.

Interpreting MMC

When applied to a tolerance value, the MMC modifier indicates that the tolerance only applies to the feature at its maximum material condition. Figure 12-23, for example, shows a part with a tolerance of .003″ at MMC. The tolerance increases as the actual size of the part feature departs from the MMC size, as shown in the chart in Figure 12-23. The tolerance increase is always identical to the amount of departure from MMC.

When the MMC modifier is used for both the tolerance value and the datum reference, the effect is similar to that shown in Figure 12-24. Here the tolerance increases as the part size and the datum size depart from their MMC sizes.

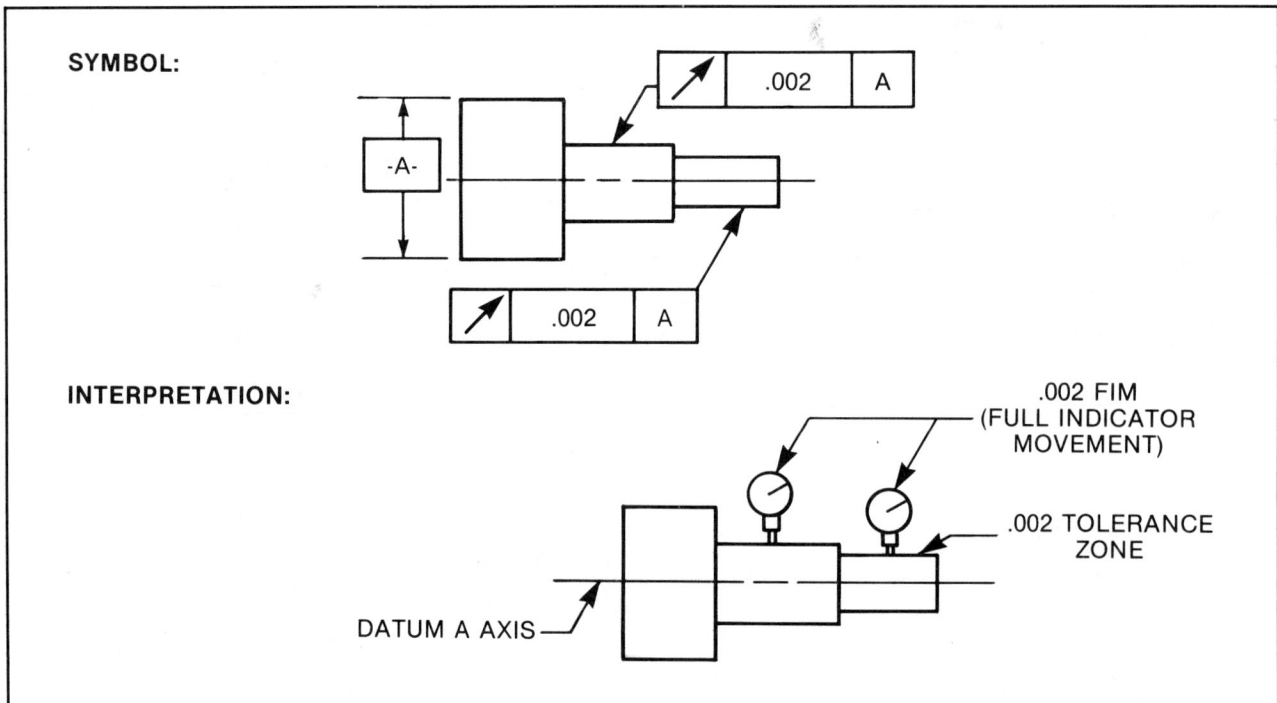

Figure 12-21. Interpreting the circular runout symbol

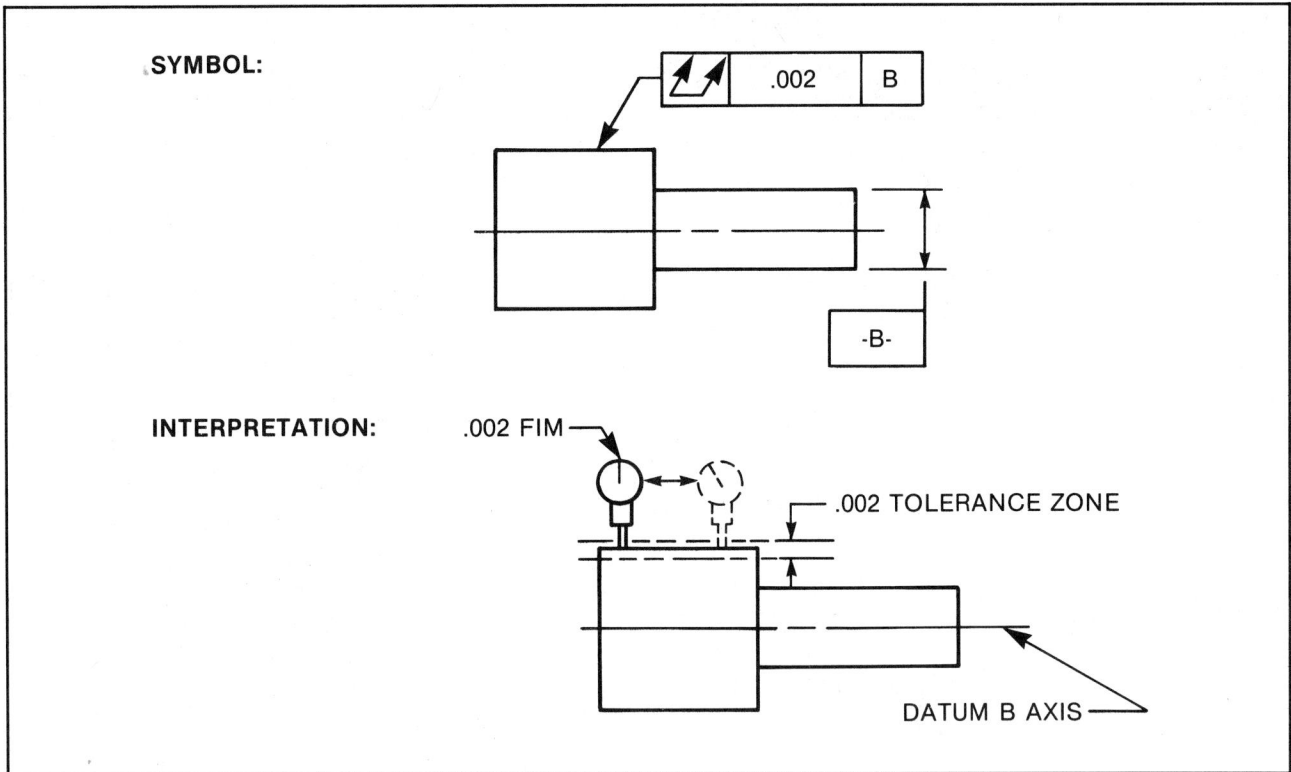

Figure 12-22. Interpreting the total runout symbol

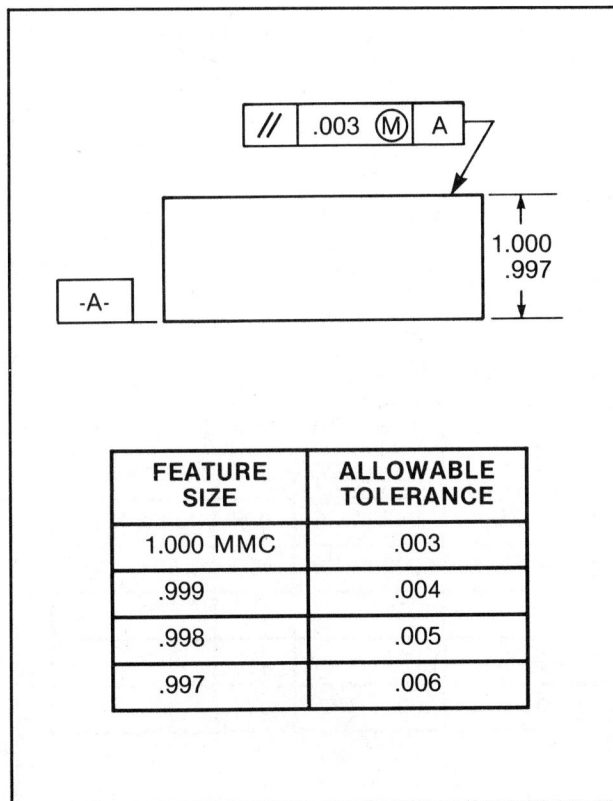

Figure 12-23. Interpreting MMC when applied to a tolerance value

FEATURE SIZE	ALLOWABLE TOLERANCE
1.000 MMC	.003
.999	.004
.998	.005
.997	.006

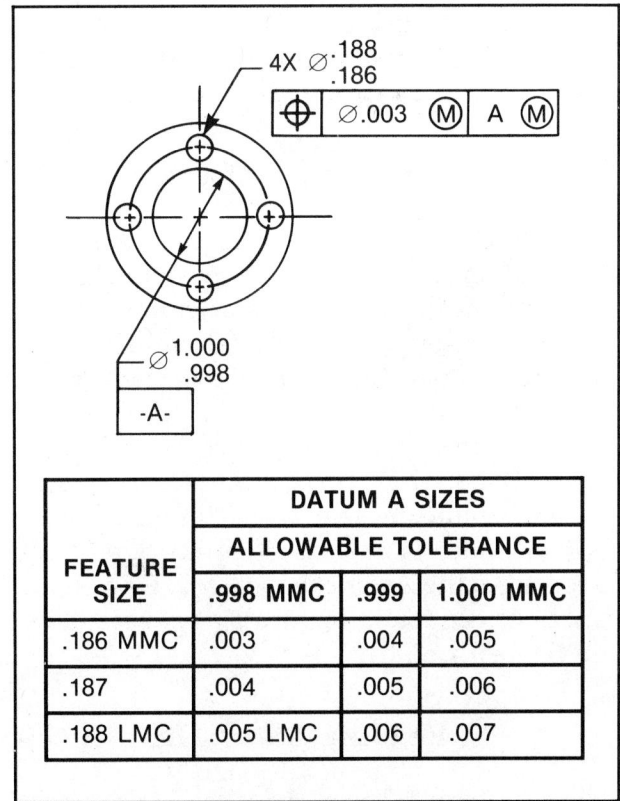

Figure 12-24. Interpreting MMC when applied to both a datum reference and a tolerance value

FEATURE SIZE	DATUM A SIZES		
	ALLOWABLE TOLERANCE		
	.998 MMC	.999	1.000 MMC
.186 MMC	.003	.004	.005
.187	.004	.005	.006
.188 LMC	.005 LMC	.006	.007

Interpreting LMC

The LMC modifier has the same effect as the MMC modifier. The only difference is in the application of the tolerance value. With the LMC modifier, the tolerance only applies at the least material condition.

Interpreting RFS

When used with a tolerance value, the RFS modifier means that the tolerance applies regardless of the actual size of the feature, Figure 12-25. When applied to both the tolerance value and the datum reference, the RFS modifier indicates that the tolerance value applies, regardless of the size of the feature or datum. See Figure 12-26.

Interpreting the Projected Tolerance Zone

The projected tolerance zone modifier is used to control the axial alignment of holes in two or more mating parts. The number shown with this modifier indicates the height of control or the thickness of the mating part. Figure 12-27 shows how this modifier is used.

Sequence of Information

The normal sequence of information contained in the feature control frame on newer prints is: (1) the geometric characteristic symbol; (2) the tolerance value; and (3) the datum reference. See Figure 12-28. This is called the *international sequence*.

A variation of this sequence is shown in Figure 12-29. Here the datum reference and tolerance value are reversed. This variation is used on older standards of prints and is referred to as the *American sequence*. Since both variations are found on prints, it is wise to be familiar with both.

Reading Datum References

Datum references can be shown as single datums or as multiple datums. When only one letter is shown in the feature control frame, it

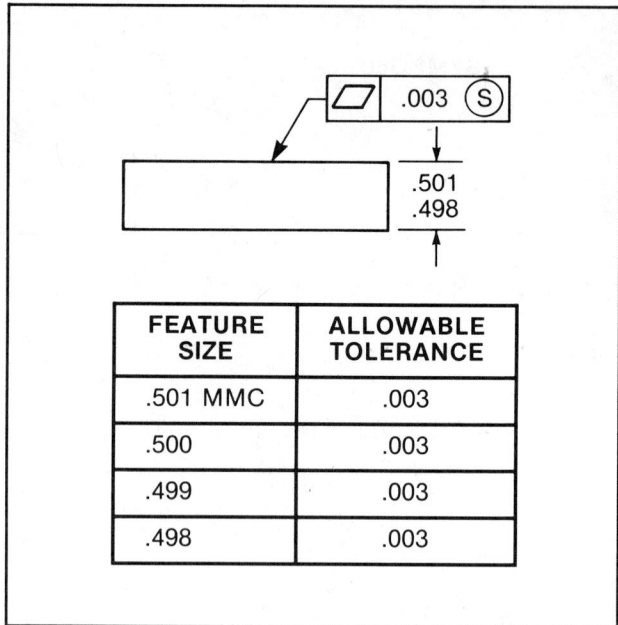

FEATURE SIZE	ALLOWABLE TOLERANCE
.501 MMC	.003
.500	.003
.499	.003
.498	.003

Figure 12-25. Interpreting RFS when applied to a tolerance value

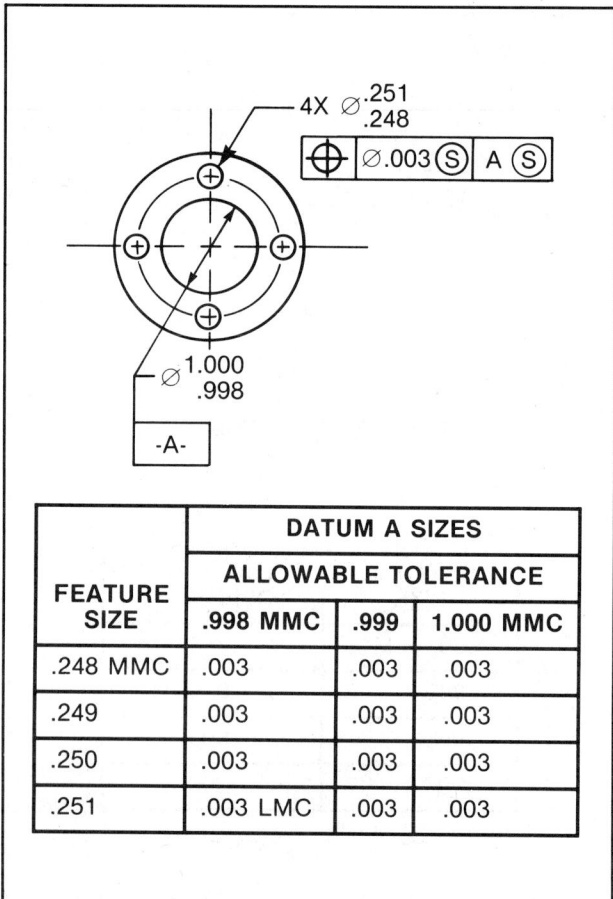

FEATURE SIZE	DATUM A SIZES		
	ALLOWABLE TOLERANCE		
	.998 MMC	.999	1.000 MMC
.248 MMC	.003	.003	.003
.249	.003	.003	.003
.250	.003	.003	.003
.251	.003 LMC	.003	.003

Figure 12-26. Interpreting RFS when applied to both a datum reference and a tolerance value

.500 – 20 UNF – 2B

⊕ | ∅.010 | Ⓜ | A

.500 | Ⓟ

-A-

INTERPRETATION:

∅.010 POSITIONAL
TOLERANCE ZONE

TRUE POSITION
AXIS

AXIS OF THREAD PD

.500 PROJECTED
TOLERANCE ZONE
HEIGHT

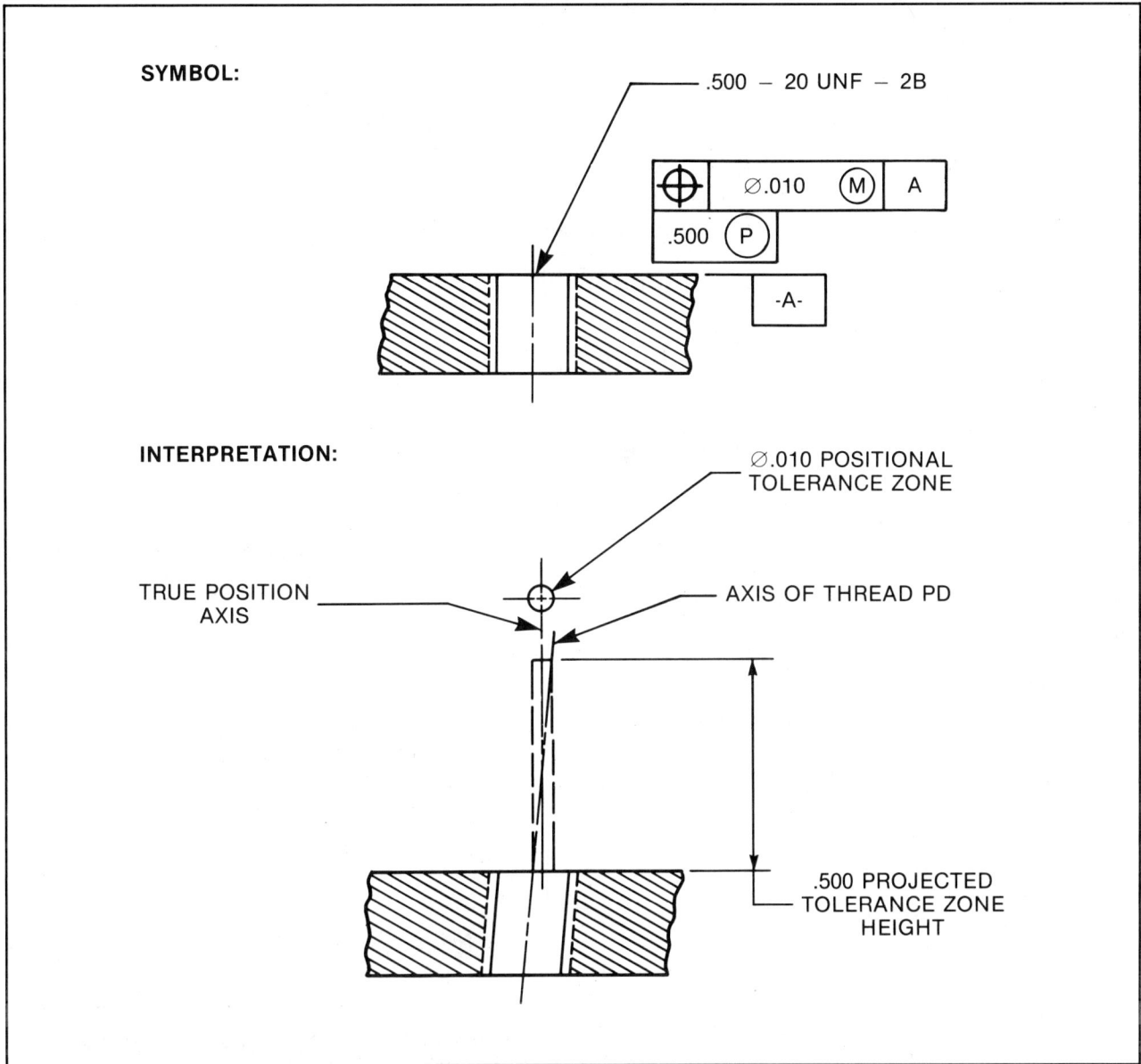

Figure 12-27. Interpreting the projected tolerance zone modifier

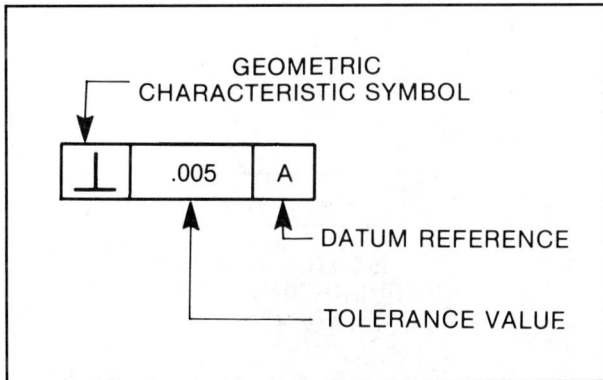

GEOMETRIC
CHARACTERISTIC SYMBOL

⊥ | .005 | A

DATUM REFERENCE

TOLERANCE VALUE

Figure 12-28. Usual (international) sequence of the feature control frame

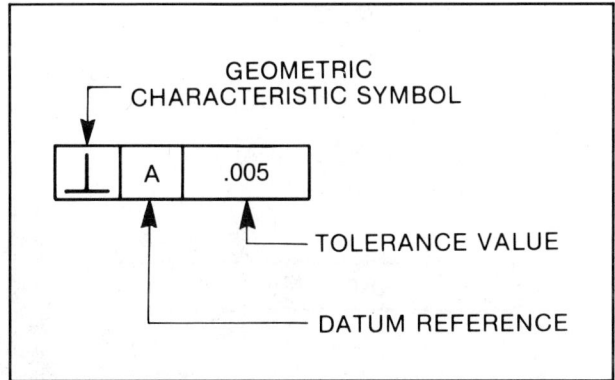

GEOMETRIC
CHARACTERISTIC SYMBOL

⊥ | A | .005

TOLERANCE VALUE

DATUM REFERENCE

Figure 12-29. A variation (American) of the sequence of the feature control frame

indicates that the feature is related only to that single datum. When a feature is related to more than one datum, multiple datums are specified.

In Figure 12-30, datums are arranged in their order of precedence. That is, the primary datum is shown first, the secondary datum is shown next, and the tertiary datum is listed last. Figure 12-31 shows how all three of these datums are related to a single part.

In cases where two datum surfaces must act as equal datums, the multiple primary, or other, datum is indicated by a dash between the datum reference letters. See Figure 12-32.

Positioning Datum References

Datum identifying symbols are generally shown alone on part features, Figure 12-33. However, when both a datum identifying symbol and a feature control frame are shown on the same plane, the symbols are combined. See Figure 12-34. When datum identifying symbols are

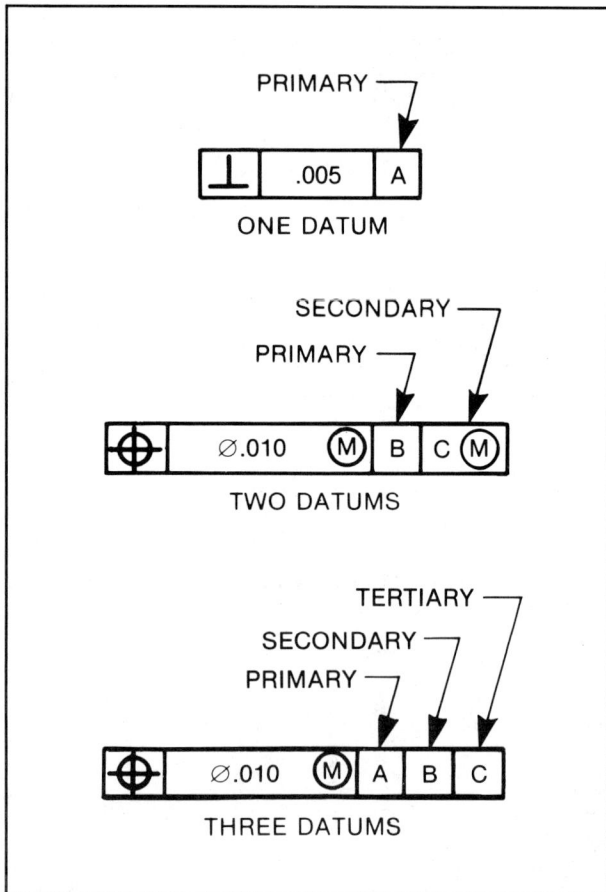

Figure 12-31. Relating datums to part features

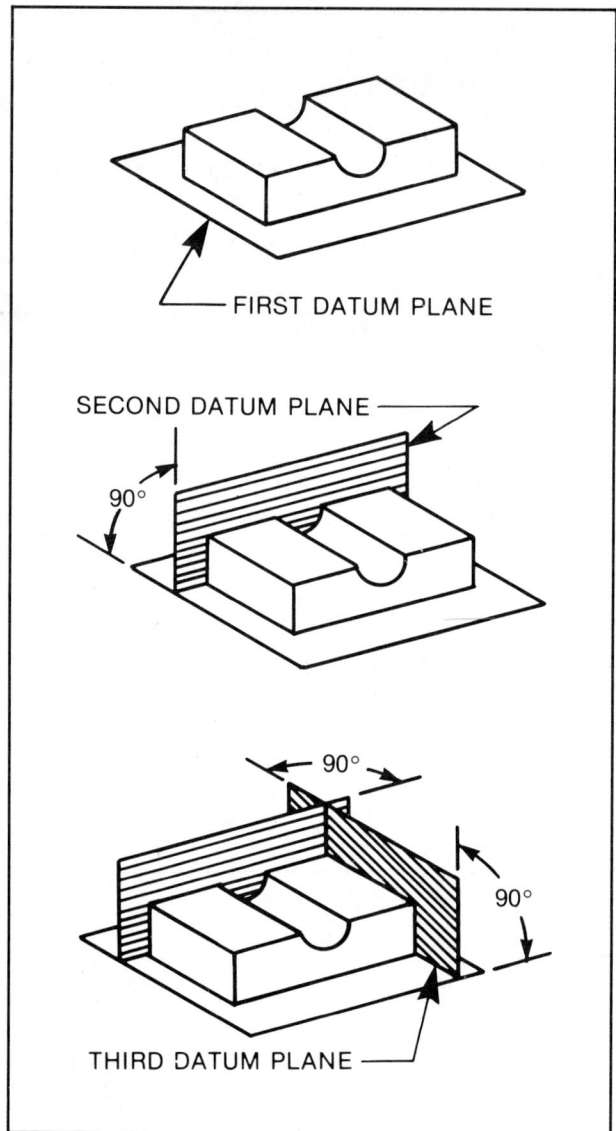

Figure 12-30. Datum order of precedence

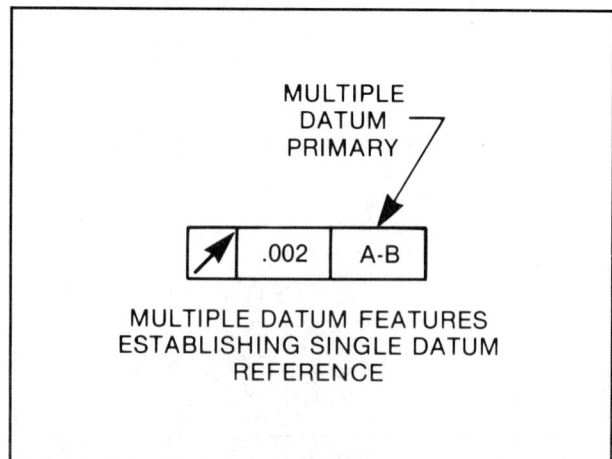

Figure 12-32. Indicating a multiple datum

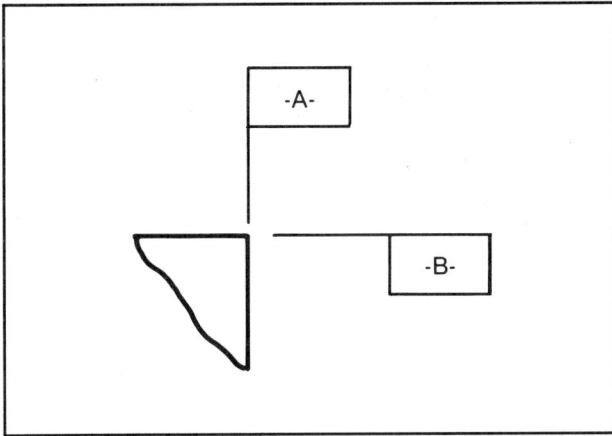

Figure 12-33. Locating the datum identifying symbol

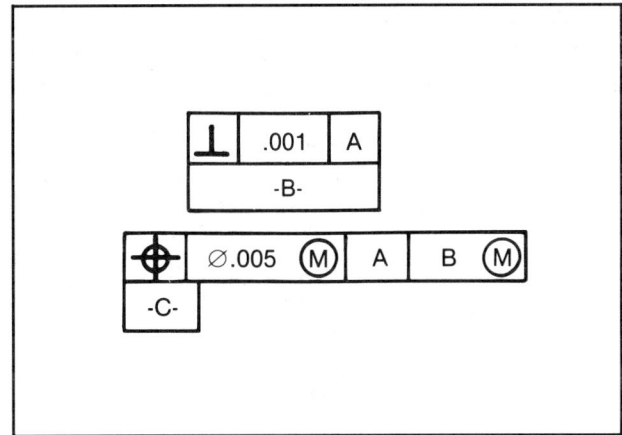

Figure 12-34. Alternate method to locate the datum identifying symbol

shown this way, remember that the datum is not part of the feature control frame. The symbols are independent and are only positioned this way to save room on the print.

To read this symbol, simply remember that the feature control frame controls the feature, and the feature created by this frame then becomes the datum shown. Therefore, in the top control frame in Figure 12-34, the surface is to be machined perpendicular to datum *A* to within .001″. This surface, once machined, becomes datum surface *B*.

ANSI *Y14.5-1973* AND *Y14.5M-1982*

The principal standard that controls the application of dimensions and tolerances in industry is American National Standards Institute (ANSI) *Standard Y14.5.* The *Y14.5-1973* standard was used from 1973 until early 1983, when the new revised standard, *Y14.5M-1982*, was adopted. The change in standards was brought about in an effort to update the older standard, and to bring American drafting practices in closer alignment with the international standards. The changes made to the standard were not dramatic. However, there could be some confusion for those who are not familiar with the 1973 version of the standard. Therefore, in this section we briefly discuss the differences between the two standards to eliminate any confusion

regarding prints produced before 1983 and those made after 1983.

Feature Control Symbols

The first, and most notable, difference between the two standards is the sequence of information contained in the feature control frame. *Y14.5-1973* uses the geometric characteristic symbol, datum reference, and tolerance value sequence. This sequence is referred to as the *American sequence.* The newer *Y14.5M-1982* standard has the geometric characteristic symbol, tolerance value, and datum reference sequence. This sequence is called the *international sequence.* The information contained within these feature control symbols is identical; the only difference is in their sequence.

Diameter Symbol

The new standard, following international practices, specifies that the diameter symbol (∅) should precede a dimensional value that indicates a diameter. The symbol *DIA* will only be seen on old prints.

Least Material Condition

The least material condition was considered a reference value in the 1973 standard, and therefore did not have an approved symbol. In the newer standard, the least material condition is shown by a circled L (Ⓛ) and is considered as a supplementary symbol. The meaning of this value is the same. Only its use in the feature control frame has been changed in the 1982 standard.

Obsolete Symbols

Some symbols that were used in the older 1973 standard have been eliminated or altered in the 1982 standard. To properly interpret the meaning of older prints, it is necessary to be familiar with these obsolete symbols as well as the new symbols.

SYMMETRY. The symbol that was used in the 1973 standard to denote symmetry ($\equiv$) is not used in the new standard. Instead, the position symbol ($\oplus$) is now used to show symmetry.

RUNOUT. The symbols that were formally used to show runout, both circular and total, were identical in the old standard. This symbol was the single arrow ($\nearrow$). The distinction between these two forms of runout was made by noting the word *TOTAL* below the feature control frame. If the note *TOTAL* was not used, it meant the runout control was circular. In

some companies, to avoid confusion, the abbreviation *CIRC* was used below the feature control frame when circular runout was specified. The new standard uses two different symbols to show runout. Circular runout is shown with a single arrow ($\nearrow$), and total runout is shown with a double arrow ($\nearrow\!\!\!\nearrow$). This eliminates the requirement of placing a note below the feature control frame.

KEY TERMS

The following key terms were introduced in this chapter. Be sure you know the meaning of each term before proceeding to the review material.

Datum
Datum Identifying Symbol
Datum Reference
Feature Control Frame
Geometric Characteristic Symbol
Least Material Condition
Maximum Material Condition
Projected Tolerance Zone
Regardless of Feature Size
Tolerance Value

Test your knowledge with this reinforcement study material. Write your answers to the questions in the spaces provided.

1. What is the symbol used most often in geometric dimensioning and tolerancing? _____

   _____

2. List the three elements normally found in a feature control frame.

   a. _____

   b. _____

   c. _____

3. What five types of tolerances are controlled with geometric dimensioning and tolerancing?

   a. _____

   b. _____

   c. _____

   d. _____

   e. _____

4. What is a datum? _____

   _____

5. Which three letters of the alphabet are not used as datum identifiers, and why? _____

   _____

   _____

6. What does the tolerance value indicate? _____

   _____

7. Identify the following geometric characteristic symbols, modifiers, and related symbols.

   a. — _____ k. ( ) _____

   b. ▱ _____ l. ∠ _____

   c. ○ _____ m. ◎ _____

   d. ⌒ _____ n. ⊥ _____

   e. ⌓ _____ o. ↗ _____

   f. ⌗ _____ p. Ⓢ _____

   g. ⌖ _____ q. Ⓟ _____

   h. ∥ _____ r. ⎯.01 _____

   i. Ⓜ _____

   j. ⌀ _____ s. ⫽ _____

8. What do the abbreviations *MMC, LMC,* and *RFS* mean? _____

_____

_____

9. How are basic dimensions shown on a print? _____

_____

_____

10. Which locational tolerance characteristic is used most often? _____

_____

11. How are the two types of runout tolerances identified on a print? _____

_____

_____

12. What do the abbreviations *TIR* and *FIM* mean? _____

_____

_____

13. What effect does the RFS modifier have on a tolerance value? _____

_____

_____

14. Study Figure R12-1 below and then fill in the information missing in the chart.

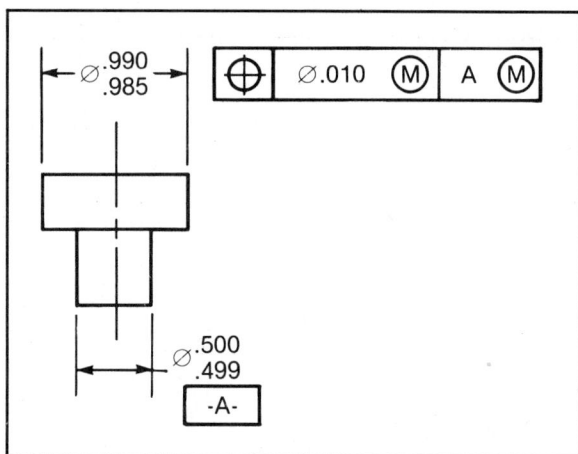

	DATUM SIZES	
FEATURE SIZE	ALLOWABLE TOLERANCES	
	.500 MMC	.499 LMC
.990 MMC		
.989		
.988		
.987		
.986		
.985 LMC		

Figure R12-1.

15. What is the American sequence of information in a feature control frame? _____

_____

_____

16. What is the international sequence of information in a feature control frame? _____

_____

_____

Refer to Figure E12-1 on page 256 to answer the following questions.

1. What is the part name? _____

2. How many datum surfaces are there on this part? _____

3. How many different geometric characteristic symbols are shown on this part? _____

4. What conditions do the geometric characteristic symbols used on this print indicate?

   _____

   _____

   _____

5. What are the boxed-in letters called? _____

   _____

6. What are the boxed-in dimensions called? _____

   _____

7. What does the symbol Ⓜ mean? _____

   _____

8. What is the MMC size of datum *B*? _____

9. What is the LMC size of datum *C*? _____

10. What does the ∅ symbol indicate? _____

    _____

11. What type of runout is indicated in symbol *A*? _____

    _____

12. Which type of sequence is shown in symbol *A*? _____

    _____

13. How does the symbol Ⓜ affect the tolerance shown in symbol *B*? _____

    _____

    _____

14. Why does symbol *C* have no datum reference? _____

    _____

    _____

15. What type of sectional view is shown? _____

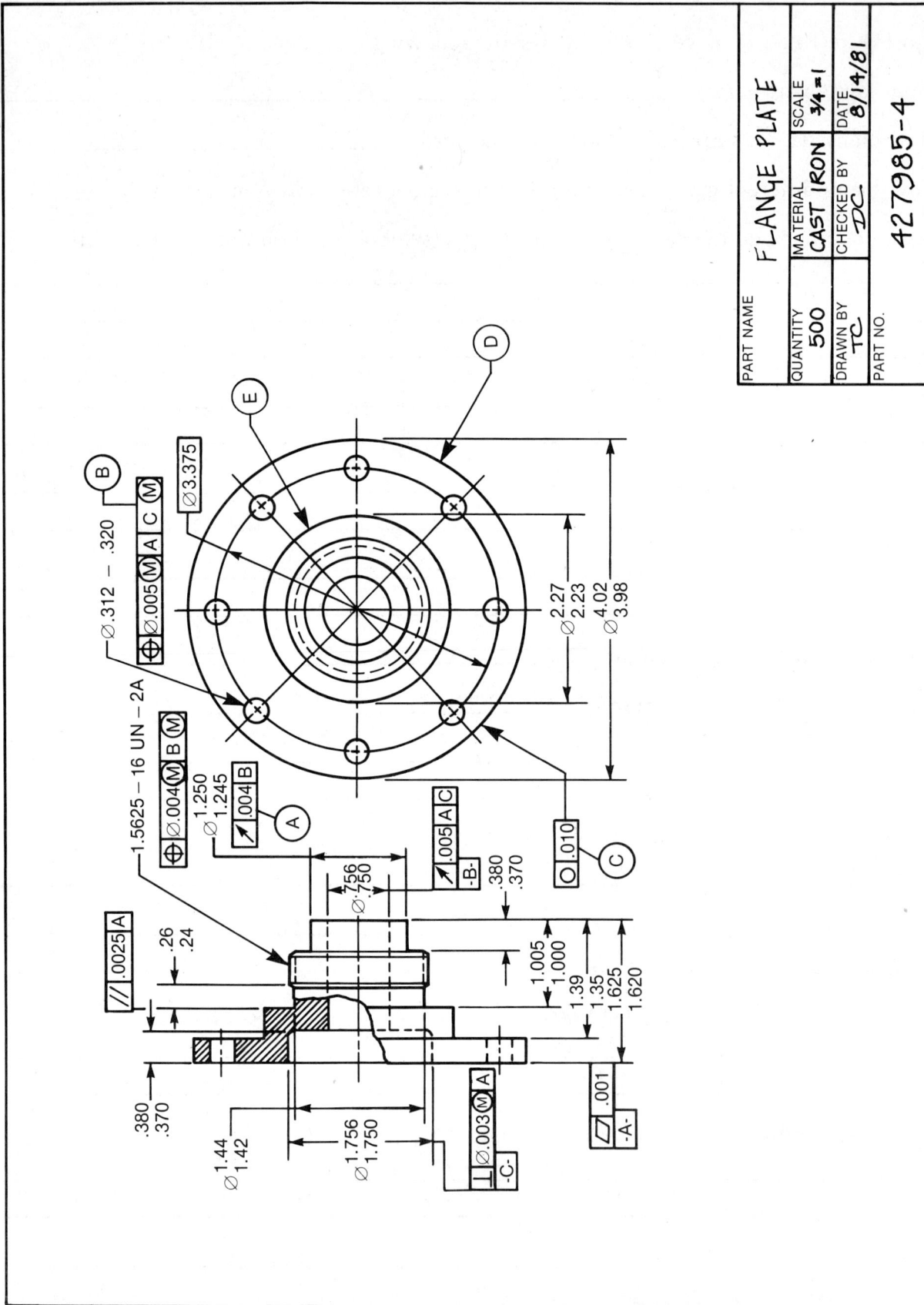

Figure E12-1.

13

PICTORIAL DRAWINGS

After studying this chapter, you will be able to:

- Identify the types of pictorial drawings.
- Describe the typical methods used to sketch pictorial drawings.
- Describe the variations of the major types of pictorial drawings.
- Draw pictorial sketches.

Pictorial drawings, while not as common as multiview drawings, are sometimes used to describe parts or assembled units. Pictorial drawings and sketches are frequently used to convey instructions for making simple parts or assemblies. They may also be used to clarify details on a multiview print.

There are three basic types of pictorial drawings: axonometric drawings, oblique drawings, and perspective drawings. Each type of pictorial drawing shows a part in approximately the same way as it would appear if photographed. Pictorial drawings are rarely dimensioned or used for production prints. Because the drawings are visually rather than proportionally correct, distortion is present. This distortion, called *foreshortening*, occurs on one or more of the part axes and makes dimensions, when applied, appear visually incorrect.

AXONOMETRIC DRAWINGS

Axonometric drawings are one-view drawings of an object and show all three dimensions—height, width, and depth—in a pictorial form. The three principal types of axonometric drawings are isometric, dimetric, and trimetric. The chief difference between the three axonometric drawing forms is the angle of the axonometric axes.

Isometric Drawings

Isometric drawings, Figure 13-1, are used for shop sketches and to clarify some multiview drawings. The popularity of isometric drawings is due largely to their ease of construction. Isometric drawings can easily be drawn or sketched with standard drawing equipment.

Isometric drawings are built around axonometric axes that have three 120° angles. See Figure 13-2. Since the angles of the axes are equal, the lengths of the sides of an isometric drawing are drawn to the same size. An isometric drawing can be drawn in any position. The four most common positions are shown in Figure 13-3.

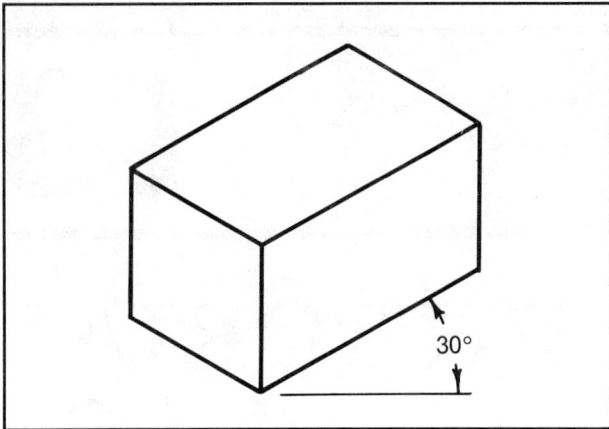

Figure 13-1. Isometric drawing

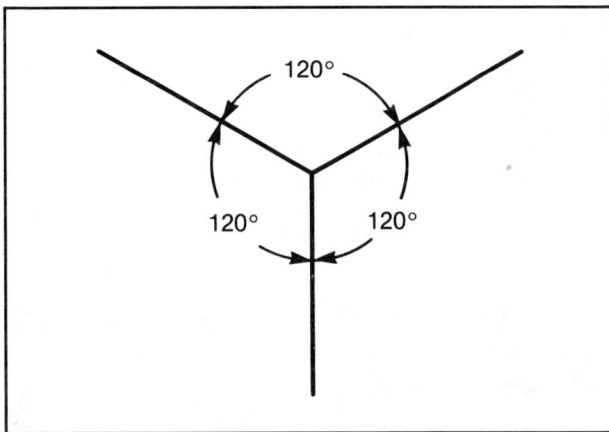

Figure 13-2. Axonometric axes for isometric drawing

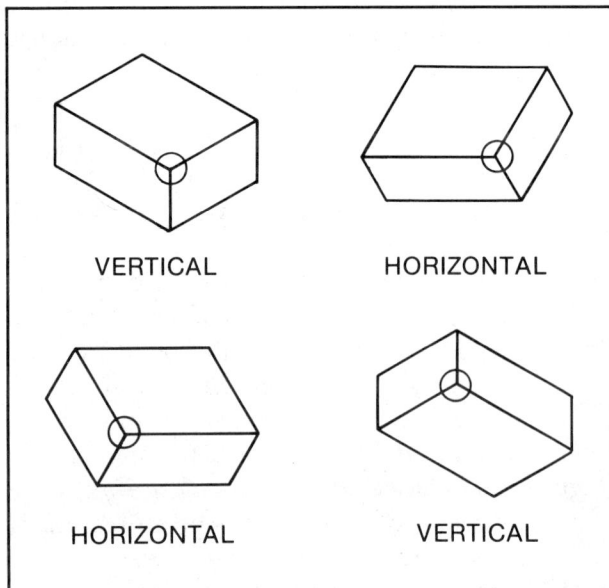

VERTICAL

HORIZONTAL

HORIZONTAL

VERTICAL

Figure 13-3. Common positions of the axonometric axes for isometric drawings

In all axonometric drawings, part details such as holes, rounds, fillets, and curved surfaces are drawn in ellipitical form, Figure 13-4. Angular details are not true length. Only vertical and 30° lines are true length. Figure 13-5 shows angular lines drawn to lengths that are proportional to the drawn object, rather than to their true lengths. This changing of line lengths is called *foreshortening* and is common in all pictorial drawings.

Dimetric Drawings

Dimetric drawings, Figure 13-6, have axonometric axes with two equal angles. The lengths of the sides that relate to these equal angles are drawn to the same size. The length of the third

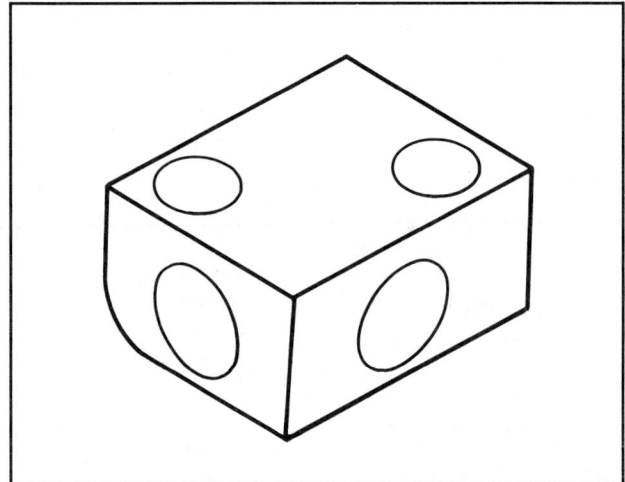

Figure 13-4. Ellipses are used to show holes and arcs in isometric drawings.

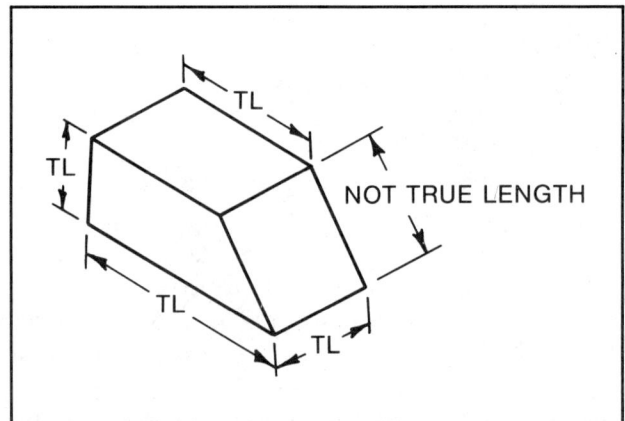

Figure 13-5. Angular lines are not true length in isometric drawings.

side is drawn to a different size. By drawing an object as a dimetric, the viewing angle is changed. This permits the part to appear more true to life than with an isometric drawing.

Trimetric Drawings

Trimetric drawings, Figure 13-7, have three different angles forming the axonometric axes. Also, since the angles of the axes are all different, the lengths of the sides are drawn to three different sizes. As with dimetric drawings, trimetric drawings are sometimes preferred to isometric drawings to make an object appear more natural.

OBLIQUE DRAWINGS

The *oblique drawing*, Figure 13-8, is another form of pictorial drawing. In oblique drawings, the principal view is drawn parallel to the viewing plane, just as in a multiview drawing. Placing the object in this position permits details such as holes, angles, and contours on the front surface to be drawn to their actual proportion and size without distortion.

In many cases, oblique drawings are preferred to other pictorial drawings because they show details and form more clearly. Oblique drawings may be used to show parts such as spacers, collars, shims, and other simple forms.

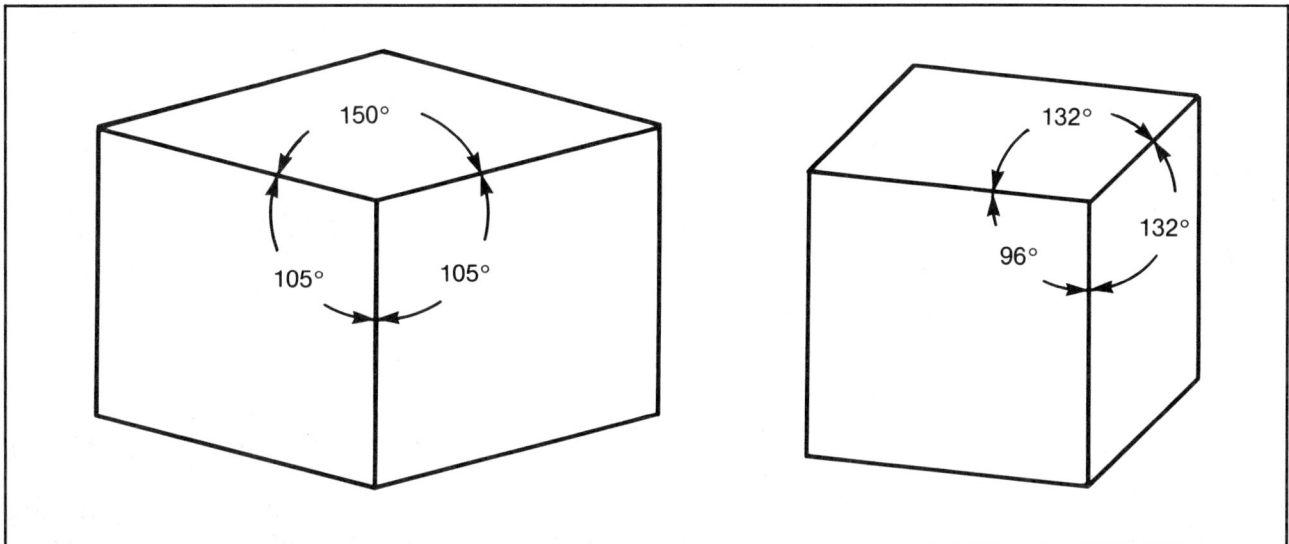

Figure 13-6. Dimetric drawings

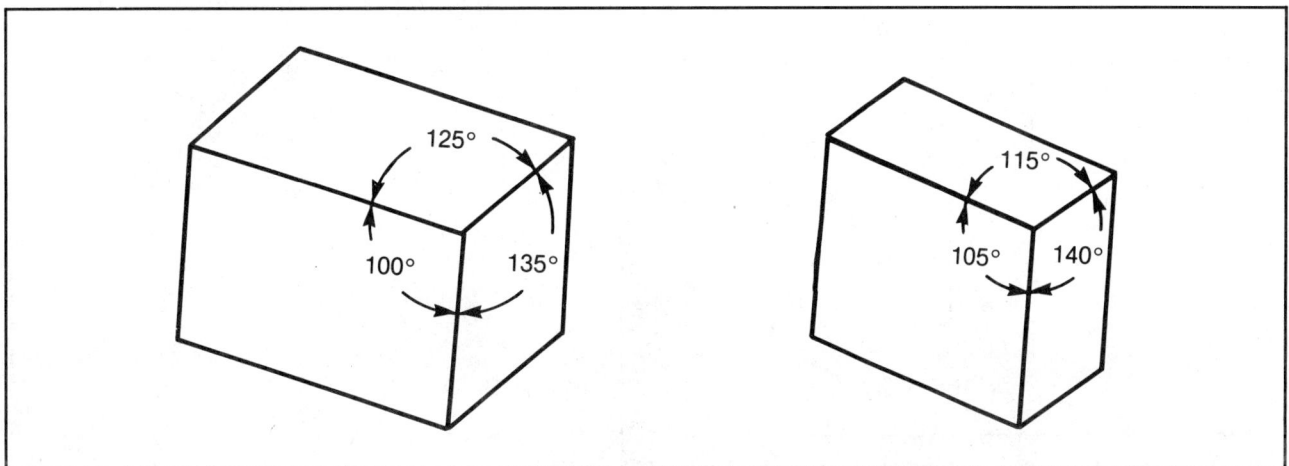

Figure 13-7. Trimetric drawings

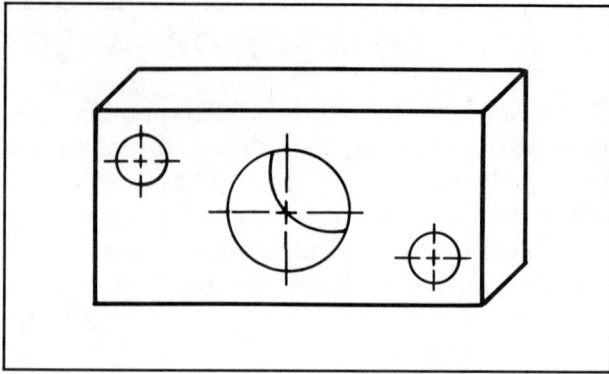

Figure 13-8. Oblique drawing

The two principal types of oblique drawings are the *cavalier* and the *cabinet*, Figure 13-9. The difference between these two types of drawings is the lengths of the receding lines. The *receding lines* are the lines that extend behind the part; they appear to give the part depth. Cavalier drawings are drawn with the receding lines drawn to full size. In cabinet drawings, the receding lines are drawn to half size. Reducing the size of the depth allows a drawing to appear more visually correct and proportional. In an oblique drawing, the receding lines can be drawn to any convenient angle. Drafters usually use 30°, 45°, or 60°. Note that all receding lines in oblique drawings are parallel.

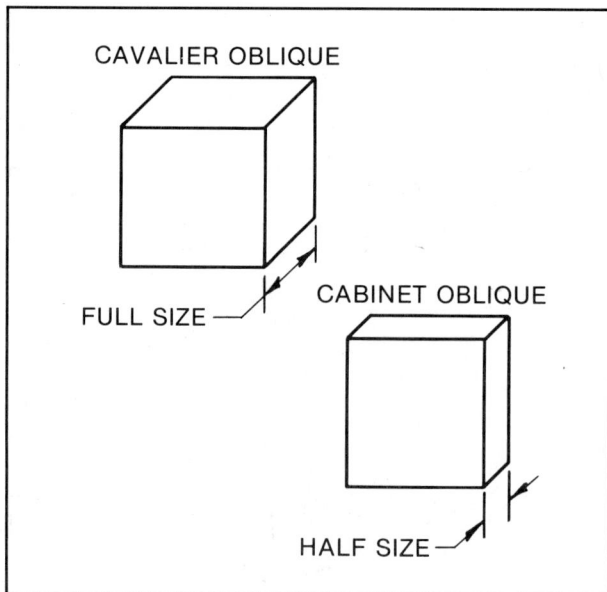

Figure 13-9. Cavalier and cabinet oblique drawings

PERSPECTIVE DRAWINGS

The *perspective drawing* is the truest pictorial method of drawing any object. A part drawn in perspective form appears as the eye would see it. Details closer to the observer appear larger, while details farther away appear smaller.

Perspective drawings are drawn by using vanishing points. *Vanishing points* are the points on the drawing where the lines forming the object come together. Each type of perspective drawing is directly related to the number of vanishing points used to construct the drawing. The three principal types of perspective views are the following:

- One-point perspective
- Two-point perspective
- Three-point perspective

One-Point Perspectives

One-point perspectives have a single vanishing point, Figure 13-10. This form of perspective normally has one face of the part drawn parallel to the viewing plane. The receding lines come together at one vanishing point on the horizon (eye level).

Two-Point Perspectives

Two-point perspectives have two vanishing points, Figure 13-11. Two-point perspectives are normally positioned at an angle to the viewing plane. The receding lines come together at two vanishing points on the horizon.

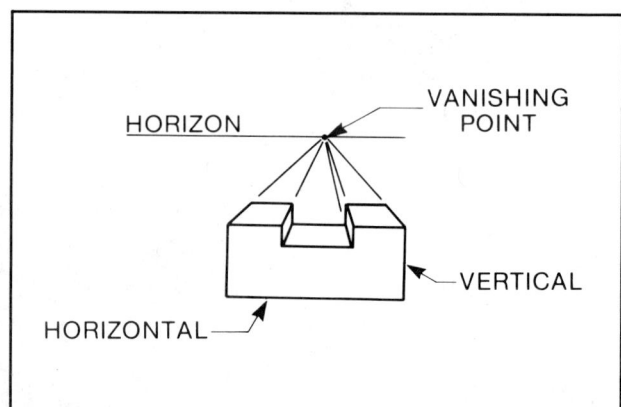

Figure 13-10. One-point perspective drawing

Three-Point Perspectives

Three-point perspectives, Figure 13-12, do not have any portion of the object parallel to the viewing plane. The receding lines on a three-point perspective meet at three vanishing points. The top two vanishing points recede to the horizon. The bottom vanishing point is placed below the drawing.

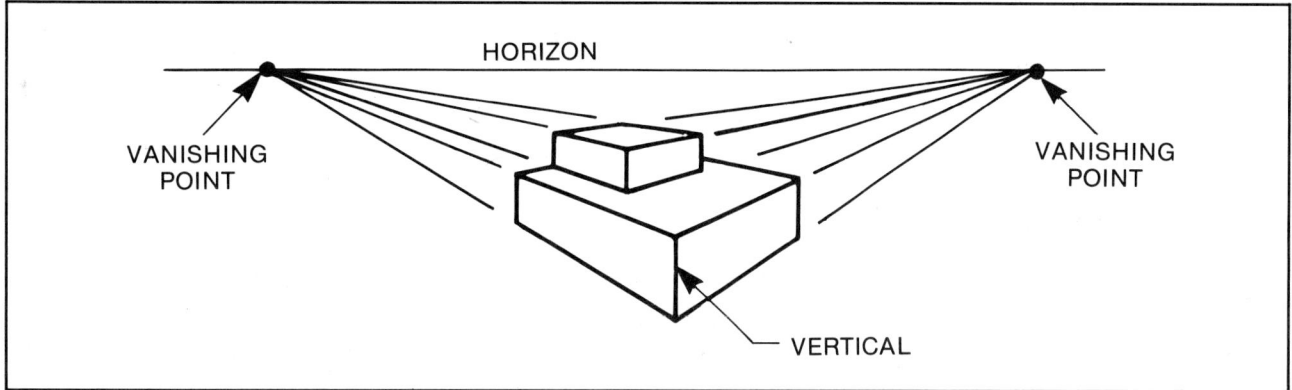

Figure 13-11. Two-point perspective drawing

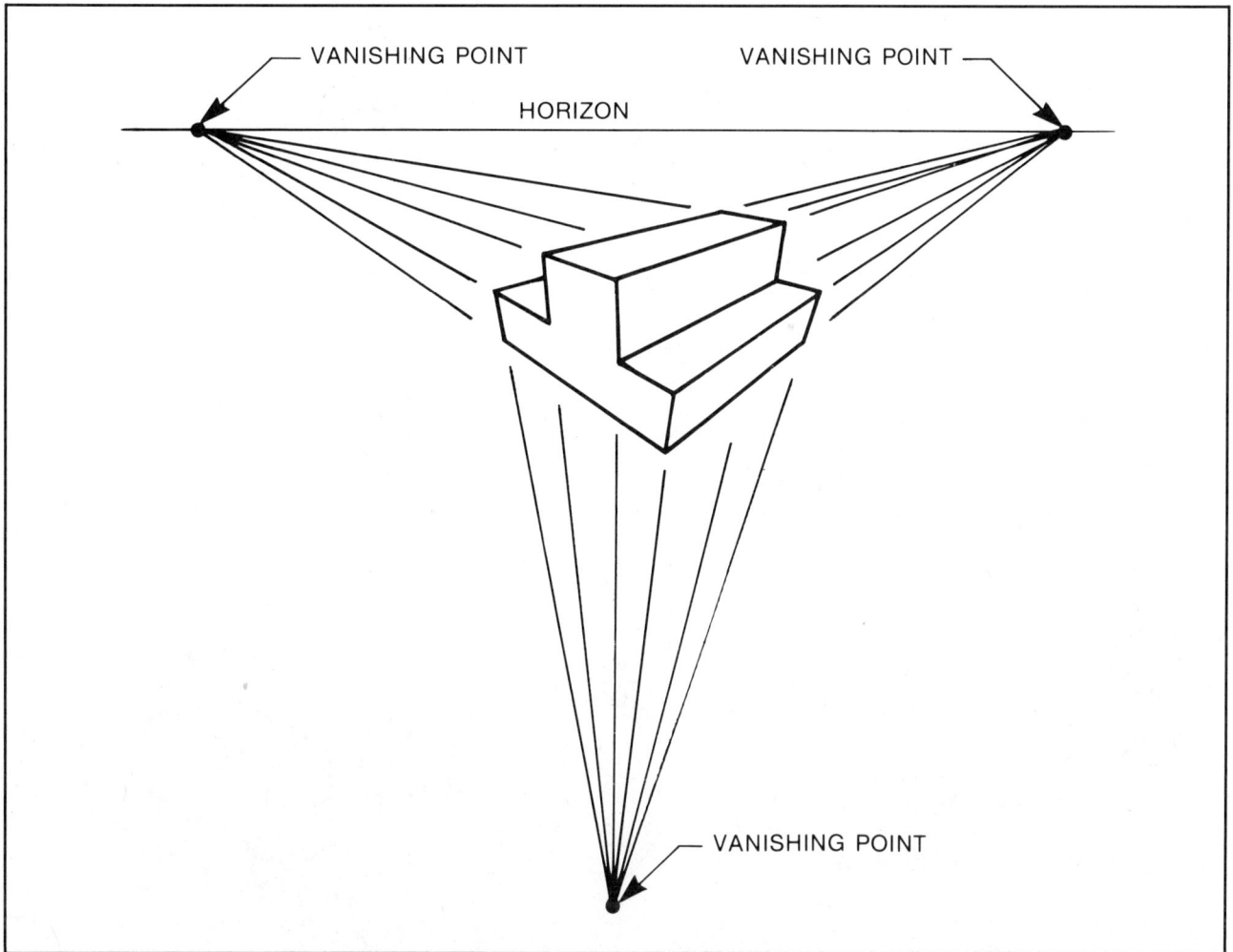

Figure 13-12. Three-point perspective drawing

MAKING PICTORIAL SKETCHES

Like other types of mechanical sketching, pictorial sketching is simply a process of sketching a frame and filling in the details. Before studying the remainder of this chapter, review the basic rules of sketching in Chapter 4.

Making Isometric Sketches

The isometric sketch is the most popular form of axonometric sketching. Several methods may be used to draw isometric sketches. Figure 13-13 shows one method that can be used to draw these sketches. The steps shown in Figure 13-13 are explained below.

STEP A. Determine the position of the part and sketch the isometric axes. Use either the lower or upper edge of the part as a reference edge. The angular lines are 30° to the horizontal.

STEP B. Estimate the true size height, width, and depth of the part and mark the sizes on the axes. Then complete the box by sketching lines parallel to the three major axes.

STEP C. Sketch in the isometric lines forming the basic shape of the object.

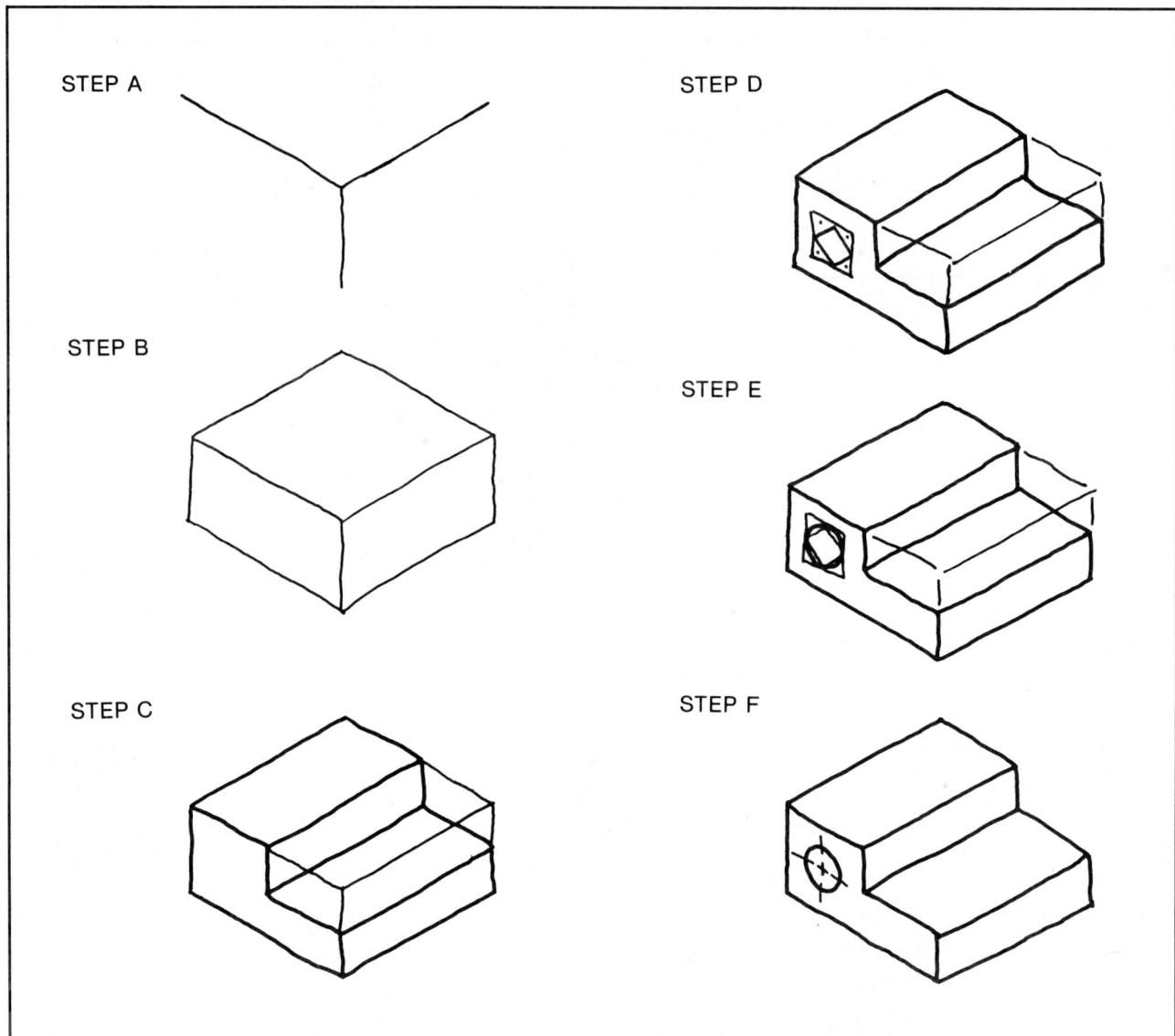

Figure 13-13. Making an isometric sketch

STEP D. To sketch circles, arcs, or rounded details, box in the area of the part where the detail is located, and mark the center point of each line. Connect these points with diagonal lines and put a dot in the approximate center of each triangle to mark the points through which the circle or arc will pass.

STEP E. Connect the dots and the points along the outside of the circle or arc.

STEP F. Erase all construction lines and darken all visible lines in the sketch.

An isometric circle will actually have an elliptical, or oval, form. To properly orient these ellipses to the part, consult the sketch in Figure 13-14. This sketch shows how the ellipses are related to the part when positioned on any of the isometric surfaces. If an isometric circle template is available it will help speed up this sketching.

The process of sketching a part with either of the other types of axonometric forms is essentially the same as for isometric sketching. The only differences are the angles of the axonometric axes and the form of the ellipses used to show circles and arcs.

To dimension axonometric sketches, simply draw the extension lines parallel to the axonometric axes and sketch the dimension lines parallel to the surfaces they control, Figure 13-15. The dimensional values can then be entered either parallel to the surface or written normally.

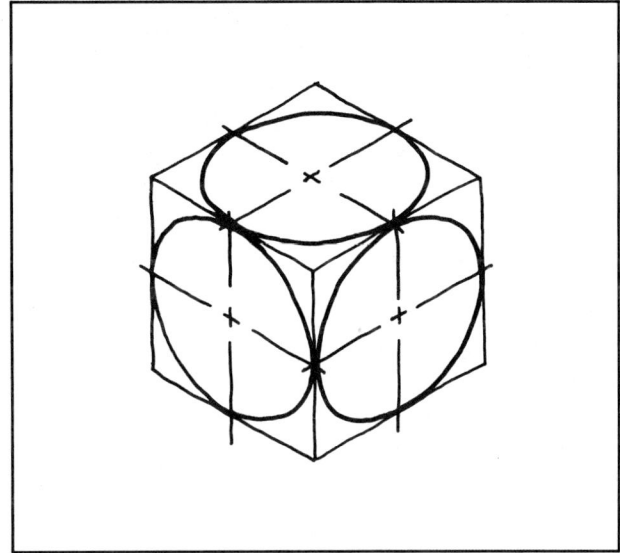

Figure 13-14. Positioning the ellipses on an isometric sketch

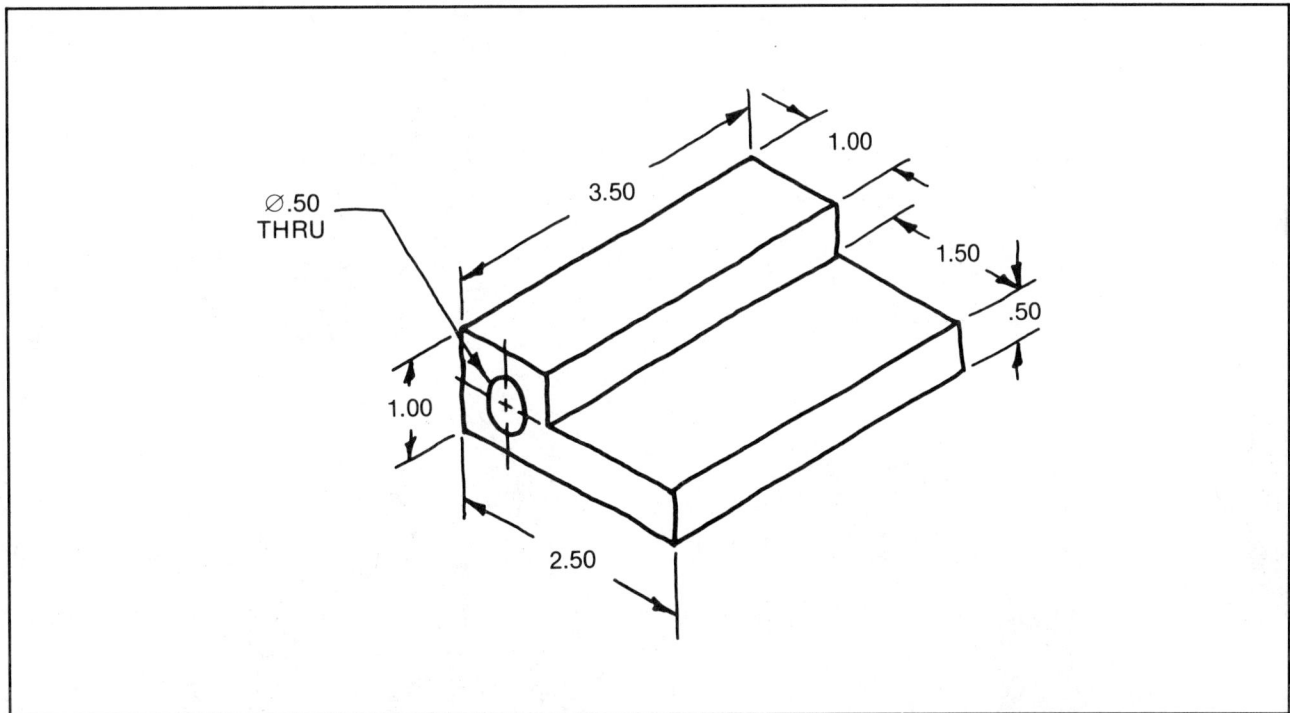

Figure 13-15. Dimensioning an isometric sketch

Making Oblique Sketches

Oblique sketches are drawn in the same manner as isometric sketches. The only difference is the alignment of the part with the viewing plane. Oblique sketches are always drawn with one side of the part parallel to the viewing plane. Figure 13-16 shows one method that can be used to draw cavalier and cabinet oblique sketches. The steps shown in Figure 13-16 are explained below.

STEP A. Determine how the part will be situated on the drawing sheet. Sketch one vertical line and one horizontal line intersecting at the corner of the part face that is to be aligned with the viewing plane.

STEP B. Estimate the height and width of the part and mark the sizes on the two lines. Next, sketch two lines parallel to the first two lines, and block in the box representing the front face of the part.

STEP C. Select the angle of the receding axes and sketch the lines showing the depth of the part. Sketch receding lines between 30° and 60° for best effect. These parallel receding lines are sketched full size for the cavalier and half size for the cabinet oblique drawing. Connect the receding lines with lines parallel to the box representing the front fact.

STEP D. Sketch the lines forming the outline of the part. Keep the lines parallel to the lines already drawn for the frame.

STEP E. To sketch arcs, circles, or curved details on the front face, siimply follow the same process as used for standard multiview

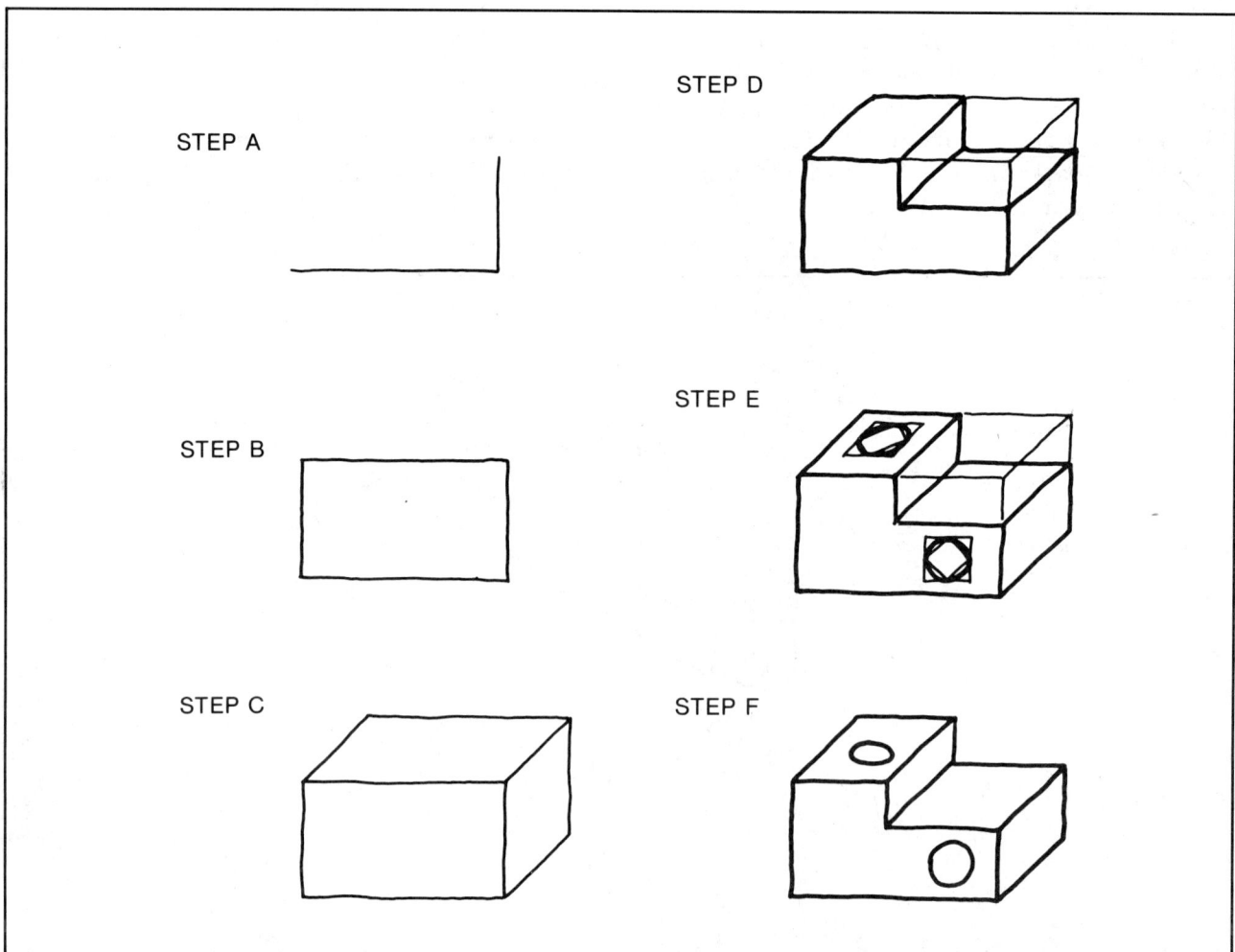

Figure 13-16. Making an oblique sketch

sketching. Follow the process used for isometric sketches if the detail is on one of the other surfaces. The angle of the receding lines determines the minor axis of the ellipse.

STEP F. Erase all construction lines and darken all visible lines on the part.

When making oblique sketches, any circles, arcs, or other curved surfaces on the part will appear either as ellipses on the receding surfaces or as normal curved lines on the front surface. The basic form of the ellipses will be determined by the shape of the part and the angle of the receding axes. The orientation of these curved surfaces is essentially the same as those on isometric sketches, Figure 13-17.

To dimension an oblique sketch, simply draw the extension lines from the part the same way as in isometric sketches. The dimension lines should be drawn parallel to the surface they control. The dimensional values can be shown parallel to the surface they control or parallel to the viewing plane.

Making Perspective Sketches

Perspective sketches, like the other types of sketches, are drawn by a simple process of blocking in the basic shape of the part and filling in the details. The only major difference is in the way the receding lines are formed. Figure 13-18 shows one method that can be

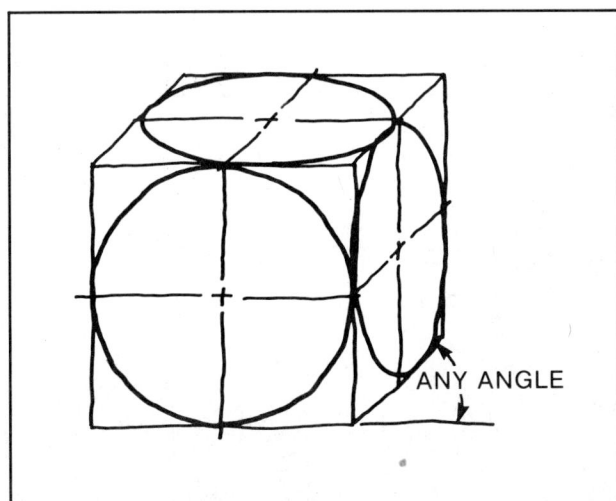

Figure 13-17. Positioning the ellipses and round holes on an oblique sketch

used to draw one-point perspective sketches. The steps shown in Figure 13-18 are explained below.

STEP A. Sketch the front view of the part just as it would be sketched for an oblique sketch. Next, select a vanishing point either behind, above, or below the part.

STEP B. Connect the edges of the front view box to the vanishing point with straight lines. Estimate the length of the part and draw a horizontal line to mark the back edge of the part. The placement of this line depends on how you want the part to appear.

STEP C. With the basic box formed, sketch in the details of the part, keeping these lines parallel to the lines forming the box.

STEP D. To sketch arcs, circles, and curved surfaces, follow the same process as that used for an oblique sketch.

STEP E. Erase all construction lines and darken all visible lines on the part.

Sketching a two-point or three-point perspective is essentially the same as sketching a one-point perspective. The only difference is in the way the part is aligned on the drawing sheet.

For two-point perspectives, Figure 13-19, first select the edge that will be in the front of the sketch, and mark the line showing this edge to its full size. Then select two vanishing points on a horizon and connect this edge to these points. Then estimate the size of the object and sketch in the details.

To sketch a three-point perspective, Figure 13-20, first select the point that will be in the front of the sketch and mark it on the drawing sheet. Next, select three vanishing points and connect the point to these three points. Then estimate the size of the object and sketch in the details.

The process of dimensioning perspective sketches is the same as that used for other pictorial sketches. Draw the extension lines from the major axes of the part. Then draw the dimension lines parallel to the surfaces they control, and insert the dimensional values parallel to the part or to the viewing plane.

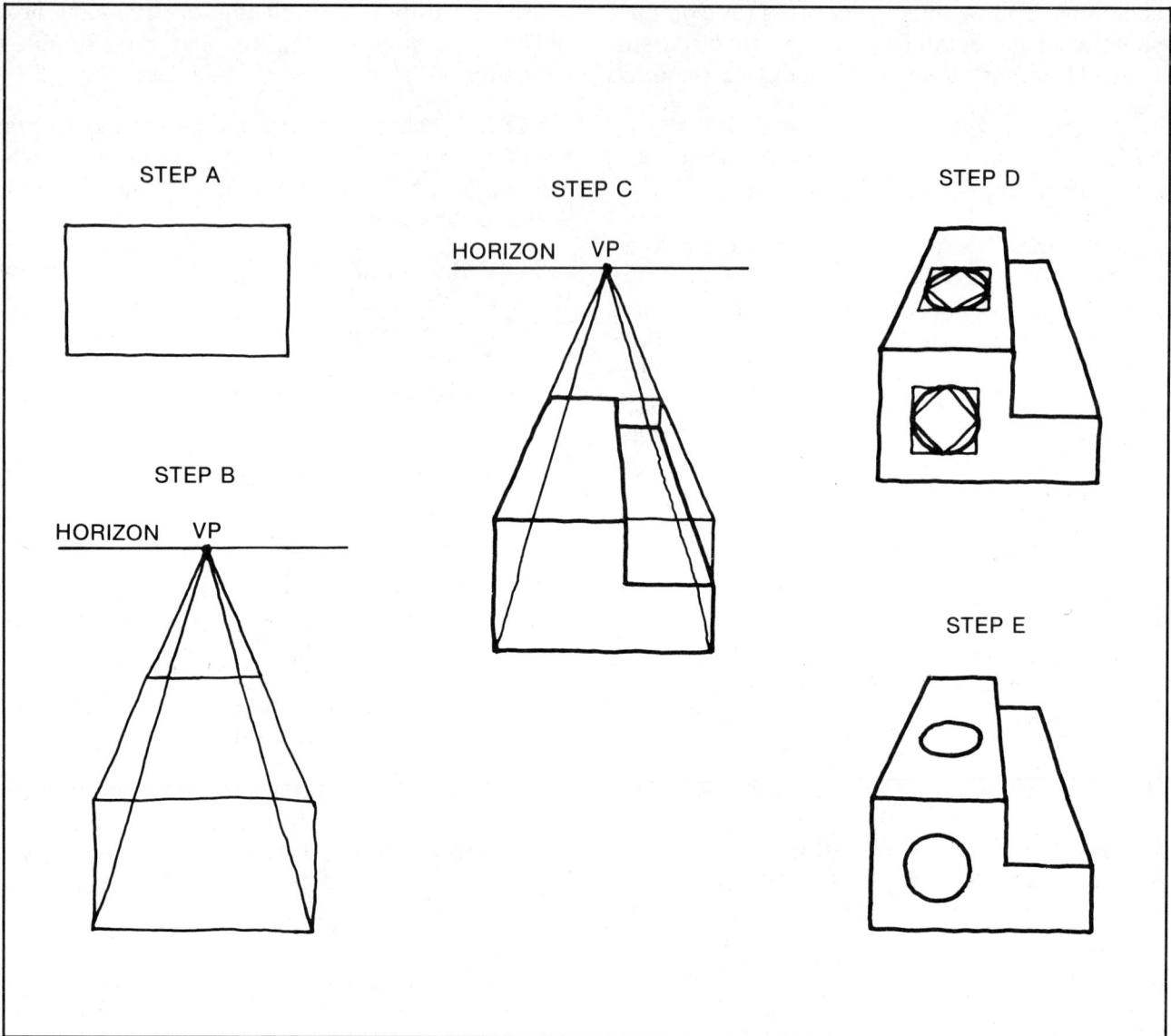

Figure 13-18. Making a one-point perspective sketch

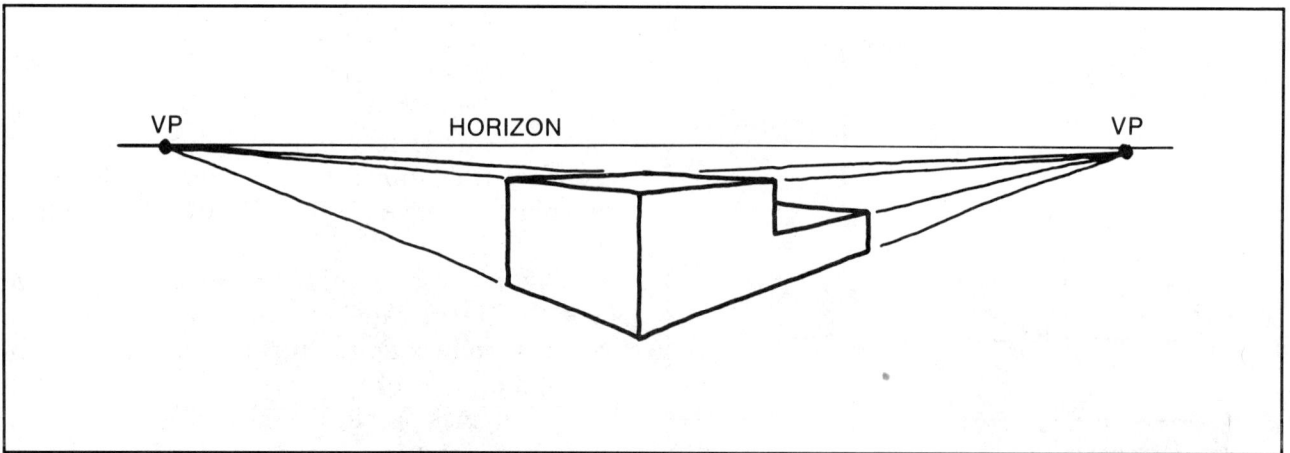

Figure 13-19. Sketching two-point perspectives

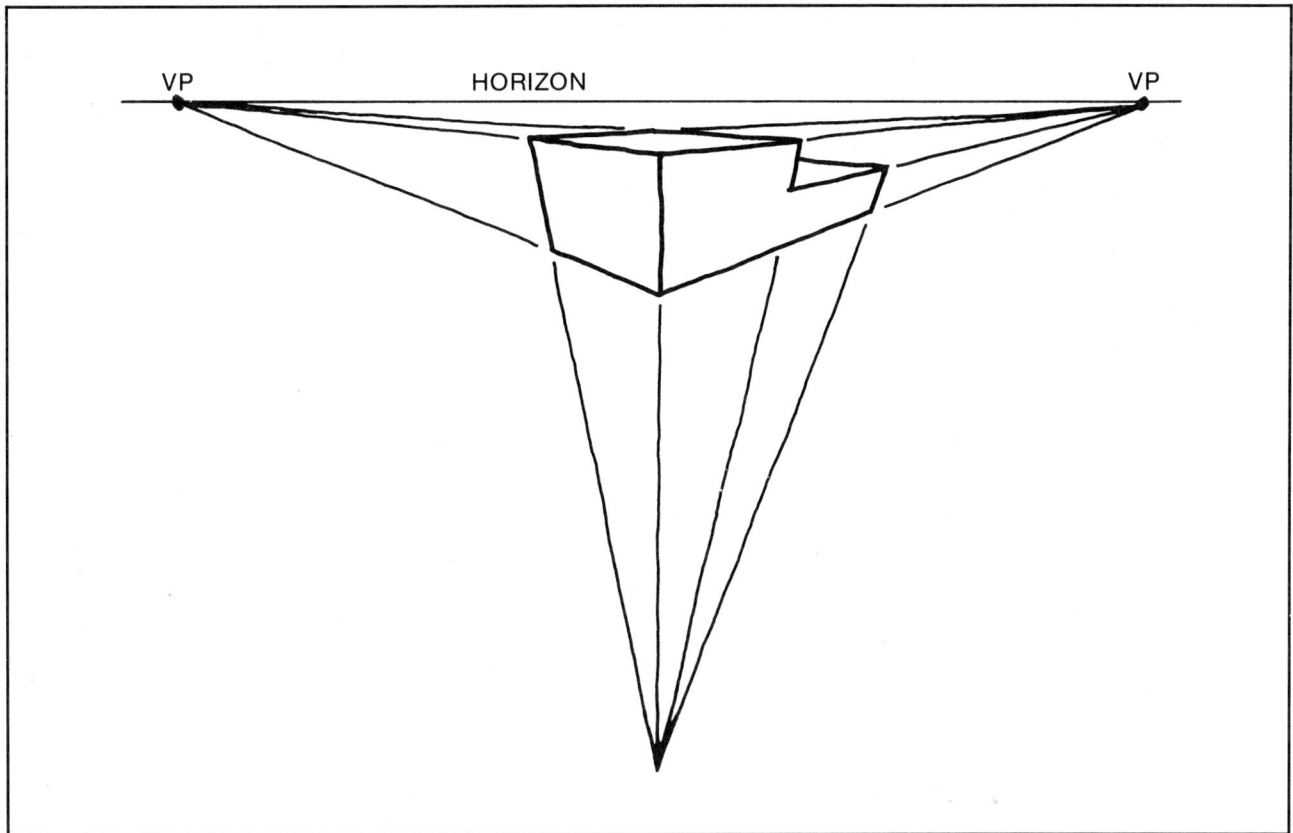

Figure 13-20. Sketching three-point perspectives

KEY TERMS

The following key terms were introduced in this chapter. Be sure you know the meaning of each term before proceeding to the review material.

Axonometric Axes
Axonometric Drawing
Cabinet Oblique Drawing

Cavalier Oblique Drawing
Dimetric Drawing
Isometric Drawing
Oblique Drawing
One-point Perspectives
Receding Lines
Three-point Perspectives
Trimetric Drawing
Two-point Perspectives
Vanishing Points

Test your knowledge with this reinforcement study material. Write your answers to the questions in the spaces provided.

1. Name the three basic forms of pictorial drawings. _____

_____

_____

2. Which type of pictorial drawing shows a part as positioned around a central axes?

_____

3. Which type of drawing has three different angles forming the axes? _____

_____

4. Which type of drawing has three identical angles forming the axes? _____

_____

5. How are holes and arcs normally shown in isometric drawings? _____

_____

6. What term describes the changing of line lengths in isometric drawings? _____

_____

7. How are angular details drawn in an isometric drawing? _____

_____

8. Which type of oblique drawing shows the part depth in half scale? _____

_____

9. Which type of oblique drawing shows the part depth in full scale? _____

_____

10. What major feature distinguishes oblique drawings from axonometric drawings?

_____

11. What are the recommended angles of the receding axes of oblique drawings? _____

_____

12. Name three different types of perspective drawings. _____

_____

13. How are holes shown in perspective sketches? _____

_____

14. For each of the following drawing characteristics, tell the form of drawing it describes. The drawing choices are *isometric, dimetric, trimetric, cavalier oblique, cabinet oblique,* and *perspective.*

    a. Depth drawn to half scale _____

    b. Three vanishing points _____

    c. Equal axonometric axes _____

    d. Two vanishing points _____

    e. No equal axonometric axes _____

    f. Two equal axonometric axes _____

    g. Full-scale depth _____

    h. Single vanishing point _____

Match the pictorial drawings shown in Figure E13-1 with their proper names.

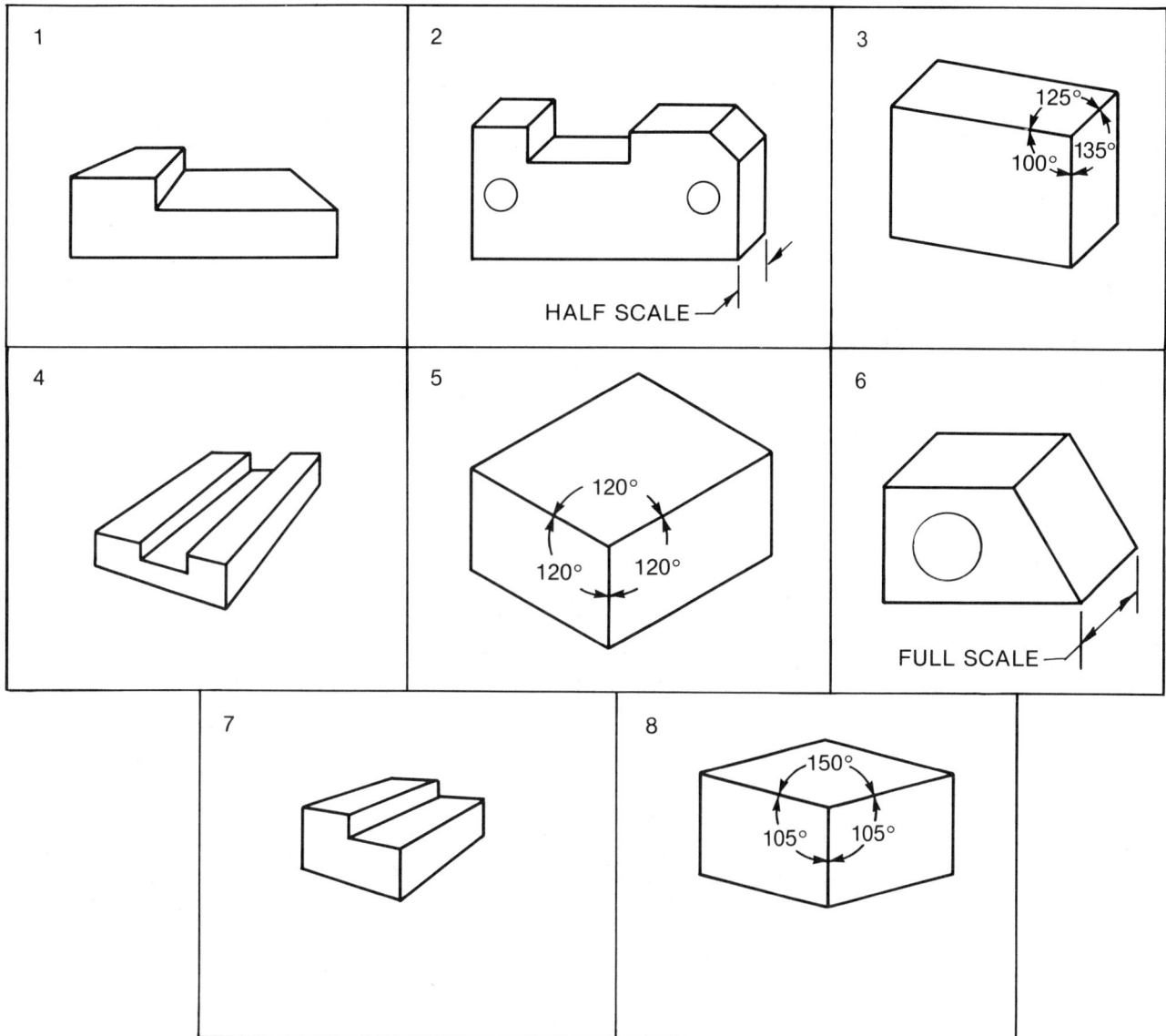

Figure E13-1.

a. Isometric _____

b. Dimetric _____

c. Trimetric _____

d. Cavalier Oblique _____

e. Cabinet Oblique _____

f. One-point Perspective _____

g. Two-point Perspective _____

h. Three-point Perspective _____

Make an isometric sketch of the part in Figure E13-2. Do not dimension the sketch.

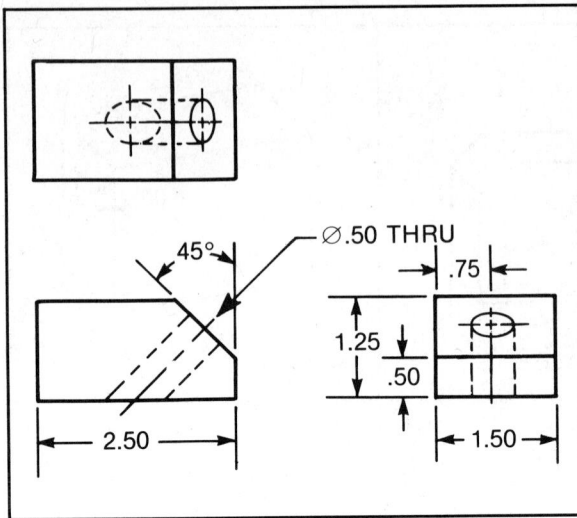

Figure E13-2.

Make a cabinet oblique sketch of the part in Figure E13-3. Do not dimension the sketch.

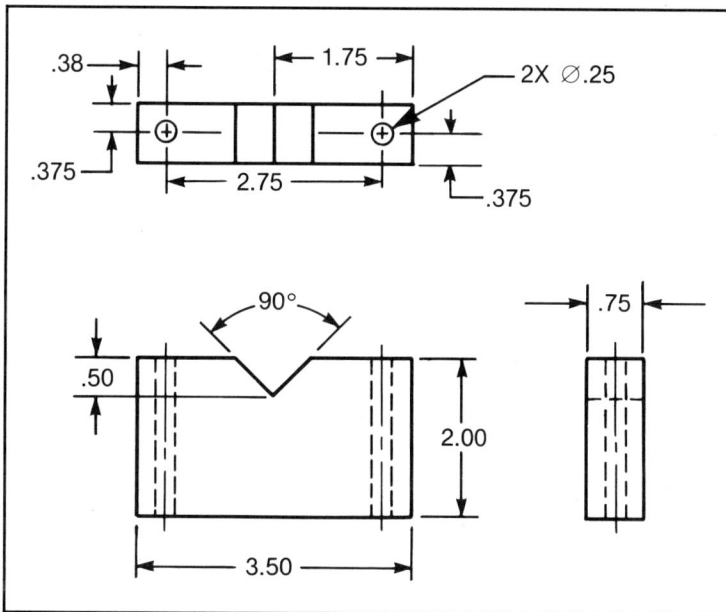

Figure E13-3.

Make a one-point perspective sketch of the part in Figure E13-4. Do not dimension the sketch.

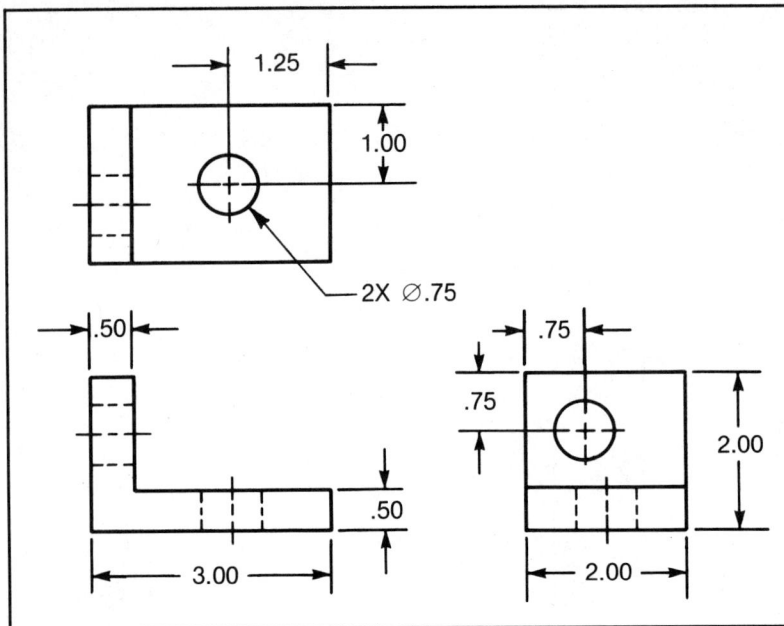

Figure E13-4.

14

WELDING AND PRECISION SHEET METAL PRINTS

OBJECTIVES

After studying this chapter, you will be able to:

■ Identify welding and sheet metal prints.

■ Interpret weld symbols.

■ Determine the required sizes of different welds.

■ Interpret a complete welding symbol.

■ Interpret precision sheet metal prints.

■ Calculate bend allowances for a variety of different bends.

■ Calculate the flat blank lengths of various parts.

Welding prints and *precision sheet metal prints* are two other forms of prints frequently used in manufacturing. These prints are basically the same as the other prints covered in this text. However, there are enough differences to make

additional study necessary. To properly interpret welding and precision sheet metal prints, it is important to understand the methods and symbols used in these prints. The print reader must also understand the formulas and processes used to calculate bend allowances.

READING WELDING SYMBOLS

Welding symbols are a form of shorthand used by welders to describe welded assemblies. These symbols provide all of the information necessary to make a weld in a compact, easy-to-read form. Properly interpreting the meaning of a welding symbol requires learning what each element of the symbol means and how it is applied to the part shown in the print.

Reference Line

The main element of the welding symbol is the *reference line*. As shown in Figure 14-1A, the welding symbol consists of a horizontal line with an arrow at one end and an optional tail

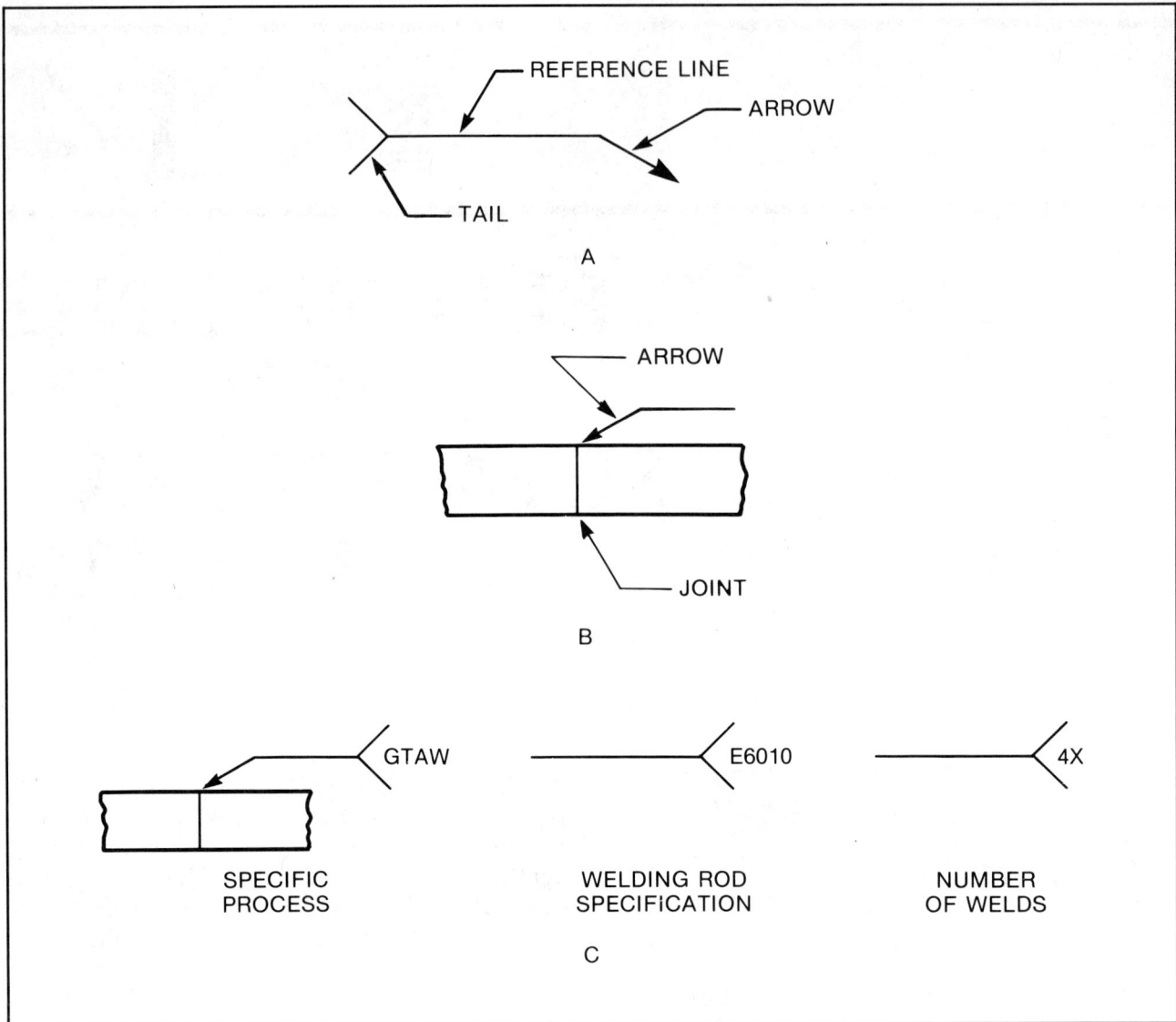

Figure 14-1. The reference line is the main element in a welding symbol (A). The arrow identifies where the weld is to be deposited (B). The tail is used to note special instructions (C).

at the other end. The horizontal line is used as a reference line. Most of the detailed information about the weld is located on the reference line. The arrow is used to identify the exact spot, or joint, where the weld is to be deposited, Figure 14-1B. The tail, when used, contains special instructions concerning the welding process, welding rod information, or any other special notes, Figure 14-1C. A list of the standard process abbreviations is included in Appendix B of this book.

When no special information is required for the weld, the tail is omitted. However, every welding symbol has the horizontal line and arrow. The arrow may extend from either end of the reference line, but the information contained above and below the reference line is always presented in the same order, regardless of the arrow's position.

Any information contained on the bottom side of the reference line applies to the arrow side of the joint. Information written on the top side of the reference line applies to the other side of the joint. The specific meaning of the arrow side and the other side is determined by the placement of the welding symbol and the location of the arrow. See Figure 14-2.

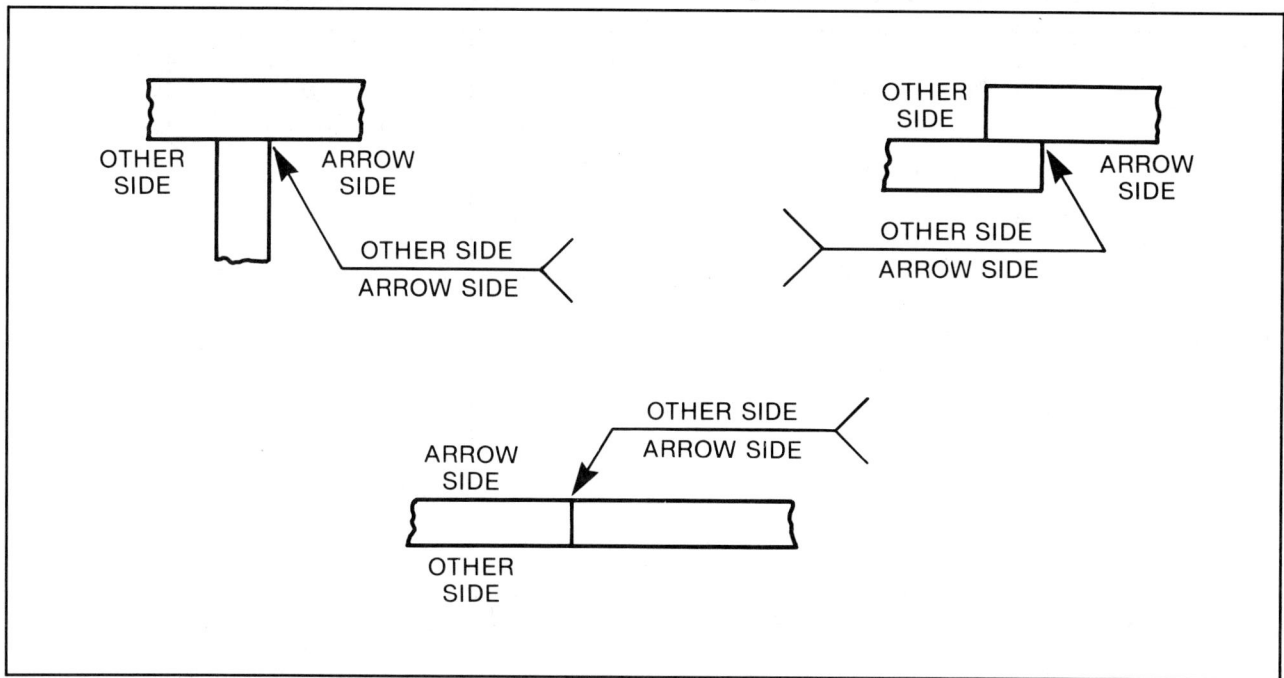

Figure 14-2. Arrow side and other side significance in a welding symbol

Reading Weld Symbols

In addition to the reference line, each welding symbol normally contains a weld symbol, Figure 14-3. *Weld symbols* are used to graphically describe the required weld.

The terms *welding symbol* and *weld symbol* should not be confused. There is a difference between these terms. A welding symbol shows the complete specifications of a welded joint. A weld symbol is only used to describe the basic shape, or form, of the specified weld.

The standard weld symbols used for welded joints are shown in Figure 14-4. These symbols are positioned on the reference line on the side of the line where the weld is to be applied. A weld intended to be on the arrow side is shown below the reference line; a weld required on the opposite side is shown above the line.

When there is no significance as to which side the weld is on, such as with spot welds, projection welds, or seam welds, the weld symbol is placed directly on the line. This allows the welder to apply the weld on the most convenient side. When weld symbols are shown on both sides of the reference line, the weld is intended to be applied to both sides of the

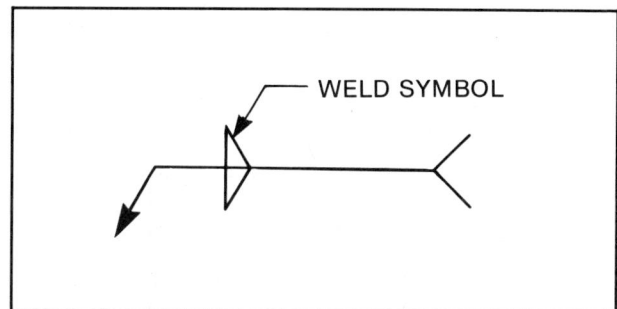

Figure 14-3. Positioning the weld symbol

joint. The weld symbol used to denote a square groove weld on both sides is the same as that used for no side significance. When this symbol is shown on a print, the meaning of the symbol is usually explained in the tail. If it is not, this normally means that the weld is to be applied to both sides.

Sometimes two different welds are specified for the same joint, Figure 14-5. In such cases, two different weld symbols are shown on the reference line. Similarly, when more than one weld is to be applied on the same side of a joint, the symbols show the required welds in their order of application. See Figure 14-6. Here the weld symbol closest to the reference line identifies the weld that is to be applied first.

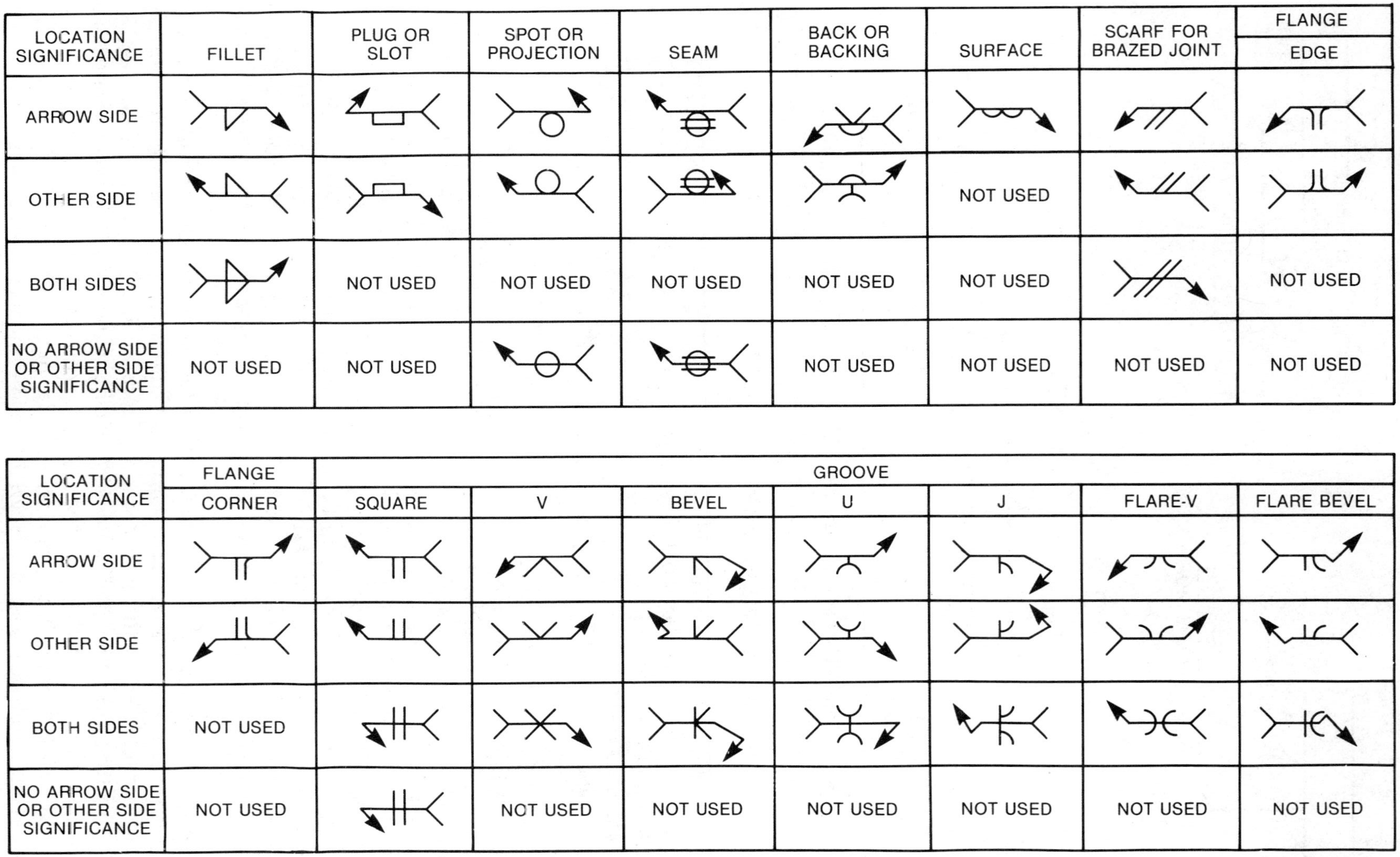

Figure 14-4. Standard weld symbols and their locational significance

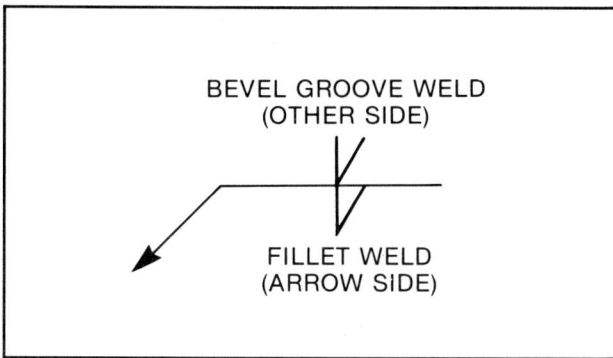

Figure 14-5. Specifying different welds for the same joint

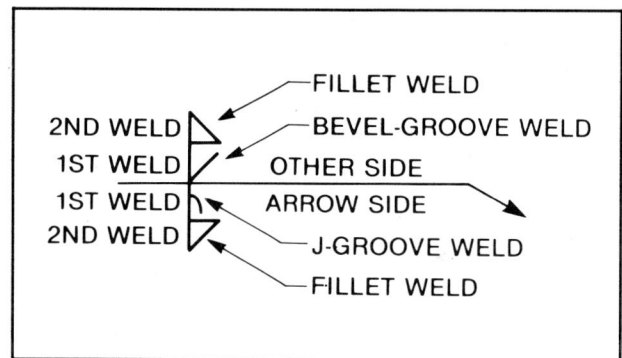

Figure 14-6. Specifying two welds in the same place

Supplementary Symbols

Supplementary symbols are used to further define the meaning of the welding symbol. As shown in Figure 14-7, these symbols include the weld-all-around symbol, field weld symbol, melt-thru symbol, backing and spacer symbols, and contour symbols.

WELD ALL AROUND. The weld-all-around symbol, Figure 14-8A, is a small circle drawn around the junction of the reference line and the arrow. This symbol means that the weld is to continue completely around the joint.

FIELD WELD. The field weld symbol, Figure 14-8B, is a small, flag-shaped identifier used to denote welds that are intended to be made away from the place of original fabrication. For example, if two different welded assemblies are made in a shop and are designed to be assembled on site, the field weld symbol would be used to note the joints that must be assembled in the field.

MELT-THRU. The melt-thru symbol is used for those welds that must be made with 100% penetration plus reinforcement. This symbol is shown on the side of the reference line opposite the weld symbol, Figure 14-8C. The melt-thru symbol should not be confused with the back-weld symbol. While they are similar in form, the melt-thru symbol is completely filled in, while the back-weld symbol is not.

BACKING AND SPACER MATERIALS. These symbols are used for welds that must have either a backup piece or a spacer inserted in the welded joint, Figure 14-8D. These symbols specify the material type. A note is also included with the backing symbol to indicate

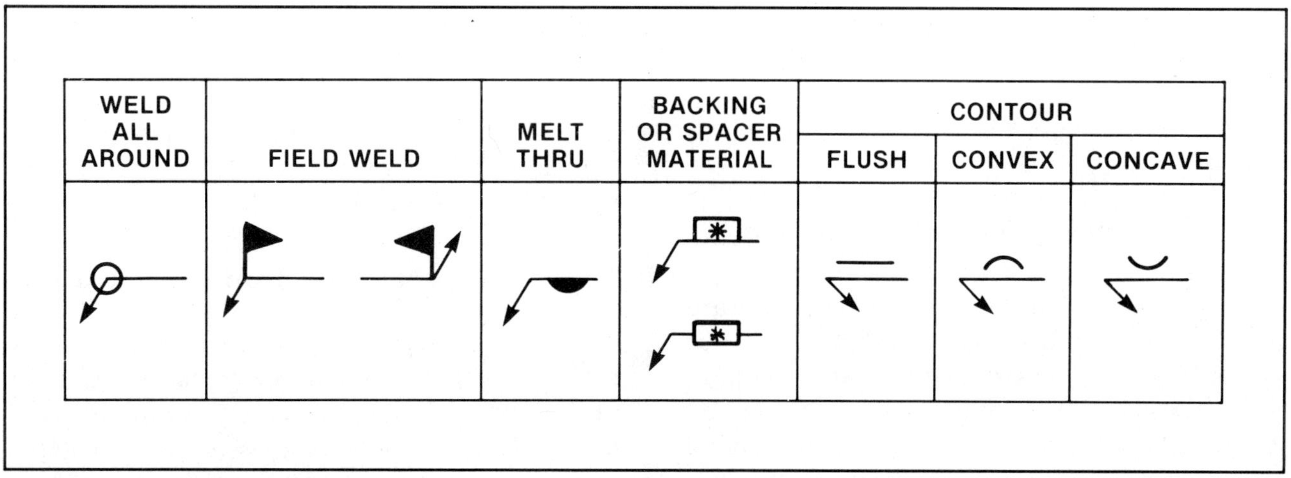

Figure 14-7. Supplementary symbols

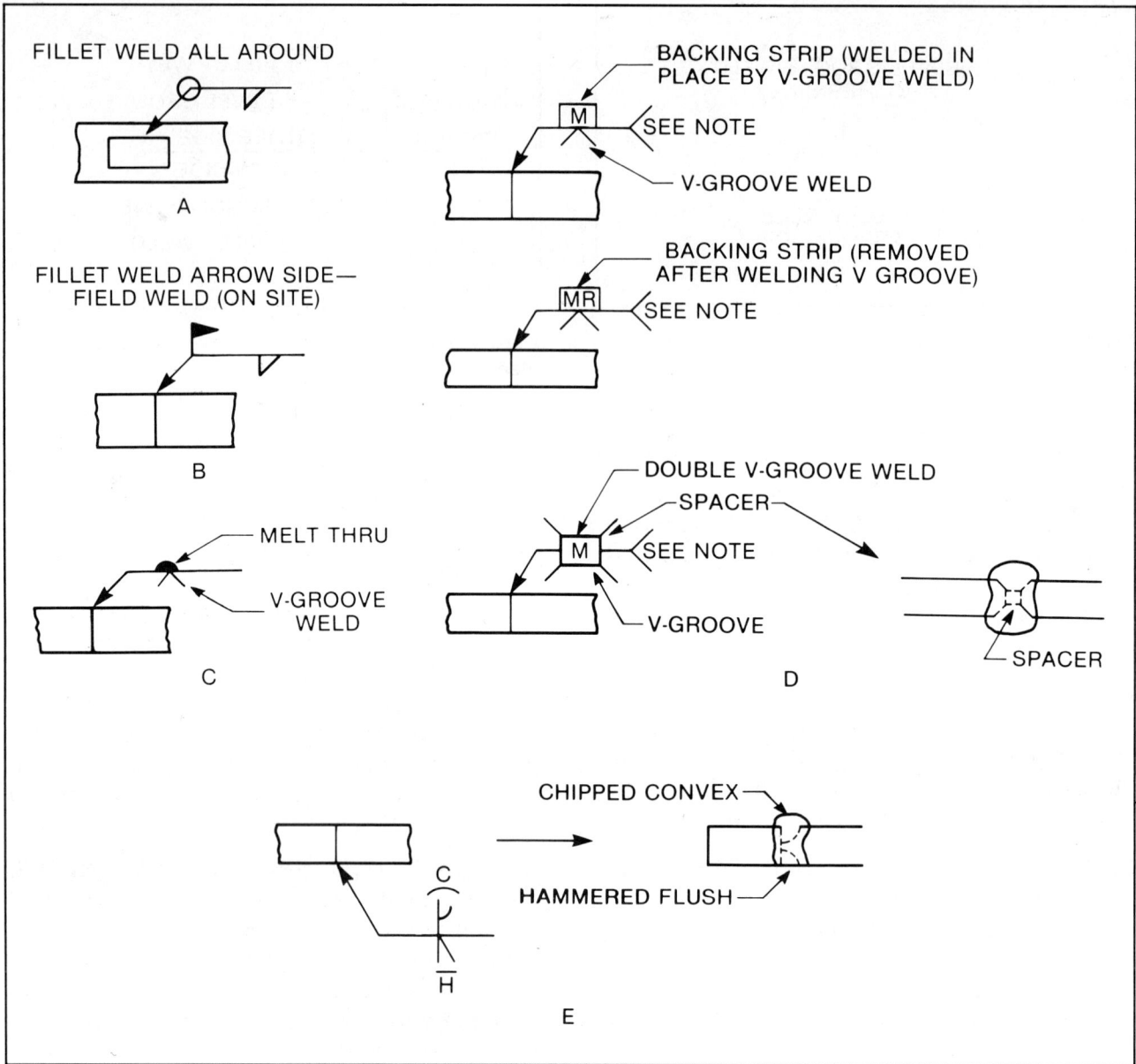

Figure 14-8. Application of (A) the weld-all-around symbol, (B) the field weld symbol, (C) the melt-through symbol, (D) the backing and spacer material symbols, and (E) the contour symbols

whether the backing is to be left in place or removed after welding.

CONTOUR SYMBOLS. Contour symbols are used to show the shape that the completed weld is to have. The three types of contours are flush, convex, and concave. When a contour symbol is used alone above the weld symbol, this means that the contour is to be finished by welding. If, however, another form of finishing is desired, a letter value is noted to identify the specific process to be used. For example, the

letter *M* means machined, *C* means chipped, *G* means ground, *R* means rolled, and *H* means hammered. See Figure 14-8E.

Specifying Joint Preparation

When groove-type welds are specified on a print, it is often necessary to specify which side of the joint is to be beveled for complete weld penetration. In these cases a broken, or bent, arrow is used to point to the side of the joint that is to be shaped or beveled, Figure 14-9.

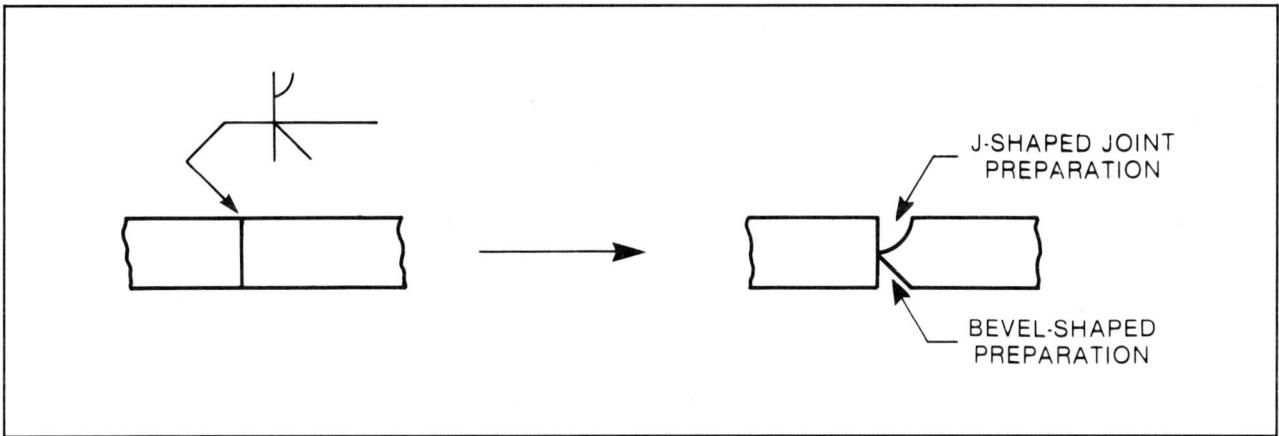

Figure 14-9. A bent arrow points to the location of the joint preparation.

SPECIFYING WELD SIZES

In addition to the graphic symbols already discussed, most welding symbols also contain numbers to denote the various weld sizes. Figure 14-10 shows the location of each size value in a typical welding symbol. Regardless of the position or location of the welding symbol, size values are always positioned in the same place with relation to the reference line, and are normally read from left to right. The only time these dimensions are not used in a welding symbol is when the values are specified in a general note on the print.

The six standard size values found in welding symbols are *size, effective throat, groove angle,*

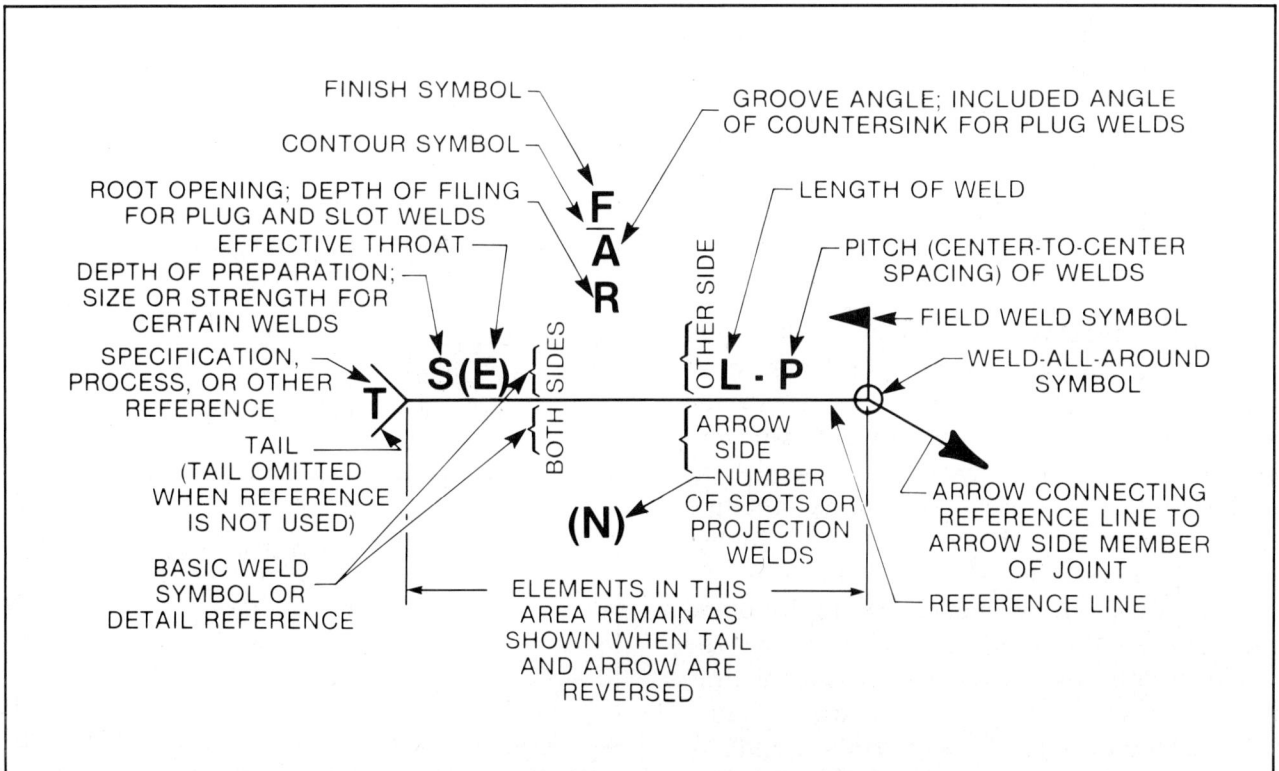

Figure 14-10. Location of size values for welding symbols

root opening, *length of weld*, and *pitch of the weld*. While not all welding symbols will contain all of these values at one time, it is necessary to be familiar with their location and meaning to properly interpret every welding symbol.

SIZE. The size designation is located on the extreme left of the reference line. It is intended to show the depth of preparation, size, or strength of the weld. The particular weld specified will normally determine the specific meaning of this value. Figure 14-11 shows several applications and meanings of this value.

EFFECTIVE THROAT. The effective throat of a weld is the minimum distance from the root to the face of the weld minus any reinforcement. As shown in Figure 14-12, this value is actually the depth of penetration of the desired weld. This value is shown within parentheses.

GROOVE ANGLE. The groove angle shows the angular size of the groove between two parts that are to be welded with a groove weld, Figure 14-13.

ROOT OPENING. The root opening specifies the amount of separation at the root of the joint between two parts to be welded, Figure 14-14.

LENGTH OF WELD. The length of weld value specifies the length of the welded area. When no value is shown in this location, it can be assumed that the weld is to be deposited across the complete length of the joint, Figure 14-15.

PITCH OF THE WELD. The pitch of a weld is the center-to-center distance between intermittent, or skip, welds, Figure 14-16. This value is read along with the length of weld value and specifies the distance between the welded areas of the joint. Therefore, if a weld is specified as *4-6*, it means that each weld is 4 inches (or millimeters) long and the centers of the welded areas are 6 inches (or millimeters) apart. When the welding symbol shows the weld symbols aligned above or below the line, a *chain intermittent weld* is specified, Figure 14-17A. However, if the weld symbols and size values are offset as in Figure 14-17B, a *staggered intermittent weld* is specified.

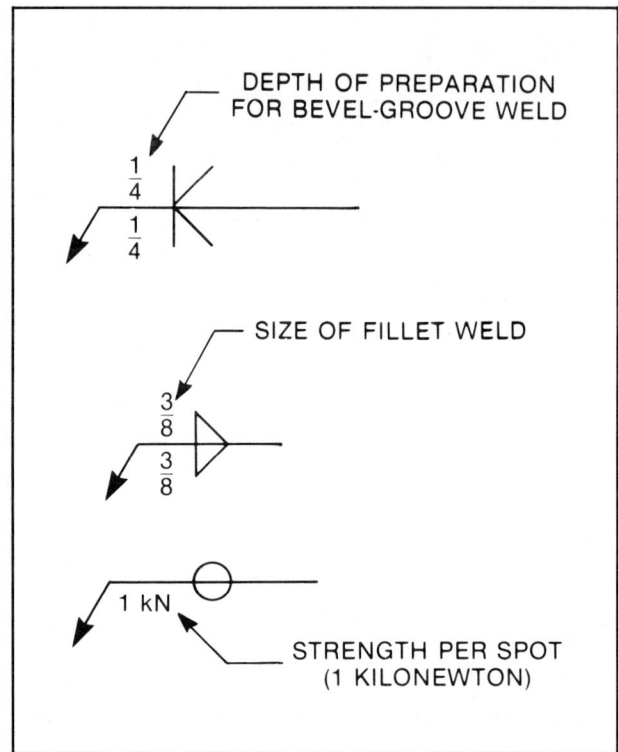

Figure 14-11. Depth of preparation, size, and strength are shown to the extreme left in a welding symbol.

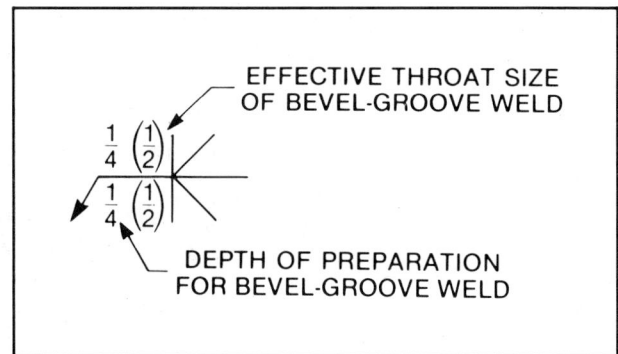

Figure 14-12. The effective throat is shown in parentheses to the right of the size value.

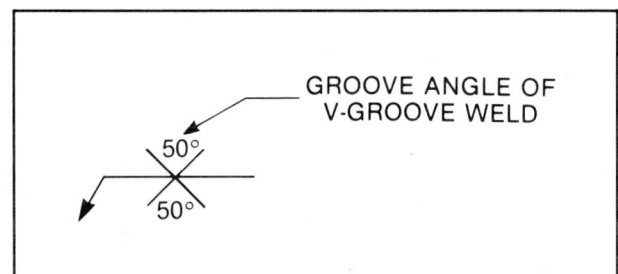

Figure 14-13. The groove angle is shown above the weld symbol.

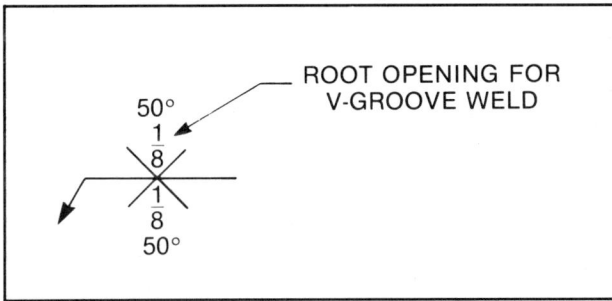

Figure 14-14. The root opening is shown below the groove angle value.

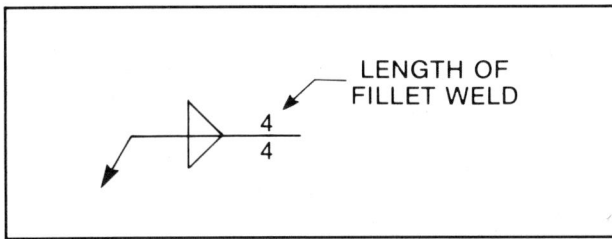

Figure 14-15. The length of weld value is shown to the right of the weld symbol.

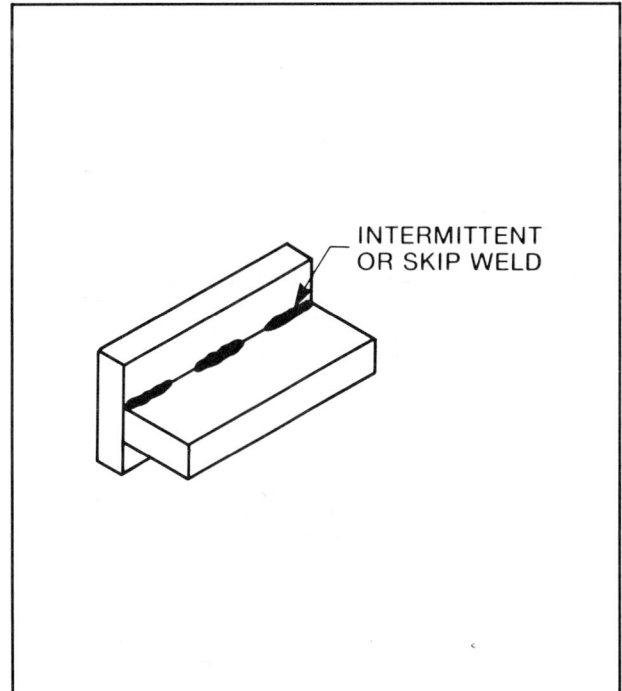

Figure 14-16. Intermittent, or skip, welds

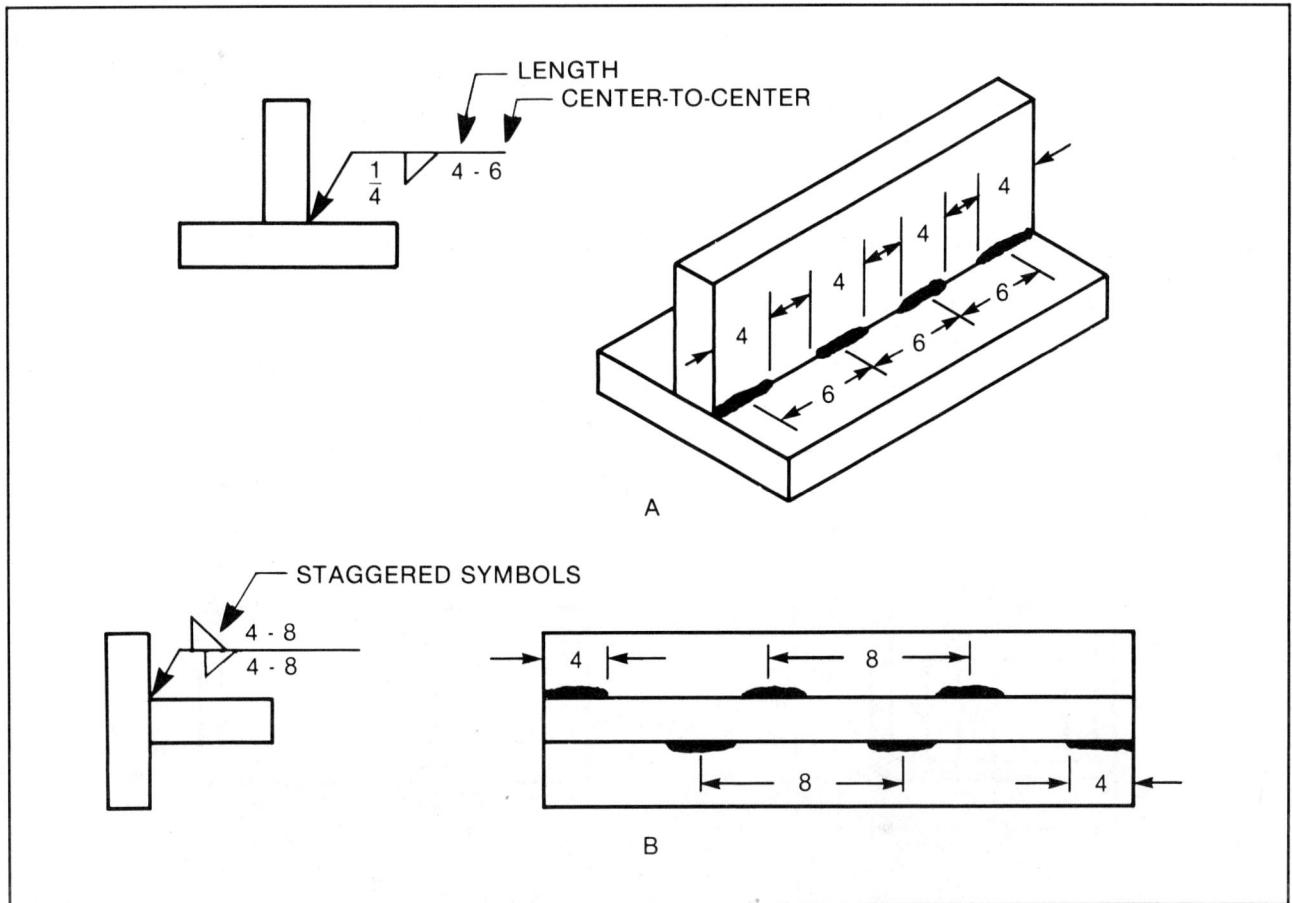

Figure 14-17. Specifying (A) a chain intermittent weld and (B) a staggered intermittent weld

READING PRECISION SHEET METAL PRINTS

Precision sheet metal parts can best be described as sheet metal parts or assemblies that are made to close tolerances. Precision sheet metal work should not be confused with the types of parts normally associated with sheet metal. While both areas involve sheet metal parts, regular sheet metal work is usually concerned with non-precision applications such as the manufacture of ducts or air vents. Precision sheet metal work, on the other hand, is primarily concerned with the fabrication of precision sheet metal parts for the automotive, computer, and electronics industries.

Most precision sheet metal prints are drawn similarly to prints used for machined parts. That is, the part is shown on the print in the same form as the final product is to appear. Sometimes, with bent or formed parts, a flat blank layout will also be included. A *flat blank layout* is a view of the part as it would appear completely flat with all of the bend allowances already calcuated. More often, however, it will be necessary to make a series of calculations to find the flat blank size of the part before it is bent.

Calculating Bend Allowances

Bend allowances must be calculated to determine the overall length of a part that is to be bent. As shown in Figure 14-18, most parts are

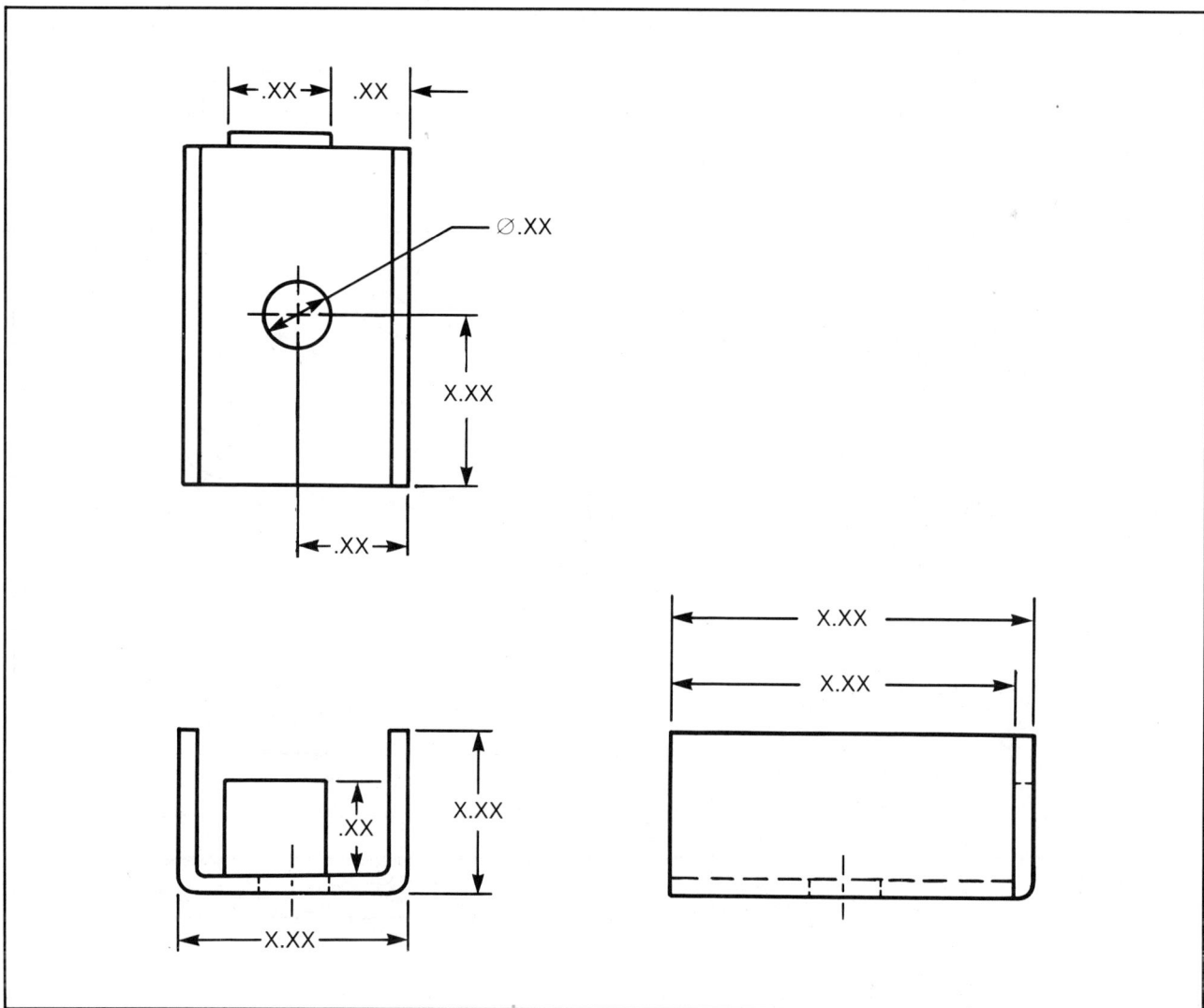

Figure 14-18. Typical dimensioning of a sheet metal part

dimensioned from their outside extremes. By dimensioning a part in this manner, the overall sizes are shown, but the exact length of the part before bending is not noted. If someone were to try to make the part based only on the dimensions shown, there would be no way to know how much material to allow for the bends. Therefore, to determine the amount of material for these bends, it is necessary to calculate the bend allowance.

When calculating the amount of material that must be added to a piece of metal for bending, several factors must be considered. First, the type of material being bent has a lot to do with the way a bend is formed. Soft materials such as soft aluminum and copper are more easily bent than harder materials such as stainless steel and steel sheet. More allowance must be provided for harder materials because they are not as readily formed as the softer types. The amount of the bend and the inside radius of the bend are also factors that must be considered. The formulas shown in this chapter are designed for general-purpose applications and may require some alteration or modification for special situations.

When bending metal, first consider the location of the neutral axis. The *neutral axis* is the axis of the part where the metal neither expands or compresses. As shown in Figure 14-19, when metal is bent, the outside area of the bend expands and the inside compresses. The neutral axis is the area where these two forces meet. It is also the area of the bend where the actual length of the bend can be measured. Therefore, calculating the bend allowance means calculating the length of the neutral axis for each bend.

This neutral axis is located at approximately one-third the metal thickness for bends having a radius less than twice the metal thickness, and one-half the thickness for bends greater than twice the metal thickness, Figure 14-20.

The general formulas used to calculate bend allowance are as follows:

- For bends less than 2t:

$$A = a \times .01745 \times (r + .33t)$$

- For bends more than 2t:

$$A = a \times .01745 \times (r + .50t)$$

Where: A = Bend allowance
      a = Desired angle of bend
      r = Inside radius of bend
      t = Metal thickness

To calculate the length of each leg of a bend when the parts are dimensioned with overall sizes, simply subtract the size of the inside radius and the metal thickness from the overall size of the leg. Expressed as a formula, this is as follows:

$$l = L - (r + t)$$

Where: l = Length of leg
      L = Overall length of leg
      r = Inside radius of bend
      t = Metal thickness

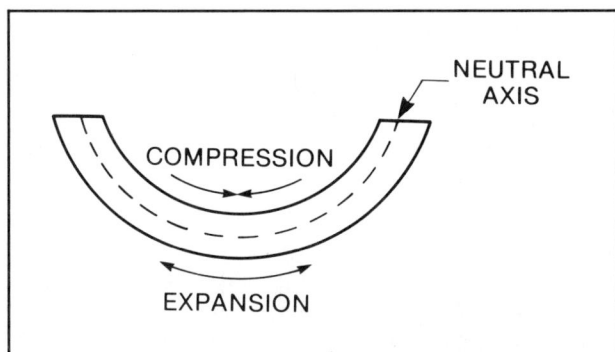

Figure 14-19. The neutral axis is the point where the compression and expansion forces meet.

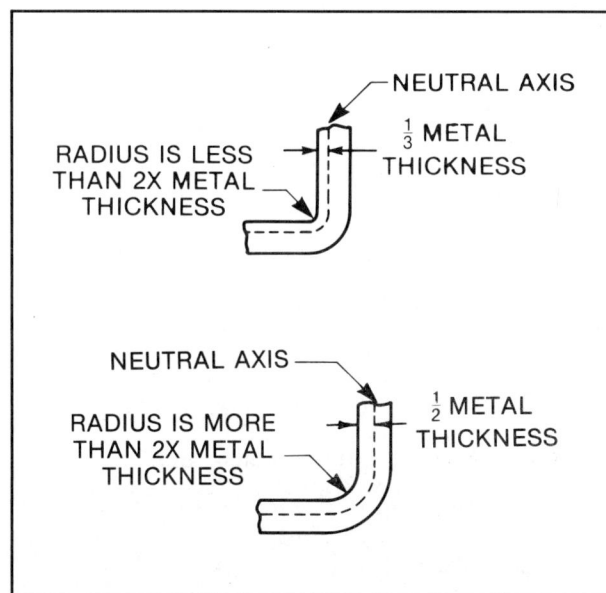

Figure 14-20. The size of the radius with relation to the metal thickness determines the location of the neutral axis.

To complete the calculations necessary to determine the size of the flat blank layout, first determine the lengths of all straight portions of the part. Then calculate all of the bend allowances and add these values to the lengths of the straight portions of the part. The easiest way to do this is to divide the part into several sections so none of the bends or straight sections are missed. Figure 14-21 shows how the flat blank sizes can be determined for a typical part.

These formulas will work equally well for thick materials and sheet metal. For problems such as those shown in Figure 14-22, were metal plate or rod must be bent, the formulas can also be used to determine the length of the metal before bending.

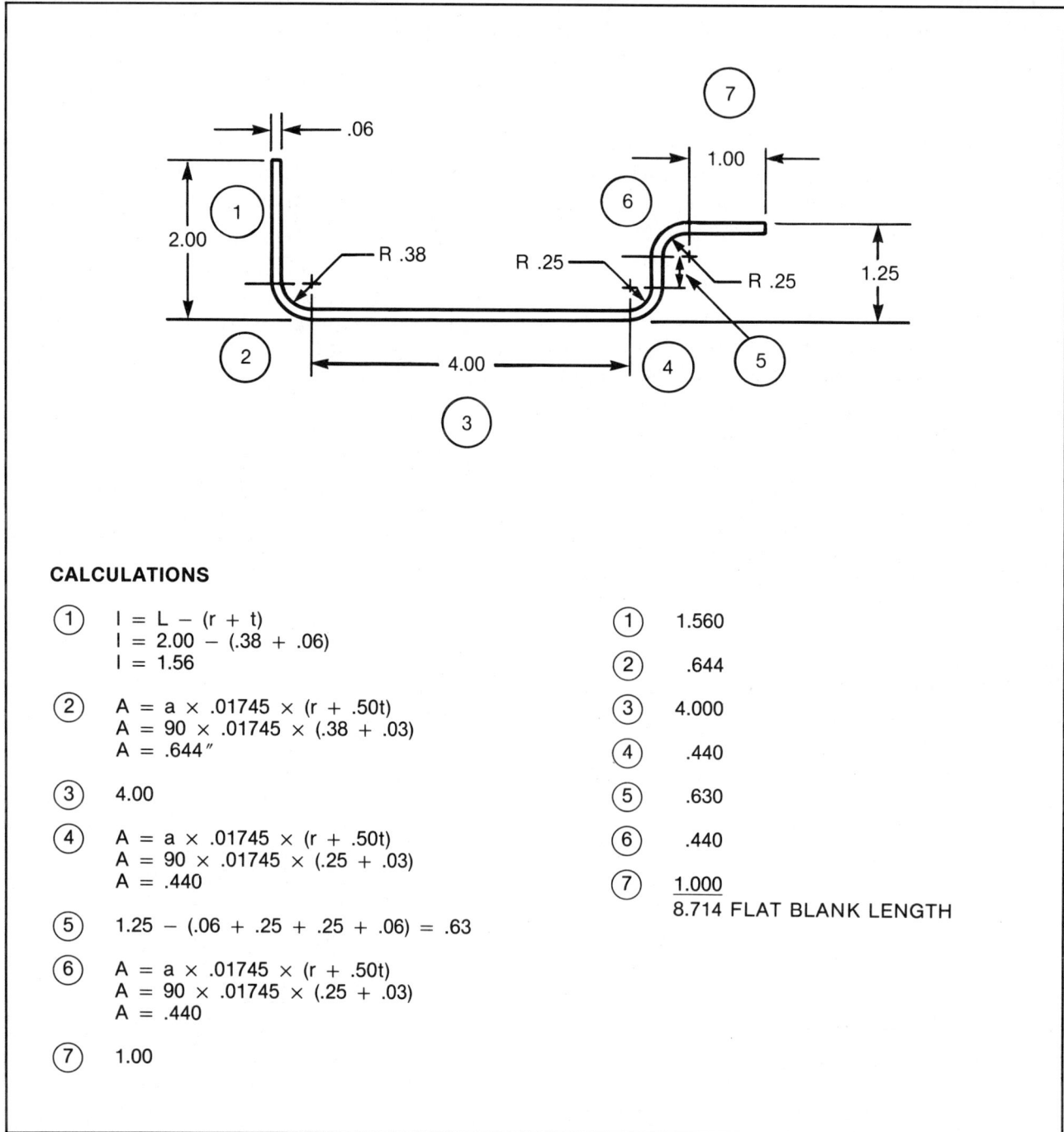

CALCULATIONS

①   I = L − (r + t)
    I = 2.00 − (.38 + .06)
    I = 1.56

②   A = a × .01745 × (r + .50t)
    A = 90 × .01745 × (.38 + .03)
    A = .644″

③   4.00

④   A = a × .01745 × (r + .50t)
    A = 90 × .01745 × (.25 + .03)
    A = .440

⑤   1.25 − (.06 + .25 + .25 + .06) = .63

⑥   A = a × .01745 × (r + .50t)
    A = 90 × .01745 × (.25 + .03)
    A = .440

⑦   1.00

①   1.560

②   .644

③   4.000

④   .440

⑤   .630

⑥   .440

⑦   <u>1.000</u>
    8.714 FLAT BLANK LENGTH

Figure 14-21. Calculating the flat blank length

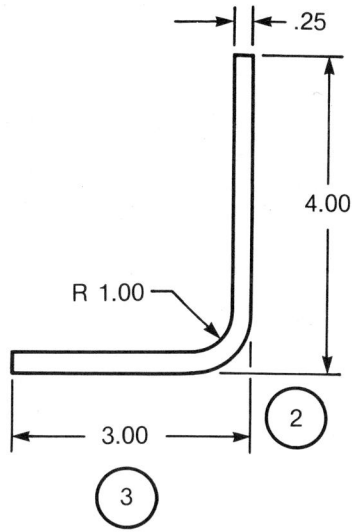

① $I = L - (r + t)$
     $I = 4.00 - (1.00 + .25)$
     $I = 2.75''$

② $A = a \times .01745 \times (r + .50t)$
     $A = 90 \times .01745 \times (1.00 + .125)$
     $A = 1.767''$

③ $I = L - (r + t)$
     $I = 3.00 - (1.00 + .25)$
     $I = 1.75''$

2.750
1.767
1.750
6.267″ FLAT BLANK LENGTH

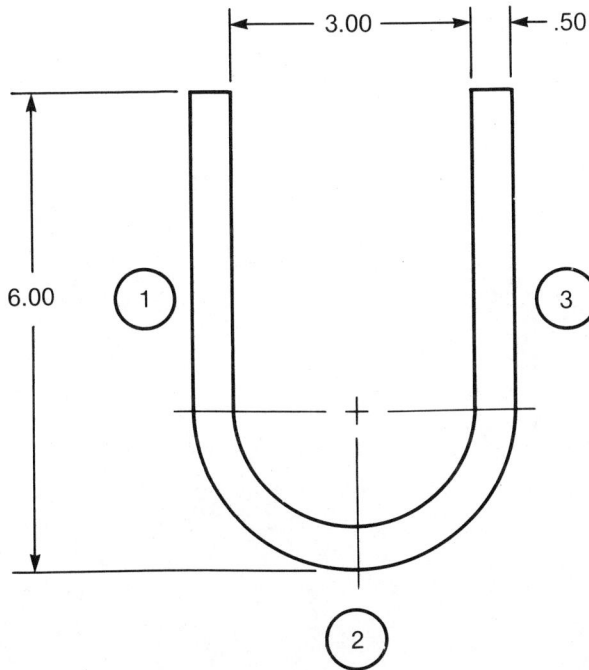

① $I = L - (r + t)$
     $I = 6.00 - (1.50 + .50)$
     $I = 4.00''$

② $A = a \times .01745 \times (r + .50t)$
     $A = 180 \times .01745 \times (1.50 + .25)$
     $A = 5.497''$

③ $I = L - (r + t)$
     $I = 6.00 - (1.50 + .50)$
     $I = 4.00''$

4.000
5.497
4.000
13.397″ FLAT BLANK LENGTH

Figure 14-22. Calculating the lengths of parts made from plate or rod

KEY TERMS

The following key terms were introduced in this chapter. Be sure you know the meaning of each term before proceeding to the review material.

Backing and Spacer Symbols
Bend Allowance
Contour Symbols
Effective Throat
Field Weld Symbol
Flat Blank Layout

Flat Blank Size
Groove Angle
Length of Weld
Melt-Thru Symbol
Neutral Axis
Pitch of the Weld
Reference Line
Root Opening
Supplementary Symbols
Weld-All-Around Symbol
Welding Symbol
Weld Size
Weld Symbol

Test your knowledge with this reinforcement study material. Write your answers to the questions in the spaces provided.

1. What is a welding symbol? _____

   _____

2. What is the main element in a welding symbol? _____

3. When is a tail used with a welding symbol? _____

   _____

4. What is a weld symbol? _____

   _____

5. What is the significance of the top and bottom of the reference line in a welding symbol?

   _____

   _____

   _____

6 What does a small circle at the intersection of the arrow and the horizontal line mean?

   _____

7. Name the three contour symbols. _____

   _____

8. What symbol is used to denote a field weld? _____

9. What does a broken, or bent, arrow mean? _____

   _____

10. What term describes the minimum distance between the root and face of a weld?

    _____

11. What three meanings may the value shown on the extreme left of a welding symbol have?

    a. _____

    b. _____

    c. _____

12. What two values are shown with a chain intermittent weld? _____

    _____

13. What purpose does a flat blank layout serve? _____

    _____

14. Along which axis is the bend allowance calculated? _____

15. Which radius is used for calculating the bend allowance? _____

    _____

Refer to Figure E14-1 on page 291 to answer the following questions.

1. What is the name of this part? _____

2. What type of weld is specified at *A*? _____

   _____

3. Which weld is applied first at *A*? _____

   _____

4. What is the effective throat of the weld shown at *A*? _____

5. What type of weld is specified at *B*? _____

   _____

6. What is the depth of preparation of the weld specified at *B*? _____

7. What type of weld is specified at *C*? _____

   _____

8. What is the effective throat of the weld at *C*? _____

9. How is the weld shown at *C* to be finished? _____

10. Why are the symbols offset in the symbol shown at *D*? _____

    _____

    _____

11. What is the length and pitch of the weld shown at *D*? _____

    _____

12. Explain the welding symbol shown at *E*. _____

    _____

    _____

13. What is indicated by line *F*? _____

    _____

    _____

14. Explain the weld shown at *G*? _____

    _____

    _____

15. How is the weld shown at *H* to be finished? _____

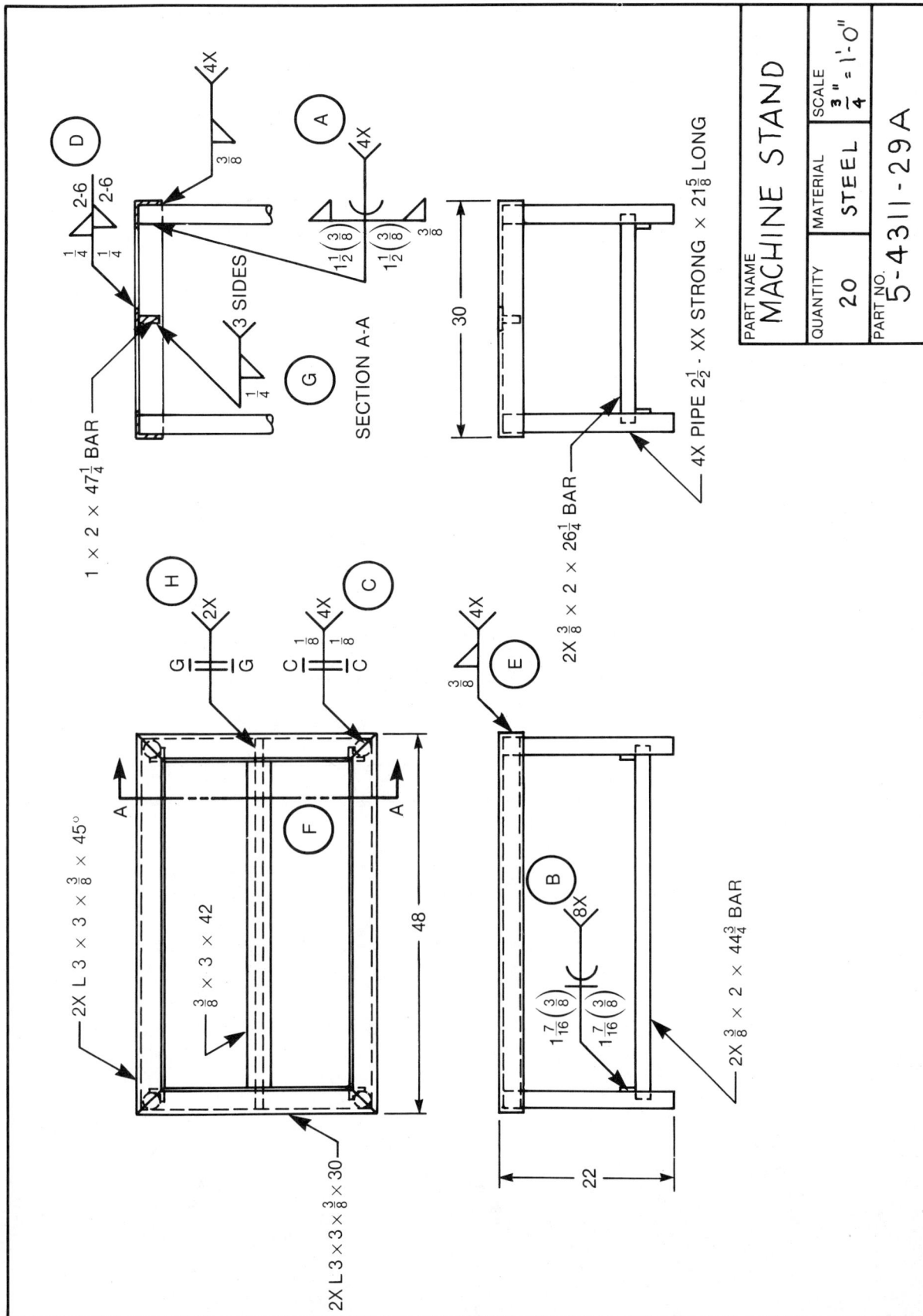

Figure E14-1.

Refer to Figure E14-2 on page 293 to answer the following questions.

1. What is the name of this part? _____

2. How many holes are in this part? What size(s) are they? _____

_____

3. What type of view is shown at *A*? _____

_____

4. What is the purpose of showing view *A*? _____

_____

_____

5. How many bends are there in this part? _____

6. What are the inside radii of these bends? _____

_____

7. What is the the thickness of the material? _____

8. What must be calculated before this part can be made? _____

_____

9. Considering the sizes of the radii, where is the neutral axis located with reference to the thickness of this part? _____

_____

10. How long must the part be before bending? _____

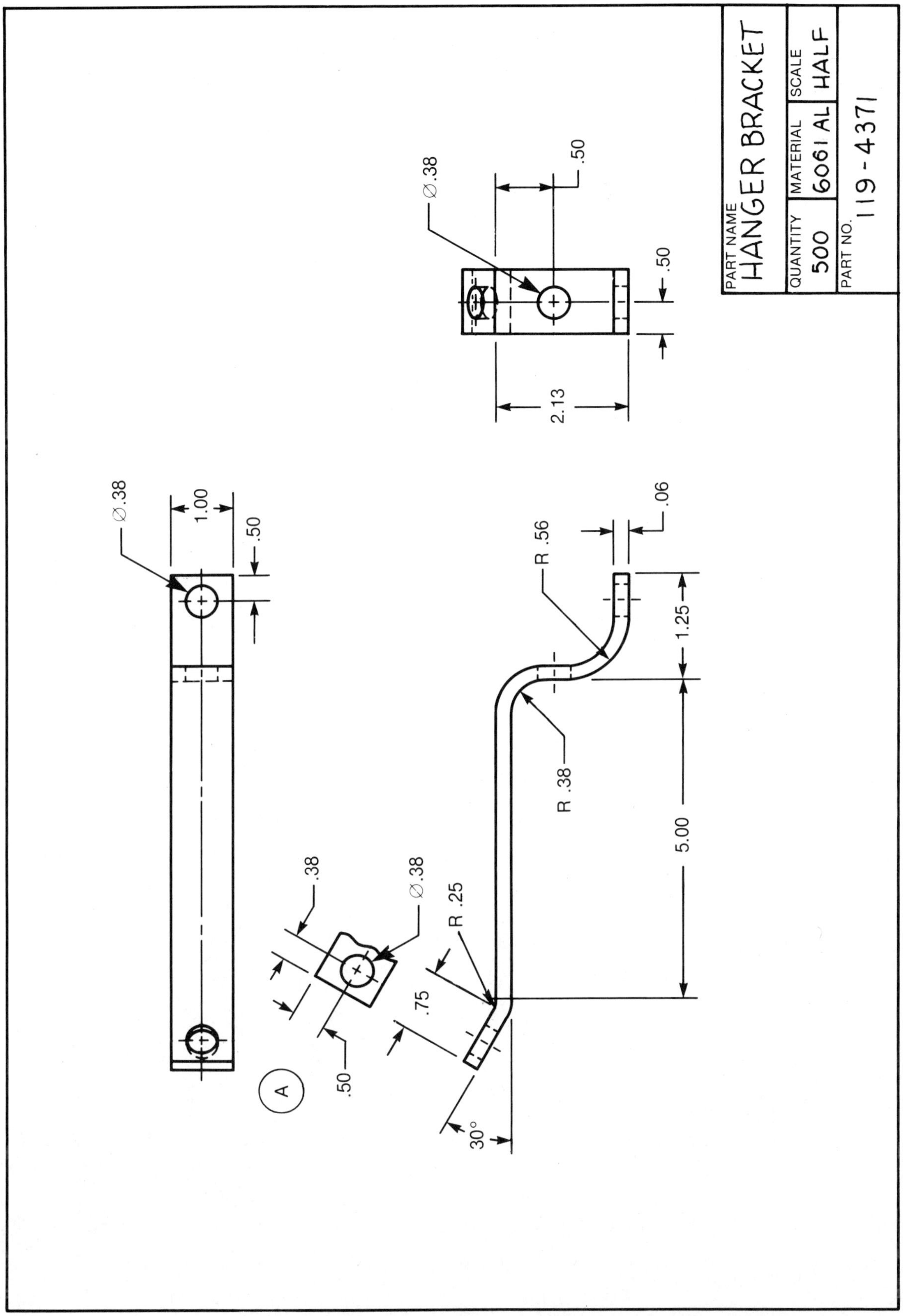

PART NAME HANGER BRACKET

	MATERIAL	SCALE
QUANTITY	6061 AL	HALF
500		

PART NO. 119-4371

∅.38

.50

.50

2.13

∅.38

1.00

.50

R.56

.06

R.38

1.25

5.00

.38

∅.38

R.25

.75

.50

A

30°

Figure E14-2.

15

METRICATION AND FIRST ANGLE PROJECTIONS

OBJECTIVES

After studying this chapter, you will be able to:

- Describe the basic rules for metric dimensioning.

- Identify and define the methods used for dual dimensioning.

- Identify the common projection systems and symbols used for prints.

The metric system was revised and updated in 1960 to meet a worldwide need for a universal measuring system. The new system is called *Le Systeme Internationale d'Unites* (The International System of Units), or *SI*. The United States adopted the SI in 1975, and is now in the gradual process of conversion. To properly interpret shop prints, it is important to be familiar with the SI and how this system is used with regard to shop prints.

The metric standards of ANSI Y14.5M-82 have several variations from the SI standards. An example is the spelling of meter (ANSI) and metre (SI).

INTERNATIONAL SYSTEM OF UNITS

The millimetre is the standard SI unit of measurement used for industrial purposes. Other units, such as the centimetre, decimetre, and metre, while proper in the SI, are not used on manufacturing prints. The following are a few of the rules that drafters follow when dimensioning metric prints.

1. Dimensions are usually specified with one-place decimals. If closer control is required, then two- or three-place decimals are used.

2. The tolerance of a dimension should never be assumed from the number of decimal places in the dimension. The title block should always be checked if no tolerance value is shown on the dimension.

3. Commas are not used to show values greater than one thousand. If necessary, a space is used for millimetre dimensions, Figure 15-1. ANSI Y14.5M-1982 does not use commas or spaces to separate digits into groups.

4. Base unit values, such as inch (in.) or millimetre (mm), are not included in a dimension. The only exception to this is when a millimetre dimension is specified on an inch-dimensioned print, or when an inch size is noted on a millimetre-dimensioned print. All prints should include the note "Unless otherwise specified, all dimensions are in inches" or "Unless otherwise specified, all dimensions are in millimetres."

5. A period is used as a decimal marker for both inch and millimetre dimensions. A comma or raised dot may also be used for a decimal marker for some millimetre dimensions (especially on European drawings). See Figure 15-2.

6. Where the dimension is less than one millimetre, a zero precedes the decimal point (0.55).

7. Where the dimension is a whole number, neither the decimal point nor a zero is shown (60).

8. Zeros are not added to the end of decimal fractions.

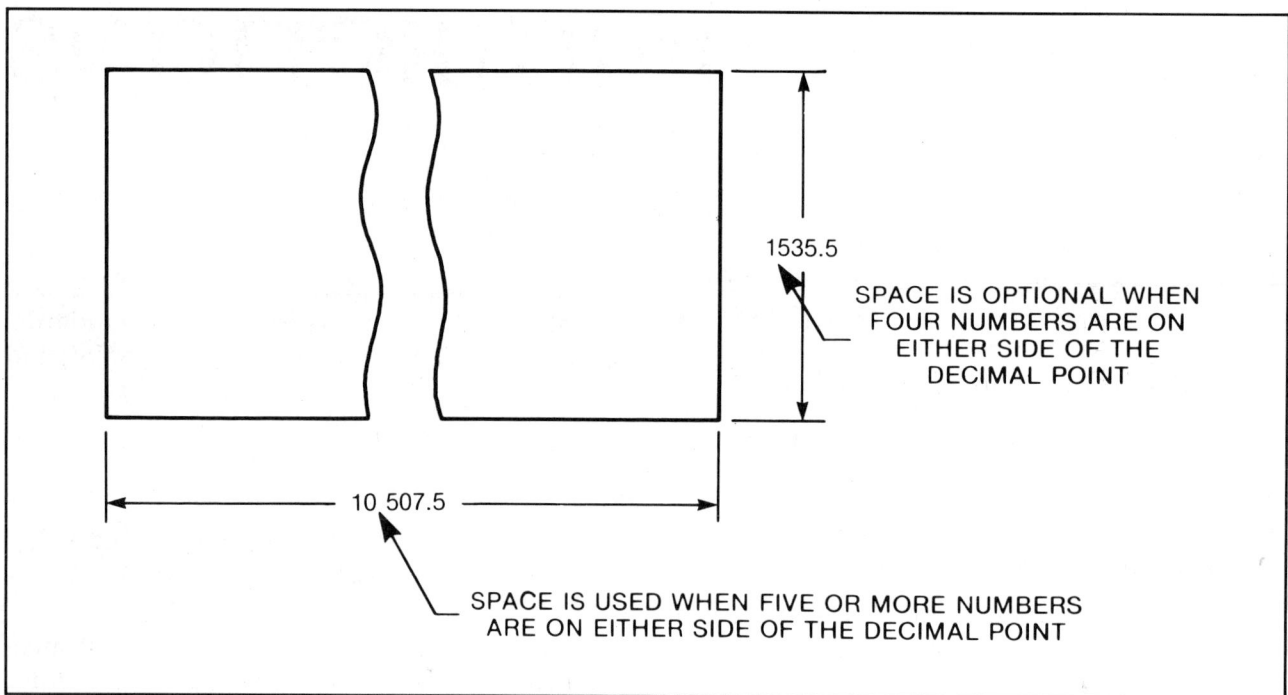

1535.5

SPACE IS OPTIONAL WHEN FOUR NUMBERS ARE ON EITHER SIDE OF THE DECIMAL POINT

10 507.5

SPACE IS USED WHEN FIVE OR MORE NUMBERS ARE ON EITHER SIDE OF THE DECIMAL POINT

Figure 15-1. Spaces are used in place of commas for numbers greater than one thousand when using ISO metric standards.

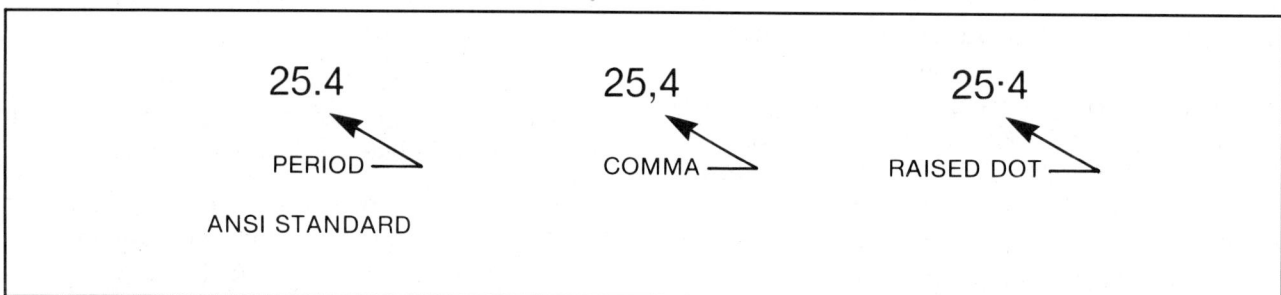

25.4
PERIOD
ANSI STANDARD

25,4
COMMA

25·4
RAISED DOT

Figure 15-2. Decimal markers

DUAL DIMENSIONING

In *dual dimensioning*, both inch and millimetre dimensional units are shown on the same print. The two methods of dual dimensioning in common use today are dual unit dimensioning and conversion chart dimensioning. Both methods will be utilized by industry until metric conversion is complete and only millimetre units are used on prints.

Dual Unit Dimensioning

Dual unit dimensioning uses both inch and millimetre units to dimension each part feature. The two principal methods used to place these dimensions on the print are the position method and the bracket method.

POSITION METHOD. The position method of dual dimensioning uses the location of the dimensions to identify the design units and the conversion units. The *design units*, shown either on the top or to the left, are the units that were originally used to design and dimension the part. The *conversion units*, shown either on the bottom or to the right, are the direct equivalent values of the design units. See Figure 15-3.

BRACKET METHOD. The bracket method of dual dimensioning uses the same format as the position method, but with one exception: the conversion values are placed within brackets. See Figure 15-4. Here again, the design units are shown on the top or to the left, and the conversion values are shown on the bottom or to the right.

Only one of these methods, either position or bracket, is shown on a single print. In addition, a note naming the method used is shown on the print, Figure 15-5.

Conversion Chart Dimensioning

Conversion chart dimensioning totally separates the design units and the conversion units on the print. With this method, the design units are shown directly on the part and the equivalent conversion values are shown in a separate chart on the print. See Figure 15-6.

The conversion chart method of dual dimensioning is rapidly replacing the dual unit method in many companies. The chief reasons for this are as follows:

• Parts are dimensioned in only one system, inches or millimetres. this permits more readable, less cluttered prints, and fewer errors.

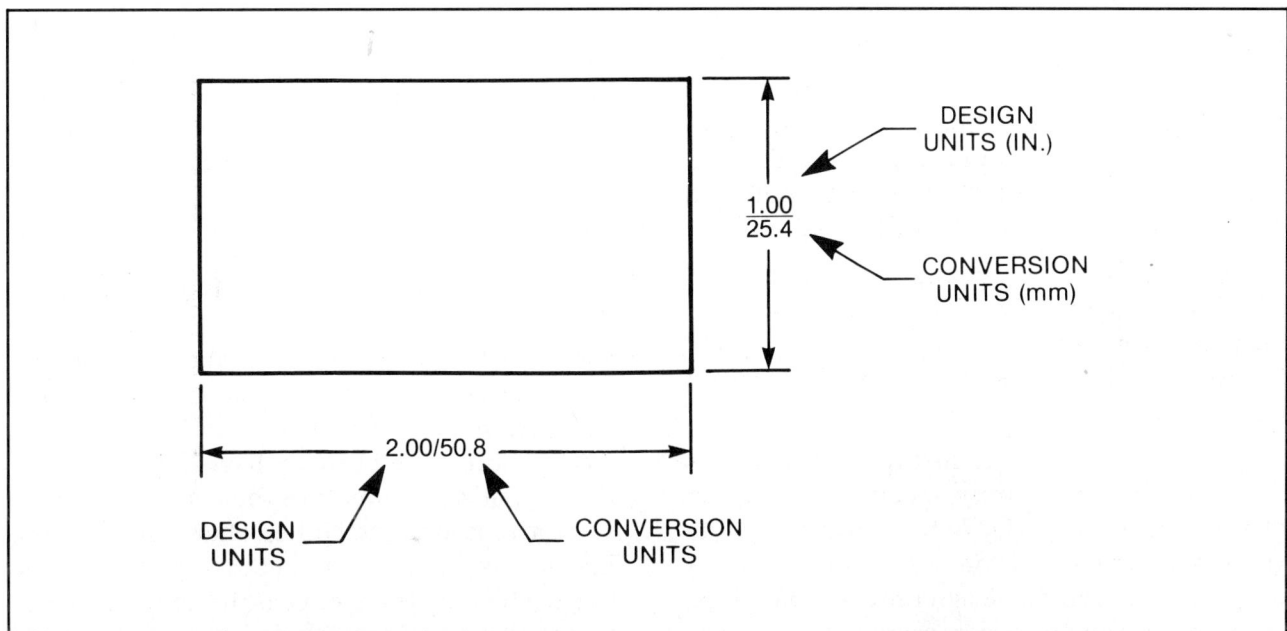

Figure 15-3. Position method of dual unit dimensioning

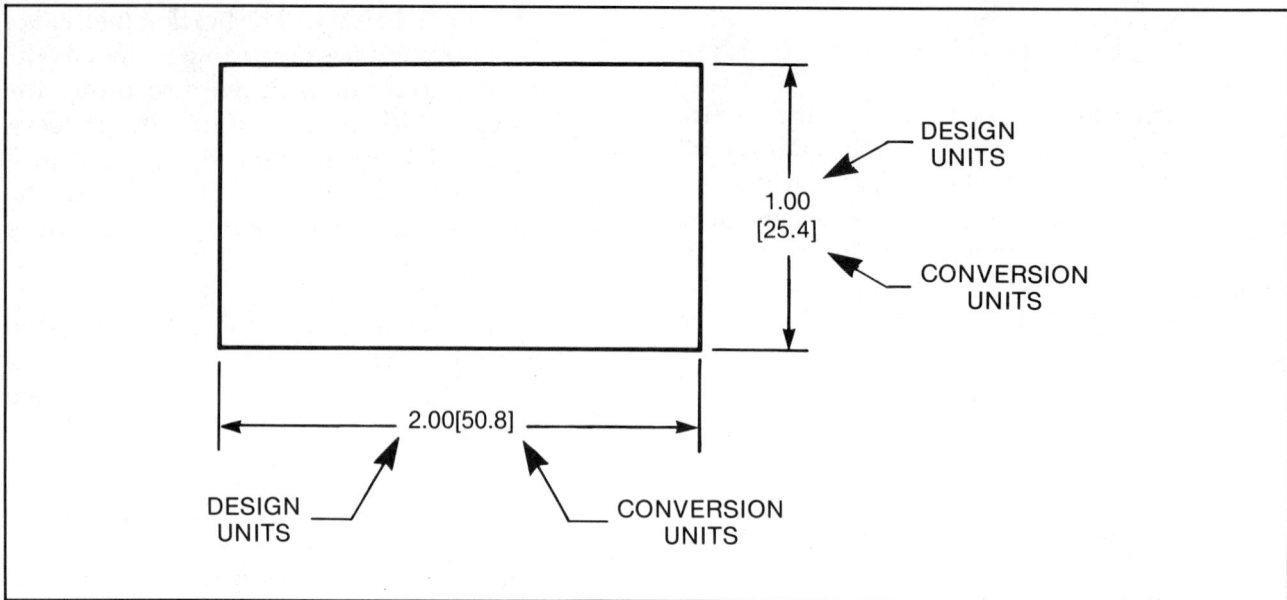

Figure 15-4. Bracket method of dual unit dimensioning

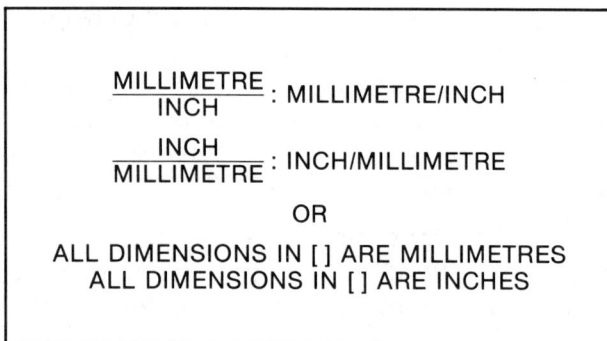

Figure 15-5. Notes to indicate dimensional placement

• With the conversion chart method, each dimension shown on the part is converted and shown in the chart. This results in less confusion and misinterpretation of hard-to-read prints.

• The conversion chart method familiarizes the shop personnel with the SI; as conversion becomes complete and shop personnel begin to "think metric," the charts can be removed from the prints.

• The conversion chart method also reduces the cost of conversion by speeding the design of manufactured parts. Rather than worrying about both design and conversion units, drafters only need to include design units on the print. A computer can be used to prepare the conversion chart.

In practice, the conversion chart is generally an "add-on." That is, the chart is made up and then placed on the original drawing. As changes to the drawing become necessary, the chart can easily be changed or replaced. In a conversion chart, the design units are usually placed to the left and the conversion units are placed to the right. The values shown in the chart usually include all tolerance values and are placed in the chart in descending order. Therefore, the smallest dimensions usually are listed first and the largest dimensions are listed last. Every dimension value that appears on the part is shown in the chart. However, all duplicate dimensions are omitted.

PROJECTION SYSTEMS

The two projection systems used throughout the world today are the first angle and third angle systems. *First angle projections* are generally used in Europe and usually have SI dimensions. *Third angle projections* are the standard projection form used in the United States and Canada. The views in both of these systems are the same. The only difference is in their placement on the print and the relationship of the views to one another.

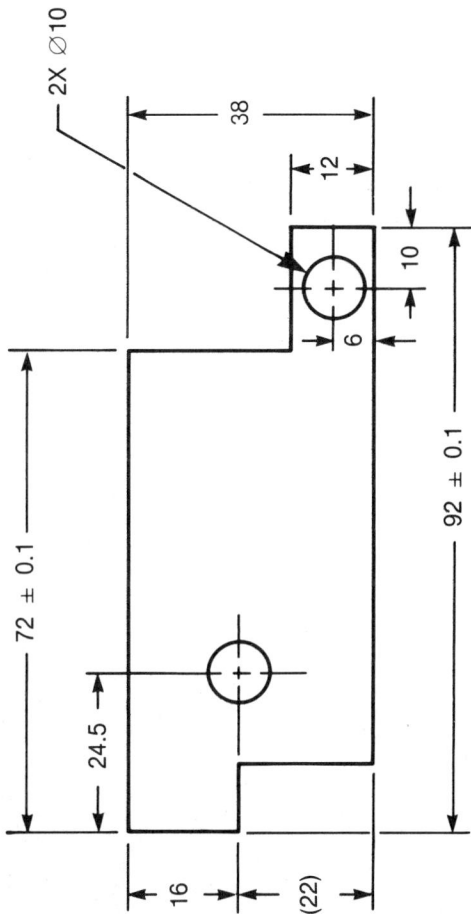

Figure 15-6. Conversion chart dimensioning

The easiest way to understand first and third angle projections is to visualize the projection planes, Figure 15-7. The vertical section represents the frontal plane and the horizontal section shows the top plane. The quadrant numbers 1, 2, 3, and 4 indicate the angle of projection.

First angle projections are drawn in the first quadrant (top front). Third angle projections are drawn in the third quadrant (bottom rear). The second and fourth quadrants are not used for industrial prints.

By placing an object in both the first and third quadrants, Figure 15-8, it can be seen how the views are projected. In first angle projection, the view is projected behind the object. The top

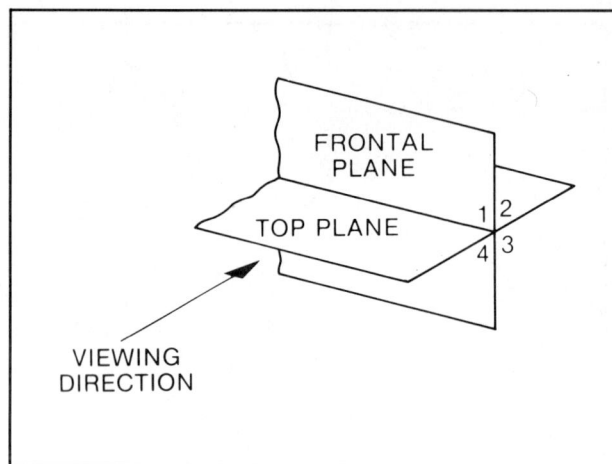

Figure 15-7. Projection planes

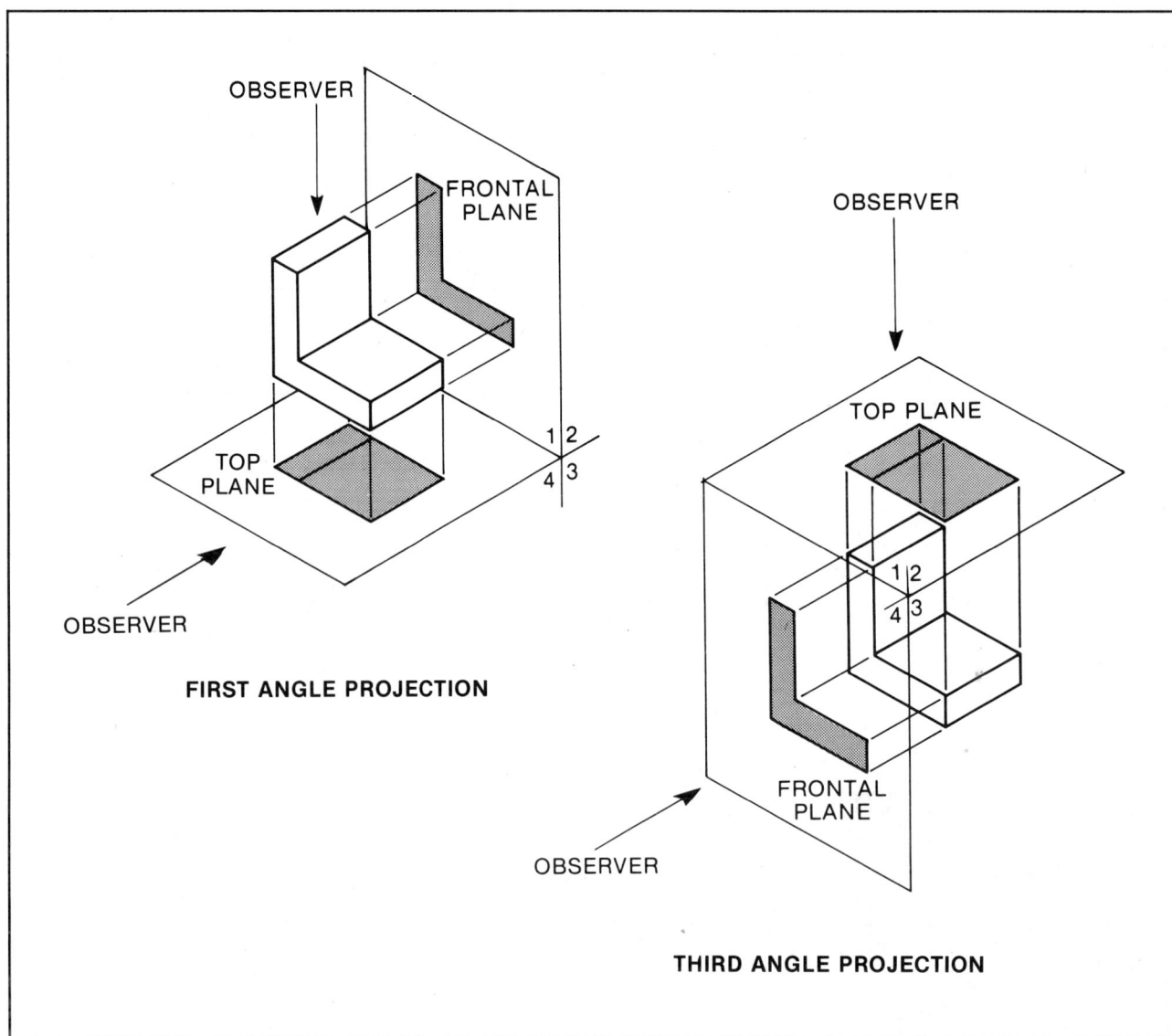

Figure 15-8. Projecting views in first and third angle projections

view is drawn below the object. In third angle projection, the front view is projected in front of the object. The top view is drawn above the object. While only the top and front views are shown here, the same principles apply to all other views as well.

As can be seen, the only difference between these systems is the relationship of the object and the projection plane. First angle projections place the object between the observer and the projection plane. Third angle projections place the projection plane between the observer and the object. See Figure 15-9.

The standard view placement used with each of these systems is shown in Figure 15-10. While third angle projections are the most common type of projection used in the United States and Canada, first angle prints are also used in some industries. With the expansion of international trade, prints using both projection systems are often exchanged internationally.

Projection Symbols

To prevent misinterpretation and to aid identification, *projection symbols* are used to indicate first and third angle projections. See Figure 15-11. These international symbols tell the reader at a glance which system was used to construct the print.

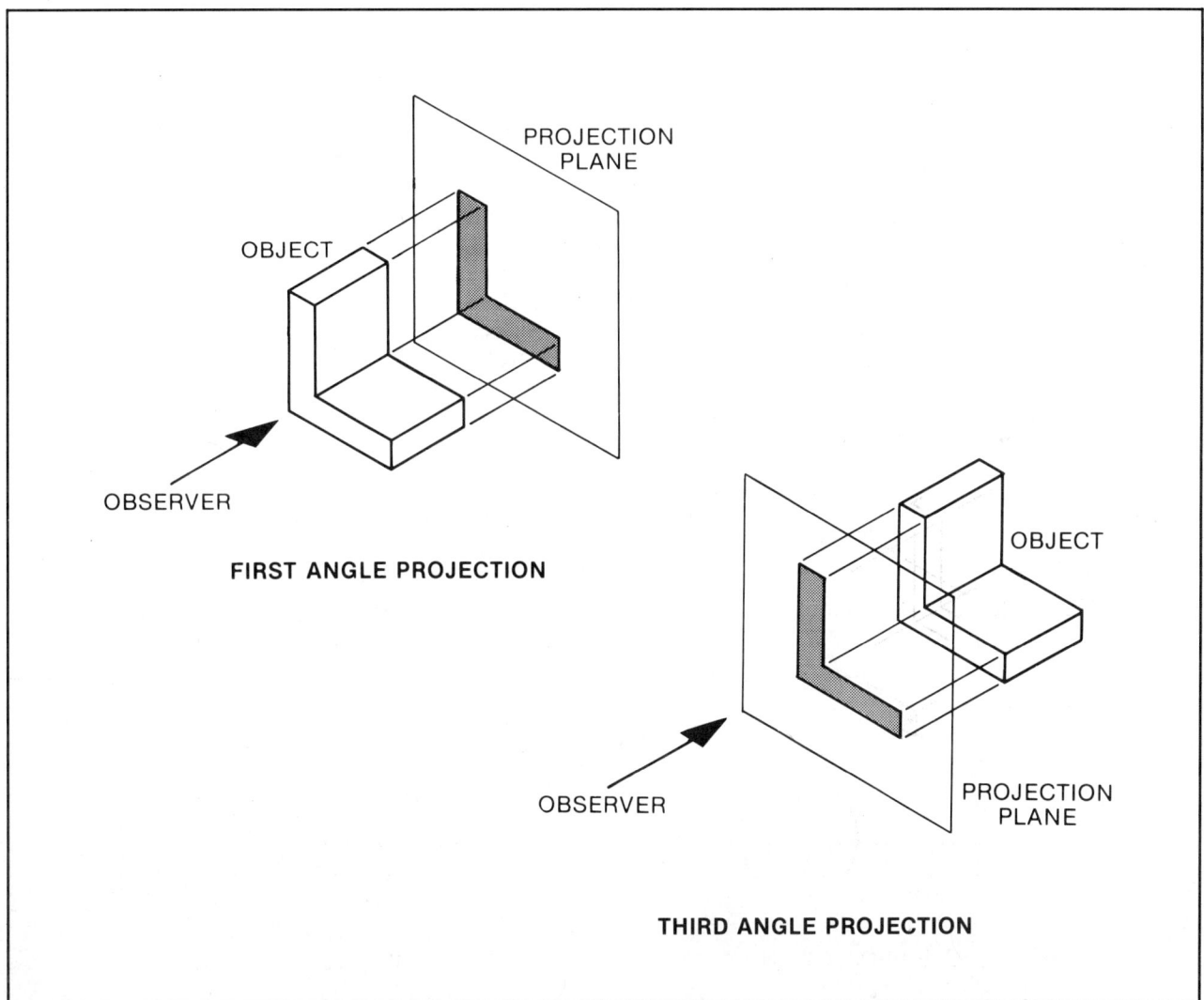

Figure 15-9. Relationships of the object, projection plane, and observer for first and third angle projections

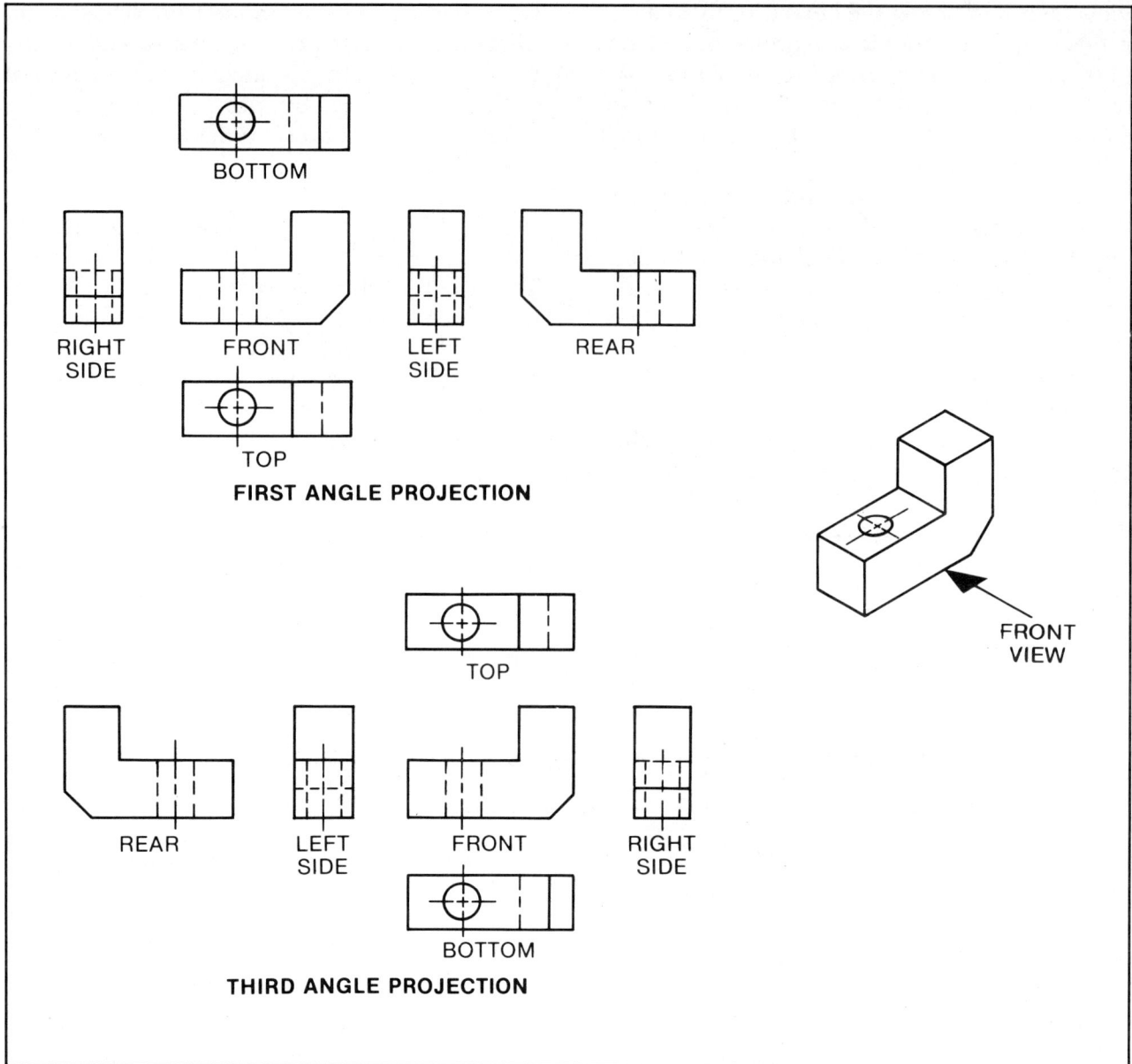

Figure 15-10. Standard placement of views for first and third angle projections

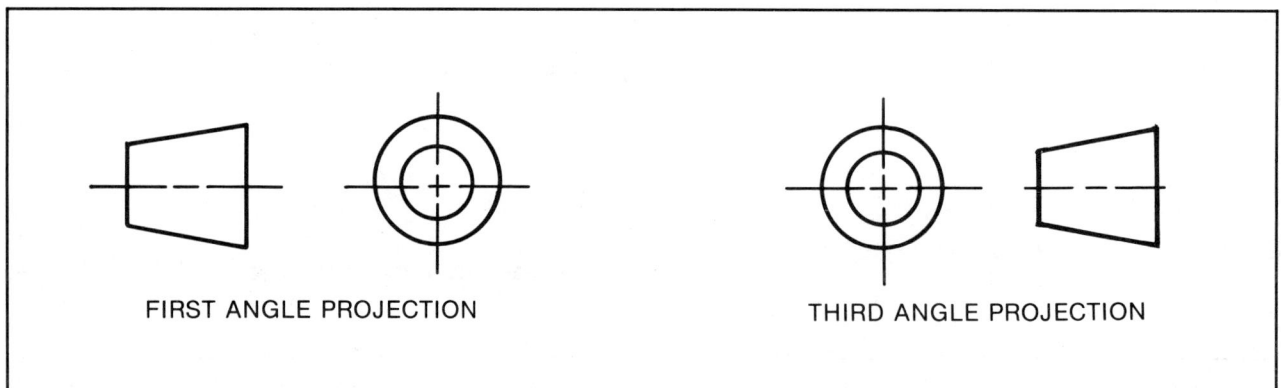

Figure 15-11. Projection symbols

Metric Designation Symbol

The *metric designation symbol*, Figure 15-12, is used to show a print dimensioned in millimetre units. The symbol is located near the title block and is shown by the word *METRIC* in large letters, framed by a dark line.

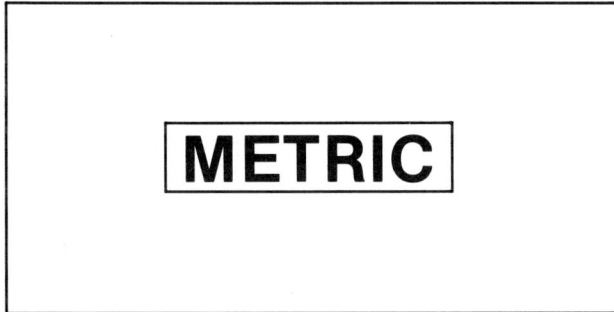

Figure 15-12. Metric designation symbol

The following key terms were introduced in this chapter. Be sure you know the meaning of each term before proceeding to the review material.

Bracket Method
Conversion Chart Dimensioning
Conversion Units
Design Units
Dual Dimensioning
Dual Unit Dimensioning
First Angle Projection
Metric Designation Symbol
Position Method
Projection Plane
Projection Symbol
Systeme Internationale d'Unites (SI)
Third Angle Projection

Test your knowledge with this reinforcement study material. Write your answers to the questions in the spaces provided.

1. What system of measurement uses the metre as its base? _____

_____

2. What is the standard metric unit used for industrial prints? _____

_____

3. How should the tolerance of a dimension be determined when no tolerance is shown on the dimensional value? _____

_____

4. How are millimetre units greater than one thousand noted on a print? _____

_____

_____

5. What is the major difference between SI and ANSI standards for metric drawings?

_____

_____

_____

6. Which system uses brackets for conversion values? _____

_____

7. Which system places the conversion values in a separate chart? _____

_____

8. What is meant by the term *design units*? _____

_____

_____

9. What is meant by the term *conversion units*? _____

_____

_____

10. What are the two projection systems used for working drawings? _____

_____

11. Which system is used mainly in the United States and Canada? _____

_____

12. Which projection system places the projection plane between the object and the observer?

_____

13. Which projection system places the object between the observer and the projection plane?

_____

14. What projection systems are indicated by the symbols shown below?

a. 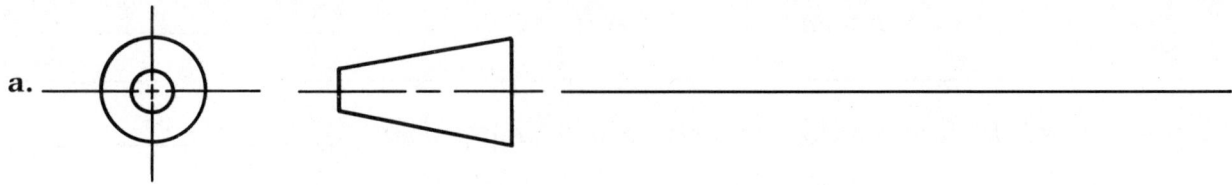 _____

b. 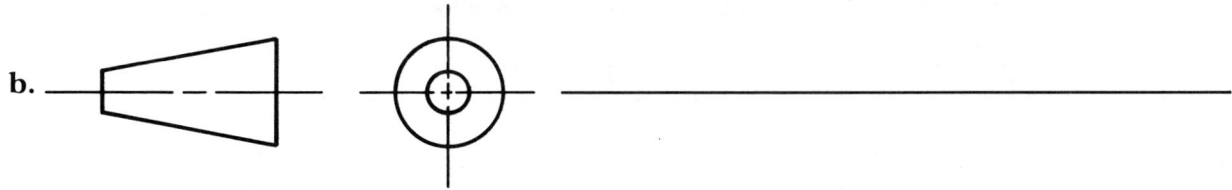 _____

Identify the views of the object shown in Figure E15-1.

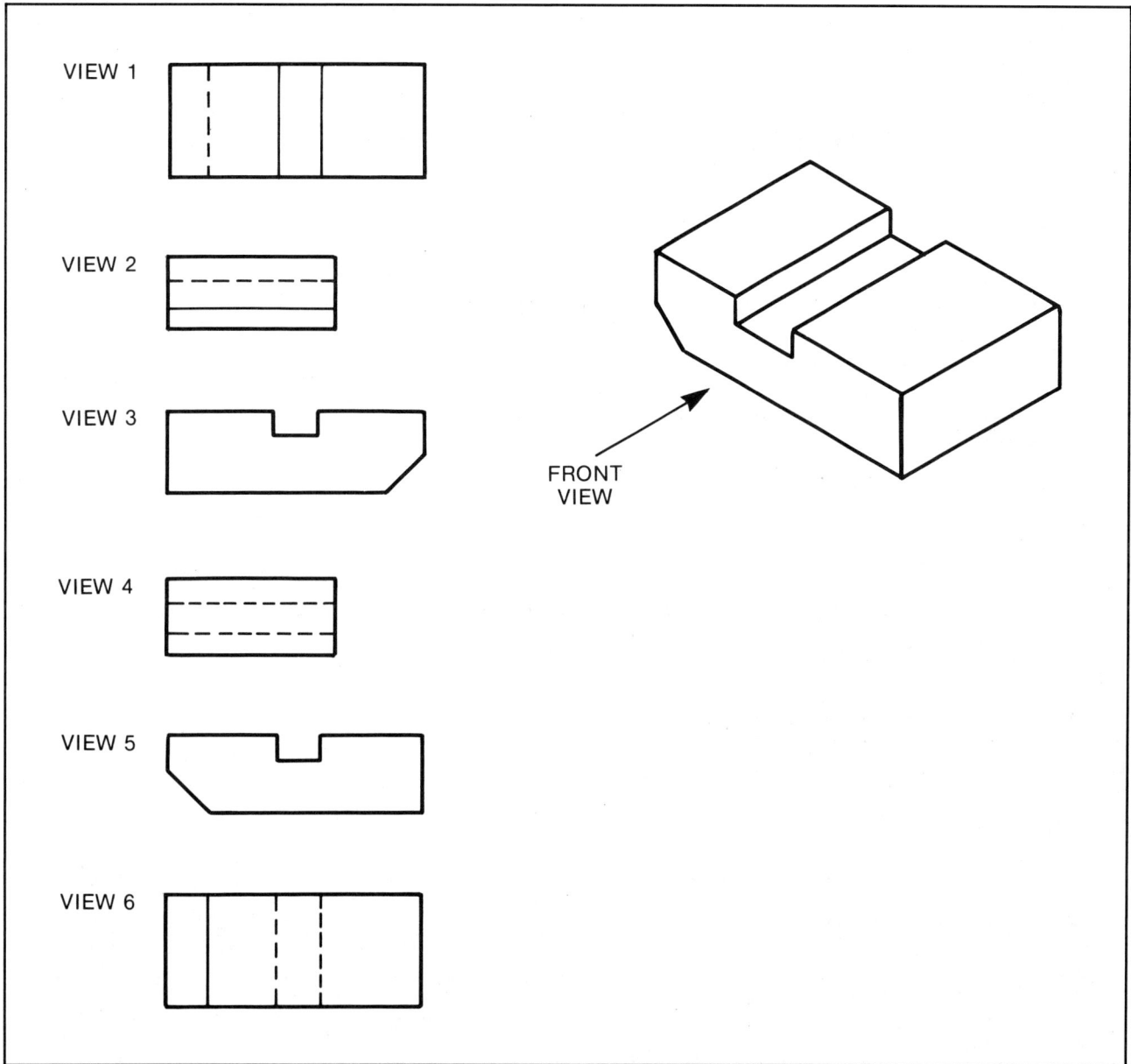

VIEW 1

VIEW 2

VIEW 3

VIEW 4

VIEW 5

VIEW 6

FRONT VIEW

Figure E15-1.

1. View #1: _____

2. View #2: _____

3. View #3: _____

4. View #4: _____

5. View #5: _____

6. View #6: _____

Refer to Figure E15-2 on page 309 to answer the following questions.

1. Which form of projection was used for this print? _____

_____

2. Identify the views shown in the print. _____

_____

3. Find the missing dimensions (in millimetres).

   a. $A =$ _____

   b. $B =$ _____

   c. $C =$ _____

4. What are the design units for this print? _____

_____

5. What are the conversion units? _____

_____

6. Which dual dimensional method is shown in this print? _____

_____

7. What thread size is specified on the print? _____

8. What is the nominal diameter of the thread? _____

9. What is the pitch of the thread? _____

10. What is the class of fit of the thread? _____

_____

11. What does the note *METRIC* mean? _____

_____

_____

12. What are the general tolerances in millimetres and in inches? _____

_____

_____

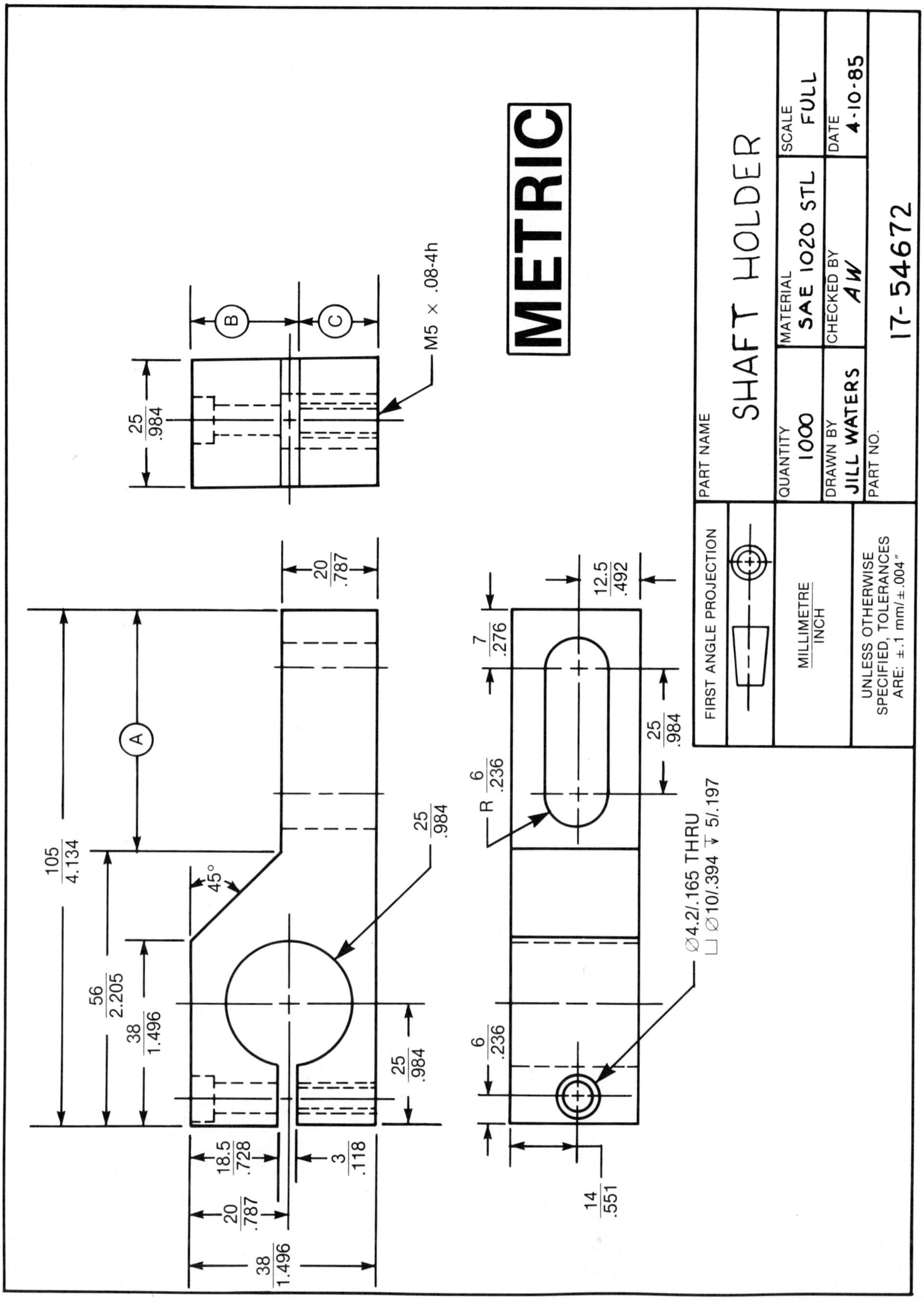

METRIC

SHAFT HOLDER

FIRST ANGLE PROJECTION	PART NAME		
	SHAFT HOLDER		SCALE FULL
	QUANTITY 1000	MATERIAL SAE 1020 STL	
	DRAWN BY JILL WATERS	CHECKED BY AW	DATE 4-10-85
MILLIMETRE INCH	PART NO. 17-54672		
UNLESS OTHERWISE SPECIFIED, TOLERANCES ARE: ±.1 mm/±.004"			

M5 × .08-4h

25 / .984

20 / .787

105 / 4.134

A

45°

25 / .984

56 / 2.205

38 / 1.496

25 / .984

18.5 / .728

20 / .787

3 / .118

38 / 1.496

12.5 / .492

7 / .276

R 6 / .236

25 / .984

6 / .236

14 / .551

Ø4.2/.165 THRU
⌴ Ø10/.394 ▽ 5/.197

B   C

Figure E15-2.

A

TABLES

Table A-1. DECIMAL EQUIVALENTS AND STANDARD DRILL SIZES

DRILL	DECIMAL	DRILL	DECIMAL	DRILL	DECIMAL	DRILL	DECIMAL	DRILL	DECIMAL	DRILL	DECIMAL
80	.0135	52	.0635	26	.1470	A	.2340	$\frac{3}{8}$	.3750	$\frac{11}{16}$	.6875
79	.0145	51	.0670	25	.1495	$\frac{15}{64}$	.2344	V	.3770	$\frac{45}{64}$	.7031
$\frac{1}{64}$	.0156	50	.0700	24	.1520	B	.2380	W	.3860	$\frac{23}{32}$	.7188
78	.0160	49	.0730	23	.1540	C	.2420	$\frac{25}{64}$	.3906	$\frac{47}{64}$	.7344
77	.0180	48	.0760	$\frac{5}{32}$	.1562	D	.2460	X	.3970	$\frac{3}{4}$	.7500
76	.0200	$\frac{5}{64}$	.0781	22	.1570	E	.2500	Y	.4040	$\frac{49}{64}$	.7656
75	.0210	47	.0785	21	.1590	$\frac{1}{4}$	.2500	$\frac{13}{32}$	.4062	$\frac{25}{32}$	.7812
74	.0225	46	.0810	20	.1610	F	.2570	Z	.4130	$\frac{51}{64}$	.7969
73	.0240	45	.0820	19	.1660	G	.2610	$\frac{27}{64}$	.4219	$\frac{13}{16}$	.8125
72	.0250	44	.0860	18	.1695	$\frac{17}{64}$	.2656	$\frac{7}{16}$	.4375	$\frac{53}{64}$	.8281
71	.0260	43	.0890	$\frac{11}{64}$	.1719	H	.2660	$\frac{29}{64}$	.4531	$\frac{27}{32}$	.8438
70	.0280	42	.0935	17	.1730	I	.2720	$\frac{15}{32}$	.4688	$\frac{55}{64}$	.8594
69	.0292	$\frac{3}{32}$	.0938	16	.1770	J	.2770	$\frac{31}{64}$	.4844	$\frac{7}{8}$	.8750
68	.0310	41	.0960	15	.1800	K	.2810	$\frac{1}{2}$	.5000	$\frac{57}{64}$	.8906
$\frac{1}{32}$	.0313	40	.0980	14	.1820	$\frac{9}{32}$	.2812	$\frac{33}{64}$	.5156	$\frac{29}{32}$	.9062
67	.0320	39	.0995	13	.1850	L	.2900	$\frac{17}{32}$	.5313	$\frac{59}{64}$	.9219
66	.0330	38	.1015	$\frac{3}{16}$	.1875	M	.2950	$\frac{35}{64}$	.5469	$\frac{15}{16}$	.9375
65	.0350	37	.1040	12	.1890	$\frac{19}{64}$	.2969	$\frac{9}{16}$	.5625	$\frac{61}{64}$	.9531
64	.0360	36	.1065	11	.1910	N	.3020	$\frac{37}{64}$	.5781	$\frac{31}{32}$	.9688
63	.0370	$\frac{7}{64}$	.1094	10	.1935	$\frac{5}{16}$	.3125	$\frac{19}{32}$	.5938	$\frac{63}{64}$	.9844
62	.0380	35	.1100	9	.1960	O	.3160	$\frac{39}{64}$	.6094	1	1.000
61	.0390	34	.1110	8	.1990	P	.3230	$\frac{5}{8}$	.6250		
60	.0400	33	.1130	7	.2010	$\frac{21}{64}$	.3281	$\frac{41}{64}$	.6406		
59	.0410	32	.1160	$\frac{13}{64}$	.2031	Q	.3320	$\frac{21}{32}$	.6562		
58	.0420	31	.1200	6	.2040	R	.3390	$\frac{43}{64}$	.6719		
57	.0430	$\frac{1}{8}$	.1250	5	.2055	$\frac{11}{32}$	.3438				
56	.0465	30	.1285	4	.2090	S	.3480				
$\frac{3}{64}$	.0469	29	.1360	3	.2130	T	.3580				
55	.0520	28	.1405	$\frac{7}{32}$	.2188	$\frac{23}{64}$	.3594				
54	.0550	$\frac{9}{64}$	.1406	2	.2210	U	.3680				
53	.0595	27	.1440	1	.2280						
$\frac{1}{16}$	.0625										

Table A-2. INCH-TO-MILLIMETER CONVERSION TABLE

DECIMALS TO MILLIMETERS

DECIMAL	mm	DECIMAL	mm
.001	0.0254	.500	12.7
.002	0.0508	.510	12.954
.003	0.0762	.520	13.208
.004	0.1016	.530	13.462
.005	0.127	.540	13.716
.006	0.1524	.550	13.97
.007	0.1778	.560	14.224
.008	0.2032	.570	14.478
.009	0.2286	.580	14.732
.010	0.254	.590	14.986
.020	0.508		
.030	0.762		
.040	1.016	.600	15.24
.050	1.27	.610	15.494
.060	1.524	.620	15.748
.070	1.778	.630	16.002
.080	2.032	.640	16.256
.090	2.286	.650	16.51
.100	2.54	.660	16.764
.110	2.794	.670	17.018
.120	3.048	.680	17.272
.130	3.302	.690	17.526
.140	3.556		
.150	3.81		
.160	4.064	.700	17.78
.170	4.318	.710	18.034
.180	4.572	.720	18.288
.190	4.826	.730	18.542
.200	5.08	.740	18.796
.210	5.334	.750	19.05
.220	5.588	.760	19.304
.230	5.842	.770	19.558
.240	6.096	.780	19.812
.250	6.35	.790	20.066
.260	6.604		
.270	6.858	.800	20.32
.280	7.112	.810	20.574
.290	7.366	.820	20.828
.300	7.62	.830	21.082
.310	7.874	.840	21.336
.320	8.128	.850	21.59
.330	8.382	.860	21.844
.340	8.636	.870	22.098
.350	8.89	.880	22.352
.360	9.144	.890	22.606
.370	9.398		
.380	9.652		
.390	9.906	.900	22.86
.400	10.16	.910	23.114
.410	10.414	.920	23.368
.420	10.668	.930	23.622
.430	10.922	.940	23.876
.440	11.176	.950	24.13
.450	11.43	.960	24.384
.460	11.684	.970	24.638
.470	11.938	.980	24.892
.480	12.192	.990	25.146
.490	12.446	1.000	25.4

FRACTIONS TO DECIMALS TO MILLIMETERS

FRACTION	DECIMAL	mm	FRACTION	DECIMAL	mm
$\frac{1}{64}$	.0156	0.3969	$\frac{33}{64}$	.5156	13.0969
$\frac{1}{32}$	.0312	0.7938	$\frac{17}{32}$	.5312	13.4938
$\frac{3}{64}$	.0469	1.1906	$\frac{35}{64}$	.5469	13.8906
$\frac{1}{16}$	.0625	1.5875	$\frac{9}{16}$	.5625	14.2875
$\frac{5}{64}$	.0781	1.9844	$\frac{37}{64}$	.5781	14.6844
$\frac{3}{32}$	.0938	2.3812	$\frac{19}{32}$	.5938	15.0812
$\frac{7}{64}$	.1094	2.7781	$\frac{39}{64}$	.6094	15.4781
$\frac{1}{8}$	.1250	3.175	$\frac{5}{8}$	.6250	15.875
$\frac{9}{64}$	.1406	3.5719	$\frac{41}{64}$	.6406	16.2719
$\frac{5}{32}$	.1562	3.9688	$\frac{21}{32}$	.6562	16.6688
$\frac{11}{64}$	.1719	4.3656	$\frac{43}{64}$	.6719	17.0656
$\frac{3}{16}$	.1875	4.7625	$\frac{11}{16}$	.6875	17.4625
$\frac{13}{64}$	.2031	5.1594	$\frac{45}{64}$	.7031	17.8594
$\frac{7}{32}$	.2188	5.5562	$\frac{23}{32}$	.7188	18.2562
$\frac{15}{64}$	.2344	5.9531	$\frac{47}{64}$	.7344	18.6531
$\frac{1}{4}$	.2500	6.35	$\frac{3}{4}$	.7500	19.05
$\frac{17}{64}$	.2656	6.7469	$\frac{49}{64}$	.7656	19.4469
$\frac{9}{32}$	.2812	7.1438	$\frac{25}{32}$	.7812	19.8438
$\frac{19}{64}$	.2969	7.5406	$\frac{51}{64}$	.7969	20.2406
$\frac{5}{16}$	.3125	7.9375	$\frac{13}{16}$	.8125	20.6375
$\frac{21}{64}$	.3281	8.3344	$\frac{53}{64}$	.8281	21.0344
$\frac{11}{32}$	.3438	8.7312	$\frac{27}{32}$	.8438	21.4312
$\frac{23}{64}$	.3594	9.1281	$\frac{55}{64}$	.8594	21.8281
$\frac{3}{8}$	.3750	9.525	$\frac{7}{8}$	.8750	22.225
$\frac{25}{64}$	.3906	9.9219	$\frac{57}{64}$	.8906	22.6219
$\frac{13}{32}$	.4062	10.3188	$\frac{29}{32}$	.9062	23.0188
$\frac{27}{64}$	.4219	10.7156	$\frac{59}{64}$	.9219	23.4156
$\frac{7}{16}$	.4375	11.1125	$\frac{15}{16}$	.9375	23.8125
$\frac{29}{64}$	.4531	11.5094	$\frac{61}{64}$	.9531	24.2094
$\frac{15}{32}$	.4688	11.9062	$\frac{31}{32}$	.9688	24.6062
$\frac{31}{64}$	.4844	12.3031	$\frac{63}{64}$	.9844	25.0031
$\frac{1}{2}$	.5000	12.7	1	1.0000	25.4

Table A-3. MILLIMETER-TO-INCH CONVERSION TABLE

mm	DECIMAL	mm	DECIMAL	mm	DECIMAL	mm	DECIMAL	mm	DECIMAL
0.01	.00039	0.41	.01614	0.81	.03189	21	.82677	61	2.40157
0.02	.00079	0.42	.01654	0.82	.03228	22	.86614	62	2.44094
0.03	.00118	0.43	.01693	0.83	.03268	23	.90551	63	2.48031
0.04	.00157	0.44	.01732	0.84	.03307	24	.94488	64	2.51969
0.05	.00197	0.45	.01772	0.85	.03346	25	.98425	65	2.55906
0.06	.00236	0.46	.01811	0.86	.03386	26	1.02362	66	2.59843
0.07	.00276	0.47	.01850	0.87	.03425	27	1.06299	67	2.63780
0.08	.00315	0.48	.01890	0.88	.03465	28	1.10236	68	2.67717
0.09	.00354	0.49	.01929	0.89	.03504	29	1.14173	69	2.71654
0.10	.00394	0.50	.01969	0.90	.03543	30	1.18110	70	2.75591
0.11	.00433	0.51	.02008	0.91	.03583	31	1.22047	71	2.79528
0.12	.00472	0.52	.02047	0.92	.03622	32	1.25984	72	2.83465
0.13	.00512	0.53	.02087	0.93	.03661	33	1.29921	73	2.87402
0.14	.00551	0.54	.02126	0.94	.03701	34	1.33858	74	2.91339
0.15	.00591	0.55	.02165	0.95	.03740	35	1.37795	75	2.95276
0.16	.00630	0.56	.02205	0.96	.03780	36	1.41732	76	2.99213
0.17	.00669	0.57	.02244	0.97	.03819	37	1.45669	77	3.03150
0.18	.00709	0.58	.02283	0.98	.03858	38	1.49606	78	3.07087
0.19	.00748	0.59	.02323	0.99	.03898	39	1.53543	79	3.11024
0.20	.00787	0.60	.02362	1.00	.03937	40	1.57480	80	3.14961
0.21	.00827	0.61	.02402	1	.03937	41	1.61417	81	3.18898
0.22	.00866	0.62	.02441	2	.07874	42	1.65354	82	3.22835
0.23	.00906	0.63	.02480	3	.11811	43	1.69291	83	3.26772
0.24	.00945	0.64	.02520	4	.15748	44	1.73228	84	3.30709
0.25	.00984	0.65	.02559	5	.19685	45	1.77165	85	3.34646
0.26	.01024	0.66	.02598	6	.23622	46	1.81102	86	3.38583
0.27	.01063	0.67	.02638	7	.27559	47	1.85039	87	3.42520
0.28	.01102	0.68	.02677	8	.31496	48	1.88976	88	3.46457
0.29	.01142	0.69	.02717	9	.35433	49	1.92913	89	3.50394
0.30	.01181	0.70	.02756	10	.39370	50	1.96850	90	3.54331
0.31	.01220	0.71	.02795	11	.43307	51	2.00787	91	3.58268
0.32	.01260	0.72	.02835	12	.47244	52	2.04724	92	3.62205
0.33	.01299	0.73	.02874	13	.51181	53	2.08661	93	3.66142
0.34	.01339	0.74	.02913	14	.55118	54	2.12598	94	3.70079
0.35	.01378	0.75	.02953	15	.59055	55	2.16535	95	3.74016
0.36	.01417	0.76	.02992	16	.62992	56	2.20472	96	3.77953
0.37	.01457	0.77	.03032	17	.66929	57	2.24409	97	3.81890
0.38	.01496	0.78	.03071	18	.70866	58	2.28346	98	3.85827
0.39	.01535	0.79	.03110	19	.74803	59	2.32283	99	3.89764
0.40	.01575	0.80	.03150	20	.78740	60	2.36220	100	3.93701

Table A-4. TAP DRILL SIZES—THREADS

COARSE THREADS (UNC)			FINE THREADS (UNF)		
NOMINAL DIAMETER	THREADS PER INCH	TAP DRILL SIZE (75% THREAD DEPTH)	NOMINAL DIAMETER	THREADS PER INCH	TAP DRILL SIZE (75% THREAD DEPTH)
$\frac{1}{4}$	20	No. 7	$\frac{1}{4}$	28	No. 3
$\frac{5}{16}$	18	"F"	$\frac{5}{16}$	24	"I"
$\frac{3}{8}$	16	$\frac{5}{16}$	$\frac{3}{8}$	24	"Q"
$\frac{7}{16}$	14	"U"	$\frac{7}{16}$	20	"W"
$\frac{1}{2}$	13	$\frac{27}{64}$	$\frac{1}{2}$	20	$\frac{29}{64}$
$\frac{9}{16}$	12	$\frac{31}{64}$	$\frac{9}{16}$	18	$\frac{33}{64}$
$\frac{5}{8}$	11	$\frac{17}{32}$	$\frac{5}{8}$	18	$\frac{37}{64}$
$\frac{11}{16}$	11N.S.	$\frac{19}{32}$	$\frac{11}{16}$	11N.S.	$\frac{19}{32}$
$\frac{3}{4}$	10	$\frac{21}{32}$	$\frac{3}{4}$	16	$\frac{11}{16}$
$\frac{13}{16}$	10N.S.	$\frac{23}{32}$	$\frac{13}{16}$	10N.S.	$\frac{23}{32}$
$\frac{7}{8}$	9	$\frac{49}{64}$	$\frac{7}{8}$	14	$\frac{13}{16}$
$\frac{15}{16}$	9N.S.	$\frac{53}{64}$	$\frac{15}{16}$	9N.S.	$\frac{53}{64}$
1	8	$\frac{7}{8}$	1	12	$\frac{59}{64}$
$1\frac{1}{8}$	7	$\frac{63}{64}$	1	14N.S.	$\frac{15}{16}$
$1\frac{1}{4}$	7	$1\frac{7}{64}$	$1\frac{1}{8}$	12	$1\frac{3}{64}$
$1\frac{3}{8}$	6	$1\frac{13}{64}$	$1\frac{1}{4}$	12	$1\frac{11}{64}$
$1\frac{1}{2}$	6	$1\frac{11}{32}$	$1\frac{3}{8}$	12	$1\frac{19}{64}$
			$1\frac{1}{2}$	12	$1\frac{27}{64}$

Table A-5. TAP DRILL SIZES—MACHINE SCREW SIZES

SCREW NUMBER	THREADS PER INCH	TAP DRILL SIZE (75% THREAD DEPTH)
0	80	$\frac{3}{64}$"
1	72	#53
1	64	#53
2	64	#50
2	56	#51
3	56	#46
3	48	$\frac{5}{64}$
4	48	#42
4	40	#43
5	44	#37
5	40	#39
6	40	#33
6	32	#36
8	36	#29
8	32	#29
10	32	#21
10	24	#25
12	28	#15
12	24	#17

Table A-6. TAP DRILL SIZES—PIPE THREAD SIZES

PIPE SIZE INCHES	THREADS PER INCH	TAP DRILL	
		TAPER NPT	STRAIGHT NPS
$\frac{1}{8}$	27	Q	$\frac{11}{32}$"
$\frac{1}{4}$	18	$\frac{7}{16}$"	$\frac{7}{16}$"
$\frac{3}{8}$	18	$\frac{9}{16}$"	$\frac{37}{64}$"
$\frac{1}{2}$	14	$\frac{45}{64}$"	$\frac{23}{32}$"
$\frac{3}{4}$	14	$\frac{29}{32}$"	$\frac{59}{64}$"
1	$11\frac{1}{2}$	$1\frac{9}{64}$"	$1\frac{5}{32}$"
$1\frac{1}{4}$	$11\frac{1}{2}$	$1\frac{31}{64}$"	$1\frac{1}{2}$"
$1\frac{1}{2}$	$11\frac{1}{2}$	$1\frac{47}{64}$"	$1\frac{3}{4}$"
2	$11\frac{1}{2}$	$2\frac{13}{64}$"	$2\frac{7}{32}$"

Table A-7. TAP DRILL SIZES—METRIC THREAD SIZES

NOMINAL SIZE (mm)	PITCH P (mm)	BASIC THREAD DESIGNATION	TAP DRILL SIZE
1.6	0.35	M1.6	1.25
1.8	0.35	M1.8	1.45
2	0.4	M2	1.60
2.2	0.45	M2.2	1.75
2.5	0.45	M2.5	2.05
3	0.5	M3	2.50
3.5	0.6	M3.5	2.90
4	0.7	M4	3.30
4.5	0.75	M4.5	3.70
5	0.8	M5	4.20
6	1	M6	5.00
7	1	M7	6.00
8	1.25 1	M8 M8 × 1	6.70 7.00
10	1.5 1.25	M10 M10 × 1.25	8.50 8.70
12	1.75 1.25	M12 M12 × 1.25	10.20 10.80
14	2 1.5	M14 M14 × 1.5	12.00 12.50
16	2 1.5	M16 M16 × 1.5	14.00 14.50
18	2.5 1.5	M18 M18 × 1.5	15.50 16.50
20	2.5 1.5	M20 M20 × 1.5	17.50 18.50
22	2.5 1.5	M22 M22 × 1.5	19.50 20.50
24	3 2	M24 M24 × 2	21.00 22.00
27	3 2	M27 M27 × 2	24.00 25.00
30	3.5 2	M30 M30 × 2	26.50 28.00
33	3.5 2	M33 M33 × 2	29.50 31.00
36	4 3	M36 M36 × 3	32.00 33.00
39	4 3	M39 M39 × 3	35.00 36.00

B

SYMBOLS AND ABBREVIATIONS

Table B-1. INTERPRETING STANDARD WELDING SYMBOLS

SLOT WELDING SYMBOL	SQUARE-GROOVE WELDING SYMBOL
Depth of filling in inches (omission indicates filling is complete) — — 3/4 — Orientation, location, and all dimensions other than depth of filling are shown on the drawing	Omission of size indicates complete joint penetration — $\frac{1}{4}$ — — Root opening
PLUG WELDING SYMBOL	**CHAIN INTERMITTENT FILLET WELDING SYMBOL**
Included angle of countersink — — Pitch (distance between centers) of welds 30° — Depth of filling in inches (omission indicates filling is complete) 1 3/8 4 Size (diameter of hole at root) —	$\frac{5}{16}$ 2-5 $\frac{5}{16}$ 2-5 — Pitch (distance between centers) of increments — Length of increments Size (length of leg) —
BACKGOUGING WELDING SYMBOL	**BACK OR BACKING WELDING SYMBOL**
$\frac{3}{8}\left(\frac{7}{16}\right)$ — Second reference line used for back gouging and welding as a second operation Back gouge $\frac{1}{2}\left(\frac{9}{16}\right)$ $\frac{1}{8}$ Note: Total effective throat not to exceed thickness of member	C Any applicable single groove weld symbol —
FLASH OR UPSET WELDING SYMBOL	**WELD-ALL-AROUND SYMBOL**
No arrow side or other side significance — FW — Process reference must be used to indicate process desired	Weld-all-around symbol indicates that weld extends completely around the joint

317

Table B-1 (Continued).

FLARE-V AND FLARE-BEVEL-GROOVE WELDING SYMBOLS	EDGE- AND CORNER-FLANGE WELDING SYMBOLS

$\frac{1}{4}$ $\frac{1}{16}$ — Root opening

Size is considered as extending only to tangent points — $\frac{1}{8}$ $\frac{1}{32}$ — Root opening

$\frac{3}{32}$
$\frac{1}{8} + \frac{1}{16}$

Radius

Size of weld — $\frac{3}{64} + \frac{1}{16}$
$\frac{1}{16}$

Height above point of tangency —

SURFACING WELDING SYMBOL INDICATING BUILT-UP SURFACE	SINGLE-V GROOVE WELDING SYMBOL INDICATING ROOT PENETRATION

Size (height of deposit) Omission indicates no specific height desired — $\frac{1}{8}$ — Orientation, location, and all dimensions other than size are shown on the drawing

Size
$\frac{1}{2}$ $(\frac{1}{2})$ 0
90°

Depth of preparation Effective throat — — Root opening — Groove angle

STAGGERED INTERMITTENT FILLET WELDING SYMBOL	DOUBLE-FILLET WELDING SYMBOL

$\frac{1}{2}$ 3-5
$\frac{1}{2}$ 3-5

Pitch (distance between centers) of increments

Size (length of leg) — — Length of increments

Size (length of leg) — IG $\frac{1}{4}$ $\frac{5}{16}$ 6 4

Specification, process, or other reference —

Length Omission indicates that weld extends between abrupt changes in direction or as dimensioned

SPOT WELDING SYMBOL	DOUBLE-BEVEL-GROOVE WELDING SYMBOL

Size (diameter of weld) Strength (in pounds per weld) may be used instead —

Process reference must be used to indicate process desired —

RSW 0.25 (5) 4

Number of welds

Pitch (distance between centers) of weld

Arrow points toward member to be prepared —

Omission of size dimension indicates a total depth of preparation equal to thickness of members —

45° $\frac{1}{8}$
1
$1\frac{1}{4}$ $\frac{1}{8}$
35°

Root opening

Groove angle

SEAM WELDING SYMBOL	PROJECTION WELDING SYMBOL

Size (width of weld) Strength (in pounds per linear inch) may be used instead —

Length of welds or increments Omission indicates that weld extends between abrupt changes in direction or as dimensioned —

0.30 3 9 RSEW

Pitch (distance between centers) of increments

Process reference must be used to indicate process desired

Size (strength in pounds per weld) Diameter of weld may be used instead for circular projection welds —

RPW 500 (7) 6

Projection welding reference must be used

Pitch (distance between centers) of welds

Number of welds

Table B-1 (Continued).

MELT-THRU SYMBOL	FIELD WELD SYMBOL
Any applicable weld symbol — 1 mm Melt-thru symbol is not dimensioned (except height) —	Field weld symbol indicates that weld is to be made at a place other than that of initial construction —

CONVEX CONTOUR SYMBOL	FLUSH CONTOUR SYMBOL
Convex contour symbol indicates face of weld to be finished to convex contour — Finish symbol (user's standard) indicates method of obtaining specified contour but not degree of finish	Flush contour symbol indicates face of weld to be made flush. When used without a finish symbol, indicates weld without subsequent finishing — Finish symbol (user's standard) indicates method of obtaining specified contour but not degree of finish

JOINT WITH BACKING	JOINT WITH SPACER
With groove weld symbol M See note Note: Material and dimensions of backing as specified	With modified groove weld symbol M See note Double bevel groove Note: Material and dimensions of spacer as specified

MULTIPLE REFERENCE LINES	COMPLETE PENETRATION
First operation shown on reference line nearest arrow — 1st Second operation, or supplementary data — 2nd 3rd Third operation, or test information —	CP Indicates complete penetration regardless of type of weld or joint preparation

LOCATION OF ELEMENTS OF A WELDING SYMBOL

FINISH SYMBOL
CONTOUR SYMBOL
ROOT OPENING; DEPTH OF FILING FOR PLUG AND SLOT WELDS
EFFECTIVE THROAT
DEPTH OF PREPARATION; SIZE OR STRENGTH FOR CERTAIN WELDS
SPECIFICATION, PROCESS, OR OTHER REFERENCE
TAIL (TAIL OMITTED WHEN REFERENCE IS NOT USED)
BASIC WELD SYMBOL OR DETAIL REFERENCE

GROOVE ANGLE; INCLUDED ANGLE OF COUNTERSINK FOR PLUG WELDS
LENGTH OF WELD
PITCH (CENTER-TO-CENTER SPACING) OF WELDS
FIELD WELD SYMBOL
WELD-ALL-AROUND SYMBOL
ARROW CONNECTING REFERENCE LINE TO ARROW SIDE MEMBER OF JOINT
REFERENCE LINE

F / A / R

T   S(E)

BOTH SIDES   OTHER SIDE

(N)

L - P

ARROW SIDE
NUMBER OF SPOTS OR PROJECTION WELDS

ELEMENTS IN THIS AREA REMAIN AS SHOWN WHEN TAIL AND ARROW ARE REVERSED

Table B-1 (Continued).

WELDING SYMBOLS FOR COMBINED WELDS	SUPPLEMENTARY SYMBOLS						
					Contour		
	Weld-All-Around	Field Weld	Melt-Thru	Backing, Spacer	Flush	Convex	Concave

BASIC JOINTS—IDENTIFICATION OF ARROW SIDE AND OTHER SIDE OF JOINT

Lap Joint	Edge Joint

Lap Joint:
- Other side member of joint
- Arrow of welding symbol
- Arrow side member of joint

Edge Joint:
- Arrow side of joint
- Arrow of welding symbol
- Joint
- 0–30

Butt Joint	Corner Joint

Butt Joint:
- Arrow of welding symbol
- Arrow side of joint
- Other side of joint

Corner Joint:
- Arrow side of joint
- Arrow of welding symbol
- Other side of joint

T-Joint	PROCESS ABBREVIATIONS

T-Joint:
- Arrow of welding symbol
- Arrow side of joint
- Other side of joint

PROCESS ABBREVIATIONS:

Where process abbreviations are to be included in the tail of the welding symbol, reference is made to *Table A*, "Designation of Welding and Allied Processes by Letters," of *AWS 2.4-79, 71*.

AMERICAN WELDING SOCIETY, INC.
2501 N.W. 7th Street, Miami, Florida 33125

Table B-2. STANDARD WELDING ABBREVIATIONS

LETTER DESIGNATION	WELDING AND ALLIED PROCESSES	LETTER DESIGNATION	WELDING AND ALLIED PROCESSES
AAC	air carbon arc cutting	HPW	hot pressure welding
AAW	air acetylene welding	IB	induction brazing
ABD	adhesive bonding	INS	iron soldering
AB	arc brazing	IRB	infrared brazing
AC	arc cutting	IRS	infrared soldering
AHW	atomic hydrogen welding	IS	induction soldering
AOC	oxygen arc cutting	IW	induction welding
AW	arc welding	LBC	laser beam cutting
B	brazing	LBW	laser beam welding
BB	block brazing	LOC	oxygen lance cutting
BMAW	bare metal arc welding	MAC	metal arc cutting
CAC	carbon arc cutting	OAW	oxyacetylene welding
CAW	carbon arc welding	OC	oxygen cutting
CAW-G	gas carbon arc welding	OFC	oxyfuel gas cutting
CAW-S	shielded carbon arc welding	OFC-A	oxyacetylene cutting
CAW-T	twin carbon arc welding	OFC-H	oxyhydrogen cutting
CW	cold welding	OFC-N	oxynatural gas cutting
DB	dip brazing	OFC-P	oxypropane cutting
DFB	diffusion brazing	OFW	oxyfuel gas welding
DFW	diffusion welding	OHW	oxyhydrogen welding
DS	dip soldering	PAC	plasma arc cutting
EASP	electric arc spraying	PAW	plasma arc welding
EBC	electron beam cutting	PEW	percussion welding
EBW	electron beam welding	PGW	pressure gas welding
EGW	electrogas welding	POC	metal powder cutting
ESW	electroslag welding	PSP	plasma spraying
EXW	explosion welding	RB	resistance brazing
FB	furnace brazing	RPW	projection welding
FCAW	flux cored arc welding	RS	resistance soldering
FLB	flow brazing	RSEW	resistance seam welding
FLOW	flow welding	RSW	resistance spot welding
FLSP	flame spraying	ROW	roll welding
FOC	chemical flux cutting	RW	resistance welding
FOW	forge welding	S	soldering
FRW	friction welding	SAW	submerged arc welding
FS	furnace soldering	SAW-S	series submerged arc welding
FW	flash welding	SMAC	shielded metal arc cutting
GMAC	gas metal arc cutting	SMAW	shielded metal arc welding
GMAW	gas metal arc welding	SSW	solid state welding
GMAW-P	gas metal arc welding— pulsed arc	SW	stud arc welding
		TB	torch brazing
GMAW-S	gas metal arc welding— short circuiting arc	TC	thermal cutting
		TCAB	twin carbon arc brazing
GTAC	gas tungsten arc cutting	THSP	thermal spraying
GTAW	gas tungsten arc welding	TS	torch soldering
GTAW-P	gas tungsten arc welding— pulsed arc	TW	thermit welding
		USW	ultrasonic welding
HFRW	high frequency resistance welding	UW	upset welding
		WS	wave soldering

Table B-3. SUFFIXES FOR OPTIONAL USE IN APPLYING WELDING AND ALLIED PROCESSES

Automatic . . . . AU
Machine . . . . ME
Manual . . . . MA
Semiautomatic . . . . SA

Table B-4. OBSOLETE WELDING SYMBOLS AND ABBREVIATIONS

TYPE OF WELD

PLUG OR SPOT	ARC SPOT OR ARC SEAM	RESISTANCE SPOT	PROJECTION	RESISTANCE SEAM	FLASH OR UPSET	FIELD
⬭	⬬	✳	✕	XXX	|	᚛

ABBREVIATIONS

Nonpressure thermit welding . . . . NTW
Pressure thermit welding . . . . PTW

Solid state welding
Die welding . . . . DW
Hammer welding . . . . HW

Table B-5. GEOMETRIC CHARACTERISTIC SYMBOLS

APPLICATION	TYPE OF TOLERANCE	CHARACTERISTIC	SYMBOL
FOR INDIVIDUAL FEATURES	FORM	STRAIGHTNESS	—
		FLATNESS	▱
		CIRCULARITY (ROUNDNESS)	○
		CYLINDRICITY	⌀
FOR INDIVIDUAL OR RELATED FEATURES	PROFILE	PROFILE OF A LINE	⌒
		PROFILE OF A SURFACE	⌓
FOR RELATED FEATURES	ORIENTATION	ANGULARITY	∠
		PERPENDICULARITY	⊥
		PARALLELISM	//
	LOCATION	POSITION	⊕
		CONCENTRICITY	◎
	RUNOUT	CIRCULAR RUNOUT	↗
		TOTAL RUNOUT	↗↗

C

RULES AND FORMULAS

Table C-1. TAPER FORMULAS

TPI = TAPER PER INCH
TPF = TAPER PER FOOT
L = LENGTH OF TAPER
D = LARGE DIAMETER
d = SMALL DIAMETER

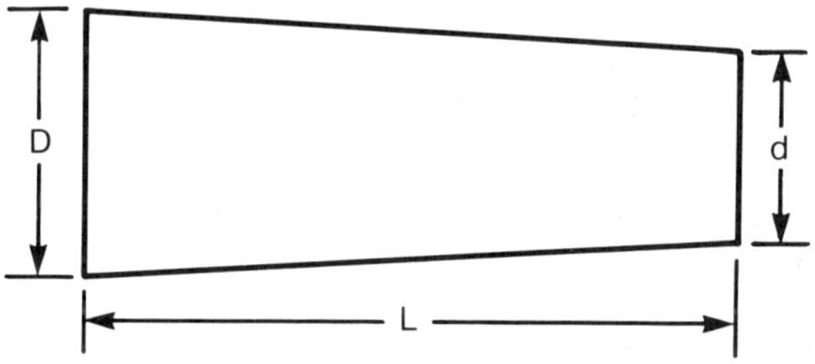

KNOWN	TO FIND	RULE
TPI	TPF	$TPF = TPI \times 12$
TPF	TPI	$TPI = \dfrac{TPF}{12}$
TPF	AMOUNT OF TAPER IN GIVEN LENGTH	$AMOUNT\ OF\ TAPER = \dfrac{TPF}{12} \times GIVEN\ LENGTH$
L, D, d	TPI	$TPI = \dfrac{D - d}{L}$
L, D, d	TPF	$TPF = \dfrac{D - d}{L} \times 12$
L, D, TPI	d	$d = D - (L \times TPI)$
L, d, TPI	D	$D = d + (L \times TPI)$
D, d, TPI	L	$L = \dfrac{D - d}{TPI}$

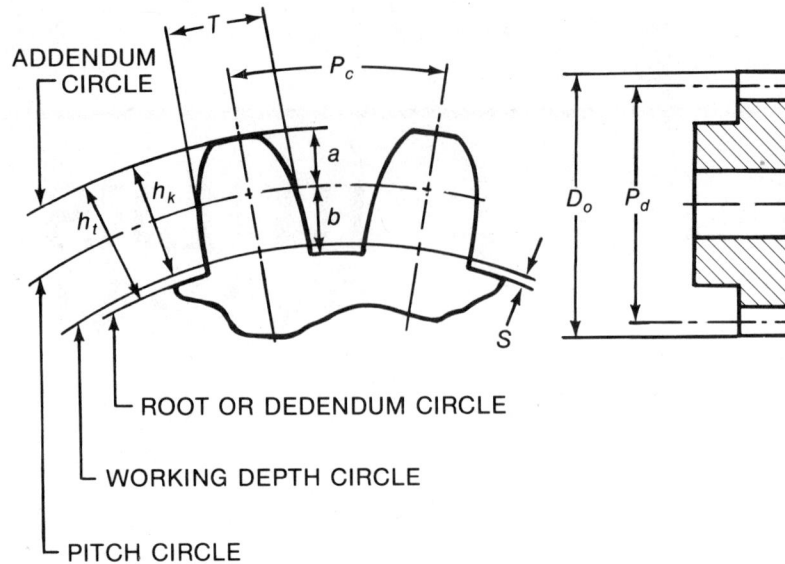

The following symbols are used in conjunction with the formulas for determining the proportions of spur gear teeth:

$$P = \text{Diametral pitch}$$
$$P_c = \text{Circular pitch}$$
$$P_d = \text{Pitch diameter}$$
$$D_o = \text{Outside diameter}$$
$$N = \text{Number of teeth in the gear}$$
$$T = \text{Tooth thickness}$$
$$a = \text{Addendum}$$
$$b = \text{Dedendum}$$
$$h_k = \text{Working depth}$$
$$h_t = \text{Whole depth}$$
$$S = \text{Clearance}$$
$$C = \text{Center distance}$$
$$L = \text{Length of rack}$$

Table C-2 (Continued).

TO FIND	RULE	FORMULA
Diametral pitch, P	Divide 3.1416 by the circular pitch.	$P = \dfrac{3.1416}{P_c}$
Circular pitch, P_c	Divide 3.1416 by the diametral pitch.	$P_c = \dfrac{3.1416}{P}$
Pitch diameter, P_d	Divide the number of teeth by the diametral pitch.	$P_d = \dfrac{N}{P}$
Outside diameter, D_o	Add 2 to the number of teeth and divide the sum by the diametral pitch.	$D_o = \dfrac{N + 2}{P}$
Number of teeth, N	Multiply the pitch diameter by the diametral pitch.	$N = P_d P$
Tooth circular thickness, T	Divide 1.5708 by the diametral pitch.	$T = \dfrac{1.5708}{P}$
Addendum, a	Divide 1.0 by the diametral pitch.	$a = \dfrac{1.0}{P}$
Dedendum, b	Divide 1.157 by the diametral pitch.	$b = \dfrac{1.157}{P}$
Working depth, h_k	Divide 2 by the diametral pitch.	$h_k = \dfrac{2}{P}$
Whole depth, h_t	Divide 2.157 by the diametral pitch.	$h_t = \dfrac{2.157}{P}$
Clearance, S	Divide .157 by the diametral pitch.	$S = \dfrac{.157}{P}$
Center distance, C	Add the number of teeth in both gears and divide the sum by two times the diametral pitch.	$C = \dfrac{N_1 + N_2}{2P}$
Length of rack, L	Multiply the number of teeth in the rack by the circular pitch.	$L = N P_c$

Table C-3. RULES AND FORMULAS FOR HELICAL GEAR CALCULATIONS

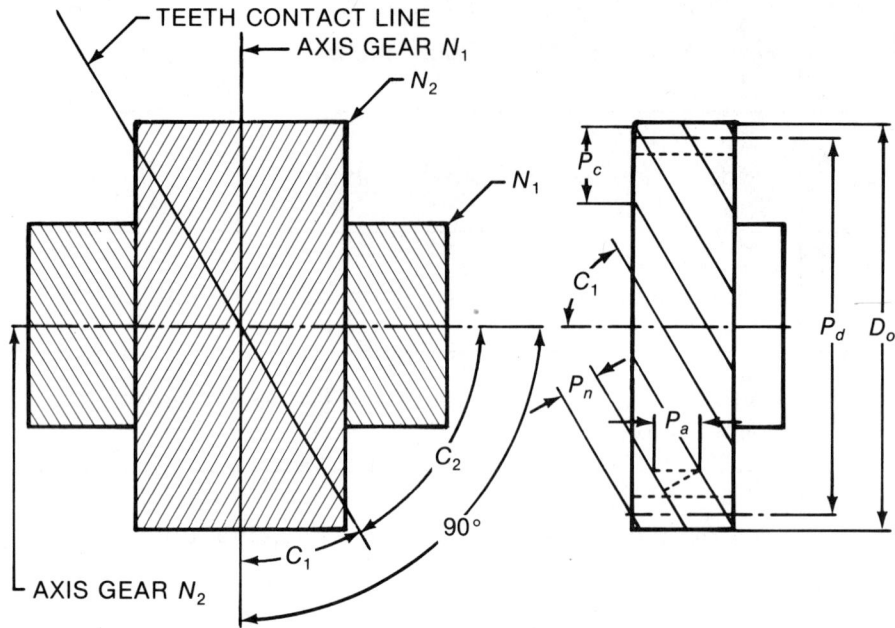

The following symbols are used in conjunction with the formulas for determining the proportions of helical gear teeth:

$$P_{nd} = \text{Normal diametral pitch (pitch of cutter)}$$
$$P_c = \text{Circular pitch}$$
$$P_a = \text{Axial pitch}$$
$$P_n = \text{Normal pitch}$$
$$P_d = \text{Pitch diameter}$$
$$S = \text{Center distance}$$
$$C, C_1, C_2 = \text{Helix angle of the gears}$$
$$L = \text{Lead of tooth helix}$$
$$T_n = \text{Normal tooth thickness at pitch line}$$
$$a = \text{Addendum}$$
$$h_t = \text{Whole depth of tooth}$$
$$N, N_1, N_2 = \text{Number of teeth in the gears}$$
$$D_o = \text{Outside diameter}$$
$$N_o = \text{Hypothetical number of teeth for which the gear cutter should be selected}$$

Table C-3 (Continued).

TO FIND	RULE	FORMULA
Normal diametral pitch, P_{nd}	Divide the number of teeth by the product of the pitch diameter and the cosine of the helix angle.	$P_{nd} = \dfrac{N}{P_d \cos C_1}$
Circular pitch, P_c	Multiply the pitch diameter of the gear by 3.1416, and divide the product by the number of teeth in the gear.	$P_c = \dfrac{3.1416 P_d}{N}$
Axial pitch, P_a	Multiply the circular pitch by the cotangent of the helix angle.	$P_a = P_c \cot C_1$
Normal pitch, P_n	Divide 3.1416 by the normal diametral pitch.	$P_n = \dfrac{3.1416}{P_{nd}}$
Pitch diameter, P_d	Divide the number of teeth by the product of the normal pitch and the cosine of the helix angle.	$P_d = \dfrac{N}{P_{nd} \cos C_1}$
Center distance, S	Divide the sum of the pitch diameters of the mating gears by 2.	$S = \dfrac{P_{d_1} + P_{d_2}}{2}$
Checking formulas (shafts at right angles)	Multiply the number of teeth in the first gear by the tangent of the tooth angle of that gear, and add the number of teeth in the second gear to the product. The sum should equal twice the product of the center distance multiplied by the normal diametral pitch, multiplied by the sine of the helix angle.	$N_1 + (N_2 \tan C_2) = 2SP_{nd} \sin C_1$
Lead of tooth helix, L	Multiply the pitch diameter by 3.1416 times the cotangent of the helix angle.	$L = 3.1416 P_d \cot C_1$
Normal circular tooth thickness at pitch line, T_n	Divide 1.571 by the normal diametral pitch.	$T_n = \dfrac{1.571}{P_{nd}}$
Addendum, a	Divide the normal pitch by 3.1416.	$a = \dfrac{P_n}{3.1416}$
Whole depth of tooth, h_t	Divide 2.157 by the normal diametral pitch.	$h_t = \dfrac{2.157}{P_{nd}}$
Outside diameter, D_o	Add twice the addendum to the pitch diameter.	$D_o = P_d + 2a$
Hypothetical number of teeth for which gear cutter should be selected, N_c	Divide the number of teeth in the gear by the cube of the cosine of the helix angle.	$N_c = \dfrac{N_1}{(\cos C_1)^3}$

Table C-4. RULES AND FORMULAS FOR BEVEL GEAR CALCULATIONS (Shafts at Right Angles)

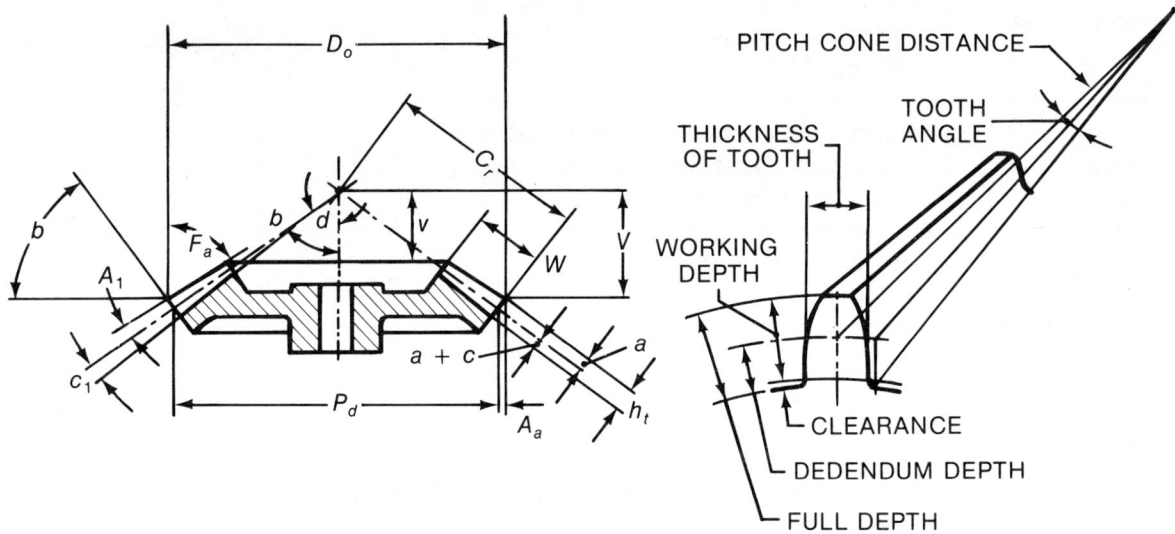

The following symbols are used in conjunction with the formulas for determining the proportions of bevel gear teeth:

P = Diametral pitch
P_c = Circular pitch
P_d = Pitch diameter
b = Pitch angle
C_r = Pitch cone distance
a = Addendum
A_1 = Addendum angle
A_a = Angular addendum
D_o = Outside diameter
c_1 = Dedendum angle
$a + c$ = Addendum plus clearance
a_s = Addendum of small end of tooth
T_L = Thickness of tooth at pitch line
T_s = Thickness of tooth at pitch line at small end of gear
F_a = Face angle
h_t = Whole depth of tooth space
V = Apex distance at large end of tooth
v = Apex distance at small end of tooth
m_g = Gear ratio
N = Number of teeth
N_g = Number of teeth in gear
N_p = Number of teeth in pinion
d = Root angle
W = Width of gear tooth face
N_c = Number of teeth of imaginary spur gear for which cutter is selected

Table C-4 (Continued).

TO FIND	RULE	FORMULA
Diametral pitch, P	Divide the number of teeth by the pitch diameter.	$P = \dfrac{N}{P_d}$
Circular pitch, P_c	Divide 3.1416 by the diametral pitch.	$P_c = \dfrac{3.1416}{P}$
Pitch diameter, P_d	Divide the number of teeth by the diametral pitch.	$P_d = \dfrac{N}{P}$
Pitch angle of pinion, $\tan b_p$	Divide the number of teeth in the pinion by the number of teeth in the gear to obtain the tangent.	$\tan b_p = \dfrac{N_p}{N_g}$
Pitch angle of gear, $\tan b_g$	Divide the number of teeth in the gear by the number of teeth in the pinion to obtain the tangent.	$\tan b_g = \dfrac{N_g}{N_p}$
Pitch cone distance, C_r	Divide the pitch diameter by twice the sine of the pitch angle.	$C_r = \dfrac{P_d}{2(\sin b)}$
Addendum, a	Divide 1.0 by the diametral pitch.	$a = \dfrac{1.0}{P}$
Addendum angle, $\tan A_1$	Divide the addendum by the pitch cone distance to obtain the tangent.	$\tan A_1 = \dfrac{a}{C_r}$
Angular addendum, A_a	Multiply the addendum by the cosine of the pitch angle.	$A_a = a \cos b$
Outside diameter, D_o	Add twice the angular addendum to the pitch diameter.	$D_o = P_d + 2A_a$
Dedendum angle, $\tan c_1$	Divide the dedendum by the pitch cone distance to obtain the tangent.	$\tan c_1 = \dfrac{a + c}{C_r}$
Addendum of small end of tooth, a_s	Subtract the width of face from the pitch cone distance, divide the remainder by the pitch cone distance and multiply by the addendum.	$a_s = a\left(\dfrac{C_r - W}{C_r}\right)$
Thickness of tooth at pitch line, T_L	Divide the circular pitch by 2.	$T_L = \dfrac{P_c}{2}$
Thickness of tooth at pitch line at small end of gear, T_s	Subtract the width of face from the pitch cone distance, divide the remainder by the pitch cone distance, and multiply by the thickness of the tooth at the pitch line.	$T_s = T_L\left(\dfrac{C_r - W}{C_r}\right)$
Face angle, F_a	Face cone of blank turned parallel to root cone of mating gear.	$F_a = b + c_1$
Whole depth of tooth space, h_t	Divide 2.157 by the diametral pitch.	$h_t = \dfrac{2.157}{P}$
Apex distance at large end of tooth, V	Multiply one-half the outside diameter by the tangent of the face angle.	$V = \left(\dfrac{D_o}{2}\right)\tan F_a$
Apex distance at small end of tooth, v	Subtract the width of face from the pitch cone distance, divide the remainder by the pitch cone distance, and multiply by the apex distance.	$v = V\left(\dfrac{C_r - W}{C_r}\right)$
Gear ratio, m_g	Divide the number of teeth in the gear by the number of teeth in the pinion.	$m_g = \dfrac{N_g}{N_p}$
Number of teeth in gear and/or pinion, N_g, N_p	Multiply the pitch diameter by the diametral pitch.	$N_g = P_d P$ $N_p = P_d P$
Cutting angle, d	Subtract the addendum plus clearance angle from the pitch angle.	$d = b - c_1$
Number of teeth of imaginary spur gear for which cutter is selected, N_o	Divide the number of teeth in actual gear by the cosine of the pitch angle.	$N_c = \dfrac{N}{\cos b}$

Table C-5. RULES AND FORMULAS FOR WORM WHEEL CALCULATIONS
(Single and Double Thread—14½° Pressure Angle)

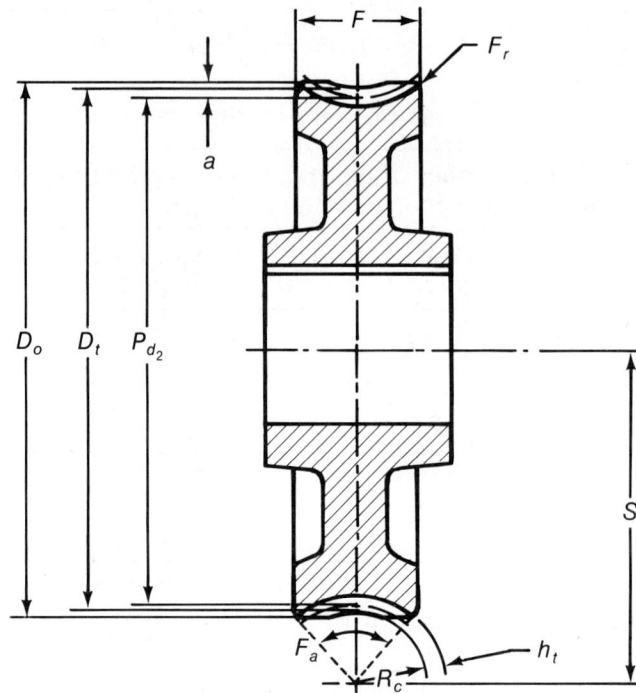

The following symbols are used in conjunction with the formulas for determining the proportions of worm wheel teeth:

P_c = Circular pitch
P_{d_2} = Pitch diameter
N = Number of teeth
D_o = Outside diameter
D_t = Throat diameter
R_c = Radius of curvature of worm wheel throat
D = Diameter to sharp corners
F_a = Face angle
F = Face width of rim
F_r = Radius at edge of face
a = Addendum
h_t = Whole depth of tooth
S = Center distance between worm and worm wheel
G = Gashing angle

Table C-5 (Continued).

TO FIND	RULE	FORMULA
Circular pitch, P_c	Divide the pitch diameter by the product of .3183 and the number of teeth.	$P_c = \dfrac{P_{d_2}}{.3183N}$
Pitch diameter, P_{d_2}	Multiply the number of teeth in the worm wheel by the linear pitch of the worm, and divide the product by 3.1416.	$P_{d_2} = \dfrac{NP_L}{3.1416}$
Outside diameter, D_o	Multiply the circular pitch by .4775 and add the product to the throat diameter.	$D_o = D_t + .4775P_c$
Throat diameter, D_t	Add twice the addendum of the worm tooth to the pitch diameter of the worm wheel.	$D_t = P_{d_2} + 2a$
Radius of curvature of worm wheel throat, R_c	Subtract twice the addendum of the worm tooth from half the outside diameter of the worm.	$R_c = \dfrac{D_o}{2} - 2a$
Diameter to sharp corners, D	Multiply the radius of curvature of the worm wheel throat by the cosine of half the face angle, subtract this quantity from the radius of curvature. Multiply the remainder by 2, and add the product to the throat diameter of the worm wheel.	$D = 2\left(R_c - R_c \cos \dfrac{F_a}{2}\right) + D_t$
Face width of rim, F	Multiply the circular pitch by 2.38 and add .25 to the product.	$F = 2.38P_c + .25$
Radius at edge of face, F_r	Divide the circular pitch by 4.	$F_r = \dfrac{P_c}{4}$
Addendum, a	Multiply the circular pitch by .3183.	$a = .3183P_c$
Whole depth of tooth, h_t	Multiply the circular pitch by .6866.	$h_t = .6866P_c$
Center distance between worm and worm wheel, S	Add the pitch diameter of the worm to the pitch diameter of the worm wheel and divide the sum by 2.	$S = \dfrac{P_{d_1} + P_{d_2}}{2}$
Gashing angle, G	Divide the lead of the worm by the circumference of the pitch circle. The result will be the cotangent of the gashing angle.	$\cot G = \dfrac{L}{3.1416d}$

Table C-6. RULES AND FORMULAS FOR WORM WHEEL CALCULATIONS (Solid Type)
(Single and Double Thread—14½° Pressure Angle)

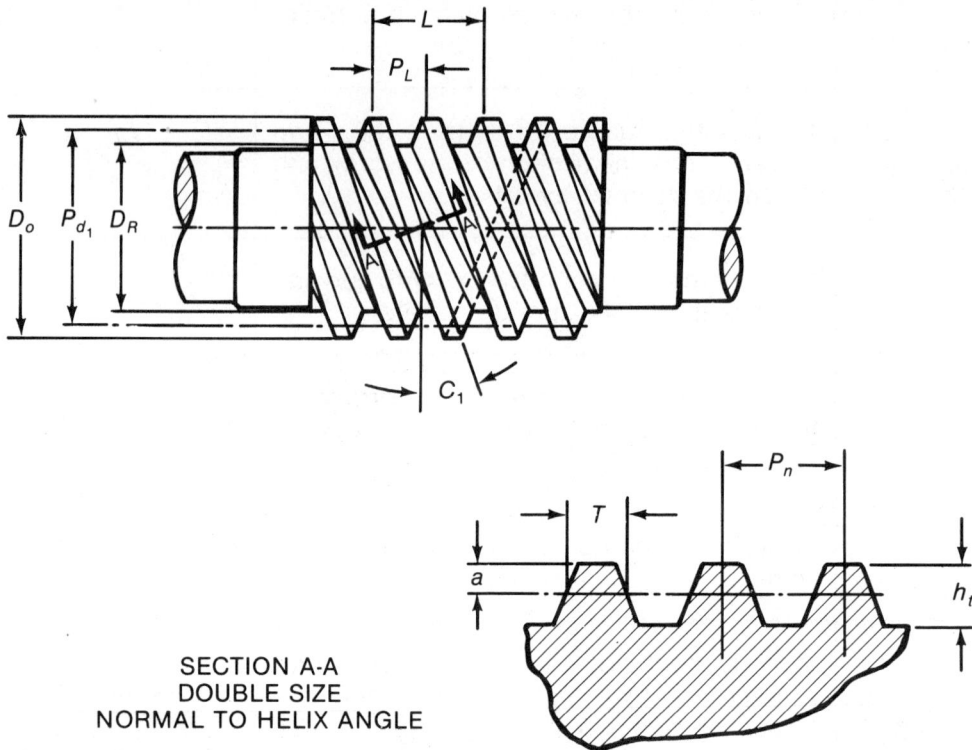

SECTION A-A
DOUBLE SIZE
NORMAL TO HELIX ANGLE

The following symbols are used in conjunction with the formulas for determining the proportions of worm wheel teeth:

P_L = Linear pitch
P_{d_1} = Pitch diameter
D_o = Outside diameter
N_w = Number of threads
D_R = Root diameter
h_t = Whole depth of tooth
C_1 = Helix angle
P_n = Normal pitch
a = Addendum
L = Lead
T = Normal tooth thickness
t = Width of thread tool at end

Table C-6 (Continued).

TO FIND	RULE	FORMULA
Linear pitch, P_L	Divide the lead by the number of threads in the whole worm; i.e., one if single threaded or four if quadruple threaded.	$P_L = \dfrac{L}{N_w}$
Pitch diameter, P_{d_1}	Subtract twice the addendum from the outside diameter.	$P_{d_1} = D_o - 2a$
Outside diameter, D_o	Add twice the addendum of the worm to the pitch diameter of the worm wheel.	$D_o = P_{d_1} + 2a$
Root diameter, D_R	Subtract twice the whole depth of the tooth from the outside diameter.	$D_R = D_o - 2h_t$
Whole depth of tooth, h_t	Multiply the linear pitch by .6866.	$h_t = .6866P_L$
Helix angle, C_1	Multiply the pitch diameter of the worm by 3.1416, and divide the product by the lead. The quotient is the cotangent of the helix angle.	$\cot C_1 = \dfrac{3.1416P_{d_2}}{L}$
Normal pitch, P_n	Multiply the linear pitch by the cosine of the helix angle of the worm.	$P_n = P_L \cos C_1$
Addendum, a	Multiply the linear pitch by .3183.	$a = .3183P_L$
Lead, L	Multiply the linear pitch by the number of threads.	$L = P_L N_w$
Normal tooth thickness, T	Multiply one-half the linear pitch by the cosine of the helix angle.	$T = \dfrac{P_L}{2} \cos C_1$
Width of thread tool at end, t	Multiply the linear pitch by .31.	$t = .31P_L$

D

COMPETENCY EXERCISES

COMPETENCY EXERCISE #1

Refer to the print on page 337 to answer the following questions.

QUESTIONS	ANSWERS
1. What is the part number?	**1.** _____
2. What is the scale of the print?	**2.** _____
3. Identify lines *A* through *E*.	**3.** *A* _____
	B _____
	C _____
	D _____
	E _____
4. Find dimensions *F* and *G*.	**4.** *F* _____
	G _____
5. What is the overall length of the part?	**5.** _____
6. How many parts are required?	**6.** _____
7. Why does line *H* end with a dot?	**7.** _____
	_____
8. What is the part name?	**8.** _____
9. How many radii are specified?	**9.** _____
10. How thick is the material specified?	**10.** _____
11. What material is specified for the part?	**11.** _____
12. Which views are shown in this print?	**12.** _____
13. What is the tolerance on the 45° angle?	**13.** _____
14. What is the tolerance on all of the other dimensions?	**14.** _____
15. Where is the general information about the part located?	**15.** _____

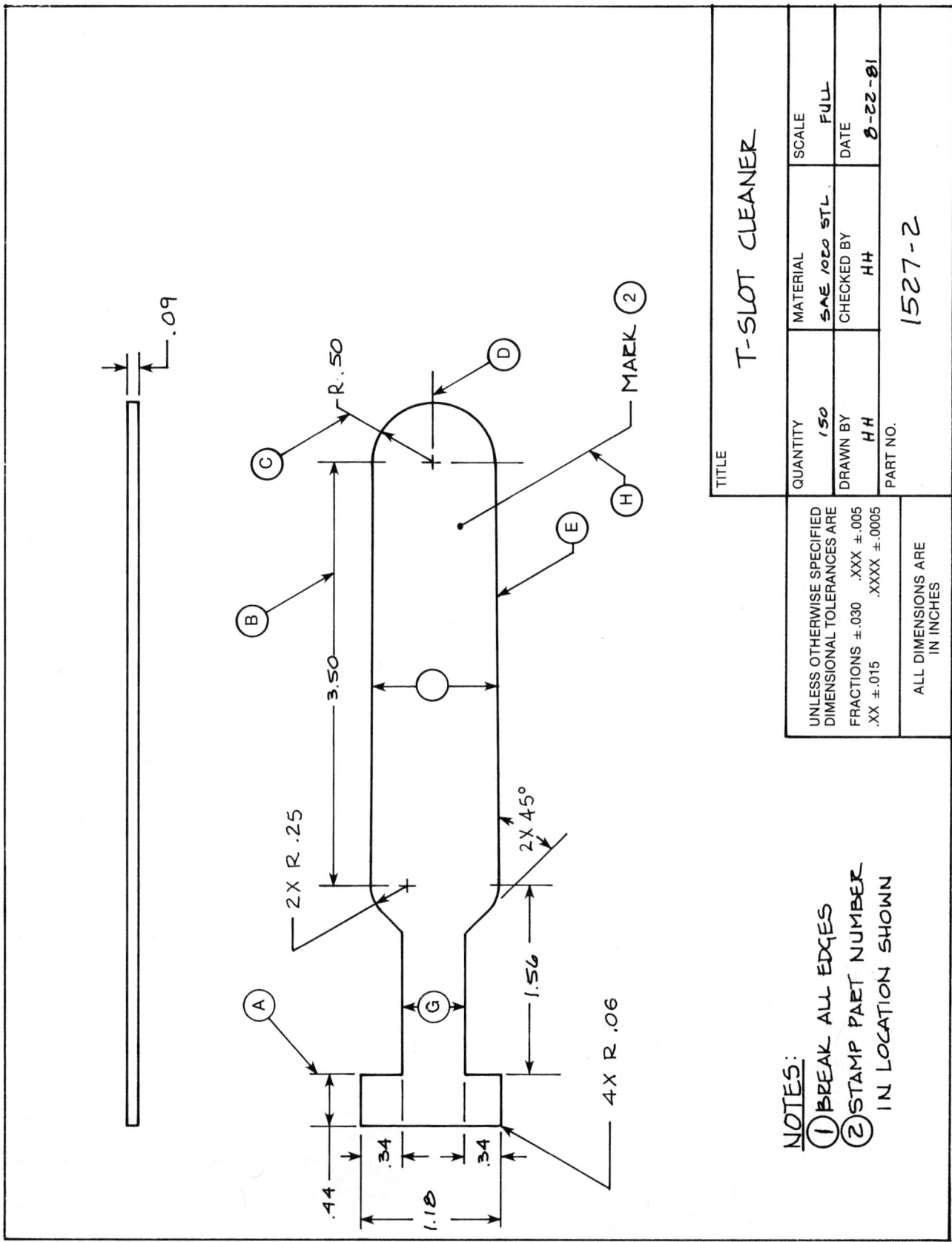

NOTES:
1. BREAK ALL EDGES
2. STAMP PART NUMBER IN LOCATION SHOWN

TITLE		
T-SLOT CLEANER		
QUANTITY	MATERIAL	SCALE
150	SAE 1020 STL.	FULL
DRAWN BY	CHECKED BY	DATE
HH	HH	8-22-81
PART NO.		
1527-2		

UNLESS OTHERWISE SPECIFIED
DIMENSIONAL TOLERANCES ARE

FRACTIONS ±.030   .XXX ±.005
.XX ±.015   .XXXX ±.0005

ALL DIMENSIONS ARE
IN INCHES

COMPETENCY EXERCISE #2

Refer to the print on page 339 to answer the following questions.

QUESTIONS	ANSWERS
1. What is the part name?	1. _____
2. What type of line is used to show the axis of the part?	2. _____
3. What type of line is used to show the visible edges?	3. _____
4. What part of the print is used to record the general information?	4. _____
5. How many parts are required?	5. _____
6. What are the part numbers of the following sizes?	
a. $\frac{1}{4}$	6. a. _____
b. $\frac{3}{8}$	b. _____
c. $\frac{7}{16}$	c. _____
7. What size are the chamfers on each part?	7. _____
8. How many chamfers are on each part?	8. _____
9. What type of knurl is specified?	9. _____
10. What are the overall lengths of the following sizes?	
a. $\frac{1}{8}$	10. a. _____
b. $\frac{1}{4}$	b. _____
c. $\frac{5}{16}$	c. _____
d. $\frac{7}{16}$	d. _____
11. What is the scale of the print?	11. _____
12. What material is specified for the parts?	12. _____
13. What is the length of the knurled area?	13. _____
14. What view is shown in the print?	14. _____
15. Identify lines (1) through (5).	15. (1) _____
	(2) _____
	(3) _____
	(4) _____
	(5) _____

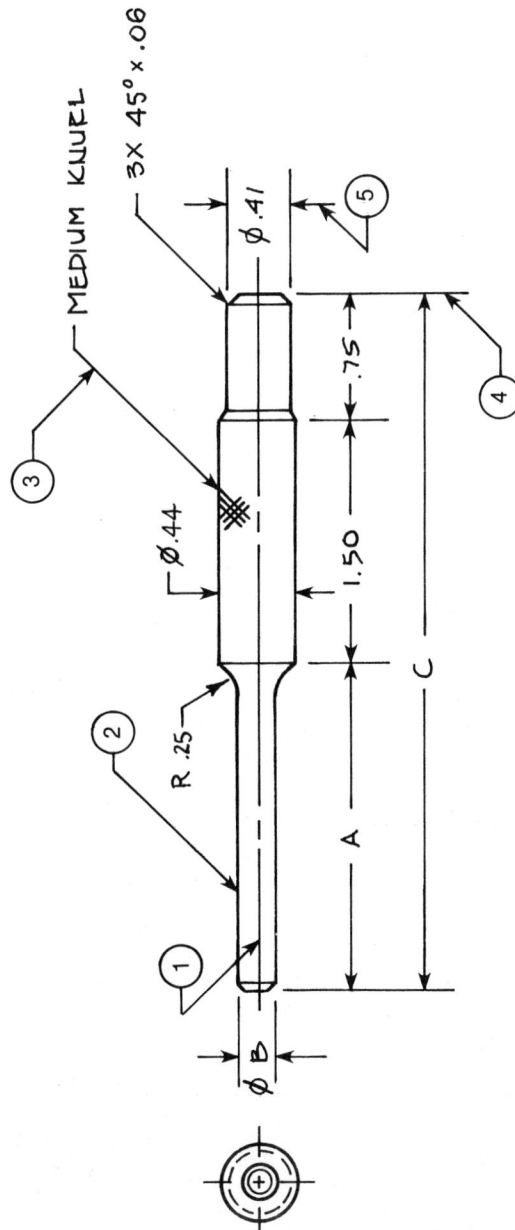

MEDIUM KNURL

3X 45° x .06

∅.41

∅.44

∅ B

R .25

1.50

.75

A

C

① ② ③ ④ ⑤

P/N	SIZE	A	B	C
–1	$\frac{1}{8}$	1.50	.120	3.75
–2	$\frac{3}{16}$	1.50	.180	3.75
–3	$\frac{1}{4}$	2.00	.240	4.25
–4	$\frac{5}{16}$	2.00	.300	4.26
–5	$\frac{3}{8}$	2.50	.365	4.75
–6	$\frac{7}{16}$	2.50	.420	4.75

TITLE

DRIVE PUNCHES

QUANTITY	MATERIAL	SCALE
10 EA.	SAE 1095 STL	FULL
DRAWN BY	CHECKED BY	DATE
TC	JB	12-16-85

PART NO.

1014-LB

COMPETENCY EXERCISE #3

Refer to the print on page 341 to answer the following questions.

QUESTIONS	ANSWERS

1. What is the part name?

1. _____

2. What material is specified?

2. _____

3. What type of line is used to show the holes in the front view?

3. _____

4. What type of line is used to show the holes in the top view?

4. _____

5. What are the overall dimensions?

5. Height _____

Width _____

Depth _____

6. How many parts are required?

6. _____

7. What size are the chamfers?

7. _____

8. How many chamfers are on each part?

8. _____

9. What is the purpose of the milled pocket?

9. _____

10. How far apart are the two rows of holes?

10. _____

11. What size are the holes?

11. _____

12. What is the scale of the drawing?

12. _____

13. What views are shown?

13. _____

14. What is the part number?

14. _____

15. Identify lines *A* and *B*.

15. *A* _____

B _____

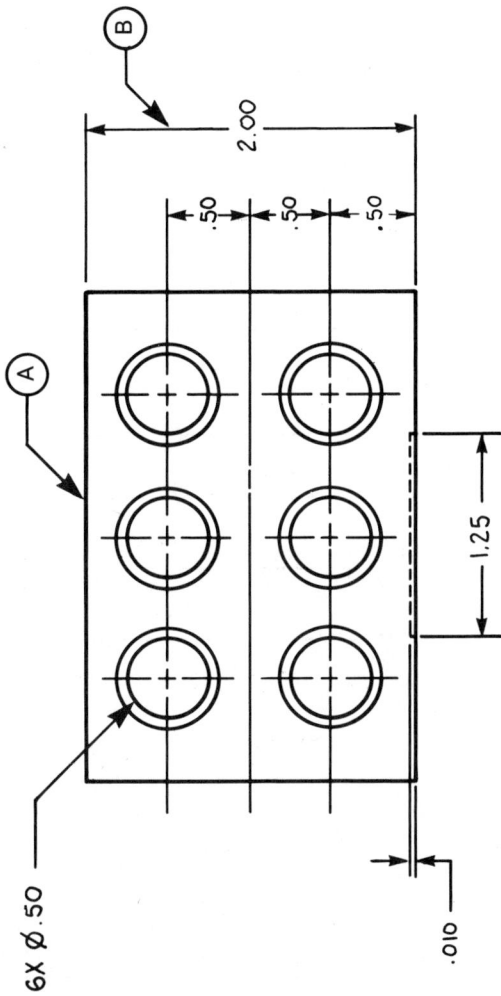

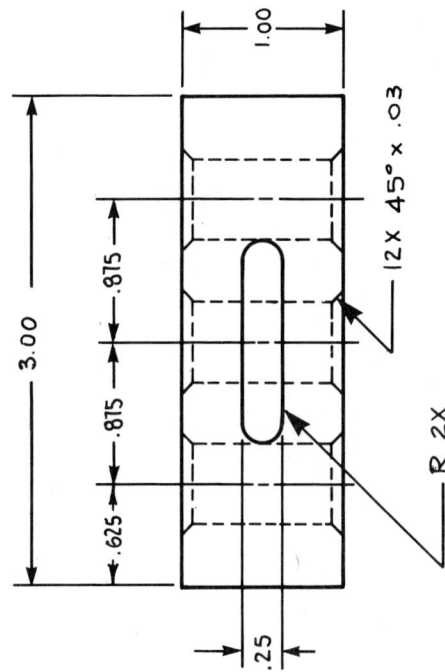

6X Ø .50

A

B

2.00

.50 .50 .50

1.25

.010

3.00

.875

.875

.625

1.00

12X 45° X .03

R 2X

.25

TITLE		
HOLE GUIDE	MATERIAL SAE 1020 STL	SCALE FULL
	CHECKED BY ANN ROSS	DATE 12-1-85
QUANTITY 5		
DRAWN BY JOHN W.		
PART NO. 4-11239-L		

COMPETENCY EXERCISE #4

Refer to the print on page 343 to answer the following questions.

QUESTIONS	ANSWERS
1. What is the name of the part?	1. _____
2. What material is specified in the print?	2. _____
3. How many tapped holes are in the part?	3. _____
4. How deep is the milled pocket?	4. _____
5. What does line *C* mean?	5. _____
	_____
6. What is the diameter of the bar?	6. _____
7. What is the part number?	7. _____
8. What is the scale of the drawing?	8. _____
9. What views are shown?	9. _____
	_____
10. How does the view arrangement differ from the standard three-view drawing?	10. _____
	_____
11. What thread size is specified for the tapped holes?	11. _____
12. How many parts are required?	12. _____
13. What size of tool bits are to be used with this tool?	13. _____
14. What angle is specified for the end of the bar?	14. _____
15. Identify lines *A* through *F*.	15. *A* _____
	B _____
	C _____
	D _____
	E _____
	F _____

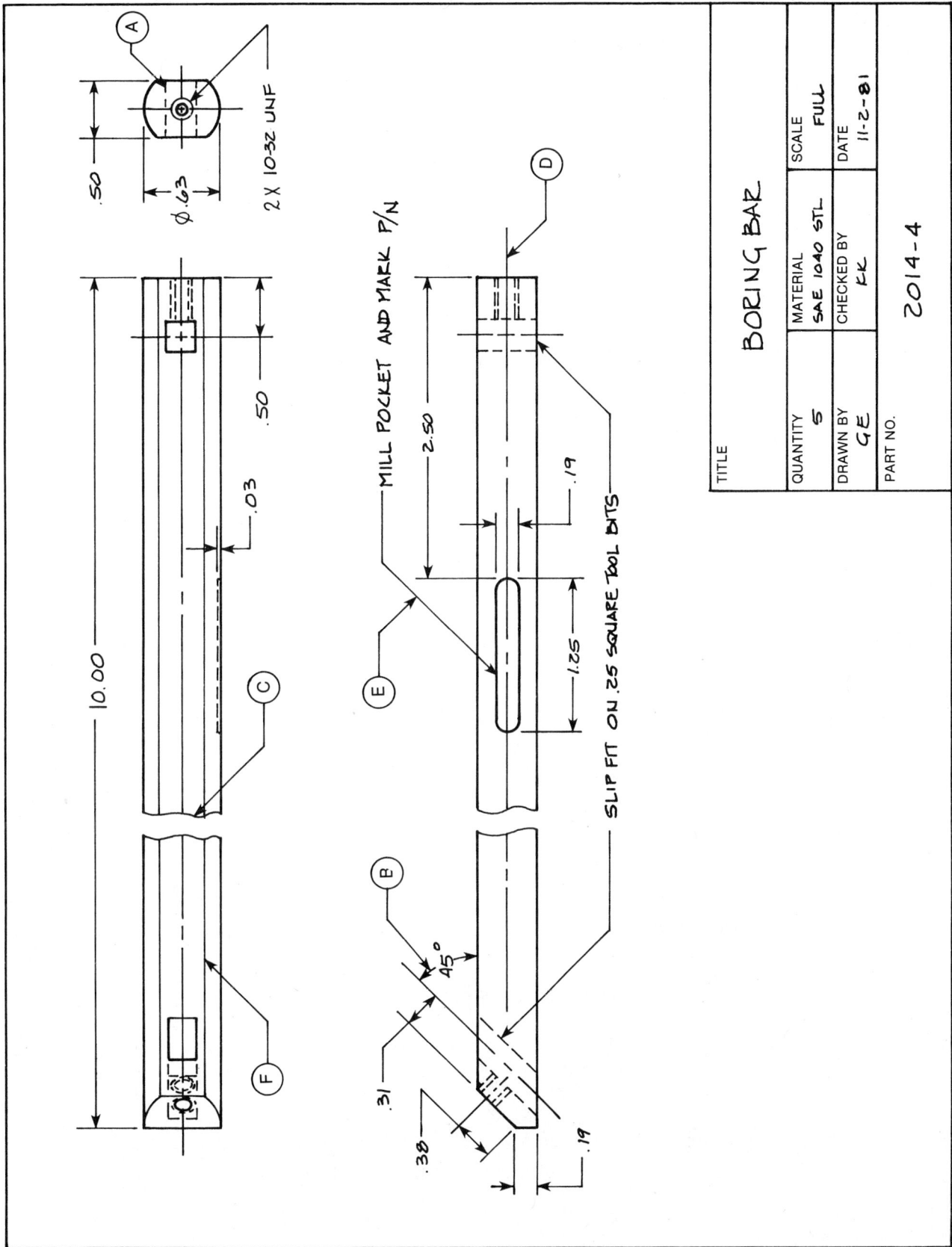

A

.50

Ø .63

2 X 10-32 UNF

.03

.50

10.00

C

F

MILL POCKET AND MARK P/N

E

2.50

.19

.125

SLIP FIT ON .25 SQUARE TOOL BITS

D

B

45°

.31

.38

.19

TITLE BORING BAR

QUANTITY 5	MATERIAL SAE 1040 STL	SCALE FULL
DRAWN BY GE	CHECKED BY KK	DATE 11-2-81
PART NO. 2014-4		

COMPETENCY EXERCISE #5

Refer to the print on page 345 to answer the following questions.

QUESTIONS	ANSWERS
1. What is the name of the part?	1. _____
2. What is the part number?	2. _____
3. What is the overall length of the part?	3. _____
4. What is the largest diameter of the part?	4. _____
5. How many tapped holes are called for in the part?	5. _____
6. What size(s) are the tapped holes?	6. _____
7. What does the dimension in parentheses mean?	7. _____
8. What material is specified?	8. _____
9. What is the angle of the slot?	9. _____
10. How wide is the slot?	10. _____
11. How deep is the slot?	11. _____
12. What is the scale of the print?	12. _____
13. What is the size of the chamfer on the part?	13. _____
14. What views are shown in this print?	14. _____
15. What is the small diameter of the part?	15. _____

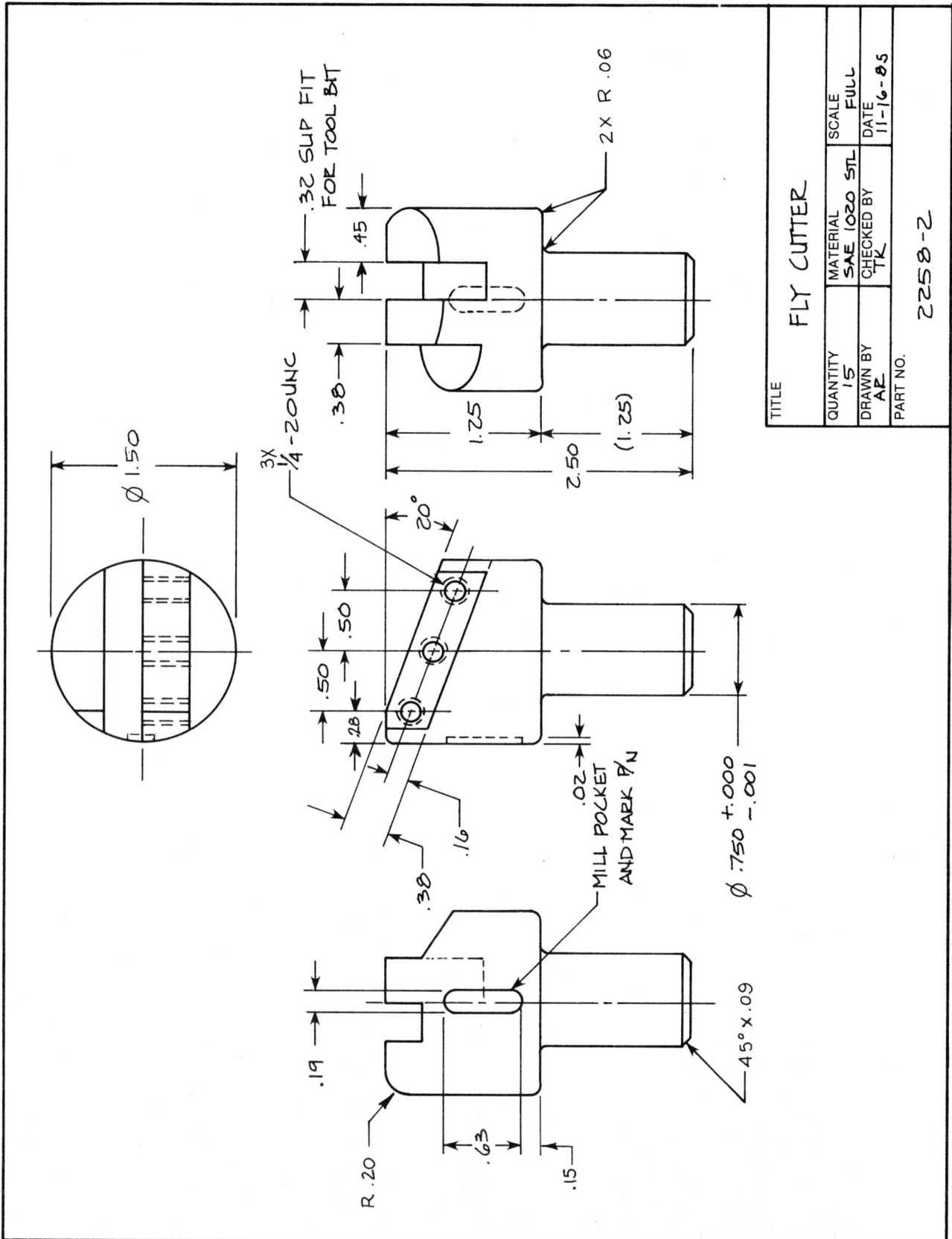

FLY CUTTER

TITLE	FLY CUTTER		
	MATERIAL SAE 1020 STL	SCALE FULL	
QUANTITY 15	CHECKED BY TK	DATE 11-16-85	
DRAWN BY AR			
PART NO.		2258-2	

.32 SLIP FIT FOR TOOL BIT

2 X R .06

.45

.38

1.75

(1.75)

2.50

∅ 1.50

3X ¼-20UNC

.20°

.50

.50

.50

.28

.16

.38

MILL POCKET AND MARK P/N

.20

∅ .750 +.000 -.001

R .20

.19

.63

.15

45° X .09

COMPETENCY EXERCISE #6

Refer to the print on page 347 to answer the following questions.

QUESTIONS	ANSWERS
1. What is the part name?	1. _____
2. How many parts are required?	2. _____
3. What type of line is shown at ①?	3. _____
4. What does line 1 mean?	4. _____ _____
5. What type of line is shown at ②?	5. _____
6. What does line 2 mean?	6. _____ _____
7. What views are used to describe this part?	7. _____ _____
8. What material is specified?	8. _____
9. What angle is specified on the part?	9. _____
10. How many radii are there on the part?	10. _____
11. Find the dimension required at ③.	11. _____
12. Find the dimension required at ④.	12. _____
13. What is the part number?	13. _____
14. What is the scale of the print?	14. _____
15. How thick is the part?	15. _____

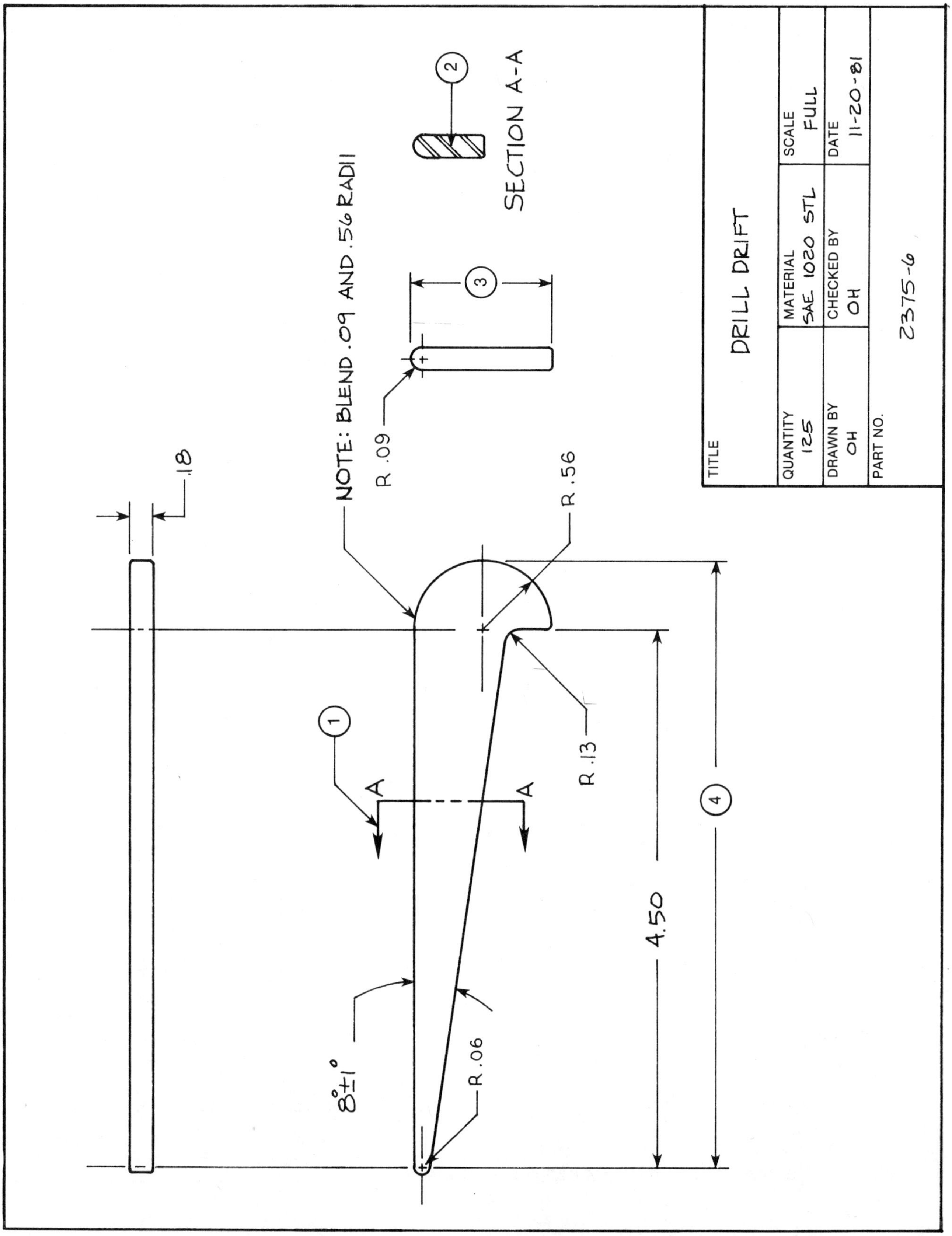
TITLE
DRILL DRIFT
QUANTITY
125
MATERIAL
SAE 1020 STL
SCALE
FULL
DRAWN BY
OH
CHECKED BY
OH
DATE
11-20-81
PART NO.
2375-6
NOTE: BLEND .09 AND .56 RADII
SECTION A-A
R .09
R .56
R .13
R .06
.18
4.50
8°±1°
A
A
1
2
3
4

COMPETENCY EXERCISE #7

Refer to the print on page 349 to answer the following questions.

QUESTIONS	ANSWERS
1. What is the part name?	1. _____
2. What views are shown in this print?	2. _____ _____
3. How much material is left on each side of the part for machining?	3. _____
4. How much material is left on the ends for machining?	4. _____
5. What type of line is shown at *A*?	5. _____
6. What does line *A* mean when used this way?	6. _____
7. What type of line is shown at *B*?	7. _____
8. Where are the rough stock sizes found?	8. _____
9. How many naval brass plates are required for each part?	9. _____
10. What is the part number?	10. _____
11. What materials are specified for the part?	11. _____
12. How many parts are required?	12. _____
13. How far apart are the hole centers?	13. _____
14. What is the size of the countersinks on the parts?	14. _____
15. What are the general tolerances for the dimensions shown on the part?	15. _____

Title Block / Material List

REVISIONS

REV	PT. NO.	DESCRIPTION	APVD	DATE
1	2	WAS .200	PW	1-2-85

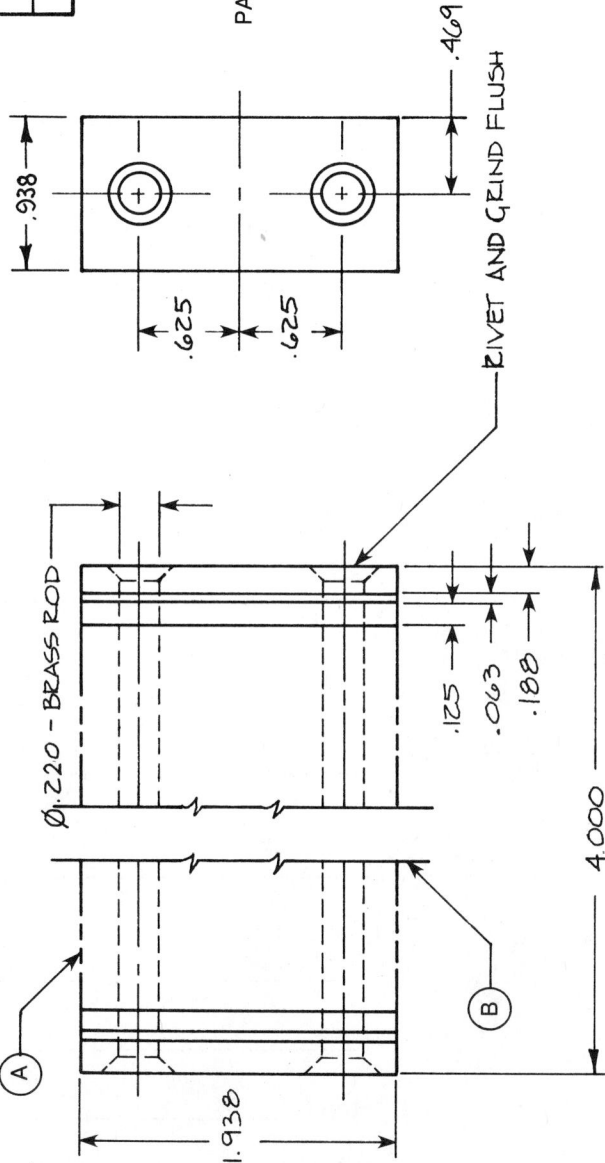

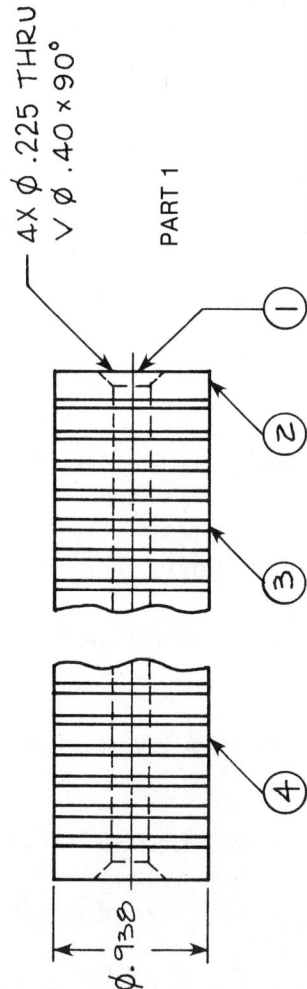

MATERIAL LIST

NO.	QTY.	PART	STOCK SIZE	MATL
4	38	PLATE	.125×1.000×2.000	SAE 1020 STL
3	40	SPACER	.063×1.000×2.000	NAVAL BRASS
2	4	PLATE	.188×1.000×2.000	SAE 1020 STL
1	4	ROD	Ø.220×4.188	NAVAL BRASS

TITLE: MAGNETIC PARALLELS

QUANTITY 2 EA.	MATERIAL NOTED	SCALE FULL
DRAWN BY CCS	CHECKED BY DAR	DATE 12-15-81

PART NO. 2673-9

UNLESS OTHERWISE SPECIFIED
DIMENSIONAL TOLERANCES ARE

FRACTIONS ±.030   .XXX ±.005
.XX ±.015   .XXXX ±.0005

ALL DIMENSIONS ARE IN
INCHES EXCEPT WHERE NOTED

PART 2

.938

.625   .625

.469

RIVET AND GRIND FLUSH

Ø.220 - BRASS ROD

.125
.063
.188

4.000

1.938

A

B

4X Ø.225 THRU
V Ø.40 × 90°

PART 1

Ø.938

NOTE: MACHINE AFTER ASSEMBLY

COMPETENCY EXERCISE #8

Refer to the print on page 351 to answer the following questions.

<table>
<tr><td>QUESTIONS</td><td>ANSWERS</td></tr>
</table>

QUESTIONS	ANSWERS
1. What is the part name?	1. _____
2. What is the part number?	2. _____
3. What does Ⓐ indicate?	3. _____
4. Find dimensions *B* through *E*.	4. *B* _____
	C _____
	D _____
	E _____
5. What is the size of the slot in the bottom of the V grooves? How many slots are there?	5. _____
	_____
6. What was the original depth of the V block?	6. _____
7. What type of thread designation is used to indicate the threads in details #2 and #3?	7. _____
8. What size thread is specified for details #2 and #3?	8. _____
9. What size threads are specified in detail #1?	9. _____
10. What is the tolerance on the overall size of the V block?	10. _____
11. What is the included angle of the V in detail #1?	11. _____
12. What are the limits of size for the included angle of the V in detail #1?	12. _____
13. What is the allowance between the mating areas of details #1 and #3?	13. _____
14. What size knurl is specified for detail #2?	14. _____
15. What is the rough stock size of detail #3?	15. Height _____
	Width _____
	Depth _____

REVISIONS

REV	PT.NO.	DESCRIPTION	DATE	APVD
—	A	WAS 2.000	1-15-82	DWS

MATERIAL LIST

NO.	QTY.	PART	STOCK SIZE	MATL
3	15	CLAMP FRAME	.63 × 1.94 × 2.22	SAE 1015 STL
2	15	SCREW	.50 RD × 2.63	SAE 1040 STL
1	15	VEE BLOCK	2.00 × 2.25 × 3.63	SAE 1095 STL

TITLE
DOUBLE "V" BLOCK & CLAMP

QUANTITY 15	MATERIAL NOTED	SCALE HALF
DRAWN BY ESR	CHECKED BY DWS	DATE 12-21-81

PART NO. 1596-3

UNLESS OTHERWISE SPECIFIED
DIMENSIONAL TOLERANCES ARE

FRACTIONS ±.030   .XXX ±.005
.XX ±.015   .XXXX ±.0005
ANGLES ±1°

ALL DIMENSIONS ARE IN
INCHES EXCEPT WHERE NOTED

Ⓘ

45°   1.000   1.063   1.313

Ⓐ   1.875   .188   .125 × .125   3 PLACES   1.063

$\frac{5}{16}$ DIA. THRU
$\frac{3}{8}$-16 UNC
.63 DEEP

Ⓑ   2.125   .875   .469   .750
$\frac{3}{8}$-16 THRU
.438   .188
1.000   1.000   1.063   3.500

Ⓒ

Ⓓ   .906 R   .688 R   .125   1.078   1.813   .875   1.094   .125

③ $\frac{1}{4}$-28 UNF   .500

② $\frac{1}{4}$-28 UNF
2.500   .125   .500   .375
.03 × .03 CHAM.
(3 PLACES)
.125 THRU
Ⓔ   MEDIUM KNURL

COMPETENCY EXERCISE #9

Refer to the print on page 353 to answer the following questions.

QUESTIONS	ANSWERS

1. What is the part name?

1. _____

2. What materials are specified for this part?

2. _____

3. How many rolls are required per part? For the entire batch?

3. _____

_____

4. What is the tolerance value applied to the length of detail #1?

4. _____

5. What is the tolerance on the diameter of the rolls?

5. _____

6. Which view shows the size of the mounting holes for the plate (detail #3)?

6. _____

7. What is the center-to-center distance between the rolls when assembled?

7. _____

8. What type of sectional view is shown in Section A-A?

8. _____

9. What are the overall finished dimensions of the bar (detail #1)?

9. Height _____

Width _____

Depth _____

10. What size cap screws are used to attach the rolls?

10. _____

11. What is the rough stock size of detail #3?

11. Height _____

Width _____

Depth _____

12. How many total parts are required to assemble one complete sine bar?

12. _____

13. What does the word *NOTED* in the title block mean?

13. _____

_____

14. What is the tolerance of the angular dimensions?

14. _____

15. What material is indicated by the section lines?

15. _____

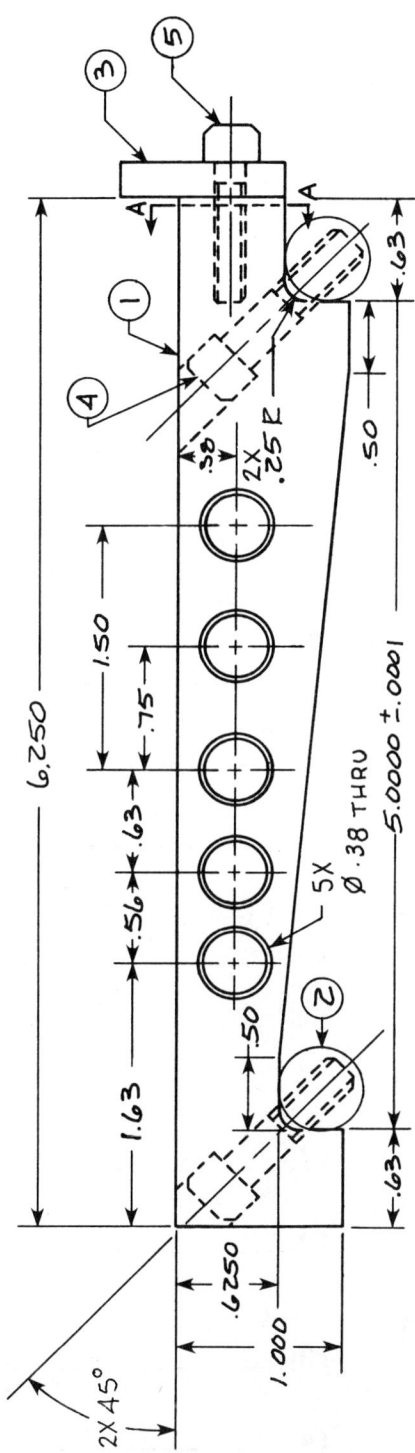

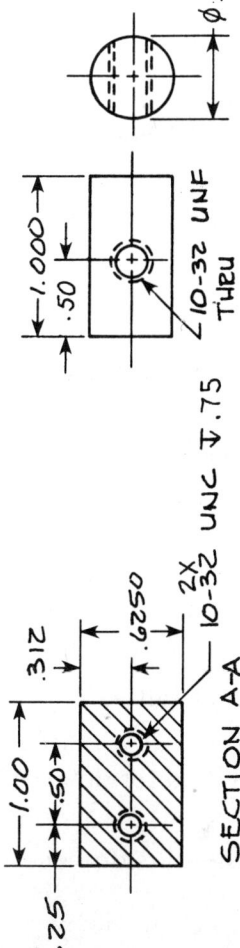

ROLL DET ②

SECTION A-A

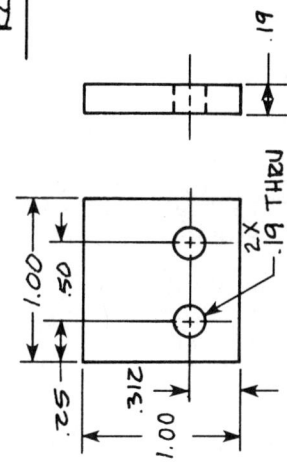

PLATE DET ③

NO.	QTY.	PART	STOCK SIZE	MATL
5	24	CAP SCR	10-32x.63	COMM
4	24	CAP SCR	10-32 x 1	COMM
3	12	PLATE	1x1x.19	SAE 1020 STL
2	24	ROLL	.5GRDx1.13	SAE 1095 STL
1	12	BAR	1x1x6.125	SAE 1095 STL

MATERIAL LIST

TITLE		
SINE BAR		
QUANTITY 12	MATERIAL NOTED	SCALE FULL
DRAWN BY DH	CHECKED BY HE	DATE 11-6-81
PART NO. 100-5-A		

UNLESS OTHERWISE SPECIFIED
DIMENSIONAL TOLERANCES ARE
FRACTIONS ±.030   .XXX ±.005
.XX ±.015   .XXXX ±.0005

ALL DIMENSIONS ARE
IN INCHES

COMPETENCY EXERCISE #10

Refer to the print on page 355 to answer the following questions.

QUESTIONS	ANSWERS
1. What is the part name?	1. _____
2. How many units are specified?	2. _____
3. What material is called for?	3. _____
4. What is the length of the threaded portion of detail #2?	4. _____
5. What is indicated by the zigzag dimension and leader lines?	5. _____ _____ _____
6. What is the overall length of the vertical slots?	6. _____
7. What is the overall length of the angular slots?	7. _____
8. How wide are the slots?	8. _____
9. What is the size of *R* in the slot dimensions?	9. _____
10. What does the Ⓐ mean? How does this change how the part is made?	10. _____ _____ _____
11. What thread size is specified for detail #2?	11. _____
12. What size is the rough stock specified for detail #2?	12. _____
13. How many radii are there on the outside edge of detail #1?	13. _____
14. What does the symbol ∅ indicate?	14. _____
15. What size chamfer is specified for detail #2?	15. _____

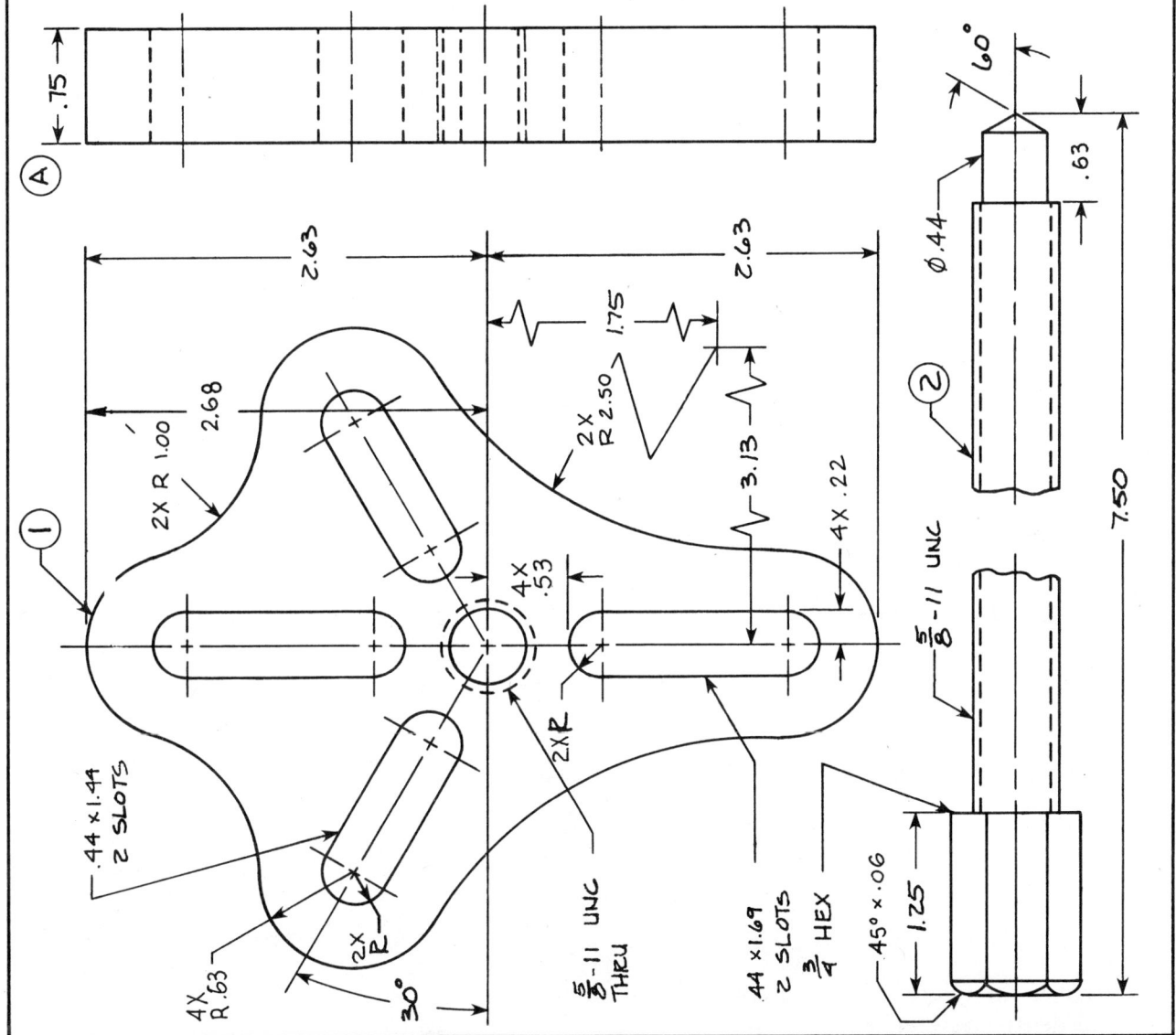

COMPETENCY EXERCISE #11

Refer to the print on page 357 to answer the following questions.

QUESTIONS	ANSWERS

1. What is the part number?

1. _____

2. What is the part name?

2. _____

3. How many units are required?

3. _____

4. What additional operation is specified for detail #1?

4. _____

5. How wide and deep is the V groove in detail #5?

5. _____

6. What type of sectional view is shown in detail #1?

6. _____

7. Find dimensions *A* and *B*.

7. *A* _____

   B _____

8. What views are shown in detail #1?

8. _____

9. What information is shown in the bottom view in detail #1?

9. _____

10. What additional treatment is specified for detail #5?

10. _____

11. What angle of projection is used for this print?

11. _____

12. What material is specified?

12. _____

13. Why is the word *NOTED* used in the title block?

13. _____

    _____

14. What tolerance is specified for the outside diameter of detail #3? For the inside diameter of detail #3?

14. Outside _____

    Inside _____

15. What is the tolerance on the counterbore diameter and depth in detail #5?

15. Diameter _____

    Depth _____

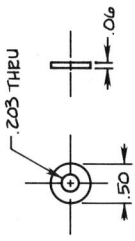

DETAIL ③

.203 THRU
.50
.06

NOTES:

DET #5 - TO BE CASE HARDENED
TO DEPTH OF .03 - HDN TO R^A-80

DET #1 - TO BE PAINTED - COLOR
NO. 537 (RED)

NO.	QTY.	PART	STOCK SIZE	MATL
5	10	PAD	1.25 RD x .94	SAE 1020
4	10	SCREW	10-32 x .75	COMM
3	10	WASHER	.50 RD x .19	SAE 1020
2	10	ELEV. SCREW	1.00 RD x 2.91	SAE 1020
1	10	BASE	1.50 RD x 2.13	SAE 1020

MATERIAL LIST

TITLE	SCREW JACK		
QUANTITY 10	MATERIAL NOTED	SCALE FULL	
DRAWN BY PJC	CHECKED BY CWD	DATE 9-7-81	
PART NO.	A-77312-2		

THIRD ANGLE

UNLESS OTHERWISE SPECIFIED
DIMENSIONAL TOLERANCES ARE
FRACTIONS ±.030 .XXX ±.002
.XX ±.015 .XXXX ±.0005

ALL DIMENSIONS ARE IN INCHES

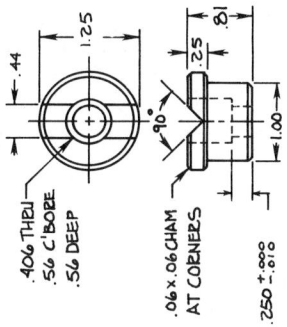

DETAIL ⑤

.44
1.25
.81
.25
90°
1.00
.250 +.000 -.010
.406 THRU
.56 C'BORE
.56 DEEP
.06 x .06 CHAM
AT CORNERS

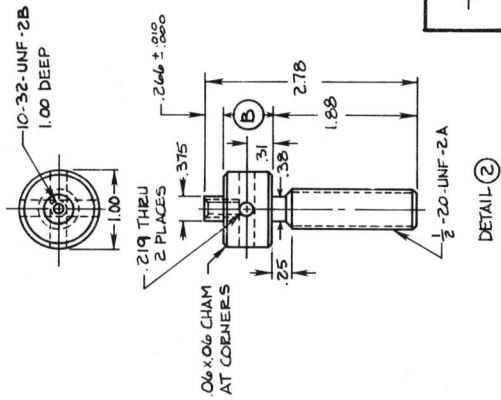

DETAIL ②

10-32-UNF-2B
1.00 DEEP
.266 +.010 -.000
.375
.31
.38
2.78
1.88
Ⓑ
½-20-UNF-2A
.219 THRU
2 PLACES
.06 x .06 CHAM
AT CORNERS
.52

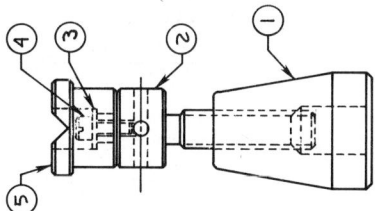

④ ③ ② ① ⑤

½-20-UNF-2B
THRU

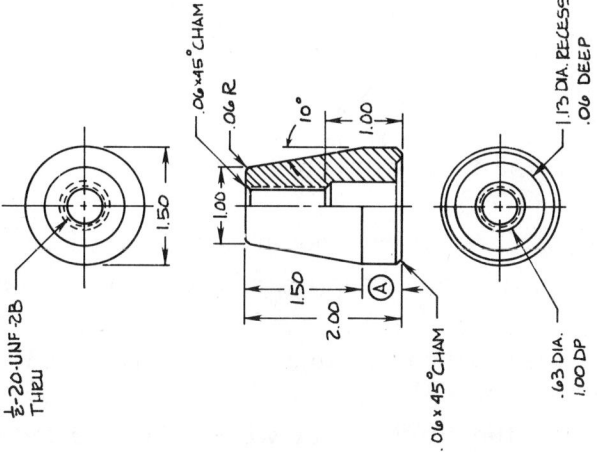

DETAIL ①

.06 x 45° CHAM
.06 R
10°
1.00
1.50
1.50
2.00
Ⓐ
.63 DIA.
1.00 DP
1.13 DIA. RECESS
.06 DEEP
.06 x 45° CHAM

COMPETENCY EXERCISE #12

Refer to the print on page 359 to answer the following questions.

QUESTIONS	ANSWERS

1. What is the part name?

2. What is the part number?

3. Identify the names of details #1 through #11.

1. _____

2. _____

3. #1 _____

   #2 _____

   #3 _____

   #4 _____

   #5 _____

   #6 _____

   #7 _____

   #8 _____

   #9 _____

   #10 _____

   #11 _____

4. What type of assembly print is this?

5. Identify the parts of the thread designations shown on detail #3, as listed to the right.

4. _____

5. $\frac{3}{8}$ _____

   16 _____

   UNC _____

   2 _____

   A _____

   10 _____

   32 _____

   UNF _____

   2 _____

   B _____

6. Find dimensions *A* and *B*.

6. *A* _____

   B _____

7. What type of fit exists between detail #4 and the hole in detail #3?

8. Why isn't detail #11 shown as a drawn detail?

9. What is the critical depth dimension of the counterbore in detail #9?

10. What is the included angle of the V grooves in details #7 and #8?

11. What is the difference between details #5 and #6?

12. What is the tolerance on the length of detail #4?

13. What are all of the individual part drawings on this print called?

14. How deep are the V grooves in details #7 and #8?

15. What is the head diameter of detail #3?

7. _____

8. _____

9. _____

10. _____

11. _____

12. _____

13. _____

14. _____

15. _____

DETAIL 10

3/8-16-UNC-2B THRU

.248 +.000 -.002

.750 +.005 -.000

.13

.13

.75

DETAIL 5

.310 ±.000 -.002

.250 ±.000 -.002

.750 +.005 -.000

.19

.19

DETAIL 6

.310 ±.000 -.002

.250 ±.000 -.002

.500 +.005 -.000

.19

.19

DETAIL 7

.312 ±.002 THRU

.06

.75

.75

.740 +.000 -.005

90°

DETAIL 8

.312 ±.002 THRU

.06

.75

.75

.490 +.000 -.005

90°

DETAIL 9

.248 ±.000 -.002

.250 ±.002 THRU C'BORE AS NOTED

.500 ±.005

.13

.13

.75

.440 +.000 -.005

.69

.38

DETAIL 1

Ⓐ

.38R

.56R

.13

1.44

.75

.31R

.250 ±.000 -.002 THRU 5 PLACES

.38R

1.75

.31R

2.50

4.50

.69

DETAIL 2

.56R

.312 THRU

.38R

1.69

.81

2.50

.75

.31R

.69

1.75

.250 THRU 3 PLACES

.31R

DETAIL 3

10-32-UNF 2B .50 DEEP

.56

.248 ±.000 -.002

.437

.38

3/8-16-UNC-2A

.249 +.000 -.001

Ⓑ

4.95

.63

.03 x .03 CHAM 2 PLACES

DETAIL 4

.750 +.001 -.000

2.75

.03 x .03 CHAM BOTH ENDS

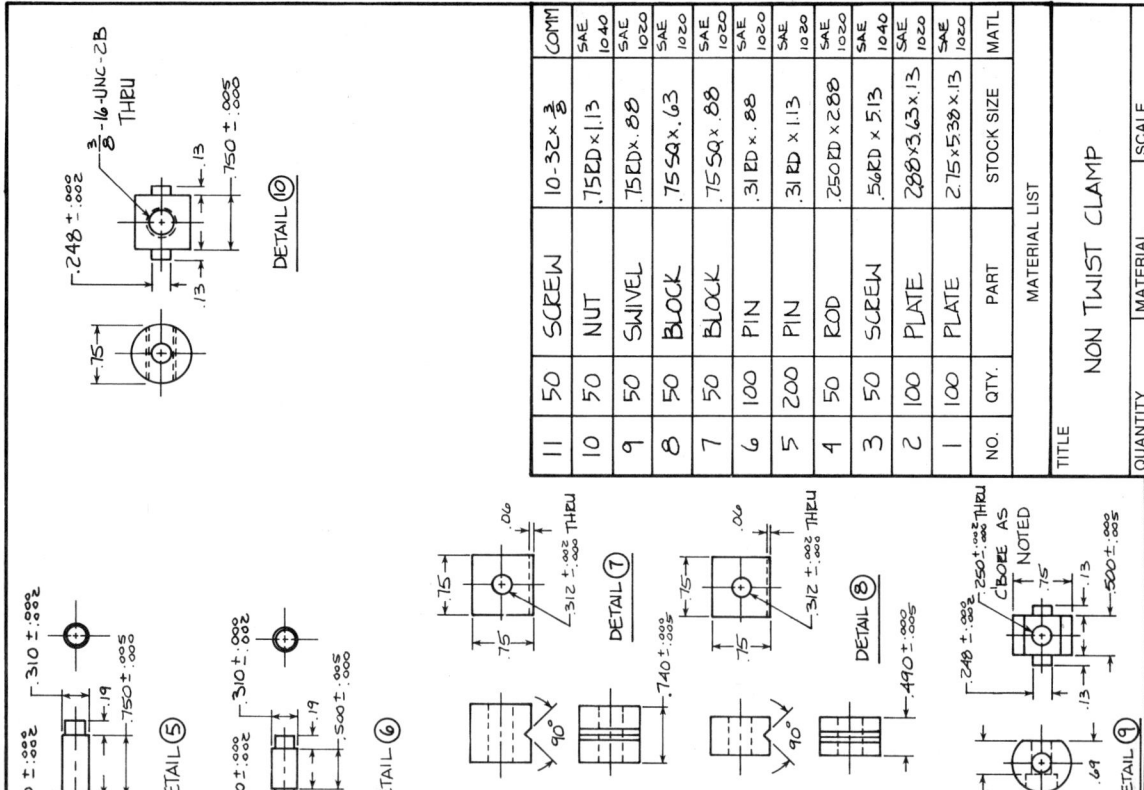

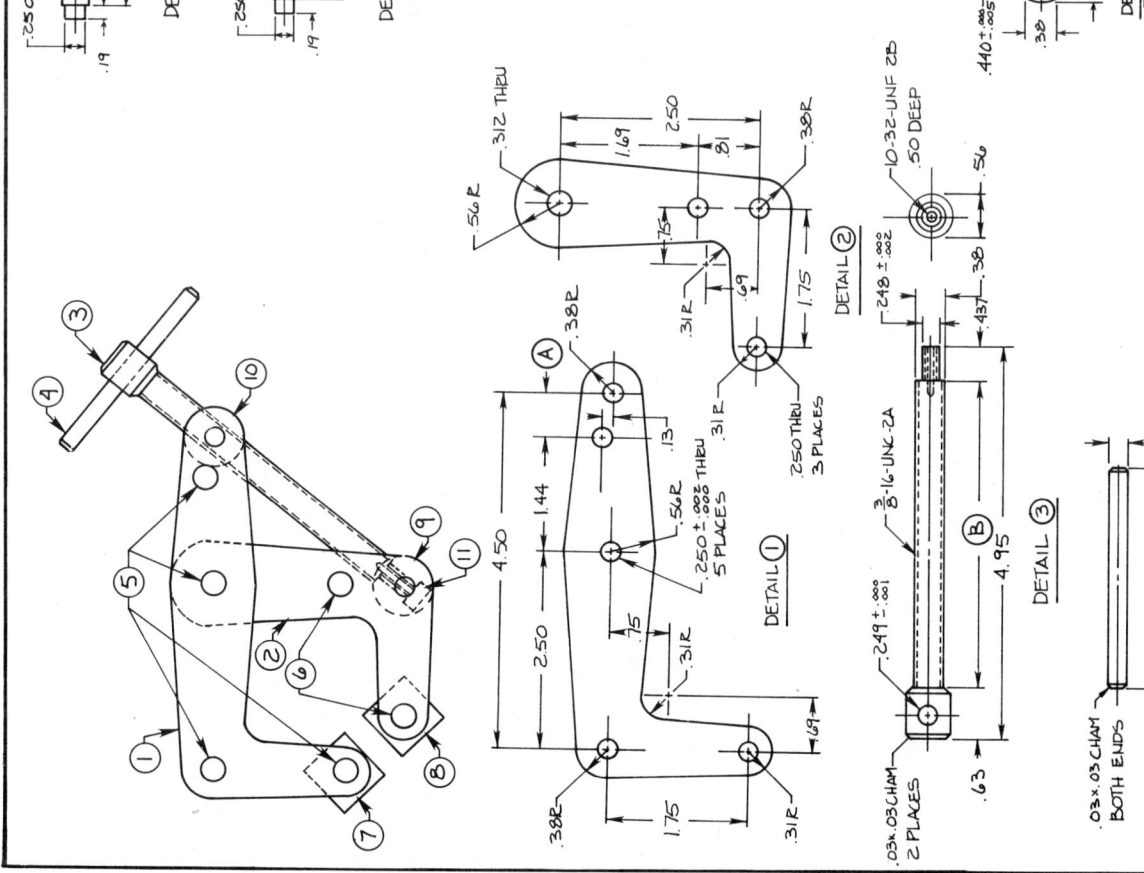

NO.	QTY.	PART	STOCK SIZE	MATL	COMM
11	50	SCREW	10-32 x 3/8		SAE 1040
10	50	NUT	.75 RD x 1.13		SAE 1020
9	50	SWIVEL	.75 RD x .88		SAE 1020
8	50	BLOCK	.75 SQ x .63		SAE 1020
7	50	BLOCK	.75 SQ x .88		SAE 1020
6	100	PIN	.31 RD x .88		SAE 1020
5	200	PIN	.31 RD x 1.13		SAE 1020
4	50	ROD	.250 RD x 2.88		SAE 1020
3	50	SCREW	.56 RD x .513		SAE 1040
2	100	PLATE	2.88 x 3.63 x .13		SAE 1020
1	100	PLATE	2.75 x 5.38 x .13		SAE 1020

MATERIAL LIST

TITLE: NON TWIST CLAMP

QUANTITY: 50

DRAWN BY: WJ

MATERIAL: NOTED

CHECKED BY: PJT

SCALE: FULL

DATE: 10-14-81

PART NO. 9-7736-C

UNLESS OTHERWISE SPECIFIED DIMENSIONAL TOLERANCES ARE
FRACTIONS ±.030   .XXX ±.005
.XX ±.015   .XXXX ±.0005

COMPETENCY EXERCISE #13

Refer to the print on page 361 to answer the following questions.

<table>
<tr><td>QUESTIONS</td><td>ANSWERS</td></tr>
<tr><td>1. What is the part name?</td><td>1. _____</td></tr>
<tr><td>2. What is the part number?</td><td>2. _____</td></tr>
<tr><td>3. What scales are used to show these parts?</td><td>3. _____</td></tr>
<tr><td>4. What materials are specified for this part?</td><td>4. _____

_____</td></tr>
<tr><td>5. What type of assembly drawing is this?</td><td>5. _____</td></tr>
<tr><td>6. What type of thread representation is used to show the threads in detail #1?</td><td>6. _____</td></tr>
<tr><td>7. What size of groove is specified on this print for thread runout?</td><td>7. _____</td></tr>
<tr><td>8. Why doesn't this print have a materials list?</td><td>8. _____

_____</td></tr>
<tr><td>9. How many revisions have been made to this print?</td><td>9. _____</td></tr>
<tr><td>10. What thread sizes are listed for detail #2?</td><td>10. _____</td></tr>
<tr><td>11. What is the taper per inch of detail #2?</td><td>11. _____</td></tr>
<tr><td>12. What size of counterbore is specified for detail #2?</td><td>12. _____</td></tr>
<tr><td>13. What size of knurl is called for on detail #1?</td><td>13. _____</td></tr>
<tr><td>14. How deep is the small hole in detail #1?</td><td>14. _____</td></tr>
<tr><td>15. What type of radius is specified for detail #3?</td><td>15. _____</td></tr>
</table>

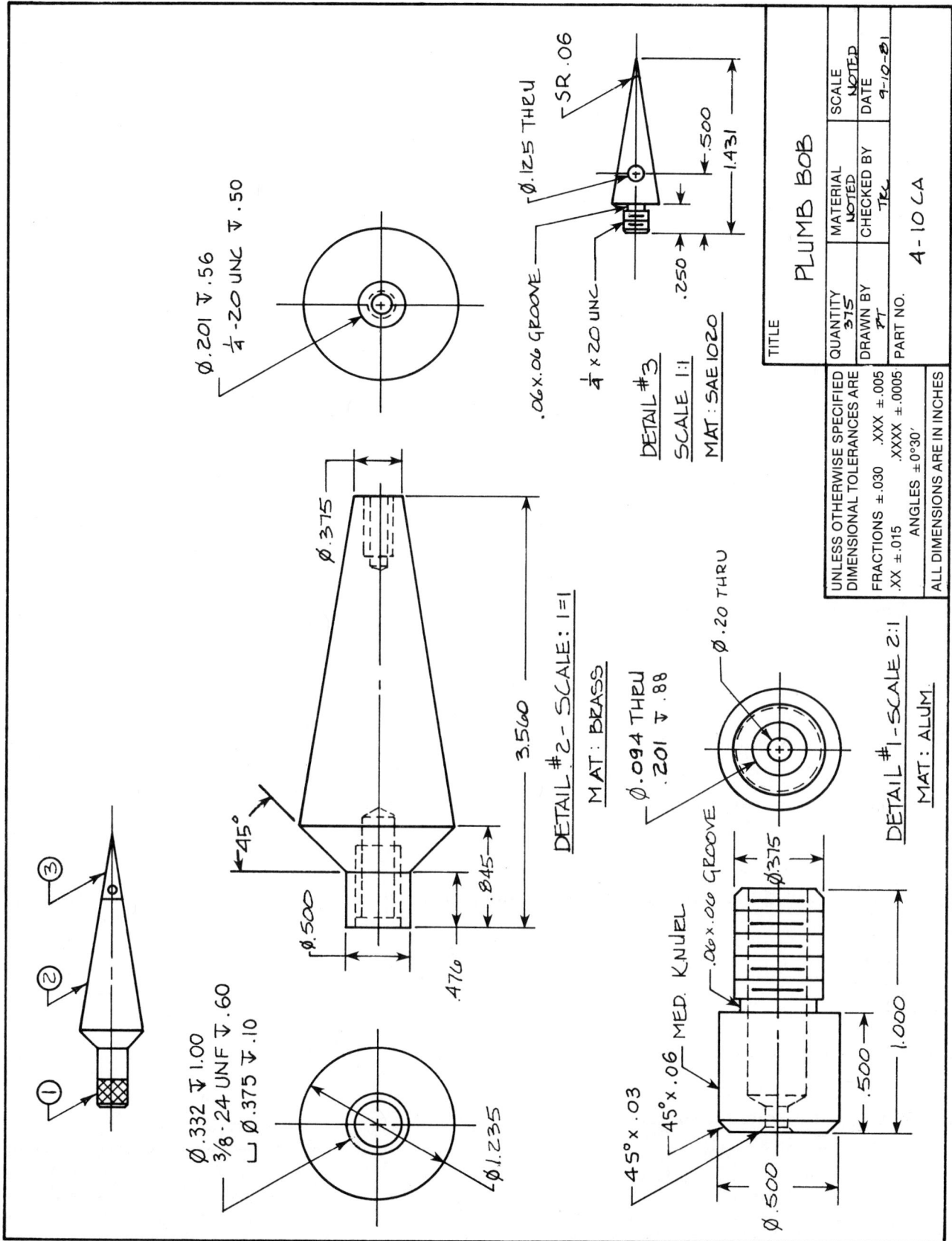

DETAIL #3
SCALE 1:1
MAT: SAE 1020

Ø.125 THRU
SR.06
.06 x .06 GROOVE
¼ x 20 UNC
.500
1.431
.250

Ø.201 ▽.56
¼ - 20 UNC ▽.50

DETAIL #2 - SCALE: 1=1
MAT: BRASS

Ø.375
3.560
45°
Ø.500
.845
.476

Ø.332 ▽1.00
3/8 - 24 UNF ▽.60
⌴ Ø.375 ▽.10

Ø1.235

DETAIL #1 - SCALE 2:1
MAT: ALUM.

Ø.20 THRU
Ø.094 THRU ▽.88
.201 ▽
Ø.375
.06 x .06 GROOVE
MED. KNURL
45° x .06
1.000
.500
45° x .03
Ø.500

TITLE		PLUMB BOB

UNLESS OTHERWISE SPECIFIED DIMENSIONAL TOLERANCES ARE
FRACTIONS ±.030   .XXX ±.005
.XX ±.015   .XXXX ±.0005
ANGLES ±0°30'
ALL DIMENSIONS ARE IN INCHES

QUANTITY 375	MATERIAL NOTED	SCALE NOTED
DRAWN BY PT	CHECKED BY TK	DATE 9-10-81
PART NO.		4-10 CA

COMPETENCY EXERCISE #14

Refer to the print on page 363 to answer the following questions.

QUESTIONS	ANSWERS
1. What is the part name?	1. _____
2. What is the part number?	2. _____
3. What views are used to describe this part?	3. _____ _____
4. What units are the design units in this print?	4. _____
5. What units are the conversion units?	5. _____
6. What dual dimensioning method is used for this print?	6. _____
7. Which units have precedence in this print?	7. _____
8. How deep is the center slot?	8. _____
9. What is the angle of the center slot?	9. _____
10. What angle is called for on the bottom surface of the part?	10. _____
11. What type of dimensioning is used for this print?	11. _____
12. What is the tap drill size specified for the tapped holes?	12. _____
13. What is the overall width of the part?	13. _____
14. What are the tolerance values that are to be applied to the hole locations?	14. Inch _____ Millimeter _____
15. What effect does the millimeter dimensioning have on the scale?	15. _____

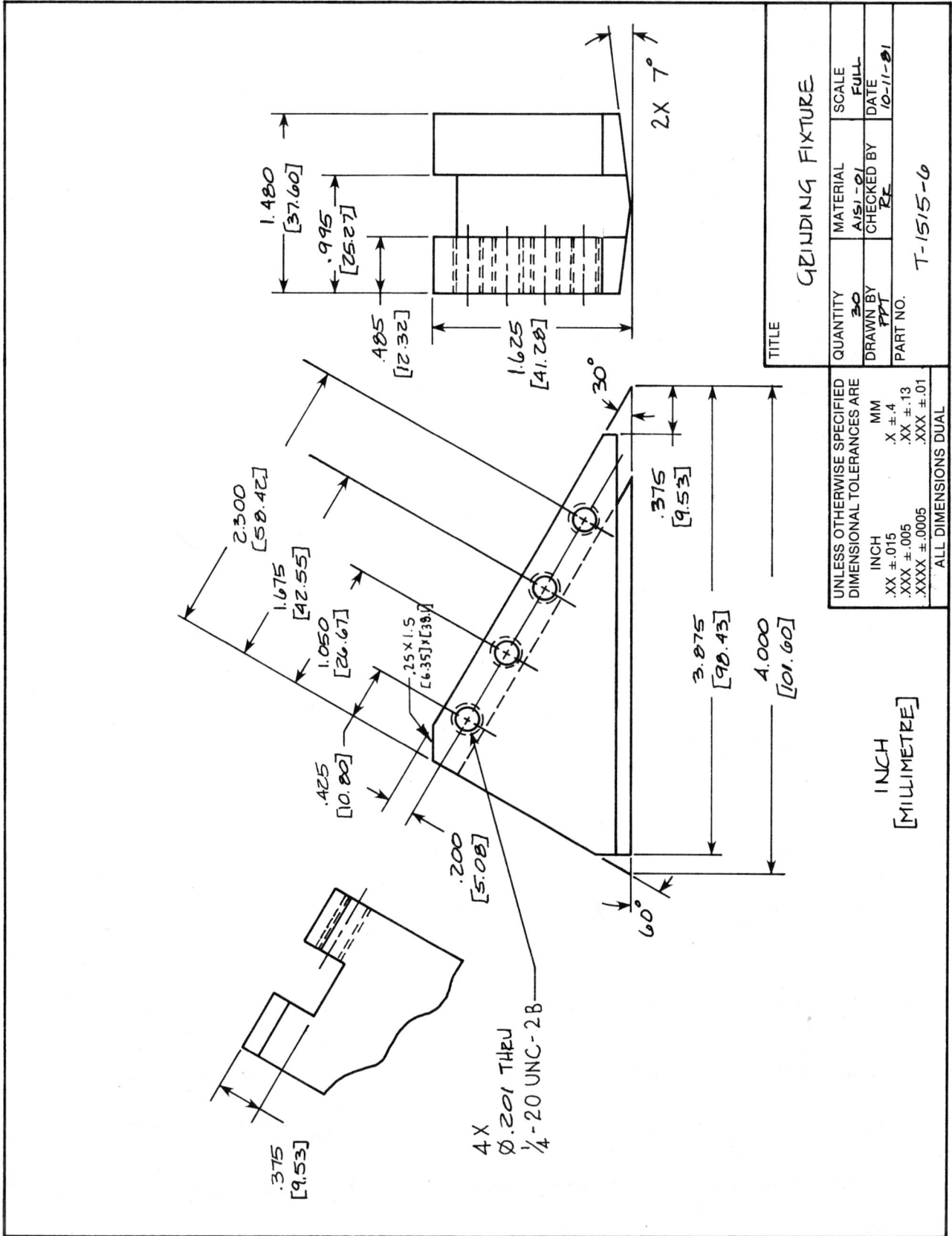

1.480
[37.60]

.995
[25.27]

.485
[12.32]

1.625
[41.28]

2X 7°

2.300
[58.42]

1.675
[42.55]

1.050
[26.67]

.25 X 1.5
[6.35] X [38.1]

.425
[10.80]

.200
[5.08]

30°

60°

.375
[9.53]

3.875
[98.43]

4.000
[101.60]

.375
[9.53]

4X
∅.201 THRU
¼-20 UNC-2B

INCH
[MILLIMETRE]

TITLE

GRINDING FIXTURE

QUANTITY
30

DRAWN BY
PPT

PART NO.

MATERIAL
AISI-01

CHECKED BY
RF

SCALE
FULL

DATE
10-11-81

T-1515-6

UNLESS OTHERWISE SPECIFIED
DIMENSIONAL TOLERANCES ARE

INCH
.XX ±.015
.XXX ±.005
.XXXX ±.0005

MM
.X ±.4
.XX ±.13
.XXX ±.01

ALL DIMENSIONS DUAL

COMPETENCY EXERCISE #15

Refer to the print on page 365 to answer the following questions.

QUESTIONS	ANSWERS
1. What is the part name?	1. _____
2. What is the part number?	2. _____
3. What is the taper per inch of the mandrel Ⓐ?	3. _____
4. What is the taper per foot of the mandrel Ⓐ?	4. _____
5. How much material is to be left on the mandrel for finish-grinding the taper?	5. _____
6. Where is the part number to be marked on the mandrel?	6. _____
7. What type of view is shown in detail "A"?	7. _____
8. What size of centerdrill is specified for the center holes?	8. _____
9. What tolerance is specified for the dimensions at the ends of the taper?	9. _____
10. What hardness is specified for the mandrel?	10. _____
11. What is the tolerance on the depth of the center holes?	11. _____
12. What material is specified for this part?	12. _____
13. What does the large dashed circle marked *A* indicate?	13. _____ _____
14. How wide is the flat at each end of the mandrel?	14. _____
15. Where are the limits of size for the overall length of the mandrel?	15. _____

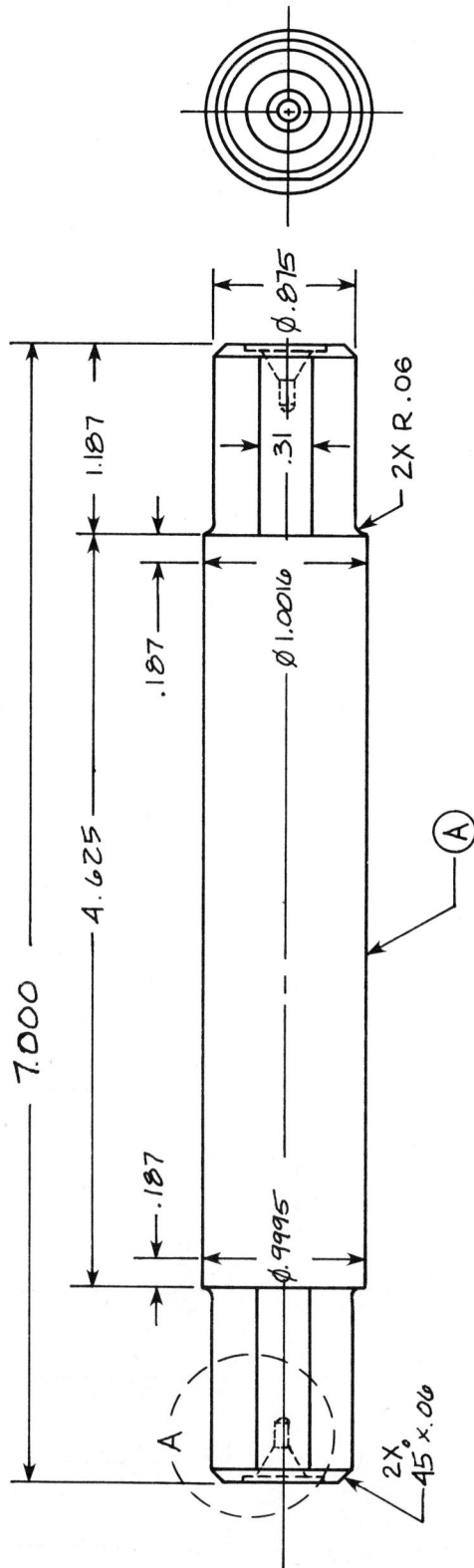

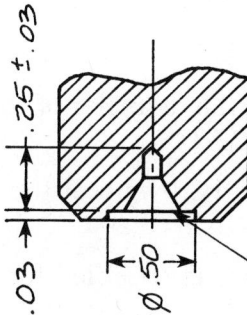

NOTES: 1. ALLOW .03 FOR
GRINDING TAPER
2. HARDEN TO R^c 58-62
3. MARK SIZE ON SMALL
END FLAT

DETAIL "A"

2X R.06

Ø.875

1.187

.31

.187

Ø1.0016

A

7.000

4.625

.187

Ø.9995

2X
45° x.06

#3 CENTERDRILL
TYPICAL-BOTH ENDS

.52 ±.03

.03

Ø.50

TITLE

LATHE MANDREL

QUANTITY	MATERIAL	SCALE
100	SAE 1095 STL	FULL
DRAWN BY	CHECKED BY	DATE
PK	DR	2-7-81
PART NO.		
6734-1		

UNLESS OTHERWISE SPECIFIED
DIMENSIONAL TOLERANCES ARE

FRACTIONS ±.030   .XXX ±.005
.XX ±.015   .XXXX ±.0002

COMPETENCY EXERCISE #16

Refer to the print on page 367 to answer the following questions.

QUESTIONS	ANSWERS
1. What is the part name?	1. _____
2. What is the part number?	2. _____
3. How many total parts are required?	3. _____
4. How many of each type are specified?	4. _____
	_____
5. What is the taper per foot, *C*?	5. _____
6. What is the purpose of the groove?	6. _____
	_____
7. Find dimensions *A* and *B*.	7. *A* _____
	B _____
8. What operation is to be performed on the headstock centers?	8. _____
9. What heat treatment is specified for the tailstock centers?	9. _____
10. What is the point angle of the center?	10. _____
11. What size of centerdrill is specified for the center holes?	11. _____
12. What is the diameter of the small end of the taper?	12. _____
13. What is the length of the tapered area?	13. _____
14. What size is specified for the groove?	14. _____
15. What is the diameter of the center at the bottom of the groove?	15. _____

#3 CENTERDRILL
⩒ .03
45° x .01

GROOVE .06 WIDE x
.06 DEEP TO IDENTIFY
HARDENED CENTER

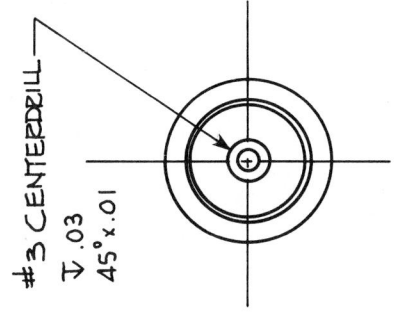

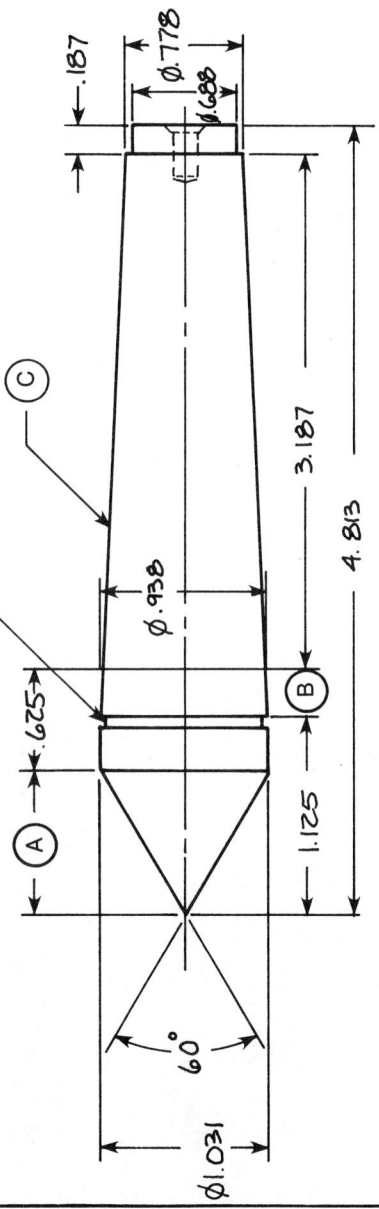

Ⓒ

Ⓑ

Ⓐ

.187

∅.778

∅.688

.625

∅.938

3.187

4.813

1.125

60°

∅1.031

NOTE: ① HEADSTOCK CENTER TO BE HARDENED
TO R꜀ 58-62.

② TAILSTOCK CENTER NOT HARDENED

③ ONLY THOSE CENTERS TO BE HARDENED
ARE TO BE GROOVED.

④ MAKE 15 HEADSTOCK & 35 TAIL STOCK

COMPETENCY EXERCISE #17

Refer to the print on page 369 to answer the following questions.

QUESTIONS	ANSWERS
1. What is the part name?	1. _____
2. What material is specified?	2. _____
3. What type of line is used to show the root circle of this part?	3. _____
4. What type of line is used to show the pitch circle?	4. _____
5. What type of view is View "A"?	5. _____
6. How many teeth are there on this gear?	6. _____
7. Find the values of *A* through *D*.	7. *A* _____
	B _____
	C _____
	D _____
8. What is the chordal thickness value?	8. _____
9. What is the chordal addendum value?	9. _____
10. What is the outside diameter of the gear?	10. _____
11. How wide is the keyway?	11. _____
12. What type of dimension is used to show the keyway depth?	12. _____
13. What is the diametral pitch of the gear?	13. _____
14. What is the minimum bore size?	14. _____
15. What tolerance is applied to the 45° angle in View "A"?	15. _____

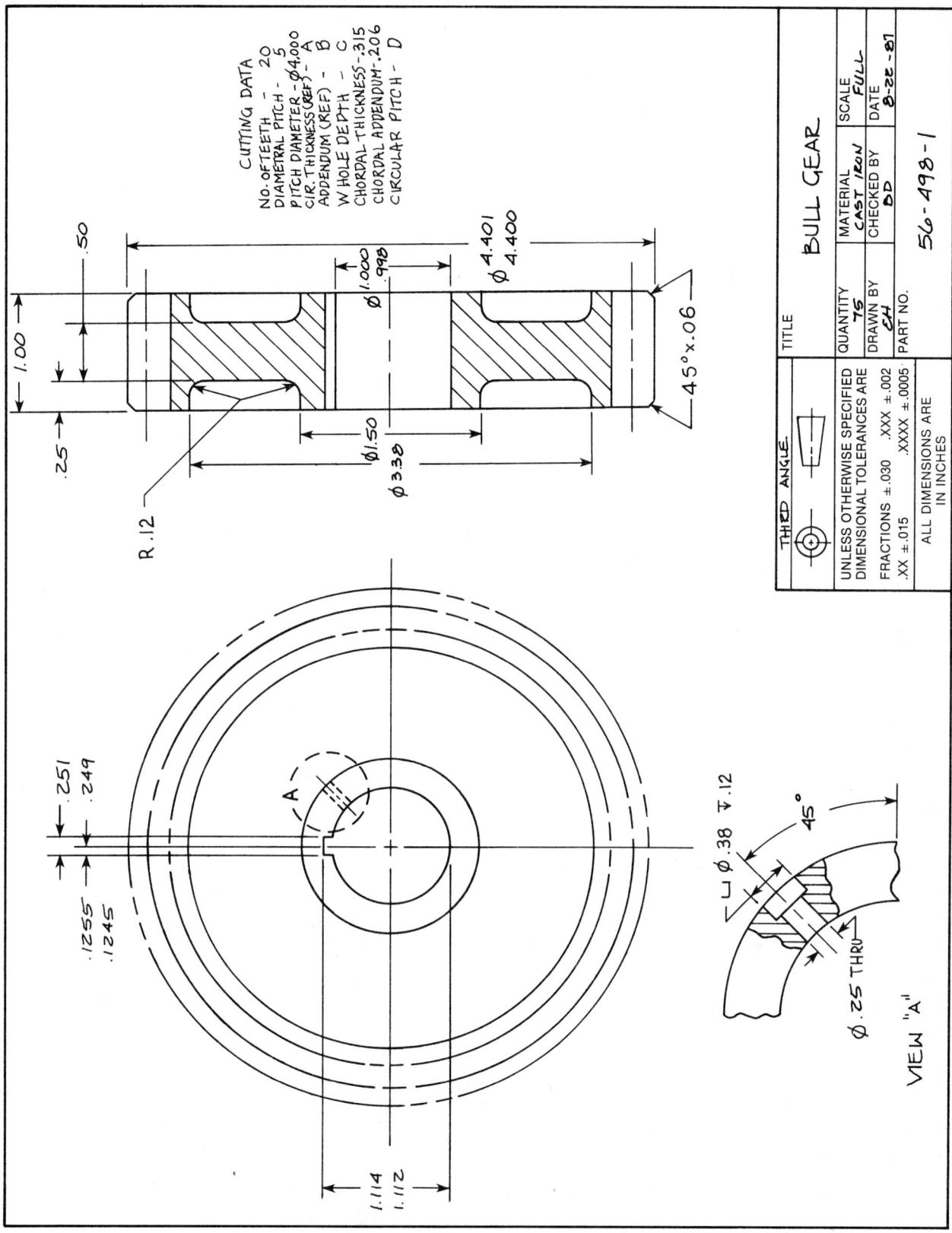

CUTTING DATA
NO. OF TEETH - 20
DIAMETRAL PITCH - 5
PITCH DIAMETER - ⌀4.000
CIR. THICKNESS (REF) - A
ADDENDUM (REF) - B
WHOLE DEPTH - C
CHORDAL THICKNESS - .315
CHORDAL ADDENDUM - .206
CIRCULAR PITCH - D

.50

1.00

.25

R .12

⌀1.000
⌀.998

⌀4.401
⌀4.400

45° x .06

⌀1.50

⌀3.38

.251
.249

1.421
1.422

1.114
1.112

A

VIEW "A"

⌀ .38 ⌴ .12

45°

⌀ .25 THRU

TITLE			
THIRD ANGLE		BULL GEAR	

		BULL GEAR	
UNLESS OTHERWISE SPECIFIED DIMENSIONAL TOLERANCES ARE	QUANTITY 75	MATERIAL CAST IRON	SCALE FULL
	DRAWN BY EH	CHECKED BY DD	DATE 8-22-81
FRACTIONS ±.030 .XXX ±.002 .XX ±.015 .XXXX ±.0005	PART NO.	56-498-1	
ALL DIMENSIONS ARE IN INCHES			

COMPETENCY EXERCISE #18

Refer to the print on page 371 to answer the following questions.

QUESTIONS

ANSWERS

1. What is the part name?

1. _____

2. What is the part number?

2. _____

3. What type of dimensioning and tolerancing is used for this part?

3. _____

4. What do the dimensions in boxes indicate?

4. _____

5. How are the datum surfaces identified?

5. _____

6. Which datum is primary?

6. _____

7. What are the symbols in the first block called?

7. _____

8. What is the entire frame around the symbol, datum, and tolerance called?

8. _____

9. Identify the symbols used on this print.

9. _____

_____

_____

10. What does the Ⓜ indicate?

10. _____

11. How are the holes in the circular pattern spaced?

11. _____

12. What tolerance is to be applied to untoleranced dimensions?

12. _____

13. What is the bolt circle diameter of the circular pattern?

13. _____

14. What significance does the order A-B-C have on the part?

14. _____

15. What tolerance is allowed for the hole diameters in the rectangular pattern?

15. _____

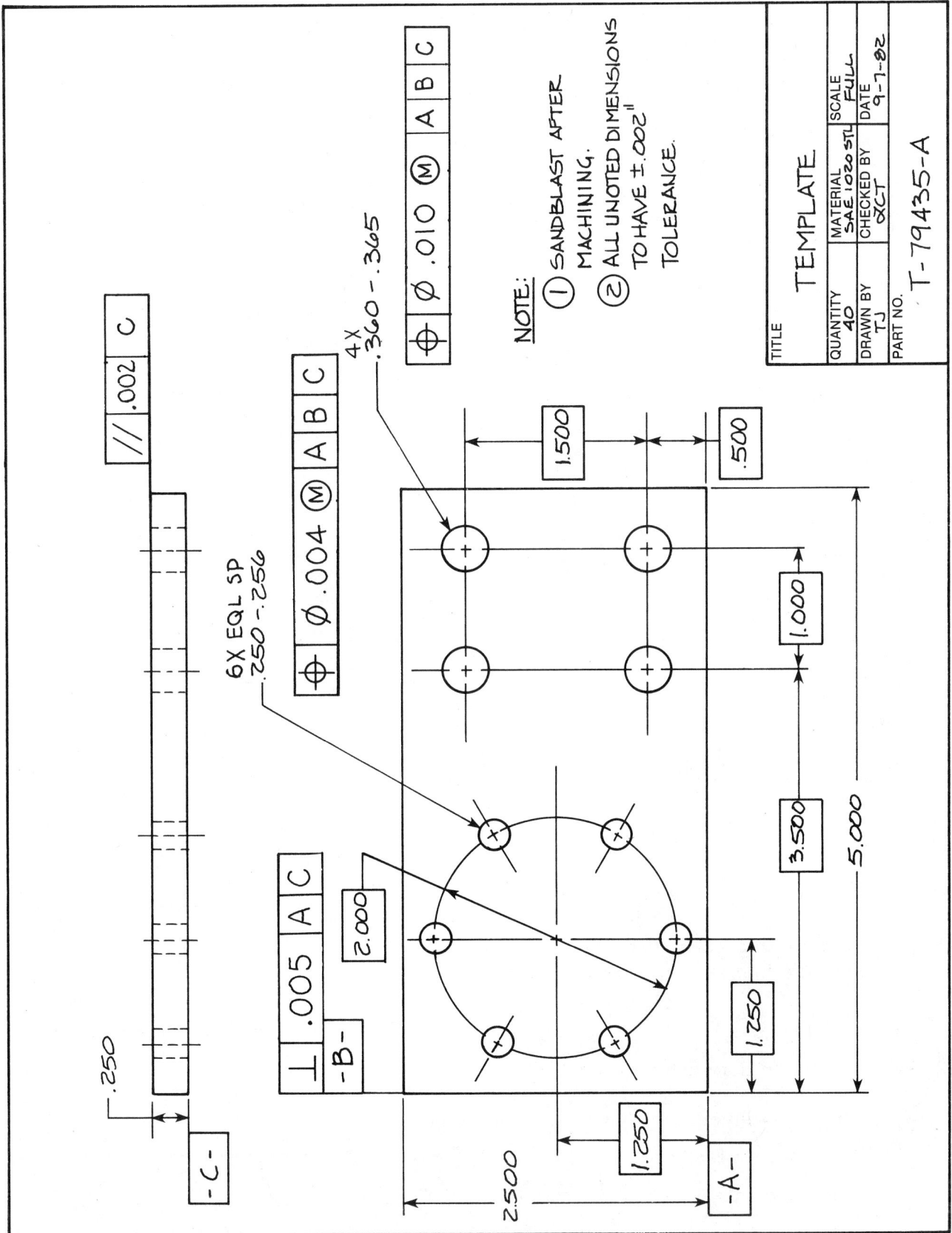

TEMPLATE

TITLE		
QUANTITY 40	MATERIAL SAE 1020 STL	SCALE FULL
DRAWN BY TJ	CHECKED BY XCT	DATE 9-7-82
PART NO. T-79435-A		

NOTE:
1. SANDBLAST AFTER MACHINING.
2. ALL UNOTED DIMENSIONS TO HAVE ±.002" TOLERANCE.

COMPETENCY EXERCISE #19

Refer to the print on page 373 to answer the following questions.

<table>
<tr><th>QUESTIONS</th><th>ANSWERS</th></tr>
</table>

QUESTIONS	ANSWERS
1. What is the part number?	1. _____
2. What is the part name?	2. _____
3. What type of projection is used for this print?	3. _____
4. What was done in revision A?	4. _____
5. Identify the following symbols.	5. // _____
	↗ _____
	▱ _____
	○ _____
	⊕ _____
	⌀ _____
6. What type of dimensioning is used for this print?	6. _____
7. What is the LMC size of the centerbore?	7. _____
8. What is the MMC size of the outside diameter?	8. _____
9. What relationship is there between the back and front faces of the flange?	9. _____ _____
10. Where is the primary datum located?	10. _____
11. What relationship exists between the outside diameter of the hub and the bore?	11. _____ _____
12. Where is the tertiary datum located?	12. _____
13. What type of dimension shows the bolt circle diameter?	13. _____
14. What is the tolerance on the thickness of the flange?	14. _____
15. What type of symbol is used to show the primary datum?	15. _____

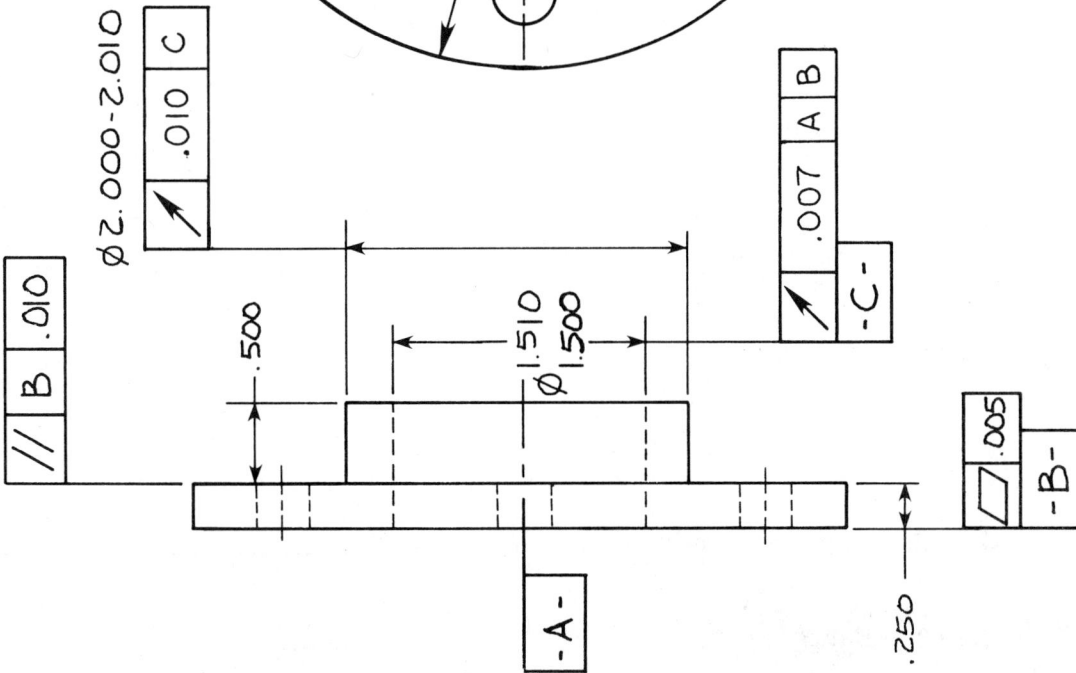

COMPETENCY EXERCISE #20

Refer to the print on page 375 to answer the following questions.

QUESTIONS	ANSWERS

1. What is the part name?

1. _____

2. What is the part number?

2. _____

3. What units are used to dimension this print?

3. _____

4. What material is specified?

4. _____

5. What is the drawing scale?

5. _____

6. What size of chamfer is shown in the print?

6. _____

7. What are the dimensions of the groove for thread runout?

7. _____

8. What is the diameter at the bottom of this groove?

8. _____

9. What size of thread is specified?

9. _____

10. What form of thread representation is used to show the threads?

10. _____

11. What does the *6g* in the thread designation indicate?

11. _____

12. How many parts are required?

12. _____

13. What is the outside diameter of the large flange?

13. _____

14. What type of dimensioning is shown in this print?

14. _____

15. How wide is the large flange?

15. _____

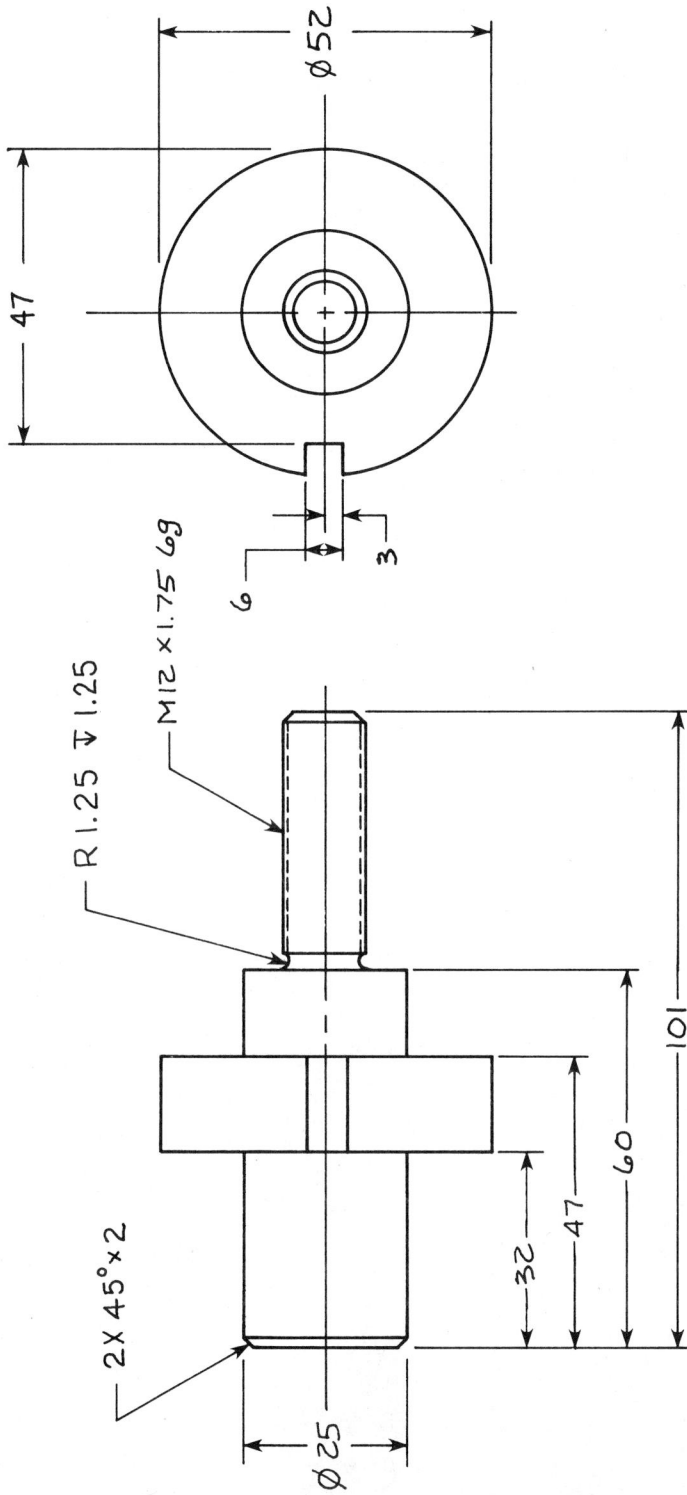

METRIC

TITLE	SHAFT		
QUANTITY 375	MATERIAL CAST IRON		SCALE FULL
DRAWN BY PPT	CHECKED BY JZ		DATE 12-7-81
PART NO.		PC-2199-22	

Ø52

47

M12 × 1.75 69

6

3

R1.25 ▽1.25

2X 45°×2

Ø25

32

47

60

101

GLOSSARY

actual size the measured size of a part

addendum the radial distance between the pitch circle and the top of each tooth

addendum angle the angular difference in the addendum from the small end to the large end of the tooth of a bevel gear

allowance the minimum intentional difference in size between two mating parts

alphabet of lines the accepted series of standard lines used for engineering drawings

angular addendum the addendum measured parallel to the axis of the bevel gear bore

angular dimensions dimensions used to show the size of angles

angular measurement the measurement of angles

angular perspectives perspective drawings that use two vanishing points; also called two-point perspectives

arc a part, or portion of a circle

assembly print a print used to show the position and relationship of the parts in an assembly

auxiliary view a view used to show angled or inclined surfaces of a part that cannot be clearly shown in any of the principal views

axonometric axes the viewing axes of axonometric drawings. The axes in isometric drawings have three equal angles, dimetric drawings have two equal angles, and trimetric drawings have three unequal angles.

axonometric drawing a form of pictorial drawing that shows an object built around axonometric axes. The three types of axonometric drawings are isometric, dimetric, and trimetric.

backing and spacer symbols the symbols used to show that a weld is to have either backing or a spacer

basic size the dimension to which the tolerance is applied

bend allowance the amount of material that must be provided to allow for any bends in the part

bilateral tolerance a tolerance that varies in both directions from the stated basic size. Bilateral tolerances may be either equal or unequal.

blocking a method of constructing sketches in which the part is first constructed using a series of boxes to form the basic shape

blueprint a copy that has white lines on a dark blue background

bolt circle an imaginary circle on which all hole centers in a circular hole pattern are located

boring a machining operation that enlarges holes. Boring is normally used for special sizes or for sizes beyond the range of normal drilling. Boring is also used to make holes straight and round prior to other machining.

bracket method dual dimensioning a method of dual dimensioning in which the conversion units on a print are shown contained within brackets

break lines lines used to show that an object was shortened or a part of the object was removed to shorten the size of the drawn object

broken section sectional view that is noted with a short break line. This view is only intended to show the internal detail of a small area of a part.

CAD Computer Aided Design; a system used to construct engineering drawings with a computer

cabinet oblique drawing an oblique drawing that shows the depth of a part in half size. This drawing has receding lines with angles of 63°26′.

cavalier oblique drawing an oblique drawing that shows the depth of a part in full size. This drawing has receding lines with angles of 45°.

center distance the distance between the center axes of mating gears

center lines lines used to show center positions of holes, radii, or part centers. Center lines are also used to show symmetry.

chain dimensioning placing dimensions in a line, one after another

chamfering a machining process that removes the sharp edge from a hole or from the end of a shaft. Chamfering is normally done at a 45° angle.

chordal addendum the length of a straight line between the top of each tooth and the chordal thickness line. This value is usually found in handbook charts.

chordal thickness the thickness of a tooth measured along a straight line, or chord, that connects the two points where the pitch circle contacts the finished contour of the tooth on both sides. This value is normally found in handbook charts.

circle a continuous curved line joined at both ends and having all points on the line equally distant from a common center point

circular pitch the distance from a point on one tooth to a corresponding point on the adjacent tooth

circular thickness the thickness of a tooth measured along the pitch circle

clearance the amount by which the dedendum of a gear is larger than the addendum of its mating gear. Or, the space between the top of one tooth and the bottom of the tooth space in meshing gears.

common fraction fractional part of a whole unit; expressed with a numerator and a denominator

complete view a view that shows every detail of a part when viewed from a particular side

contour symbols the symbols used to show the finished shape of a weld

conventions the standard rules drafters follow when constructing engineering drawings

conversion chart dimensioning a system of dual dimensioning in which the converted values are shown in a separate chart

conversion units the direct equivalent conversions of the design units

coordinate dimensions dimensions that use a single reference point, called a datum, for all dimensions

counterboring a machining operation used to produce a cylindrical recessed area around the edge of a hole for recessing bolt heads or nuts

countersinking a machining operation used to produce a conical recess on the edge of a hole for recessing flat head screws

cutting angle the angle on which the teeth of a bevel gear are actually cut. This angle is measured from the center axis of the gear bore.

cutting plane lines lines used to show the path of an imaginary cut that forms a sectional view

datum any specific plane, surface, line, center point, or feature used as a reference point for dimensioning

datum dimensioning positioning dimensions so they are all referenced from a single point, or datum

datum identifying symbol the symbol used in a print to identify a datum on the part

datum reference the letter value that is used to identify a datum in the feature control symbol

decimal fraction fractional part of a whole unit; expressed with a decimal number

dedendum the radial distance between the pitch circle and the bottom of each tooth

dedendum angle the angular difference in the dedendum from the small end to the large end of the tooth of a bevel gear

degree (°) the primary measurement unit used to measure angles

denominator the bottom number of a common fraction

design units the units used to make the original design of a part

detail assembly print a print used to show both the position and size of the parts in an assembly

detail print a multiview print of a single part

diagonal an angular line that connects the corners of a square or rectangle

dial caliper a variation of a vernier caliper; used to make linear measurements to an accuracy of .001″ or .02 mm

diameter the distance across the center of a circle

diametral pitch the ratio of the number of teeth to the pitch diameter. This is also the factor that determines the size of the teeth on a gear; the larger the diametral pitch number, the smaller the size of the teeth.

dimetric drawing an axonometric drawing with two equal angles; for example, two 105° angles and one 150° angle forming the axonometric axes

diazo print a copy that has dark lines on a white background

dimensional and process notes notes used to show information that cannot be shown or dimensioned on the part

dimension lines lines used to show the direction and extent of a dimension

dimension numerical description of size and location of part details

direct dimensioning positioning dimensions directly between two part features

discrimination the smallest, or finest, division of a scale that can be read reliably

dual dimensioning a process of dimensioning in which both inch and millimeter units are used to dimension part features

dual unit dimensioning a system of dual dimensioning in which both inch and millimeter units are shown on the part

effective throat the minimum distance from the root to the face of a weld, minus any reinforcement

engineering drawing original mechanical drawing from which prints are made

enlarged detailed view a view that is drawn to show more detail than shown in any principal view. This view identifies and isolates special features of a part and shows them enlarged so that all details are understandable.

extension lines lines used to establish the limits of dimension lines

face angle the angle of the face of the teeth measured from the center axis of the bevel gear bore

face width the width of the bevel gear teeth measured along the pitch cone radius

feature control symbol the combined symbol that shows the geometric characteristic symbol, tolerance value, and datum reference for geometrically toleranced dimensions

field weld symbol a symbol used to denote that the indicated joint is to be welded somewhere away from the place where the assembly was originally built

fillet the rounded area on the inside corner of a cast or welded part

first angle projection a method of projecting the views of an object; used primarily in Europe

fit the term used to describe the amount of tightness between two assembled parts

flat blank layout a drawn view of a bent sheet metal part, shown flat, with all bend allowances already calculated

flat blank size the size of a part before bending

flats machined details that are used for seats for set screws on shafts

full section sectional view that shows the complete internal details of a part

general oblique drawing an oblique drawing that has an angle other than 45° or 63°26″

geometric characteristic symbol the symbol used to indicate the characteristic to be controlled

graduations the individual lines found on each edge of a steel rule

groove angle the angular size of the groove between two parts that are to be welded with a groove weld

half section sectional view that is generally used with symmetrical parts to show both internal and external features

helix angle the angle of the teeth on a helical gear to the axis of the gear bore

hidden lines lines used to show surfaces and details not directly visible in a particular view

hole pattern a grouping of holes in either a straight (linear) or a circular pattern

improper fraction a common fraction in which the numerator is greater than the denominator

isometric drawing an axonometric drawing with three equal angles, each 120°, forming the axonometric axes

key the solid coupler used between the keyway and keyseat in a keyed assembly

keyseat the slot, or groove, cut into the shaft, or internal part, of a keyed assembly

keyway the slot, or groove, cut into the external part of a keyed assembly

knurling a machining operation that produces a raised surface on a shaft to provide a better hand grip or to enlarge a shaft for a press fit. Knurling can be either straight or diamond pattern.

lay symbols the symbols used to identify the desired pattern of surface roughness

lead the distance a thread advances in one revolution. With single lead worms, the lead is equal to the linear pitch. With multiple lead worms, the lead is a multiple of the number of the lead. For example, a double lead thread has a lead equal to twice the linear pitch. A triple lead worm has a lead that is three times the size of the linear pitch.

lead angle the angle of the thread measured perpendicular to the worm thread. This value is also called the helix angle.

lead of helix the linear distance a helical gear tooth would travel in one 360° revolution if it were free to move axially

leader lines lines used to show one-way dimensions and dimensional notes or to point out process and material specifications

least common denominator the smallest denominator value that is equally divisible by the other possible denominator values

least material condition the smallest size of an external feature or the largest size of an internal feature on a part

length of weld the length of the welded area

limit dimensioning a method of tolerancing in which the dimensions show the upper and lower limits of size

limits the upper and lower sizes of a dimension when the tolerance is applied

linear dimensions straight line dimensions

linear measurement the measurement of length

linear pitch the distance between corresponding points on adjacent threads on the worm. This value is equal to the circular pitch of the worm gear.

materials list a list used to identify the materials, sizes, and part numbers of each item in an assembly

maximum clearance the maximum intentional difference in size between two mating parts

maximum material condition the largest size of an external feature or the smallest size of an internal feature on a part

measured point the end point of a measurement

melt-thru symbol a symbol used to denote welds that must be made with 100 percent penetration plus reinforcement

metric designation symbol the symbol used to show a print dimensioned in millimeters

microinch one millionth of an inch; .000001″

micrometer caliper an instrument used to make linear measurements to an accuracy of .001″, .0001″, .01 mm, or .002 mm, depending on the particular micrometer caliper used.

minute (′) a subdivision of a degree. There are 60 minutes in 1 degree.

mixed number a combination of a whole number and a common fraction

necking the process of cutting an external groove on a part

neutral axis the axis of a bent part in which neither compression nor expansion occurs. This is the axis that determines the length of the bend allowance.

nominal size the size designation used for general identification

normal circular pitch the distance from the center of one tooth to the center of the next, measured at the pitch line at a right angle to the tooth

normal diametral pitch the diametral pitch of the cutter needed to cut the teeth on a helical gear to the proper size and form

notes used on a print to give specific information about making a part

numerator the top number of a common fraction

oblique drawing a form of pictorial drawing in which the front of the object is drawn parallel to the viewing plane and the depth is shown at an angle along receding lines

oblique perspectives perspective drawings that use three vanishing points; also called three-point perspectives

offset section sectional view that uses an offset cutting plane line to show an internal detail

one-point perspectives perspective drawings that use a single vanishing point; also called parallel perspectives

orthographic projection the method used by drafters to project the views parallel to the surfaces they represent

orthographic view a view drawn by orthographic projection

outside diameter the distance across the center of a gear from one edge to the other. This is also the diameter of the addendum circle.

parallel perspectives perspective drawings that use a single vanishing point; also called one-point perspectives.

partial view a view used to show specific information about a limited area of a part

phantom lines lines used to show alternate positions of moving parts, adjacent positions of mating or related parts, or used to eliminate unnecessary repeated details

pitch circle an imaginary circle formed by the pitch diameter

pitch cone angle the angle of the bevel gear in relation to the mating gear

pitch cone distance the radius of the imaginary cone formed by the gear face of a bevel gear

pitch diameter the distance across the center of the gear from the pitch point on one side to the pitch point on the opposite side. This is the diameter from which most of the other gear values are calculated.

pitch of the weld the center-to-center distance between welds in either chain or staggered intermittent welding

pitch point the point where meshing (running together) gears contact each other

plus and minus dimensioning a method of tolerancing in which the tolerances are shown in a plus and minus form

position method dual dimensioning a method of dual dimensioning in which the position of the values determines the design and conversion units

precedence of lines the line priority used to specify which line is shown when two or more different lines occupy the same space or location

primary auxiliary view an auxiliary view that is shown adjacent to, or aligned with, the surface it represents in one of the principal views. The three major types of primary auxiliary views are the front, top, and side auxiliary views.

print an exact copy of an original drawing

print body the portion of the print where the drawn views are shown

projected tolerance zone the area projected above a part to permit alignment of a mating hole or hole pattern

projection plane the plane on which the view is shown. In most cases this plane is represented by the drawing sheet.

projection symbol the international symbol used to show the projection system used to construct a print

projectors the imaginary lines of sight used to construct an orthographic view

proper fraction a common fraction in which the numerator is smaller than the denominator

radial dimensions dimensions used to show the size of radii

radii rounded areas on either the inside or outside corners of a machined part. Radii are normally machined onto a part.

radius of curvature the size of the radius formed by the curvature of the worm gear tooth

reaming a machining operation used to enlarge holes to an exact size

receding lines the lines used to show the depth of an object when drawn in pictorial form

reducible fraction a common fraction that can be reduced to lower terms

reference dimensions dimensions that are placed on a print for convenience. These dimensions can normally be derived from the working dimensions shown.

reference line the horizontal line in a welding symbol that organizes all the relevant data about the weld

reference point the starting point of a measurement

regardless of feature size the symbol that is used to indicate that the tolerance value must be followed regardless of the actual size of the feature

removed section sectional view that is placed away from the detail it refers to; generally identified by SECTION A-A, SECTION B-B, etc.

revolved section sectional view that is used to show cross sections of spokes, ribs, or arms. Revolved sections are generally turned and are either superimposed on the object or shown in a broken area.

revisions list a list used to record any changes to an engineering drawing since the drawing was originally made

right angle an angle that has vertical and horizontal members at exactly 90°

root diameter the distance across the center of a gear from the bottom of the tooth on one side to the bottom of a tooth on the opposite side. This is also the diameter of the dedendum circle.

root opening the amount of separation at the root of the joint between two parts to be welded

roughness the fine irregularities in the surface finish caused by the machining process used to produce the finish

roughness average the arithmetical average of the roughness of a surface within a certain roughness sampling length; expressed in microinches

roughness sampling length the length of the sample used to determine the roughness average

roughness spacing the average spacing between adjacent peaks of the roughness profile

round the rounded area on the outside corner of a cast or welded part

scale the proportional relationship of the part in the print to the actual object

scales the series of graduation lines along each edge of a steel rule

second (") a subdivision of a minute. There are 60 seconds in 1 minute.

secondary auxiliary view an auxiliary view that is shown perpendicular to a primary auxiliary view

section lines lines used to show the areas of an object cut by the cutting plane line

section detail view sectional view that is drawn to identify, isolate, and enlarge special internal part details that are not clearly visible in a principal view or another form of sectional view

sectional view a view used to show the internal details of a part

shop sketching a quick and inexpensive way to convey instructions graphically

spotfacing a machining process used to provide flat seats for bolts or nuts on an irregular surface

steel protractor an instrument used to make angular measurements to an accuracy of 1°

steel rule a flat, thin strip of steel with evenly spaced graduations used for measuring parts to an accuracy of $\frac{1}{64}''$, $\frac{1}{100}''$, or .5 mm, depending on the rule used

supplementary symbols the symbols used to further define the meaning of a welding symbol

symmetry equality on both sides of a center line

System Internationale d'Unities (SI) the International System of Units

tabular dimensions dimensions in chart form

taper a uniform change in width or diameter over a specific length

taper per foot the amount of difference in end diameters between the large and small ends of a part per foot of taper length

taper per inch the amount of difference in end diameters between the large and small ends of a part per inch of taper length

third angle projection a method of projecting the views of an object; used primarily in the United States and Canada

thread class the class of fit of a thread. Designations include 1, 2, and 3 with 1 the loosest and 3 the tightest for Unified threads. The letters used with these designations are "A" for external threads and "B" for internal threads.

thread designation the system used to dimension screw threads

thread form the shape and singular characteristics of a screw thread

thread series the group of thread sizes that a particular thread belongs to. The most common are Unified Coarse (UNC), Unified Fine (UNF), and Unified Extra Fine (UNEF).

three-point perspectives perspective drawings that use three vanishing points; also called oblique perspectives

throat diameter the diameter of an imaginary circle connecting the tops of the worm gear teeth at their center point

title block a block where all the general information related to the part is shown

tolerance the total amount a dimension is allowed to vary from the stated size

trimetric drawing an axonometric drawing that has no equal angles; for example, one 105° angle, one 140° angle, and one 115° angle forming the axonometric axes

two-point perspectives perspective drawings that use two vanishing points; also called angular perspectives

undercutting a term used to describe the process of cutting an internal groove in a part

unilateral tolerance a tolerance that varies in only one direction from the stated basic size

unit assembly print a print used to show the assembled relationship of two or more parts

vanishing points the points used to construct perspective drawings

vernier bevel protractor an instrument used to make angular measurements to an accuracy of 5′

vernier caliper an instrument used to make linear measurements to an accuracy of .001″ or .02 mm

views individual drawings of each side of an object

visible lines lines used in a print to show visible edges, surfaces, and corners of an object

visualization the process of studying the individual views of an object to form a mental picture of how the finished workpiece should appear

waviness the large, widely spaced irregularities of surface texture

waviness height the peak-to-valley height of the waviness within the waviness spacing

waviness spacing lay the direction of the predominant marks of the surface pattern

weld-all-around symbol a small circle drawn around the intersection of the arrow and reference line that is used to indicate that the specified weld is to go completely around a joint

weld size the value shown in a welding symbol to denote the depth of preparation, actual weld size, or weld strength

weld symbol the symbol used to graphically describe the required weld

welding symbols the symbols used to specify all the information needed to make any welds

whiteprint a copy that has dark lines on a white background

whole depth the depth at which the gear is cut or the total value of the addendum plus the dedendum

working depth the depth of engagement between meshing gears. The working depth is equal to the addendum plus the dedendum minus the clearance

working dimensions dimensions that control the sizes of a part

INDEX

E

representation of screw threads, 193; *illus.*, 194
    pictorial, 193
    schematic, 193
    simplified, 193
revisions list, 161-162; *illus.*, 162
revolved section, 173-174; *illus.*, 174
right angle, 48
right-side view, 61-62; *illus.*, 61, 62
root diameter, 197; *illus.*, 198
root opening, 282; *illus.*, 283
roughness, 207
roughness average, 208; *illus.*, 208
    for common manufacturing processes, 211
roughness sampling length, 208; *illus.*, 208
roughness spacing, 208; *illus.*, 208
round, 46, 134; *illus.*, 46
    dimensioning, 134; *illus.*, 135
round holes, 128
    dimensioning, 128; *illus.*, 128
runout tolerances, 246-247; *illus.*, 234
    circular, 246
    total, 246, 247

S

scale
    discrimination of, 95
    on print, 11, 119; *illus.*, 119
    on steel rule, 95
    vernier, 99; *illus.*, 100
schematic representation, 193; *illus.*, 194
screws, 194
    common head styles, 195
    specifying, 194; *illus.*, 194
screw threads, 188-193
    designations of, 189-193
    forms of, 188-189
    representing, 193; *illus.*, 194
secondary auxiliary views, 67-68; *illus.*, 68
secondary datum, 250; *illus.*, 250
seconds, 106, 121; *illus.*, 123
sectional detailed views, 175; *illus.*, 176
sectional views, 167-174
    indicating, 168-170
    types of, 170-174
section lines, 24; *illus.*, 20, 26
    in sectional views, 170; *illus.*, 171, 172
sections
    broken, 170
    full, 170
    half, 170
    offset, 170-173
    removed, 173
    revolved, 173-174
self-holding tapers, 183; *illus.*, 184

self-releasing tapers, 183-184; *illus.*, 184
sharp V thread, 188; *illus.*, 188
shop mathematics, 76-89
shop measuring tools, 95-108
    micrometers, 99-102
    protractors, 106-108
    steel rules, 95-98
    verniers, 103-106
shop sketching, 39-50
    practical techniques for, 47-49
    procedures for, 49-50
    techniques for 40-47
    tools for, 39-40
shop sketching techniques, 40-47
    for circles and arcs, 44-47; *illus.*, 44, 46
    for isometric sketches, 262-263; *illus.*, 262
    for oblique sketches, 264-265; *illus.*, 264
    for perspective sketches, 265-267; *illus.*, 266
    practical, 47-49
    square and triangle method as, 44
    for straight lines, 40-44; *illus.*, 40, 41
shortened radius dimensions, 124
side view, 12, 61-63
    left, 61-62
    right, 61-62
simplified representation, 193; *illus.*, 194
size
    actual, 137
    basic, 137
    estimating, in sketching, 48; *illus.*, 48
    limits of, 137
    nominal, 136
    specifying weld, 281-283; *illus.*, 281
    in welding symbol, 282; *illus.*, 282
size dimensions, 117; *illus.*, 117
size values for welding symbols, 281-283;
        illus., 281
    effective throat, 282
    groove angle, 282
    length of weld, 282, 283
    pitch of the weld, 282, 283
    root opening, 282, 283
    size, 282
sketching arcs, 46-47; *illus.*, 46
sketching circles, 44-47; *illus.*, 45
sketching half-circles, 46; *illus.*, 46
sketching procedures, 49-50
    for isometric sketches, 262-263
    for oblique sketches, 264-265
    for perspective sketches, 265-267
sketching straight lines, 40-44; *illus.*, 40, 41
sketching tools, 39-40
    erasers, 40
    paper, 40
    pencils, 39